The Biology of Neotropical Scorpions

André Felipe de Araujo Lira
Edmundo González-Santillán
Editors

The Biology of Neotropical Scorpions

From Taxonomy and Genetics to Ecology and Ethnobiology

Springer

Editors
André Felipe de Araujo Lira
Colección Nacional de Arácnidos
Departamento de Zoología
Instituto de Biología
Universidad Nacional Autónoma de México
Coyoacán, Mexico City, Mexico

Edmundo González-Santillán
Colección Nacional de Arácnidos
Departamento de Zoología
Instituto de Biología
Universidad Nacional Autónoma de México
Coyoacán, Mexico City, Mexico

ISBN 978-3-032-23193-2 ISBN 978-3-032-23194-9 (eBook)
https://doi.org/10.1007/978-3-032-23194-9

This Springer imprint is published by the registered company Springer Nature Switzerland AG
The registered company address is: Gewerbestrasse 11, 6330 Cham, Switzerland

To all those passionate about the Neotropical scorpion biodiversity, whose dedication to study and conservation inspires this work.

Preface

This book focuses on comprehensive aspects of scorpions of the Neotropical region, from public health to biogeographic history, synthesised in a single work. Therefore, this innovative book is devoted to tackling that lack of knowledge and bringing together a bundle of studies to offer students, colleagues, professionals, and policy-makers a comprehensive background to start their research with scorpions as model organisms. Since 1990, with the publication of the *Biology of Scorpions* edited by Gary Polis, the study of scorpions has thrived following the tradition focusing on the Palearctic and Nearctic fauna, leaving behind ancient and important scorpion fauna localised in the Neotropics. The Neotropical region encompasses the tropical areas from South America to central Mexico, including the Antilles, comprising different biomes, such as deserts, dry forests, savannahs, mountain ranges, and rainforests. These myriad environments influence scorpions' venom composition, morphology, and their human interactions. To this date, many researchers from different areas and countries have generated a growing body of information about Neotropical scorpions, but this is still fragmentary. Our proposal seeks to innovate in the following fields of knowledge: the biology of scorpion fluorescence, chromosomal and molecular genetics, conservation ecology, and ethnobiology of scorpions from the perspective of traditional medicine.

Coyoacán, Mexico City, Mexico André Felipe de Araujo Lira
Coyoacán, Mexico City, Mexico Edmundo González-Santillán

Acknowledgements

We gratefully acknowledge the Universidad Nacional Autónoma de México (UNAM) for institutional support throughout this research. We thank the Dirección General de Asuntos del Personal Académico (DGAPA) for the postdoctoral fellowship from the UNAM to André Felipe de Araujo Lira. Special thanks to the experts who generously contributed data, photographs, and review. Heartfelt gratitude also goes to our family and friends for their unwavering encouragement.

Competing Interests The authors have no competing interests to declare that are relevant to the content of this manuscript.

About the Book

The Neotropical scorpions represent a remarkable group within one of the richest zoogeographical regions on Earth. This work integrates decades of accumulated knowledge with new discoveries, highlighting their diversity, ecology, and evolutionary significance. It is a valuable resource for researchers and conservationists working to understand and protect these fascinating animals.

Contents

About the Editors

André Felipe de Araujo Lira is a biologist (PhD in Animal Biology, 2018) who has been dedicated to the study of arachnids in the Neotropical region. His main interests are ecology and the behaviour of scorpions, pseudoscorpions, and harvestmen. He has conducted research focused on the conservation ecology, landscape ecology, macroecology, and defensive and reproductive behaviour of these arachnids for more than 10 years, resulting in more than 100 published peer-reviewed articles and 1000 citations.

Edmundo González-Santillán received a PhD in Ecology, Behavior, and Evolutionary Biology from CUNY (2013) and focuses on the origin, evolution, and maintenance of arachnids. He published over 30 peer-reviewed contributions on the Order Scorpiones and other arachnids, focusing mainly on Mexican fauna. He has extensive experience in teaching, outreach, and speaking at local and international meetings and congresses. He is the curator of the National Collection of Arachnids in the Biology Institute, UNAM, a full-time researcher, and group leader of the research group. He has devoted more than 20 years to scorpion systematics research, venoms, and other salient biological features of scorpions.

Chapter 1
What Are the Neotropical Scorpions?

André Felipe de Araujo Lira

Abstract The Neotropical region, extending from northern Argentina to central Mexico and including the Antilles, is one of the world's most biologically diverse zoogeographic zones, harboring exceptional species richness and endemism driven by complex geological and climatic histories. Among its diverse fauna, Neotropical scorpions represent an ecologically important and morphologically diverse group adapted to various habitats ranging from caves and forests to deserts and urban environments. Despite their notoriety as medically significant arthropods responsible for envenomation incidents, these scorpions contribute essential ecosystem functions and offer valuable insights into evolutionary biology, biogeography, and ecology. However, key knowledge gaps persist, particularly related to species diversity and distribution (Darwinian and Wallacean shortfalls), which are exacerbated by accelerated habitat loss and biodiversity decline in the region. The loss of Neotropical scorpion biodiversity not only threatens ecological balance but also jeopardizes potential pharmacological resources derived from their venoms. This chapter provides an updated overview of Neotropical scorpion biology, emphasizing their taxonomy, habitat diversity, morphological adaptations, and ecological roles, while situating these within the broader context of conservation challenges facing Neotropical biodiversity.

Keywords Arachnids · Biodiversity · Zoogeographical region

A. F. de Araujo Lira (✉)
Colección Nacional de Arácnidos, Departamento de Zoología, Instituto de Biología, Universidad Nacional Autónoma de México, Coyoacán, Mexico City, Mexico
e-mail: andre.lira@st.ib.unam.mx

A. F. de Araujo Lira, E. González-Santillán (eds.), *The Biology of Neotropical Scorpions*, https://doi.org/10.1007/978-3-032-23194-9_1

1.1 Introduction

The Neotropical region represents one of the most biologically diverse zoogeographical zones in the world (e.g., Leal et al. 2016; Rull and Carnaval 2020), stretching from northern Argentina to central Mexico (Morrone et al. 2022) (Fig. 1.1) and encompassing numerous ecosystems such as tropical rainforests, savannas, wetlands, and deserts. This immense diversity of habitats has fostered the evolution of a rich and unique fauna, marked by high levels of species richness and endemism (Morawetz and Raedig 2007; Andrade-Díaz et al. 2017; Noguera-Urbano and Escalante 2018). Understanding the drivers behind this biodiversity is crucial not only for scientific knowledge but also for the conservation of these fragile ecosystems, many of which face increasing threats from human activities.

Fig. 1.1 Neotropical region extension. Sensu Morrone (2014)

Among the remarkable groups inhabiting the Neotropical region, scorpions stand out both for their ecological roles and their diversity (see Chaps. 8 and 13). These arthropods have adapted to a range of environments, from caves and forests to urban areas, showcasing a variety of morphologies and life strategies (Polis 1990). Despite their medical significance due to the potential for causing envenomation incidents, Neotropical scorpions are much more than just a public health concern—they are integral components of their ecosystems and serve as valuable models for studying evolutionary biology, biodiversity patterns, and ecological interactions.

However, substantial gaps in our knowledge about scorpion diversity and distribution persist, linked to the so-called Darwinian and Wallacean shortfalls (Diniz-Filho et al. 2023). These gaps are especially alarming given the accelerated rates of habitat loss and species extinctions affecting the Neotropics today (e.g., Antonelli 2021; Pires and Galetti 2023; Boron et al. 2023). Addressing these challenges is imperative not only for preserving scorpion biodiversity and their ecosystem functions but also for safeguarding the potential benefits their unique venom compounds may offer in pharmacology. This chapter aims to explore the introductory aspects of biology, diversity, and ecological significance of Neotropical scorpions within the broader context of the region's biodiversity.

Since the iconic book *The Biology of Scorpions*, published in 1990 by Gary A. Polis, which marked a fundamental milestone in scorpion studies, research involving these animals has advanced significantly in various biological fields. However, one of the regions least addressed by Polis was precisely the Neotropical region, whose richness and complexity have demanded more in-depth studies. Therefore, this book emerges as an updated and comprehensive contribution to the state of the art of Neotropical scorpions, covering not only classical aspects such as taxonomy, biogeography, and ecology, but also emerging and innovative topics including ethnobiology and recent advances in understanding the fluorescence of these animals. This book reflects recent progress and broadens knowledge about a group essential to the biodiversity and ecosystems of the Neotropical region.

1.2 Neotropical Region

The Neotropical region corresponds to one of the eight zoogeographical zones, extending from northern Argentina to central Mexico, including the Antilles (Morrone 2018). The region encompasses a vast range of biomes and habitat types (Fig. 1.2), each with a particular history of landscapes and biotic evolution (e.g., Muller 1972; Silvério et al. 2013; Cabanne et al. 2016). The Neotropics are home to outstandingly high levels of species richness and endemism for many taxa (Fernández et al. 2021; Palma-Silva et al. 2022; Machado et al. 2024; Guedes et al. 2025), including most of the biodiversity *hotspots* (Myers et al. 2000).

Several hypotheses have been proposed to try to explain the biodiversity found in the Neotropical region. Among them, the main ones are: the separation between the South American and African continents (Raven and Axelrod 1974; Gentry 1982),

Fig. 1.2 Different habitat types found in the Neotropical region. (**a**) Wetland (Pantanal), (**b**) Seasonal Dry Tropical Forest (Caatinga), (**c**) Savanna-like vegetation (Cerrado), (**d**) Cave, (**e**) Tropical rainforest (Brazilian Atlantic Forest), and (**f**) Desert. (Photo by (**a**) César Correa, (**b**) Gabriel Celante, (**c**) Everton Tizo-Pedroso, (**d**) Eder Barbier, (**e**) Rafael Ramos, and (**f**) Alejandro Valdez-Mondragón)

the emergence of the Isthmus of Panama, allowing migration between South and North American faunas (Leigh et al. 2013; Meyer et al. 2019), the climatic variations that occurred during the Pleistocene which resulted in expansion and contraction events of habitats (e.g., Prance 1982; Cabanne et al. 2016; Camps et al. 2018), and finally, tectonic events such as the uplift of the Andes (Antonelli et al. 2009; Mendoza-Henao and Garcia-R 2021) and the Sierra Madre (Mastretta-Yanes et al. 2015; Bryson Jr et al. 2018) in south and north America, respectively.

The high biodiversity found in the Neotropical region has attracted the interest of the scientific community over the years (e.g., Ceballos and García 1995; Escribano-Avila et al. 2017; Valente-Neto et al. 2025). Despite this, some gaps still persist, such as the Darwinian shortfall and Wallacean shortfall. These two shortfalls are related to our lack of knowledge about species diversity and distribution, respectively (Diniz-Filho et al. 2023). This is particularly concerning given the current scenario of accelerated biodiversity extinction (Antonelli 2021) and the higher endemism rate for the Neotropical region (e.g., Morawetz and Raedig 2007; Nogueria et al. 2011; Nogué et al. 2012). A large part of species loss is associated with the high rate of native vegetation destruction and its replacement by pastures, agriculture, and cities (e.g., Durães et al. 2013; Pelicice et al. 2021; Curtis et al. 2022). This loss of Neotropical scorpion biodiversity may be not only harmful to the habitat where these animals live due to the loss of ecosystem functions (see Chap. 13), but also financially negative for humans since these animals are considered "living pharmacies" and many toxins with pharmaceutical properties would be lost (see Chap. 15), and finally, it may lead to the proliferation of species capable of causing accidents with humans (see Chap. 16).

Fig. 1.3 Scorpions in different microhabitats. Leaf-litter dwelling species *Tityus pusillus* (**a**) and *Megacormus seductus* (**b**), ground dwelling species *Hoffmanihadrurus gerstchi* (**c**), and *Bothriurus asper* (**d**), vegetation dwelling species *Physoctonus debilis* (**e**), and cave dwelling species *Typhlochactas* sp. (**f**). (Photos by Matheus Leonydas (**a**), Kleyton Rezende (**b**), Alejandro Valdez-Mondragón (**c** and **f**), Gabriel Celante (**d**), and André Otávio (**e**)

Neotropical scorpions are particularly known for their capacity to cause accidents, with scorpionism being a serious public health problem in many countries of the region (see Chap. 16). However, Neotropical scorpions are much more diverse than the few synanthropic species. For example, Brito and Borges (2015) reported 47 scorpion species from Ecuador of which only two being considered as synanthropic (*Teuthraustes atramentarius* Simon, 1878 and *Centruroides margaritatus* (Gervais, 1841)). Scorpions are found in virtually every terrestrial habitat of the Neotropical region, occurring from caves to mountains, deserts to rainforests, and even in urban environments (e.g., Ojanguren-Affilastro and Mattoni 2017; Ochoa and Rojas-Runjaic 2019; Lira et al. 2019; Amado et al. 2021). The diversity of habitats where scorpions can be found directly reflects the diversity of microhabitats used by these animals (Fig. 1.3). For example, there are lineages specialized in living in galleries, such as bothriurids and diplocentrids, while others are adapted to life on sandy soil, like hadrurids, and even in vegetation, such as some buthids.

1.3 Scorpions

Scorpions were one of the first groups of invertebrates to migrate from the aquatic to the terrestrial environment (see Chap. 7). Despite the morphological changes necessary to conquer the terrestrial environment, the *bauplan* of these animals has

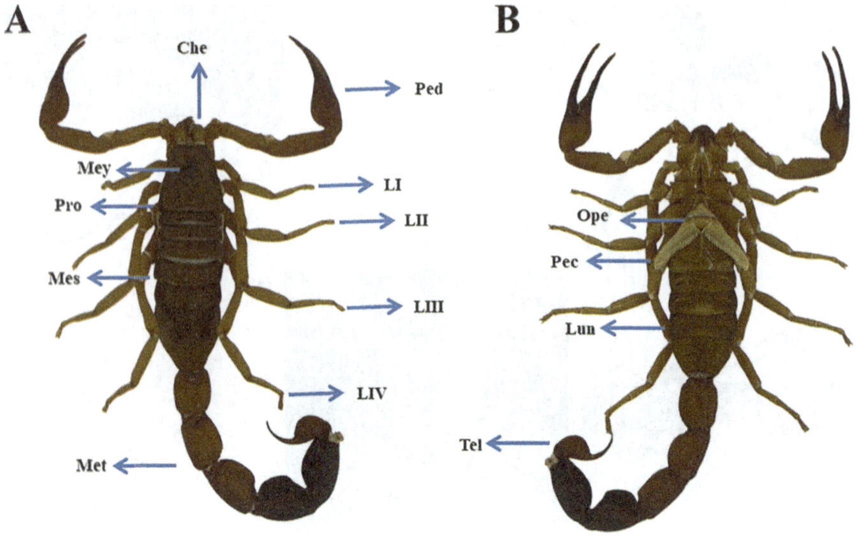

Fig. 1.4 *Bauplan* of scorpions. *Jaguajir agamemnon* A) dorsal view and B) ventral view. Che—chelicerae, Ped—pedipalp, Mey—median eyes, Pro—prosoma, Mes—mesosoma, Met—metasoma, LI—leg I, LII—leg II, LIII—leg III, LIV—leg IV, Ope—genital operculum, Pec—pectines, Lun—book lung, Tel—telson. (Photo by Randyson Pinheiro)

been little altered. The body of scorpions is divided into two parts: prosoma, containing all the appendages, and opisthosoma, which is further divided into mesosoma and metasoma (Fig. 1.4).

Dorsally on the prosoma are two median eyes that may or may not be present (see Chap. 9) and a variable number of lateral eyes (0–6) (Loria and Prendini 2014). Ventrally are the chelicerae, pincer-like structures used both for crushing food and for reproduction (see Chap. 3). Next are the pedipalps, pincer-like structures characteristic of scorpions. The pedipalps have important functions and can be used in reproductive contexts (see Chap. 3), defense, food capture (see Chap. 10), and male-male competition (Tang 2023; Wyman et al. 2025). Additionally, the pedipalp is one of the structures that can exhibit different allometric growth between the sexes (see Chap. 2). Its morphology can vary, for example, in species where males have elongated pedipalp chelae, while in other species they are more robust (Fig. 1.5).

Scorpions have four pairs of legs, which increase in size from the anterior to the posterior. The legs have tarsal setae that assist the animals according to their habitat (Fet et al. 1998; Loria and Prendini 2021); for example, species living in sandy dunes exhibit many setae to walk on the sand (Fet et al. 1998) while species that inhabit wet habitats possess a well-developed pseudonychium that allows them to climb in vegetation (Prendini 2001). On the ventral side of the prosoma are the pectines, organs with a mechano-chemoreceptive function (reviewed in Wolf 2017). The pectines are fundamentally important for these animals, being involved in the detection and recognition of sexual partners and predators (e.g., Pordeus et al. 2019; Feitosa et al. 2022; Dionisio-da-Silva et al. 2025), location of potential prey (Mineo

Fig. 1.5 Scorpion species with males showing different pedipalp morphology. *Tityus dinizi* with gracile pedipalp (**a**) and *Tityus metuendus* with robust pedipalp (**b**). (Photos by Jonas Martins)

and Del Claro 2006; Drozd et al. 2022), and navigation (Gaffin and Brayfield 2017; Prévost and Stemme 2020; Hladik et al. 2025). Finally, on the ventral face of the prosoma is the genital operculum, a structure used for reproduction.

The opisthosoma is subdivided into the mesosoma (composed of seven segments) and the metasoma (composed of five segments). On the ventral side of the mesosoma are the openings of the book lungs; additionally, some species have glandular tissues whose function is still unknown. The metasoma is a characteristic structure of scorpions, also known as the tail. This structure can also exhibit different allometric values (see Chap. 2) (Fig. 1.6). At the end of the metasoma are the anus and the telson. Due to its terminal position and the pasty type of excreta produced by scorpions, these animals need to clean it after excretion to avoid blockage (Armas 1980; Pordeus et al. 2020). Finally, the telson is the structure where the venom glands and the stinger are located, and some species also have glands on the telson dorsal side used in reproduction (e.g., Peretti 1997; Teruel et al. 2015).

1.4 Conclusion

Neotropical scorpions represent a highly diverse and ecologically important group within one of the world's richest zoogeographical regions. Their remarkable adaptability across a wide range of habitats—from rainforests and deserts to caves and urban environments—reflects the complex evolutionary and biogeographical history of the Neotropics. Despite their medical significance, these organisms contribute vital ecosystem functions and offer unique insights into biodiversity patterns, evolutionary processes, and ecological interactions.

Fig. 1.6 Neotropical scorpion species with different metasoma size. (**a**) *Tityus neglectus*, (**b**) *Jaguajir rochae*, (**c**) *Opistacanthus cayaporum*, and (**d**) *Tityopsis canizaresorum*. (Photos by Adeilson Melo (**a**), Gabriel Celante (**b**), Fernando J.M. Rojas-Runjaic (**c**), and Rolando Teruel (**d**))

However, significant knowledge gaps regarding their diversity and distribution persist, compounded by increasing habitat loss and species decline across the region. Addressing these gaps is critical not only to preserve the ecological roles of Neotropical scorpions but also to safeguard potential benefits arising from their unique venom compounds. This chapter sets the foundation for further research, emphasizing the urgent need for comprehensive studies to support conservation efforts and enhance our understanding of these fascinating arthropods within the broader context of Neotropical biodiversity.

Acknowledgments I would like to thank the Dirección General de Asuntos del Personal Académico (DGAPA) for the postdoctoral fellowship from the Universidad Nacional Autónoma de México. I would also like to thank Dr. Adriano DeSouza for the map preparation and Doctors César Correa, Gabriel Celante, Everton Tizo-Pedroso, Eder Barbier, Rafael Ramos, Alejandro Valdez-Mondragón, Matheus Leonydas, Kleyton Rezende, André Otávio, Randyson Pinheiro, Jonas Martins, Adeílson Melo, Fernando J.M. Rojas-Runjaic, and Rolando Teruel for kindly providing the photos used in the chapter.

References

Amado TF, Moura TA, Riul P, Lira AFA, Badillo-Montaño R, Martinez PA (2021) Vulnerable areas to accidents with scorpions in Brazil. Trop Med Int Health 26(5):591–601. https://doi.org/10.1111/tmi.13561

Andrade-Díaz MS, Hibbard TN, Díaz-Gómez JM (2017) Identifying endemism areas: an example using Neotropical lizards. S Am J Herpetol 12(1):61–75. https://doi.org/10.2994/SAJH-D-16-00038.1

Antonelli A (2021) The rise and fall of Neotropical biodiversity. Bot J Linn Soc 199(1):8–24. https://doi.org/10.1093/botlinnean/boab061

Antonelli A, Nylander JAA, Persson C, Sanmartín I (2009) Tracing the impact of the Andean uplift on Neotropical plant evolution. PNAS 106(24):9749–9754. https://doi.org/10.1073/pnas.0811421106

Armas LF (1980) Aspecto de la biología de alguns escorpiones cubanos. Poeyana 211:1–28

Boron V, Deere NJ, Hyde M, Bardales R, Stasiukynas D, Payán E (2023) Habitat modification destabilizes spatial associations and persistence of Neotropical carnivores. Curr Biol 33(17):3722–3731. https://doi.org/10.1016/j.cub.2023.07.064

Brito G, Borges A (2015) A checklist of the scorpions of Ecuador (Arachnida: Scorpiones), with notes on the distribution and medical significance of some species. J Venom Anim Toxins Incl Trop Dis 21:1–17. https://doi.org/10.1186/s40409-015-0023-x

Bryson RW Jr, Zarza E, Grummer JÁ, Parra-Olea G, Flores-Villela O, Klicka J, McCormack JE (2018) Phylogenomic insights into the diversification of salamanders in the *Isthmura belli* group across the Mexican highlands. Mol Phylogenet Evol 125:78–84. https://doi.org/10.1016/j.ympev.2018.03.024

Cabanne GS, Xalderón L, Arias NT, Flores P, Pessoa R, d'Horta FM, Miyaki CY (2016) Effects of Pleistocene climate changes on species ranges and evolutionary processes in the Neotropical Atlantic Forest. Biol J Linn Soc 119(4):856–872. https://doi.org/10.1111/bij.12844

Camps GA, Martínez-Meyer E, Verga AR, Sérsic NA, Cosacov A (2018) Genetic and climatic approaches reveal effects of Pleistocene refugia and climatic stability in an old giant of the Netropical dry Forest. Biol J Linn Soc 125(2):401–420. https://doi.org/10.1093/biolinnean/bly115

Ceballos G, García A (1995) Conserving Neotropical biodiversity: the role of dry forest in western Mexico. Conserv Biol 9(6):1349–1353

Curtis JR, Robinson WD, Rompré G, Austin SH (2022) Urbanization is associated with unique community simplification among birds in a neotropical landscape. Landsc Ecol 37:209–231. https://doi.org/10.1007/s10980-021-01344-1

Diniz-Filho JAF, Jardim L, Guedes JJM, Meyer L, Stropp J, Frateles LEF, Pinto RB, Lohmann LG, Tessarolo G, Carvalho CJB, Ladle RJ, Hortal J (2023) Macroecological links between the Linnean, Wallacean, and Darwinian shortfall. Front Biogeogr 15(2):e59566. https://doi.org/10.21425/F5FBG59566

Dionisio-da-Silva W, Rocha-da-Silva KLA, Veloso HMG, Silva MB (2025) Hide and seek: chemical cues drive site preference among potential mates and intraguild competitors. Biol J Linn Soc 144(4):blae058. https://doi.org/10.1093/biolinnean/blae058

Drozd D, Wolf H, Stemme T (2022) Mechanosensory pathways of scorpion pectin hair sensillae-adjustment of body height and pectin position. J Comp Neurol 530(16):2918–2937. https://doi.org/10.1002/cne.25384

Durães R, Carrasco L, Smith TB, Karubian J (2013) Effects of forest disturbance and habitat loos on avian communities in a Neotropical biodiversity hotspot. Biol Conserv 166:203–211. https://doi.org/10.1016/j.biocon.2013.07.007

Escribano-Avila G, Cervera L, Ordóñez-Delgado L, Jara-Guerrero A, Amador L, Paladines B, Briceño J, Parés-Jiménez V, Lizcano DJ, Duncan DH, Espinosa CI (2017) Biodiversity patterns and ecological processes in Neotropical dry forest: the need to connect research and

management for long-term conservation. Neotrop Biodivers 3(1):107–116. https://doi.org/10.1080/23766808.2017.1298495
Feitosa MLB, Dionisio-da-Silva W, Lira AFA, Teles-Pontes WJ (2022) Fear as an enemy? Behavioral changes of *Ananteris mauryi* (Scorpiones: Buthidae) triggered by chemical cues from an intraguild predator. Can J Zool 100(8):488–493. https://doi.org/10.1139/cjz-2022-0027
Fernández F, Guerrero RJ, Sánchez-Restrepo AF (2021) Systematic and diversity of Neotropical ants. Rev Colomb Entomol 47(1):e11082. https://doi.org/10.25100/socolen.v47i1.11082
Fet V, Polis GA, Sissom WD (1998) Life in sandy deserts: the scorpion model. J Arid Environ 39:609–622
Gaffin DD, Brayfield BP (2017) Exploring the chemo-textural familiarity hypothesis for scorpion navigation. J Arachnol 45(3):265–270. https://doi.org/10.1636/JoA-S-16-070.1
Gentry A (1982) Neotropical floristic diversity: phytogeographical connections between Central and South America. Pleistocene climatic fluctuations or an accident of Andean orogeny? Ann Mo Bot Gard 69:557–593
Guedes GHS, Santangelo JM, Pires APF, Araújo FG (2025) β-diversity scaling patterns across different bioregionalisations for a megadiverse Neotropical fish family. J Biogeogr 52(4):e15088. https://doi.org/10.1111/jbi.15088
Hladik J, Bailer Y, Wolf H, Stemme T (2025) Shelter selection in females of two scorpion species depends on shelter size and scent. J Comp Physiol A 211:163–183. https://doi.org/10.1007/s00359-024-01721-6
Leal BSS, Silva CP, Pinheiro P (2016) Phylogeographic studies depict the role of space and time scales of plant speciation in a highly diverse Neotropical region. Crit Rev Plant Sci 35(4):215–230. https://doi.org/10.1080/07352689.2016.1254494
Leigh EG, O'Dea A, Vermeij GJ (2013) Historical biogeography of the Isthmus of Panama. Biol Rev 89:148–172. https://doi.org/10.1111/brv.12048
Lira AFA, Salomao RP, Albuquerque CMR (2019) Pattern of scorpion diversity across a bioclimatic dry-wet gradient in Neotropical forests. Acta Oecol 96:10–17. https://doi.org/10.1016/j.actao.2019.02.004
Loria SF, Prendini L (2014) Homology of the lateral eyes of Scorpiones: a six-ocellus model. PLoS One 9(12):e112913. https://doi.org/10.1371/journal.pone.0112913
Loria SF, Prendini L (2021) Burrowing into the forest: phylogeny of the Asian forest scorpions (Scorpionidae: Heterometrinae) and the evolution of ecomorphotypes. Cladistics 37(2):109–161
Machado VR, Pereira MJR, Tirelli FP, Bennet D, Ferrari A (2024) Endemism and regionalization of Neotropical mammals: a multi-taxon analysis. J Mammal 105(4):792–808. https://doi.org/10.1093/jmammal/gyae055
Mastretta-Yanes A, Moreno-Letelier A, Piñero D, Jorgensen TH, Emerson BC (2015) Biodiversity in the Mexican highlands and the interaction of geology, geography and climate within the trans-Mexican Volcanic Belt. J Biogeogr 42(9):1586–1600. https://doi.org/10.1111/jbi.12546
Mendoza-Henao AM, Garcia-R JC (2021) Neotropical biodiversity: hypotheses of species diversification and dispersal in the Andean mountain forests. In: Myster RW (ed) The Andean cloud forest. Springer, Cham. https://doi.org/10.1007/978-3-030-57344-7_9
Meyer NFV, Moreno R, Sutherland C, de la Torre JA, Esser HJ, Jordan CA, Olmos M, Ortega J, Reyna-Hurtado R, Valdes S, Jansen PA (2019) Effectiveness of Panama as an intercontinental land bridge for large mammals. Conserv Biol 34(1):207–219. https://doi.org/10.1111/cobi.13384
Mineo MF, Del Claro K (2006) Mechanoreceptive function of pectines in the Brazilian yellow scorpion *Tityus serrulatus*: perception of substrate-borne vibrations and prey detection. Acta Ethologica 9(2):79–85. https://doi.org/10.1007/s10211-006-0021-7
Morawetz W, Raedig C (2007) Angiosperm biodiversity, endemism and conservation in the Neotropics. Taxon 56(4):1245–1254. https://doi.org/10.2307/25065916
Morrone JJ (2014) Biogeographic regionalisatoin of the Neotropical region. Zootaxa 3782:1110
Morrone JJ (2018) The spectre of biogeographical regionalization. J Biogeogr 45(2):282–288. https://doi.org/10.1111/jbi.13135

Morrone JJ, Escalante T, Rodríguez-Tapia G, Carmona A, Arana M, Mercado-Gómez JD (2022) Biogeographic regionalization of the Neotropical region: new map and shapefile. An Acad Bras Cienc 94:e20211167. https://doi.org/10.1590/0001-3765202220211167
Muller P (1972) Centres of dispersal and evolution in the neotropical region. Stud Neotrop Fauna 7(2):173–185. https://doi.org/10.1080/01650527209360442
Myers N, Mittermeier RA, Mittermeier CG, da Fonseca GA, Kent J (2000) Biodiversity hotspots for conservation priorities. Nature 403(6772):853–858
Nogué S, Rull V, Vegas-Vilarrúbia T (2012) Elevational gradients in the neotropical table mountains: patterns of endemism and implications for conservation. Divers Distrib 19(7):676–687. https://doi.org/10.1111/ddi.12017
Noguera-Urbano EA, Escalante T (2018) The Neotropical region sensu the areas of endemism of terrestrial mammals. Aust Syst Bot 30(6):470–484. https://doi.org/10.1071/SB16053
Nogueria C, Ribeiro S, Costa GC, Colli GR (2011) Vicariance and endemismo in a Neotropical savana hotspot: distribution patterns of Cerrado squamate reptiles. J Biogeogr 36(10):1907–1922. https://doi.org/10.1111/j.1365-2699.2011.02538.x
Ochoa JA, Rojas-Runjaic FJ (2019) Scorpions. In: Biodiversity of Pantepui. Academic, London, pp 223–246
Ojanguren-Affilastro AA, Mattoni CI (2017) *Mauryius* n. gen. (Scorpiones: Bothriuridae) a new neotropical scorpion genus. Arthropod Syst Phylogeny 75(1):125–139
Palma-Silva C, Turchetto-Zolet AC, Fay MF, Vasconcelos T (2022) Drivers of exceptional Neotropical biodiversity: an update view. Bot J Linn Soc 199(1):1–7. https://doi.org/10.1093/botlinnean/boac005
Pelicice FM, Bialetzki A, Camelier P, Carvalho FR, García-Berthou E, Pompeu OS, Mello FT, Pavanelli CS (2021) Human impacts and the loss of Neotropical freshwater fish diversity. Neotrop Ichthyol 19(3):e210134. https://doi.org/10.1590/1982-0224-2021-0134
Peretti AV (1997) Relación de las glándulas caudales de machos de escorpiones Bothriuridae con el comportamiento sexual (Scorpiones). Revue Arachnologique 12(3):31–41
Pires MM, Galetti M (2023) Beyond the "empty forest": the defaunation syndromes of Neotropical forests in the Anthropocene. Glob Ecol Conserv 41:e02362. https://doi.org/10.1016/j.gecco.2022.e02362
Polis GA (1990) The biology of scorpions. Stanford University Press, Stanford
Pordeus LM, Lira AFA, Albuquerque CMR (2019) Male courtship behavior is triggered by female chemical cues in the scorpion *Tityus pusillus* (Scorpiones: Buthidae). Can J Zool 97(12):1122–1125. https://doi.org/10.1139/cjz-2019-0020
Pordeus LM, Dionisio-da-Silva W, Lira AFA, Albuquerque CMR (2020) Excreta grooming behavior in a litter-dwelling scorpion species (Scorpiones: Buthidae) from Brazil. J Arachnol 48(1):98–99. https://doi.org/10.1636/0161-8202-48.1.98
Prance GT (1982) A review of the phytogeographic evidences for Pleistocene climate change in the Neotropics. Ann Mo Bot Gard 69(3):594–624. https://doi.org/10.2307/2399085
Prendini L (2001) Substratum specialization and speciation in southern African scorpions: the effect hypothesis revisited. In: Fet V, Selden PA (eds) Scorpions 2001. In memoriam Gary A. Polis. British Arachnological Society, Burnham Beeches, U.K., pp 113–138
Prévost EM, Stemme T (2020) Non-visual homing and the current status of navigation in scorpions. Anim Cogn 23:1215–1234. https://doi.org/10.1007/s10071-020-01386-z
Raven PH, Axelrod DI (1974) Angiosperm biogeography and past continental movements. Ann Mo Bot Gard 61:539–673
Rull V, Carnaval AC (2020) Neotropical diversification: patterns and processes. Springer, Cham, p 820
Silvério DV, Brando PM, Balch JK, Putz FE, Nepstad DC, Oliveira-Santos C, Bustamante MMC (2013) Testing the Amazon savannization hypothesis: fire effects on invasion of a neotropical forest by native Cerrado and exotic pasture grasses. Philos Trans R Soc B 368(1619):e20120427. https://doi.org/10.1098/rstb.2012.0427

Tang V (2023) Non-aggressive competition between males of *Srilankametrus yaleensis* (Kovařík et al., 2019) (Scorpionidae), and other types of agonistic behavior observed in scorpions. Euscorpius 2023(368):1–17

Teruel R, Lowe G, Fet V, Daniel KA (2015) Occurrence of a telson gland in the genus *Superstitionia* Stahnke, 1940 (Scorpiones: Superstitioniidae). Euscorpius 204:1–5

Valente-Neto F, Mello JLS, Pestana GC, Shimabukuro EM, Siqueira AS, Covich AP, Saito VS (2025) Ecological perspectives on the organization of biodiversity in Neotropical streams. Hydrobiologia 852:3025–3047. https://doi.org/10.1007/s10750-024-05631-1

Wolf H (2017) Scorpions pectines—idiosyncratic chemo- and mechanosensory organs. Arthropod Struct Dev 46(6):753–764. https://doi.org/10.1016/j.asd.2017.10.002

Wyman JT, Wright-Ueda J, Agnew Q, Castellano I, Simone Y (2025) First report of arm-span competition in buthid scorpions: male-male contest in *Tityus* cf. *rosenbergi* Pocock, 1898. The. J Arachnol 52(3):210–213. https://doi.org/10.1636/JoA-S-23-009

Chapter 2
Allometry and the Evolution of Sexual Size Dimorphism in Scorpions: Insights from Neotropical Species

Stênio Ítalo Araújo Foerster

Abstract Sexual size dimorphism (SSD) and sex-related differences in body parts arise when intersexual genetic correlations are attenuated, thereby allowing morphology to evolve more freely. As much of the morphological variation results from changes in the relative proportions of body parts rather than absolute size, quantifying allometry is essential to understanding the evolution of morphology and the influence of sexual selection on SSD. Although SSD and static allometry have received considerable attention in scorpions, most research to date has focused on intraspecific aspects, leaving macroevolutionary patterns largely unexplored. In this chapter, I use phylogenetic comparative methods to analyse 10 morphometric traits measured in both sexes of 95 Neotropical buthid species, investigating SSD, evolutionary allometry, and sex-specific evolutionary dynamics of body size. The results confirm modest female-biased SSD, with females averaging 8% larger than males. However, SSD declined with increasing body size, in line with Rensch's rule. Despite limited differences in allometric slopes between sexes, males exhibited positive allometry for chela length and depth, traits likely influenced by sexual selection due to their role in courtship and mating. Evolutionary model fitting indicated that male size evolved under an Ornstein-Uhlenbeck process, consistent with stabilising selection towards an adaptive optimum, whereas female size followed a Brownian motion model, suggesting weaker or absent directional constraints. Overall, the findings suggest that the subtle SSD in Neotropical buthids is shaped by differing selective pressures acting with varying intensity on each sex and primarily results from evolutionary changes in male size. I emphasise that caution must be exercised when generalising these findings to scorpion families that differ radically in morphology from buthids.

Keywords Phylogenetic comparative methods · Sexual size dimorphism · Allometry

S. Í. A. Foerster (✉)
Department of Zoology, University of Tartu, Tartu, Estonia

Department of Ecology, Swedish University of Agricultural Sciences, Uppsala, Sweden
e-mail: stenio.foerster@slu.se

A. F. de Araujo Lira, E. González-Santillán (eds.), *The Biology of Neotropical Scorpions*, https://doi.org/10.1007/978-3-032-23194-9_2

2.1 Introduction

Scorpions are morphologically diverse, particularly in tropical regions where numerous species and forms thrive. The family Buthidae, the most species-rich, and the one that includes many of the most noxious scorpions, exhibits a wide range of morphological adaptations that reflect their ecological niches and life-histories. The morphological variability observed in these animals extends beyond body size, encompassing other traits that are inherently correlated with body size. Investigating the covariation patterns between body size and these traits lies at the core of allometric studies (Pélabon et al. 2014, 2018) and can provide substantial insights on the evolution of the morphological diversity seen today.

As a fundamental concept in biology, allometry describes the observed relationship linking morphological, behavioural, physiological, and life-history traits with the size of the organisms (Bonduriansky 2007; Pélabon et al. 2014, 2018). Allometric relationships can be described at the intraspecific level, during growth (ontogenetic allometry), among individuals at a specific developmental stage (static allometry), or at the interspecific level (evolutionary allometry, the focus of this chapter), the last two approaches typically focusing on adults (Esquerré et al. 2017; Pélabon et al. 2018; Painting et al. 2024). Understanding evolutionary allometry is crucial for elucidating the mechanisms of morphological diversification and adaptation among species. It allows researchers to identify evolutionary constraints and selective pressures that influence the development of specific traits, which can be particularly important in understanding phenomena such as sexual size dimorphism (Tsuboi and Takahashi 2024). For instance, variations in allometric scaling can indicate how traits evolve differently in males and females due to distinct reproductive strategies or ecological roles (Bolstad et al. 2015; Huijsmans et al. 2024).

Allometric relationships can be categorised into isometry, positive allometry, and negative allometry. Isometry occurs when the scaling of a trait is proportional to the scaling of the organism as a whole. This means that as body size increases, the trait size increases at the same rate, maintaining a constant ratio; the allometric slope (β) in such cases equals unity. In contrast, the steeper allometric slope ($\beta > 1$) seen in positive allometry indicates that as body size increases, the trait increases disproportionately larger. This phenomenon is often observed in sexually selected traits, where larger individuals develop more pronounced features that enhance their attractiveness (Rodríguez-Escobar et al. 2020) or competitive ability (Fromhage and Kokko 2014). Examples include the positive allometric slopes of secondary sexual traits in the case of the Aotearoa/New Zealand kelp flies (Coelopidae), as well as some noctuid moths (Rabieh et al. 2015; LeGrice and Holwell 2022). Such findings of positive allometry suggest that certain structures may evolve to become more prominent relative to overall body size due to evolutionary pressures. Negative allometry ($\beta < 1$), on the other hand, indicates that as body size increases, the trait scales to it at a slower rate, resulting in a relatively smaller size for the trait relative to the body. This pattern is often observed in traits that may incur higher costs or are less critical for survival or reproduction. For example, studies have found that male genitalia may exhibit negative allometry, suggesting that larger males do not necessarily have proportionately larger genitalia (Teder 1998; Bonduriansky and Day 2003; Mutanen et al. 2006), implying no clear functional advantages in growing larger genital structures.

Previous studies on scorpions have yielded mixed results in terms of allometric relationships between sexes, but with a certain tendency of morphometric traits to display isometric scaling (Sánchez-Quirós et al. 2012, but see Visser and Geerts 2021a, Carbonari et al. 2024). Selective forces acting on fecundity and sexual selection of secondary sexual traits have commonly been invoked to explain allometric patterns and sexual size dimorphism (SSD) in scorpions at the intraspecific level (Fox et al. 2015; Lira et al. 2018; Visser and Geerts 2021a, b). According to these studies, selection acting on female fecundity is thought to explain the general pattern of female-biased SSD, where females are larger than males, commonly observed in scorpions (summarised in Giménez Carbonari et al. 2024), and more broadly, in invertebrates (García-Navas et al. 2017; Rohner et al. 2018). Conversely, sexual selection acting on males likely accounts for the presence of certain traits in males that are larger than those in females and/or scale positively with overall body size (Fox et al. 2015; Visser and Geerts 2021b). This apparent dichotomy between selection on female size due to fecundity and sexual selection on male size has been widely discussed in studies on SSD and is not unique to scorpions—see the so-called "equilibrium model of SSD" in Blanckenhorn et al. (2007) and references therein.

A compelling concept for identifying drivers of SSD at the comparative level is Rensch's rule (Rensch 1950, 1959), which predicts that SSD increases with body size in species exhibiting male-biased SSD (i.e., males larger than females) or decreases with SSD in female-biased taxa (i.e., females larger than males). In such cases, the sex with greater body size variation (e.g., males) is considered the primary driver of size divergence (Ceballos et al. 2013) and is supposedly under stronger selection. This assumption can be further supported if, over the evolutionary history of a clade, males exhibit a faster rate of body size evolution compared to females (e.g., Hendry et al. 2014). Conversely, if female body size positively correlates with SSD and evolves faster than male body size, it would suggest that females are under stronger selection, likely driven by fecundity, since reproductive output is often positively associated with body size in scorpions and many other invertebrates (Outeda-Jorge et al. 2009; McLean et al. 2018). Therefore, analysing the dynamics of body size evolution within a clade and testing whether it conforms to Rensch's rule can provide valuable insights into the selective pressures shaping SSD and the allometric scaling of morphological traits.

Although substantial efforts have been made to describe SSD and allometric patterns in scorpions at the intraspecific level (i.e., static allometry) (e.g., Sánchez-Quirós et al. 2012; Carlson et al. 2014; Fox et al. 2015; Visser and Geerts 2021a, b; Carbonari et al. 2024), relatively little attention has been given to characterising these patterns in a comparative context (evolutionary allometry). Furthermore, no studies have yet specifically examined the tempo and mode of body size evolution in scorpions due to the previously limited availability of phylogenetic information and representative comparative datasets.

In this chapter, I present a concise yet comprehensive analysis of evolutionary allometries in 10 morphometric traits measured in male and female Neotropical buthid scorpions (95 species), using phylogenetic comparative methods. I also examine the mode and rate of body size evolution for both sexes, testing whether Neotropical buthids conform to the predictions of Rensch's rule. Based on previous findings on static allometry in scorpions (citations in the previous paragraph), I formulated the following hypotheses to be addressed here:

1. Sexual size dimorphism (SSD) in Neotropical scorpions is female-biased; however, size differences between sexes will tend to decrease as species get larger as predicted by Rensch's rule.
2. Body size will evolve at a faster rate in males than in females, resulting in more size variation in males. Thus, SSD in Neotropical species will be primarily driven by variation in male size.
3. At average body size, certain traits will be larger in males and exhibit positive allometry, particularly traits essential for courtship and mating such as chela measurements (e.g., chela length, width, and depth).

The findings and methodologies outlined here enhance our understanding of the ecological and evolutionary processes driving the morphological diversity observed in Neotropical scorpions (primarily buthids). They also provide a foundation for future comparative research on body size evolution, while offering new perspectives for interpreting previous findings on allometry and sexual size dimorphism in terrestrial arthropods.

2.2 Methods

Data on linear measurements of 10 morphometric traits were obtained from Foerster (2025), which were primarily scanned from specialised literature (see Supplementary material in Foerster 2025 for a complete list of sources). Carapace length was imputed for a small subset of species ($n = 16$) missing this measurement in the original dataset, using the predictive models developed by Foerster (2025). The list of traits analysed here includes length (L), width (W), and depth (D) measurements of chela (cheL, cheW, cheD), telson (teL, teW, teD), plus the length and width of the mesosoma (mesL and mesW, respectively), the length of the metasoma (metL) as the sum of metasomal segments I-V, and the width of the metasomal segment V (met5W), which can be used to represent the width of the metasoma. The selected traits are all involved in biological functions related to feeding, hunting, defence, and reproduction, and their respective allometric patterns are expected to covary, to a certain degree, with such biological functions (Prendini 2001; Coelho et al. 2017; Loria and Prendini 2021; Simone and van der Meijden 2021). Carapace length—the linear distance between anterior and posterior margins of the carapace—was taken as the proxy for body size, as its measurement is more straightforward and presumably less affected by measurement error than total length (McLean et al. 2018; Foerster 2025).

Allometric relationships are the expression of both the size of a trait (y) and the overall size of the organism (z) being controlled by a common growth factor (Huxley 1932). In this setting, the relative growth of y and z can be described by a power function in the form $y = \alpha z^{\beta}$, where α is a constant estimated from the data and β is the allometric exponent expressing the growth of trait y relative to the overall size z (Huxley 1932; Pélabon et al. 2018). Conveniently, allometric relationships are studied with log-transformed data (Glazier 2013; Pélabon et al. 2018) since the power function described above becomes linear on a log-log scale, so that the allometric relationship is expressed by $\log(y) = \log(\alpha) + \beta \times \log(z)$, with $\log(\alpha)$ and β being the allometric intercept and the allometric slope, respectively (Pélabon et al. 2018). Therefore, all analyses performed in this study (not only the allometric analyses) were carried out with log-transformed (natural logarithm) values of both dependent and independent variables.

To calculate the allometric slopes of the 10 morphometric traits, I used phylogenetic linear mixed models fitted under Bayesian settings with the R package MCMCglmm (Hadfield 2010). During the modelling, shared ancestry among species was taken into account by adding the inverse of the variance-covariance matrix calculated from the ultrametric molecular phylogeny of Buthidae proposed by Foerster (2025). The models were fitted separately for each trait and included a species-level random intercept factor to account for the repeated measures nature of the data (i.e., one measurement per sex). The models included an interaction term with sex to examine whether allometric slopes differed between males and females. All models were fitted with default priors through a single MCMC chain with 50,000 iterations sampled every 100; a burn-in period of 5000 iterations was applied. Such parameters were enough to ensure proper convergence of MCMC chains (effective sample sizes > 200). The posterior mean and 95% credible intervals (CIs) were then obtained for each trait's allometric slope (β) and interpreted in the following way: isometry if the 95% CIs of the β parameter overlaps with 1, positive allometry if the lower CI > 1, or negative allometry if the upper CI < 1. Mean differences in trait values between males and females were investigated using a similar approach, except that the mixed models did not include an interaction with sex. Instead, each trait was modelled at time in a model that had sex as the independent variable and carapace length as covariate (ANCOVA-like model). In this way, it was possible to access and compare trait sizes at the average body size across sexes, which was achieved using the R package marginaleffects (Arel-Bundock et al. 2024).

Following Abouheif and Fairbairn (1997) and Ceballos et al. (2013), phylogenetic major axis regression implemented in phytools (Revell 2024) was used to test whether the studied species conform to Rensch's rule by regressing log female size on log male size. In this setting (i.e., female size is the dependent variable), a slope less than unity indicates allometric scaling consistent with Rensch's rule, while a slope higher than unity indicates allometry inconsistent with Rensch's rule (Abouheif and Fairbairn 1997; Webb and Freckleton 2007). A slope of 1 would be interpreted as absence of allometry (Abouheif and Fairbairn 1997). The analysis explicitly includes shared evolutionary history among the species and estimates the degree of phylogenetic signal in model residuals (λ) by maximum likelihood (Revell 2024); a statistical test on the null hypothesis that the slope is equal to unity is calculated based on Clarke (1980) (hereafter "Clarke's test") and reported together with the regression results.

Lastly, the mode and rate of phenotypic evolution in body size were analysed using models of continuous trait evolution implemented in the geiger R package (Pennell et al. 2014). Three models were fitted for each sex. The first, the Brownian motion (BM) model, is described by two parameters: θ, the ancestral trait value at the start of the simulation (also termed phylogenetic mean), and σ^2, the rate parameter that determines the speed at which trait values move randomly within the trait space over time. The BM model assumes that traits evolve through a random walk, with changes accumulating over time without a specific direction (Beaulieu et al. 2012; Cooper et al. 2016). The second model, the Ornstein-Uhlenbeck (OU) model, extends the BM framework by incorporating an additional parameter, α, which represents the strength of attraction towards a central tendency (θ), as expected under a stabilising selection process (Beaulieu et al. 2012). The OU model also includes the evolutionary rate parameter σ^2. The parameter α, measuring the absolute strength of selection, can be expressed in more intuitive terms such as the phylogenetic half-life, calculated as $t_{1/2} = \ln 2/\alpha$, which indicates the time required for a trait to move halfway from an ancestral value towards the selective optimum θ (Hansen et al. 2008). Additionally, the metric ρ (Cornuault 2022) quantifies the percentage reduction in trait variance over a specified time period (T) due to selection, relative to the variance expected under pure drift. It is calculated as:

$$\rho T = 1 - \frac{(-2\alpha T)}{2\alpha T}$$

where α comes from the OU model and T is the time period of interest (e.g., the tree height). The third model was the white noise (WN) model, a non-phylogenetic approach that assumes trait values are drawn from a normal distribution with no covariation structure among species (Pennell et al. 2014). The three models were compared within each sex using Akaike weights calculated with the phytools package. Subsequently, using the same packages, I simulated 1000 trait values conditioned on the tree and the parameters estimated from the best-fit models for each sex. These simulations were used to calculate 95% confidence intervals for the phylogenetic mean trait value (θ) and the evolutionary rate parameter (σ^2), also allowing the calculation of P-values in Monte Carlo tests.

2.3 Results

Body size, measured as carapace length, ranged from 2 to 11 mm in females and 1.6 to 11 mm in males. On average, females were 8% larger than males (pMCMC < 0.01), with posterior means and 95% credible intervals estimated at 4.1 mm (2.7–6.7) for

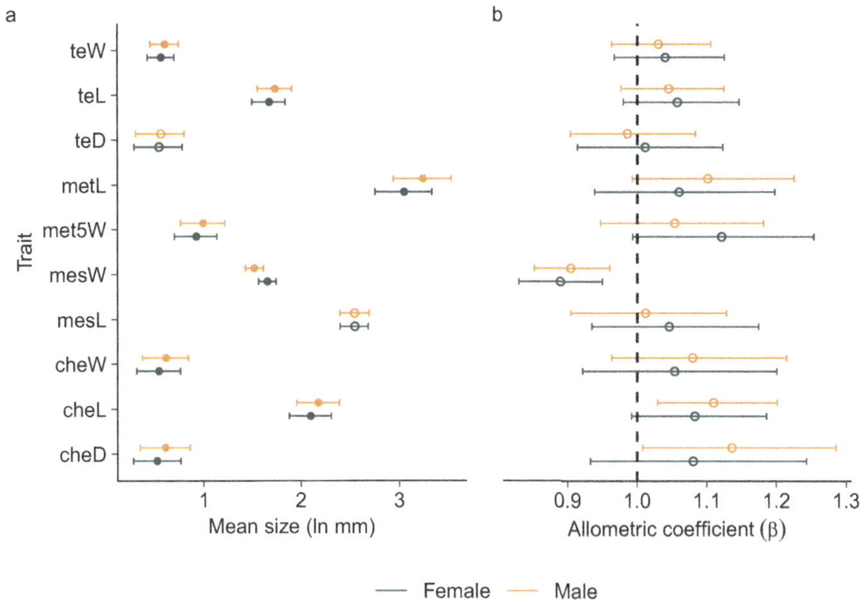

Fig. 2.1 Sex differences in mean trait values (**a**) and allometric coefficients (**b**) across 95 species of Neotropical buthid scorpions, estimated using Bayesian phylogenetic linear mixed models. Points represent mean posterior estimates, with surrounding 95% credible intervals (CIs). Solid points indicate statistically significant differences (pMCMC < 0.05) between males and females, while open points denote non-significant differences. In panel (**a**), coefficients were estimated including carapace length (a proxy for body size) as a covariate. In panel (**b**), allometric coefficients were calculated using carapace length as a measure of body size. Allometric patterns are classified as: isometric (CI includes 1), positive (lower CI > 1), or negative (upper CI < 1). The allometric coefficients and their CIs are provided in Appendix 1

females and 3.8 mm (2.5–6.1) for males. Trait-specific comparisons revealed that males were, on average, slightly larger than females for seven out of the 10 traits when controlling for body size (Fig. 2.1a). The exceptions were telson depth and mesosoma length, where no significant differences between sexes were observed, and mesosoma width, where male mesosomas were, on average, 13% narrower than those of females (Fig. 2.1a).

None of the traits exhibited statistically significant differences in allometric coefficients (slopes) between sexes (Fig. 2.1b). However, some differences in allometric patterns were observed. In males, chela length and chela depth displayed positive allometry, while in females, these traits exhibited isometry (Fig. 2.1b). For chela length, the lower bound of the 95% credible interval for the female allometric slope was slightly below one—a pattern also observed for the length of the fifth metasomal segment in females and for metasoma length in males (Fig. 2.1b).

The phylogenetic major axis regression analysis demonstrated that the relationship between female and male body size in the Neotropical species examined conforms to Rensch's rule (intercept = 0.29, $\beta = 0.87$). This conclusion is supported by

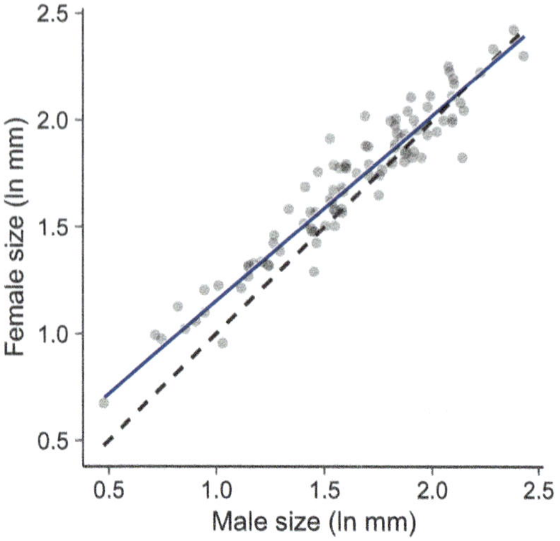

Fig. 2.2 Phylogenetic major axis regression of female size on male size in 95 Neotropical scorpion species. The observed regression slope (blue line) was significantly shallower than the expected slope under the assumption of isometry (dashed line). Body size was represented by the natural logarithm of carapace length

the observed regression slope being shallower than the theoretical expectation of isometry (i.e., $\beta = 1$) across the studied species (Fig. 2.2). The discrepancy between the observed and expected regression slopes was further validated by Clarke's test ($r^2 = 0.85$, $T = 3.5$, $df = 68$, $P = 0.001$). Additionally, the residuals of the female vs. male regression showed a relatively high degree of phylogenetic signal ($\lambda = 0.73$).

Evidence for different evolutionary processes describing the evolution of body size between female and male scorpions could be observed. For females, the evolution of body size was better described by a random walk (BM) process, although the OU model received considerable support (Table 2.1). Conversely, body size in males was better explained by an OU process, with the BM model being the second-best-fit model (Table 2.1).

The examination of the best-fit models for each sex (Table 2.1) revealed that the phylogenetic mean (θ), representing the trait value at the root, was similar between sexes. Ancestral body sizes (carapace length) for females and males were estimated at 4.2 mm and 3.7 mm, respectively (Table 2.1). Monte Carlo simulations indicated no significant differences between female and male ancestral sizes (Fig. 2.3a; $P = 0.87$). The 95% confidence intervals for the phylogenetic mean across 1000 Monte Carlo simulations ranged from 2.7 to 6.5 mm in females and from 3.1 to 5.4 mm in males.

The average rate of body size evolution (σ^2) for males was nearly three times higher than that for females, as estimated by the best-fit models of continuous trait evolution (Table 2.1). Monte Carlo simulations confirmed that this difference in evolutionary rate was statistically significant (Fig. 2.3b; $P = 0$). The 95% confidence intervals for the average σ^2 ranged from 0.01 to 0.06 log mm My^{-1} in males and from 0.003 to 0.005 log mm My^{-1} in females.

In absolute terms, the alpha parameter from the OU models indicated that the strength of selection in male size was three times higher than that estimated for female size (Table 2.1). Accordingly, male size is expected to approach its

Table 2.1 Models of continuous trait evolution estimated for male and female body size in 95 Neotropical scorpion species. The models include Brownian Motion (BM), Ornstein-Uhlenbeck (OU), and White Noise (WN) (see Methods for details). Analyses were conducted using log--transformed carapace length as a proxy for body size, incorporating measurement error in the trait measurements

	Female				Male			
Model	AICc (weight)	θ	σ^2	α	AICc (weight)	θ	σ^2	α
BM	21.53 (**0.68**)	1.44	0.003	–	82.73 (0.19)	1.32	0.005	–
OU	23.04 (0.32)	1.46	0.004	0.007	79.85 (**0.81**)	1.40	0.008	0.021
WN	76.30 (0.00)	1.70	0.122	–	98.42 (0.00)	1.61	0.153	–

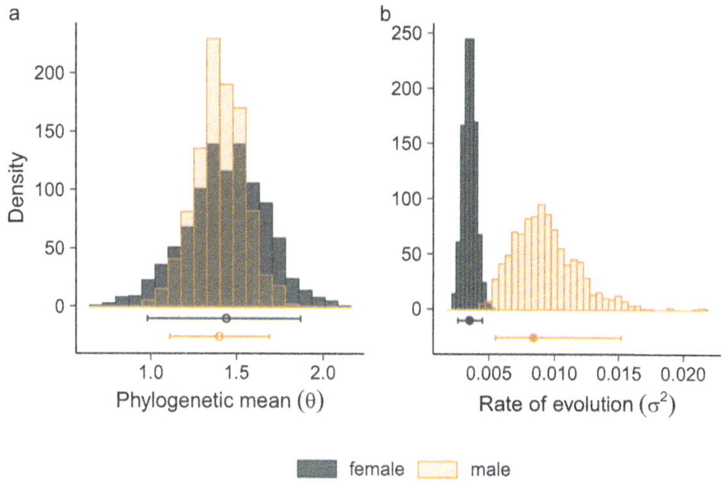

Fig. 2.3 Distributions of the (**a**) phylogenetic mean and (**b**) the rate of phenotypic evolution in body size for male and female Neotropical scorpion species ($n = 95$), based on 1000 Monte Carlo simulations. Distributions were generated using parameters from the best-fit models of continuous trait evolution (Brownian motion for females and Ornstein–Uhlenbeck for males; see Methods). Points represent the maximum likelihood estimates of θ and σ^2 from the best-fit models, with 95% confidence intervals derived from the Monte Carlo simulations. Solid points indicate parameters for which differences between sexes were statistically significant ($P < 0.05$), whereas open points indicate non-significant differences

phenotypic optimum more rapidly than female size, as revealed by the phylogenetic half-lives. The time required for male size to move halfway from the ancestral size towards the phenotypic optimum was estimated at 33.2 My, whereas for females, the phylogenetic half-life was estimated at 105.3 My. In relative terms, such numbers translate to 50% and 160% of the tree height (66 My) for average male and female body size, respectively. Furthermore, the percent reduction in trait variance due to selection (ρ) was twice as high for males (66%) compared to females (33%).

2.4 Discussion

Sexual size dimorphism in animals arises from differential selection pressures on males and females, leading to distinct evolutionary trajectories for body size, and scorpions are not an exception to this rule. To gain more insights on that, this chapter provides a comparative analysis of body size and evolutionary allometries in 10 morphometric traits in male and female Neotropical scorpions, framed within an explicit phylogenetic context.

In absolute terms, SSD was modest, with females being, on average, 8% larger than males. It is worth saying that carapace length was used as the reference trait for body size in all analyses, although previous research has highlighted potential ambiguities in selecting a reference character (Fox et al. 2015; Coelho et al. 2017). For instance, Fox et al. (2015) found that SSD in the North American species *Hadrurus arizonensis* Ewing, 1928, varied depending on the trait chosen to represent body size. Their study identified the width of the first metasomal segment as the most appropriate metric. While Fox et al. (2015) focused on a single species, Foerster (2025) recently demonstrated that the length of the fifth metasomal segment provided the best predictions for total length (though not substantially better than carapace length) in a phylogenetic comparative study of 195 buthids. These findings contrast with the nearly universal use of carapace length or total length as proxies for body size in scorpions (e.g., Brown 2003, Outeda-Jorge et al. 2009, McLean et al. 2018; Moreira et al. 2022). Yet, the preference for linear measurements of metasomal segments as body size proxies may be more statistically than biologically motivated, particularly in terms of allometric scaling. In other words, the biological meaning of the scaling of a trait relative to the length or width of a specific metasomal segment might be difficult to conceptualise. Moreover, all traits examined in this study, including common body size proxies, are strongly intercorrelated (Foerster 2025). Thus, the choice between total length and carapace length as a representative body size metric is less critical for the analyses presented here.

In qualitative terms, the female-biased pattern of SSD observed in this study aligns with intraspecific SSD patterns previously described (Fox et al. 2015; Lira et al. 2018; Visser and Geerts 2021a, b; Carbonari et al. 2024). Mesosoma length and width were the only characters where males did not significantly exceed females when accounting for body size (carapace length). Females showed significantly wider mesosomas than males, suggesting greater mesosomal areas (assuming a simplistic length × width calculation). A plausible explanation for the enlargement of compartmental and superficial mesosomal areas in females may rely, at least partially, on pre- and post-embryonic developmental features in scorpions. From a pre-embryonic perspective, a larger compartmental area could accommodate greater reserves stored in the hepatopancreas, which is an organ that constitutes nearly 40% of total body weight and plays a direct role in allocating resources to oogenesis and vitellogenesis (Warburg 2011). From a post-embryonic standpoint, larger and wider mesosomas in females may provide more surface area for carrying scorpionlings after birth, potentially enabling them to sustain larger broods in terms of either offspring size or number. This explanation, however, should be taken with caution because a precise anatomical quantification of compartmental and superficial mesosomal areas has yet to be developed and applied to support (or not) the explanations above.

The observed female-biased SSD supports the common notion that reproductive benefits linked to larger body size may favour selection for larger females in scorpions (Visser and Geerts 2021a, b, Carbonari et al. 2024). Interestingly, the pattern of SSD in Neotropical buthids follows Rensch's rule, indicating that selection for larger females may vary across the size spectrum. SSD is more pronounced in smaller species and nearly absent in larger ones. This pattern indirectly suggests that selection on female fecundity may be stronger in smaller species, favouring larger female size relative to males, or alternatively, that selective pressures may have driven a reduction in male size in these species (see below). Proximate causes of SSD may involve sex-specific mortality, energy acquisition, and energy expenditure related to maintenance, reproductive tissues, and reproductive behaviour (Mollet et al. 2023). However, disentangling the relative contribution of these factors to the observed SSD patterns in scorpions remains largely speculative due to the paucity of comparative studies addressing these mechanisms. Nevertheless, potential explanations for female-biased SSD in Neotropical buthids can still be approximated from experimental work on some of the (female-biased) species included in this study. For instance, males of *Tityus neibae* Armas, 1999, can reach maturity by the fourth instar when fed weekly, whereas females do not mature by this stage under identical conditions (Seiter et al. 2020). This suggests that males develop more rapidly than females, potentially as a trade-off that favours earlier maturity at smaller body size, presumably reducing predation risk. In contrast, females grow larger, which may enhance reproductive success by increasing offspring number or size (Seiter et al. 2020). Higher mortality risk associated with male behaviour (e.g., males being more active and thus more exposed to predators) has also been proposed to explain earlier maturation at smaller body size in *Tityus pusillus* Pocock, 1893 (Albuquerque and Lira 2016; Lira et al. 2018). These observations imply that individuals adopt strategic growth decisions in response to trade-offs between growth and mortality risk, and between attaining large size and emerging early (Tuni and Berger-Tal 2012; Albuquerque and Lira 2016). Findings from other arachnid groups also support this logic. In some spiders, females may reach larger sizes by extending their development, either by increasing the number of juvenile instars (Higgins and Goodnight 2010, 2011; Kuntner et al. 2012; Agnarsson et al. 2024) or by prolonging the duration of each instar without altering their number (Higgins and Goodnight 2011). Additionally, female spiders may grow more efficiently through greater energy assimilation and reduced activity levels relative to males (Chelini et al. 2019). It is worth noting, however, that these findings are drawn primarily from spider species exhibiting extreme female-biased SSD, which is not the case in buthid scorpions, including those from other biogeographical regions. In addition, the general pattern of longer developmental times leading to larger adult sizes is well-known in invertebrates (Blanckenhorn 2000; Blanckenhorn et al. 2007; Tammaru and Teder 2025), but despite the examples discussed above, it cannot be taken as an uncontested rule in Neotropical buthids. Lira et al. (2021) demonstrated that in *Ananteris mauryi* Lourenço, 1982, for instance, males take slightly more time to mature than females, yet females still attain larger final (adult) size on average.

Yet another perspective to understand the female-biased SSD in scorpions could be focused on the average size reduction in males, rather than size increase in females. This is consistent with the "Ghiselin-Reiss small-male hypothesis" (Ghiselin 1974; Reiss 1989), which posits that smaller males are likely to evolve when reproductive success is largely determined by scramble competition for access to females (Blanckenhorn et al. 1995). Under such conditions, males are expected to prioritise reproductive investment over somatic growth, potentially resulting in reduced survival and a corresponding relaxation of male-male competition (Blanckenhorn et al. 1995; Mollet et al. 2023). Looking at other arachnids, smaller male size has been associated with reproductive success in mating systems that involve scramble competition (Danielson-François et al. 2012; Jordan et al. 2014), where males race to find and mate with females as quickly as possible. According to this scenario, different sex-related functional roles may underlie the evolution of SSD in Neotropical buthids, with male morphology shaped by selection for sprinting-based foraging strategies, and female morphology optimised for maximising reproductive output and parental care, as suggested by intraspecific studies on non-Neotropical buthids (Carlson et al. 2014; Miller et al. 2016). As the pattern of SSD reported here is consistent with Rensch's rule, reduction in absolute male size occurs principally in comparatively smaller species—e.g., *Microtityus* sp. Kjellesvig-Waering, 1966, *Ananteris* sp. Thorell, 1891, *Chaneke* sp. Francke, Teruel & Santibanez-Lopez, 2014, and some few species in the genera *Centruroides* Marx, 1890 and *Tityus* C. L. Koch, 1836. This implies that males of larger species benefit from increased absolute size. Such benefits may correlate with pre-copulatory sexual selection mechanisms like male-male competition and territoriality, which have been reported for large scorpions (Tang 2023), including Neotropical buthids (Wyman et al. 2025), but less likely to play a significant role in the morphology of smaller-bodied ones. Nevertheless, the evolutionary drivers behind the subtle but significant size differences observed in Neotropical buthids are difficult to predict, but just like in other arthropods (e.g., Higgins and Goodnight 2011; Quiñones-Lebrón et al. 2021; Agnarsson et al. 2024), constraints such as intersexual genetic correlations and the physiological and ecological costs of attaining and maintaining larger body sizes in both sexes (Blanckenhorn 2000; Blanckenhorn et al. 2007; Higgins and Goodnight 2011; Chelini et al. 2019) certainly apply.

When looking at the allometric scaling of the studied traits, I found no evidence of sex-specific differences in allometric slopes, although certain traits exhibited distinct allometric patterns between sexes. Chela length and chela depth displayed positive allometry in males but were isometric in females. Based on prior intraspecific syntheses (Carbonari et al. 2024), I anticipated these traits to exhibit positive allometry in males due to their functional role in sexual interactions. Scorpion reproduction involves a complex courtship ritual in which the male uses his pedipalps (chelae) to grasp the female and performs coordinated walking movements to induce copulation (Foerster et al. 2021; Simone and van der Meijden 2021). This mating behaviour, which can last from a few minutes to hours depending on the female's resistance, demands significant pinch force from males, particularly in species with female-biased SSD. Indeed, size-corrected analyses have shown a strong positive correlation between chela depth and pinch force in scorpions (van der Meijden et al. 2010). The presence of positive allometry in male chela depth (but not in females) thus suggests that males of smaller species possess disproportionately greater pinch force relative to their body size, aligning with the greater degree of SSD observed among these species. Similarly, positive allometry was found in male chela length. However, in this case, mechanical performance related to pinch force may not be the primary selective driver. Instead, positive allometry in chela length might reflect an adaptation to enhance closing speed, as previously suggested (van der Meijden et al. 2010). If so, this functional shift would also be more pronounced in smaller-bodied species, which showed the highest degree of SSD. However, precise descriptions of potential associations between male pinch force and SSD in scorpions remain to be tested using comparative datasets.

Regarding the models of continuous trait evolution, I found significant differences not only in the rate of body size evolution between sexes (roughly three times faster in males) but also differences in evolutionary mode: with OU and BM processes providing the best fit to the male and female body size data, respectively. These findings support the hypothesis of different selective pressures acting on each sex, with varying intensities. Although this may sound obvious at first, no studies have yet identified and quantified such evolutionary processes acting on body size evolution in scorpions. The results suggest that the evolution of body size in females is better described by a drift (or other stochastic) process without directional selection towards a specific adaptive peak. While female size is not constrained to evolve towards a particular body size optimum, this finding should not be taken as a general lack of constraints on female size, as their low evolutionary rate (σ^2) indicates an overall low evolvability, or ability to explore the trait space. In addition, it should be noticed that the fit of an OU process to the female size cannot be completely ruled out, as there was some statistical support for this model in terms of AIC weights. More or less the same can be said for male size, although their statistical support favouring the OU model was more convincing. In the case of males, the best fit of an OU process suggests the presence of stabilising selection towards an optimal size (estimated to be at roughly 4.06 mm in carapace length). The higher rate of size evolution in males also indicates faster exploration of size spectrum during the evolution of the species, but as in females, this does not imply an unconstrained exploration of trait space. Instead, the absolute strength of selection (α) was three times higher for males than for females (using the OU model for females as a didactic example). Accordingly, the much lower phylogenetic half-life and much higher percentual decrease in trait variance (ρ) estimated for male size indicate intense selective pressures—potentially sexual selection (Carbonari et al. 2024)—on male size, even while permitting (or perhaps leading to) greater size variation than observed in females. Thus, despite the limitations of making direct conclusions about sex-specific evolutionary constraints on body size based solely on singe-rate or single-peak models, the results presented here mark an important foundation for future phylogenetic comparative studies of body size evolution in scorpions and other arthropods.

2.5 Conclusions

By using phylogenetic comparative methods and a representative morphological dataset, this chapter provides an insightful introduction to the understanding of SSD and evolutionary allometry in scorpions at the interspecific level. I confirmed the presence of (a modest) female-biased pattern of SSD in Neotropical buthid scorpions at the comparative level, and that it conforms to the predictions established by the Rensch's rule. The findings support the idea that size variation in males is the main contributor for the pattern of SSD observed in the studied species, which agrees with recent conclusions reported in a more representative study with buthids (Foerster 2025). Sexual selection appears to play a limited but non-negligible role in explaining size variation in males of Neotropical buthids, with foraging and mating behaviour probably influencing sex-specific morphologies in the studied taxa. Finally, I emphasise the need for further phylogenetic comparative research into scorpion morphology and ecology to improve the current understanding of trait evolution and to avoid extrapolating intraspecific patterns to macroevolutionary contexts. In this regard, I also acknowledge the limitations of my own study by cautioning against generalising the results presented here to scorpion families with radically different morphologies than buthids.

References

Abouheif E, Fairbairn DJ (1997) A comparative analysis of allometry for sexual size dimorphism: assessing Rensch's rule. Am Nat 149(3):540–562. https://doi.org/10.1086/286004

Agnarsson I, Goodnight C, Higgins L (2024) Developmental asynchrony: a potential cost of extreme sexual size dimorphism in seasonal environments. J Arachnol 52(1). https://doi.org/10.1636/JoA-S-22-058

Albuquerque CMR, Lira AFA (2016) Insights into reproductive strategies of *Tityus* (*Archaeotityus*) *pusillus* Pocock, 1893 (Scorpiones, Buthidae). C R Biol 339(5–6):179–184. https://doi.org/10.1016/j.crvi.2016.03.003

Arel-Bundock V, Greifer N, Heiss A (2024) How to interpret statistical models using marginaleffects for R and Python. J Stat Softw 111(9):1–32. https://doi.org/10.18637/jss.v111.i09

Beaulieu JM, Jhwueng D-C, Boettiger C, O'Meara BC (2012) Modeling stabilizing selection: expanding the Ornstein-Uhlenbeck model of adaptive evolution: expanding the Ornstein Uhlenbeck model. Evolution 66(8):2369–2383. https://doi.org/10.1111/j.1558-5646.2012.01619.x

Blanckenhorn WU (2000) The evolution of body size: what keeps organisms small? Q Rev Biol 75(4):385–407

Blanckenhorn WU, Preziosi RF, Fairbairn DJ (1995) Time and energy constraints and the evolution of sexual size dimorphism? To eat or to mate? Evol Ecol 9(4):369–381. https://doi.org/10.1007/BF01237760

Blanckenhorn WU, Dixon AFG, Fairbairn DJ, Foellmer MW, Gibert P, Linde K, van der Meier R, Nylin S, Pitnick S, Schoff C, Signorelli M, Teder T, Wiklund C (2007) Proximate causes of Rensch's rule: does sexual size dimorphism in arthropods result from sex differences in development time? Am Nat 169(2):245–257. https://doi.org/10.1086/510597

Bolstad GH, Cassara JA, Márquez E, Hansen TF, van der Linde K, Houle D, Pélabon C (2015) Complex constraints on allometry revealed by artificial selection on the wing of *Drosophila melanogaster*. Proc Natl Acad Sci 112(43):13284–13289. https://doi.org/10.1073/pnas.1505357112

Bonduriansky R (2007) Sexual selection and allometry: a critical reappraisal of the evidence and ideas. Evolution 61(4):838–849. https://doi.org/10.1111/j.1558-5646.2007.00081.x

Bonduriansky R, Day T (2003) The evolution of static allometry in sexually selected traits. Evolution 57(11):2450. https://doi.org/10.1554/03-213

Brown CA (2003) Offspring size-number trade-offs in scorpions: an empirical test of the Van Noordwijk and De Jong model. Evolution 57(9):2184–2190. https://doi.org/10.1111/j.0014-3820.2003.tb00397.x

Carbonari JJG, Oviedo-Diego MA, Peretti AV, Mattoni CI (2024) Sexual dimorphism and functional allometry in scorpions: a comparative study from a neotropical species. Zoology:126208. https://doi.org/10.1016/j.zool.2024.126208

Carlson BE, McGinley S, Rowe MP (2014) Meek males and fighting females: sexually-dimorphic antipredator behavior and locomotor performance is explained by morphology in bark scorpions (*Centruroides vittatus*). PLoS One 9(5):e97648. https://doi.org/10.1371/journal.pone.0097648

Ceballos CP, Adams DC, Iverson JB, Valenzuela N (2013) Phylogenetic patterns of sexual size dimorphism in turtles and their implications for Rensch's rule. Evol Biol 40(2):194–208. https://doi.org/10.1007/s11692-012-9199-y

Chelini M-C, Delong JP, Hebets EA (2019) Ecophysiological determinants of sexual size dimorphism: integrating growth trajectories, environmental conditions, and metabolic rates. Oecologia 191(1):61–71. https://doi.org/10.1007/s00442-019-04488-9

Clarke MRB (1980) The reduced major axis of a bivariate sample. Biometrika 67(2):441–446. https://doi.org/10.1093/biomet/67.2.441

Coelho P, Kaliontzopoulou A, Rasko M, van der Meijden A (2017) A 'striking' relationship: scorpion defensive behaviour and its relation to morphology and performance. Funct Ecol 31(7):1390–1404. https://doi.org/10.1111/1365-2435.12855

Cooper N, Thomas GH, Venditti C, Meade A, Freckleton RP (2016) A cautionary note on the use of Ornstein Uhlenbeck models in macroevolutionary studies. Biol J Linn Soc 118(1):64–77. https://doi.org/10.1111/bij.12701

Cornuault J (2022) Bayesian analyses of comparative data with the Ornstein–Uhlenbeck model: potential pitfalls. Syst Biol 71(6):1524–1540. https://doi.org/10.1093/sysbio/syac036

Danielson-François A, Hou C, Cole N, Tso I-M (2012) Scramble competition for moulting females as a driving force for extreme male dwarfism in spiders. Anim Behav 84(4):937–945. https://doi.org/10.1016/j.anbehav.2012.07.018

Esquerré D, Sherratt E, Keogh JS (2017) Evolution of extreme ontogenetic allometric diversity and heterochrony in pythons, a clade of giant and dwarf snakes. Evolution 71(12):2829–2844. https://doi.org/10.1111/evo.13382

Foerster SÍA (2025) Body size prediction in scorpions: a phylogenetic comparative examination of linear measurements of individual body parts. PeerJ 13:e18621. https://doi.org/10.7717/peerj.18621

Foerster SÍA, Dionisio-da-Silva W, Santos AB, Albuquerque CMR, Lira AFA (2021) Notes on the courtship behavior of the parthenogenetic scorpion *Tityus stigmurus*. Acta Arachnol 70(2):83–86. https://doi.org/10.2476/asjaa.70.83

Fox GA, Cooper AM, Hayes WK (2015) The dilemma of choosing a reference character for measuring sexual size dimorphism, sexual body component dimorphism, and character scaling: cryptic dimorphism and allometry in the scorpion *Hadrurus arizonensis*. PLoS One 10(3):e0120392. https://doi.org/10.1371/journal.pone.0120392

Fromhage L, Kokko H (2014) Sexually selected traits evolve positive allometry when some matings occur irrespective of the trait. Evolution 68(5):1332–1338. https://doi.org/10.1111/evo.12349

García-Navas V, Noguerales V, Cordero PJ, Ortego J (2017) Ecological drivers of body size evolution and sexual size dimorphism in short-horned grasshoppers (Orthoptera: Acrididae). J Evol Biol 30(8):1592–1608. https://doi.org/10.1111/jeb.13131

Ghiselin MT (1974) The economy of nature and the evolution of sex. University of California Press. https://books.google.ee/books?id=S55CJYKKEUoC

Glazier DS (2013) Log-transformation is useful for examining proportional relationships in allometric scaling. J Theor Biol 334:200–203. https://doi.org/10.1016/j.jtbi.2013.06.017

Hadfield JD (2010) MCMC Methods for multi-response generalized linear mixed models: the MCMCglmm R Package. J Stat Softw 33(2):1–22

Hansen TF, Pienaar J, Orzack SH (2008) A comparative method for studying adaptation to a randomly evolving environment. Evolution:1965–1977. https://doi.org/10.1111/j.1558-5646.2008.00412.x

Hendry CR, Guiher TJ, Pyron RA (2014) Ecological divergence and sexual selection drive sexual size dimorphism in new world pitvipers (Serpentes: Viperidae). J Evol Biol 27(4):760–771. https://doi.org/10.1111/jeb.12349

Higgins L, Goodnight C (2010) *Nephila clavipes* females have accelerating dietary requirements. J Arachnol 38(1):150–152. https://doi.org/10.1636/Sh08-92.1

Higgins L, Goodnight C (2011) Developmental response to low diets by giant *Nephila clavipes* females (Araneae: Nephilidae). J Arachnol 39(3):399–408. https://doi.org/10.1636/B11-18.1

Huijsmans TERG, Courtiol A, Van Soom A, Smits K, Rousset F, Wauters J, Hildebrandt TB (2024) Quantifying maternal investment in mammals using allometry. Commun Biol 7(1):475. https://doi.org/10.1038/s42003-024-06165-x

Huxley J (1932) Problems of relative growth. Johns Hopkins University Press, Methuen/London

Jordan LA, Kokko H, Kasumovic M (2014) Reproductive foragers: male spiders choose mates by selecting among competitive environments. Am Nat 183(5):638–649. https://doi.org/10.1086/675755

Kuntner M, Zhang S, Gregorič M, Li D (2012) *Nephila* female gigantism attained through post-maturity molting. J Arachnol 40(3):345–347. https://doi.org/10.1636/B12-03.1

LeGrice RJ, Holwell GI (2022) Size variation, allometry and mating success in AotearoalNew Zealand kelp flies (Coelopidae). Evol Ecol 36(4):683–700. https://doi.org/10.1007/s10682-022-10159-8

Lira AFA, Pordeus LM, Rego FNAA, Iannuzzi K, Albuquerque CMR (2018) Sexual dimorphism and reproductive behavior in the Brazilian scorpion *Tityus pusillus* (Scorpiones, Buthidae). Invertebr Biol 137(3):221–230. https://doi.org/10.1111/ivb.12221

Loria SF, Prendini L (2021) Burrowing into the forest: phylogeny of the Asian forest scorpions (Scorpionidae: Heterometrinae) and the evolution of ecomorphotypes. Cladistics 37(2):109–161. https://doi.org/10.1111/cla.12434

Lira AFA, Correia JA, Dionisio-da-Silva W, Albuquerque CMR (2021) Life-history traits of the Brazilian litter-dwelling scorpion: post-embryonic development and reproductive behaviour in Ananteris mauryi Lourenço, 1982 (Scorpiones: Buthidae). J Nat Hist 55(21–22):1323–1334. https://doi.org/10.1080/00222933.2021.1925766

McLean CJ, Garwood RJ, Brassey CA (2018) Sexual dimorphism in the Arachnid orders. PeerJ 6:e5751. https://doi.org/10.7717/peerj.5751

Miller DW, Jones AD, Goldston JS, Rowe MP, Rowe AH (2016) Sex differences in defensive behavior and venom of the striped bark scorpion *Centruroides vittatus* (Scorpiones: Buthidae). Integr Comp Biol 56(5):1022–1031. https://doi.org/10.1093/icb/icw098

Mollet FM, Enberg K, Boukal DS, Rijnsdorp AD, Dieckmann U (2023) An evolutionary explanation of female-biased sexual size dimorphism in North Sea plaice, *Pleuronectes platessa* L. Ecol Evol 13(1):e8070. https://doi.org/10.1002/ece3.8070

Moreira MOM, Araújo VLN, Foerster SÍA, Moura GJB, Lira AFA (2022) Relationship between body size and habitat heterogeneity on cannibalism and intraguild predation in scorpions. Biologia 77(10):2867–2873. https://doi.org/10.1007/s11756-022-01154-z

Mutanen M, Kaitala A, Mönkkönen M (2006) Genital variation within and between three closely related *Euxoa* moth species: testing the lock-and-key hypothesis. J Zool 268(2):109–119. https://doi.org/10.1111/j.1469-7998.2005.00029.x

Outeda-Jorge S, Mello T, Pinto-da-Rocha R (2009) Litter size, effects of maternal body size, and date of birth in South American scorpions (Arachnida: Scorpiones). Zoologia 26:45–53

Painting CJ, Symonds MRE, Holwell GI (2024) The evolution of positive allometry and exaggerated traits in a diverse beetle clade. Funct Ecol 38(5):1265–1283. https://doi.org/10.1111/1365-2435.14532

Pélabon C, Firmat C, Bolstad GH, Voje KL, Houle D, Cassara J, Rouzic AL, Hansen TF (2014) Evolution of morphological allometry. Ann N Y Acad Sci 1320(1):58–75. https://doi.org/10.1111/nyas.12470

Pélabon C, Tidière M, Lemaître J-F, Gaillard J-M (2018) Modelling allometry: statistical and biological considerations—a reply to Packard. Biol J Linn Soc. https://doi.org/10.1093/biolinnean/bly141

Pennell MW, Eastman JM, Slater GJ, Brown JW, Uyeda JC, FitzJohn ZG, Alfaro ME, Harmon LJ (2014) geiger v2.0: an expanded suite of methods for fitting macroevolutionary models to phylogenetic trees. Bioinformatics 30(15):2216–2218. https://doi.org/10.1093/bioinformatics/btu181

Prendini L (2001) Substratum specialization and speciation in southern African scorpions: The Effect Hypothesis revisited. In: Scorpions 2001: in memoriam Gary A. Polis. British Arachnological Society, Burnham Beeches, pp 113–138

Quiñones-Lebrón SG, Kuntner M, Kralj-Fišer S (2021) The effect of genetics, diet, and social environment on adult male size in a sexually dimorphic spider. Evol Ecol 35(2):217–234. https://doi.org/10.1007/s10682-020-10097-3

Rabieh MM, Esfandiari M, Seraj AA, Bonduriansky R (2015) Genital and body allometry in two species of noctuid moths (Lepidoptera: Noctuidae). Biol J Linn Soc 116(1):183–196. https://doi.org/10.1111/bij.12550

Reiss MJ (1989) The allometry of growth and reproduction, 1st edn. Cambridge University Press. https://doi.org/10.1017/CBO9780511608483

Rensch B (1950) Die Abhängigkeit der relativen sexualdifferenz von der Körpergrösse. Bonner Zoologische Beiträge 1:58–69

Rensch B (1959) Evolution above the species level. Columbia University Press, New York

Revell LJ (2024) phytools 2.0: an updated R ecosystem for phylogenetic comparative methods (and other things). PeerJ 12:e16505. https://doi.org/10.7717/peerj.16505

Rodríguez-Escobar FE, Carrillo-Muñoz AI, Serrano-Meneses MA (2020) Seasonal variation in the allometry of wing pigmentation in adult males of the territorial damselfly *Hetaerina vulnerata* (Insecta Odonata). Ethol Ecol Evol 32(2):148–161. https://doi.org/10.1080/03949370.2019.1693432

Rohner PT, Teder T, Esperk T, Lüpold S, Blanckenhorn WU (2018) The evolution of male-biased sexual size dimorphism is associated with increased body size plasticity in males. Funct Ecol 32(2):581–591. https://doi.org/10.1111/1365-2435.13004

Sánchez-Quirós C, Arévalo E, Barrantes G (2012) Static allometry and sexual size dimorphism in *Centruroides margaritatus* (Scorpiones: Buthidae). J Arachnol 40(3):338–344. https://doi.org/10.1636/B12-14.1

Seiter M, Mosetig L, Milasowszky N (2020) The trade-off between adult size and development time due to different feeding regimes in the scorpion *Tityus neibae*. Invertebr Reprod Dev 64(4):274–280. https://doi.org/10.1080/07924259.2020.1806119

Simone Y, van der Meijden A (2021) Armed stem to stinger: a review of the ecological roles of scorpion weapons. J Venomous Anim Toxins Incl Trop Dis 27:e20210002. https://doi.org/10.1590/1678-9199-jvatitd-2021-0002

Tammaru T, Teder T (2025) Body size and timing of maturation. In: Life history evolution: traits, interactions, and applications. Wiley, Hoboken, p 496

Tang V (2023) Non-aggressive competition between males of *Srilankametrus yaleensis* (Kovařík et al., 2019) (Scorpionidae), and other types of agonistic behavior observed in scorpions. Euscorpius 2023(368):1–17

Teder T (1998) Limited variability of genitalia in the genus *Pimpla* (Hymenoptera: Ichneumonidae): inter-or intraspecific causes? Neth J Zool 48(4):335–347

Tsuboi M, Takahashi T (2024) Sexually divergent selection, allometric constraints, and the evolution of sexual dimorphism in cichlids from Lake Tanganyika. J Evol Biol 37(12):1563–1575. https://doi.org/10.1093/jeb/voae101

Tuni C, Berger-Tal R (2012) High mortality and female-biased operational sex ratio result in low encounter rates and moderate polyandry in a spider: ecological factors shape polyandry. Biol J Linn Soc 107(4):910–919. https://doi.org/10.1111/j.1095-8312.2012.01990.x

van der Meijden A, Herrel A, Summers A (2010) Comparison of chela size and pincer force in scorpions; getting a first grip. J Zool 280(4):319–325. https://doi.org/10.1111/j.1469-7998.2009.00628.x

Visser JH, Geerts S (2021a) Static allometry and sexual dimorphism in the Striped Lesser-thicktail Scorpion *Uroplectes lineatus*. Arachnology 18(7). https://doi.org/10.13156/arac.2020.18.7.700

Visser JH, Geerts S (2021b) Sexual dimorphism and static allometry in the South African scorpion *Opistophthalmus karrooensis*. Arachnology 18(9). https://doi.org/10.13156/arac.2021.18.9.1057

Warburg MR (2011) Scorpion reproductive strategies, allocation and potential; a partial review. Eur J Entomol 108(2):173–181. https://doi.org/10.14411/eje.2011.023

Webb TJ, Freckleton RP (2007) Only half right: species with female-biased sexual size dimorphism consistently break Rensch's rule. PLoS One 2(9):e897. https://doi.org/10.1371/journal.pone.0000897

Wyman JT, Wright-Ueda J, Agnew Q, Castellano I, Simone Y (2025) First report of arm-span competition in buthid scorpions: male-male contest in *Tityus* cf. *rosenbergi* Pocock, 1898. J Arachnol 52(3). https://doi.org/10.1636/joa-s-23-009

Chapter 3
Mating or Not? Reproductive Aspects of Sexual Reproduction and Parthenogenesis in Neotropical Scorpions

Gracielle de Fátima Braga-Pereira and André Felipe de Araujo Lira

Abstract The reproduction of scorpions is a complex process that involves specific behaviors and evolutionary adaptations to ensure the continuity of the species. These arachnids exhibit a predominantly sexual mode of reproduction, with well-defined sexual dimorphism in some species. The mating process begins with courtship, which involves a series of interactions between the pair, including specific body movements such as the *dance* or *promenade à deux*, aimed at finding a suitable location for the deposition of the spermatophore, a structure containing the sperm cells. The transfer of sperm occurs indirectly, followed by internal fertilization. Many scorpion species use chemical signals for sexual communication, with females being the main producers of pheromones. Embryonic development occurs through viviparity. Scorpions exhibit parental care; after birth, the young climb onto the mother's back and remain there until they undergo their first molt, after which they may disperse into the environment. In addition to sexual reproduction, some species can reproduce by parthenogenesis, a form of reproduction in which the offspring develop from an unfertilized gamete.

Keywords Sexual selection • Non-sexual reproduction • Litter size

G. de Fátima Braga-Pereira
Laboratório de Etologia de Artrópodes, Departamento Acadêmico de Tecnologia em Saúde e Biologia, Instituto Federal da Bahia, Salvador, Bahia, Brazil
e-mail: gracielle.pereira@ifba.edu.br

A. F. de Araujo Lira (✉)
Colección Nacional de Arácnidos, Departamento de Zoología, Instituto de Biología, Universidad Nacional Autónoma de México, Coyoacán, Mexico City, Mexico
e-mail: andre.lira@st.ib.unam.mx

A. F. de Araujo Lira, E. González-Santillán (eds.), *The Biology of Neotropical Scorpions*, https://doi.org/10.1007/978-3-032-23194-9_3

3.1 Introduction

During the nineteenth century and the first half of the twentieth century, the first records on scorpion mating were made (e.g., Maccary 1810; Brongniart and Gaubert 1891; Fabre 1923); at that time, researchers believed that sperm transfer occurred directly, through a copulatory organ. However, in the 1950s, studies showed that sperm transfer in scorpions is carried out indirectly, via a spermatophore, which is deposited on a substrate (Bücherl 1956; Zolessi 1956; Alexander 1956, 1959). Only from the 1990s onwards were detailed descriptive analyses of scorpion courtship conducted (e.g., Peretti 1991, 1995; Tallarovic et al. 2000), including species of scorpions found in South America, such as the buthid *Zabius fuscus* Thorell, 1876 (Peretti 1991). In the last 20 years, several Neotropical species have had their courtship described in detail, including medically important species such as *Tityus bahiensis* (Perty, 1833) (Outeda-Jorge 2010), *Tityus serrulatus* Lutz & Mello, 1922 (Braga-Pereira 2021), and *Tityus stigmurus* (Thorell 1876) (Foerster et al. 2021).

Scorpion mating is performed through behavioral acts that share common aspects among different species (Polis and Sissom 1990). Sexual recognition between individuals occurs through chemical signaling, and after this through small changes during pre-courtship, such as the duration of certain behavioral acts (Peretti 1993) employed to mitigate the female aggressivity (Polis and Farley 1979, Simone and Meijden 2021). One of the aspects that varies most among species is the total duration of courtship, which can last from a few minutes to several days; this time is related to the female's receptivity, which depends on the reproductive period and the necessary time for the males to find a suitable place to deposit the spermatophore (e.g., Tallarovic et al. 2000; Ross 2009a; Sánchez-Quirós et al. 2012; Chantall-Rocha and Japyassú 2017; Olivero et al. 2017).

Scorpion courtship can be divided into three phases: initiation, dance, and sperm transfer (Fig. 3.1). In the initiation phase, recognition occurs between individuals, both at the species and sexual level; in the dance phase, known as *promenade à deux*, the male, guiding the female by the pedipalps, searches for a suitable substrate to deposit the spermatophore; finally, during the sperm transfer phase, the male deposits the spermatophore in the environment and leads the female over it for sperm transfer to occur (Polis and Sissom 1990). Some species may exhibit a fourth phase, called post-transfer, in which various behaviors can be observed (Alexander 1959; Tallarovic et al. 2000), including the consumption of the spermatophore by the female, as in *T. bahiensis* (Outeda-Jorge 2010), or by the male, as in *T. serrulatus* (Braga-Pereira 2021). Below are the main behavioral acts observed in the different phases of mating in Neotropical scorpion species.

Fig. 3.1 Courtship and mating phases in *Tityus serrulatus* (**a**) Initiation, (**b**) dance, and (c) spermatophore deposition. (Photos by Gracielle Braga-Pereira)

3.2 Sexual Recognition

Initially, studies evaluated the scorpion mating recognition proposed that males recognized their partners by tactile cues, perceiving anatomic differences after their first contact (Alexander 1959; Polis and Farley 1979). Particularly, anatomic differences in the metasoma (sexual dimorphism) would be key to recognition, because males would initiate contact by preferably grasping this structure (Alexander 1959). However, recently growing evidence has suggested that sexual recognition in these animals occurs by the presence of the pheromones (e.g., Pordeus et al. 2019; Braga-Pereira 2021; Oviedo-Diego et al. 2021; Barbosa-da-Silva et al. 2025).

Sex pheromones are chemical compounds that play a fundamental role in reproduction. In the animal kingdom, various species use pheromones for locating partners, sexual recognition, and selecting individuals for mating (Johansson and Jones 2007; Wyatt 2009). In the case of scorpions, this form of chemical communication is also of great importance, facilitating the location and recognition of sexual partners in their environment (Steinmetz et al. 2004; Taylor et al. 2012; Pordeus et al. 2019). In neotropical scorpions, the presence of chemical signals used in mating has been recorded in some species, such as *Tityus pusillus* Pocock, 1893 (Pordeus et al. 2019) and *T. serrulatus* (Braga-Pereira 2021).

Female arthropods are generally the main emitters of sex pheromones (Greenfield 2002; Tabata et al. 2012). In scorpions, it is believed that these compounds are released into the soil or nearby surfaces, where males may detect them with their pectines, sensory structures located on the ventral side, which play a fundamental role in detecting these chemical signals (Polis and Sissom 1990; Gaffin and Brownell 1992; Trabalon and Bagnères 2010). By moving their pectines over the soil, males may capture pheromone molecules and track their origin, thus being able to find the female (Barbosa-da-Silva et al. 2025). For example, in a Y-maze test, males of *Urophonius* spp. were able to locate the female pheromones (Oviedo-Diego et al. 2021). After detecting the female pheromones, males begin a series of specific movements and interactions, characterizing the initiation stage of mating (Pordeus et al. 2019; Braga-Pereira 2021; Barbosa-da-Silva et al. 2025). Sex pheromones, besides attracting partners, also provide crucial information about the receptivity and reproductive status of females (Johansson and Jones 2007; Thomas 2011; Steiger et al. 2012). However, research on pheromones in scorpions is still limited compared to other groups of invertebrates. Despite this, initial studies have shown that the chemical composition of these pheromones can vary between species, among different populations of the same species, and between sexes (e.g., Olivero et al. 2015; Braga-Pereira 2021; Barbosa-da-Silva et al. 2025). This suggests that pheromones not only facilitate mating but also play a role in speciation and reproductive isolation, since populations that produce different pheromones may not recognize each other as potential partners (Grimaldi and Engel 2005).

3.3 Partner Choice

Mate choice is expected to occur when there is variability in the quality of potential mates, resulting in direct benefits for the chooser (Møller and Jennions 2001) or indirect benefits, such as enhanced genetic diversity, increased offspring attractiveness and variability, improved paternal care, or greater compatibility between sperm and the female reproductive tract and/or eggs (Jennions and Petrie 2000; Tomkins et al. 2004; Parker and Birkhead 2013). Sexual selection may favor mechanisms that act both before and after mating, reflecting a range of reproductive strategies in which the costs of mating can be substantial for both sexes (Cunningham and Birkhead 1998; Edward and Chapman 2011; Han et al. 2020).

In scorpions, females are the decisive sex regarding the timing of copulation. Females screen male reproductive attempts (Chantall-Rocha and Japyassú 2017), which typically results in a lower rate of male mating success (e.g., Lira et al. 2018a; Foerster et al. 2021; Oviedo-Diego et al. 2025). Generally, females react aggressively to male advances; however, when courted by males in better physical condition, this aggression decreases (Olivero et al. 2019; Moreira 2021; Oviedo-Diego et al. 2025). According to Olivero et al. (2019), better condition provides males with a mating advantage and serve as an indicator of courtship performance to females. For example, males in better condition may perform certain courtship behaviors, such as spending less time in the *promenade à deux*, prolonging the sexual sting, and exhibiting more persistent conduct (e.g., Olivero et al. 2015; Moreira 2021). This cryptic selection by females may pressure males to compete for the opportunity to copulate. Recently, Tang (2023) described a behavioral display called arm-span in various Heterometrinae species. In the Neotropical context, this same behavior has been described in the species *Tityus rosenbergi* Pocock, 1898, where individuals face off and extend their pedipalps laterally during male-male contests (Wyman et al. 2025). Additionally, previous studies have shown that males are equally capable of assessing female physical condition (Olivero et al. 2015; Moreira 2021; Oviedo-Diego et al. 2025). Thus, males may focus their reproductive efforts on females in better condition, optimizing their chances of reproductive success.

3.4 Pre-courtship (Initiation)

The initiation phase begins with the encounter between the pair; the male initiates a series of behavioral acts that culminate in grasping the female's pedipalps with his own pedipalps. The behavioral acts performed by the male are probably related to sexual and species recognition between the couple, as well as to calming the female when she shows aggression or low receptivity (Peretti and Carrera 2005; Braga-Pereira 2021). Among the behaviors displayed by males, pedipalp shaking, body shaking, or metasoma swagging may occur (e.g., Lira et al. 2018a, 2021; Foerster et al. 2021). Furthermore, it has been suggested that male behavioral acts from the initiation phase may be related to cryptic female choice (Briceño and Bonilla 2009), a process in which the female selects the male during copulation, and this choice may be based on behavioral acts performed by the male (Eberhard 1996). However, recent studies also indicate that females can choose males with better body condition (Olivero et al. 2019; Moreira 2021; Oviedo-Diego et al. 2025).

A striking behavior presented by males of some scorpion families, such as Buthidae, is the production of body vibrations, which begin after detecting the female, before there is physical contact between the pair (e.g., Chantall-Rocha and Japyassú 2017; Lira et al. 2021; Foerster et al. 2021). The vibrations can extend throughout the entire mating process. The vibration behavior has been observed and described in several Neotropical species, such as *Jaguajir rochae* (Borelli, 1910) (Chantall-Rocha and Japyassú 2017), *Ananteris mauryi* Lourenço, 1982 (Lira et al. 2021), *T. serrulatus* (Braga-Pereira 2021), and *T. stigmurus* (Foerster et al. 2021). Briceño and Bonilla (2009) identified that in *Centruroides margaritatus* (Gervais, 1841), the vibration may be transmitted through the substrate. The authors discuss possible functions for this behavior, such as sexual and species recognition, as well as calming females that show aggression or low receptivity.

In some species, before grasping the pedipalps, the male may rub the metasoma on different parts of the female's body, as well as touch different parts of her body with his pedipalps (e.g., Peretti and Acosta 1999; Toscano-Gadea 2010). At this stage, the female may show aggression if she is not receptive to mating. Resistance and receptivity behaviors have been described in species such as *Z. fuscus* (Peretti and Carrera 2005), *J. rochae* (Chantall-Rocha and Japyassú 2017), *T. serrulatus* (Braga-Pereira 2021), and *T. pusillus* (Moreira et al. 2024). Characteristics indicating a lack of female receptivity include moving away from the male, arching and moving the metasoma, and attempts to sting (e.g., Chantall-Rocha and Japyassú 2017; Braga-Pereira 2021; Moreira et al. 2024). Passive females show submission and accept the male's advances, allowing him to hold their pedipalps after little or no insistence. Generally, virgin females are more receptive to mating, not exhibiting resistance behaviors.

3.5 Dance

Considered one of the most striking characteristics of the reproductive biology of scorpions, the dance phase begins with the male holding the female's pedipalps with his own pedipalps. In this phase, the female may still display resistance behaviors to copulation, such as attempts to release her pedipalps and arching her metasoma, moving it laterally (e.g., Lira et al. 2018a; Moreira et al. 2024; Oviedo-Diego et al. 2025). The male then continues the courtship, until she no longer shows resistance. The male then guides the female across the substrate while exhibiting some behavioral acts, such as rubbing his chelicerae against the female's chelicerae, projecting and moving them, and opening and closing the chelae (e.g., Outeda-Jorge 2010; Lira et al. 2018a, 2021). During the dance, the male moves his pectines over the substrate in search of a suitable place to deposit the spermatophore. Another behavior observed during the dance is the frequent movement of the male's first pair of legs in the ventral region of the female, which may also be related to cryptic female choice, as recorded in *T. bahiensis* (Outeda-Jorge 2010) and *T. serrulatus* (Braga-Pereira 2021).

3.6 Sperm Transfer

After the male finds a suitable place to deposit the spermatophore, both stop moving around the environment. The male raises his metasoma and moves it while the spermatophore exits from his genital opening (e.g., Alexander 1959; Ross 2009a; Outeda-Jorge 2010). After the spermatophore is expelled, the male guides the female underneath him until it fits into the female genital opening. The sperm transfer can last from a few seconds to several minutes. After the transfer, the male and female separate their pedipalps through shaking movements.

3.7 Post Sperm Transfer

After the separation of the couple, some species of scorpions may exhibit certain behavioral acts, one of the most notable is the consumption of the spermatophore. Ingestion may occur by the female, as in *T. bahiensis* (Outeda-Jorge 2010), or by the male, as observed in *T. serrulatus* (Braga-Pereira 2021). The consumption of the spermatophore is likely linked to energy gain, where the female may use the energy to generate offspring, while the male may use the energy gain to reconstitute a new spermatophore (Francke 1979; Peretti and Acosta 1999; Tallarovic et al. 2000). After spermatophore deposition, scorpion males take between 1 and 14 days to regenerate the hemispermatophores and show sexual interest in new partners (e.g., Peretti and Acosta 1999; Ross 2009b; Lira et al. 2018a). These data suggest that males are capable of mating with more than one female per reproductive period. Scorpion males exhibit high mortality compared to females, rarely reaching a second reproductive period (Polis and Sissom 1990). A recent study shows that senescence negatively impacts sperm quality and reproductive capacity, gradually declining testicular mass, sperm quantity, and viability (Vrech et al. 2024). Similarly, females may copulate with more than one partner during the reproductive period (Warburg 2012). In addition, particularly buthids, females possess the capacity to store sperm and allow them to produce multiple broods throughout the year (3–4 broods) (e.g., Dupre 1992; Outeda-Jorge et al. 2009; Albuquerque and Lira 2016). However, in some species, after insemination females retain a mating plug adhered to their genital operculum, preventing further copulations (e.g., Mattoni and Peretti 2004; Oviedo-Diego et al. 2020). This mating plug is absorbed by the female's body during gestation, thus allowing the birth of the offspring.

3.8 Viviparity and Parental Care

Scorpions are viviparous animals, meaning the embryos develop inside the female's body and the offspring are born fully formed (Francke 1982; Polis and Sissom 1990). The embryos can be nourished in the ovariuterus or from a diverticulum connected to the mother (Polis and Sissom 1990; Lourenço 2000; Warburg 2012). The gestation period can vary between 3 and 18 months, depending on the species and environmental conditions (Brownell and Polis 2001, Outeda-Jorge et al. 2009) similarly to litter size (Table 3.1).

A remarkable characteristic in scorpions is parental care, which is exercised by the female immediately after the offspring are born. As soon as they are born, scorpion offspring immediately climb onto the mother's back, where they remain until they undergo their first molt (Peretti 1994; Brownell and Polis 2001). This behavior is crucial for the survival of newborns, since at this early stage they are still vulnerable and unable to adequately defend themselves or seek food on their own. Parental care in scorpions involves protection against predators and physical support while the offspring complete their cuticular development, as they are born with a soft cuticle that does not provide the same resistance as adults (Ugolini et al. 1986; Benton 1991; Guo et al. 2024). During this phase, the mother may adopt highly protective behavior, even reduce her hunting and feeding activities to prioritize the safety of the offspring (Mahsberg 2010). Upon reaching the second instar, the offspring become independent and are ready to disperse into the environment.

Table 3.1 Litter size in Neotropical scorpion species. For a complete list of references please see Dupre (2020)

Scorpion species	Litter size	References
Ananteridae		
Ananteris balzanii	10–34	Outeda-Jorge et al. (2009)
Ananteris canalera	4–31	Lezcano et al. (2022)
Ananteris coineaui	15	Lourenço (2007)
Ananteris mauryi	9–31	Lira et al. (2021)
Ananteris platnicki	6–39	Lezcano et al. (2022) and Triana and Bonilla (2023)
Ananteris terueli	13	Kovařík (2006)
Bothriuridae		
Bothriurus araguayae	28–53	Outeda-Jorge et al. (2009)
Bothriurus bonariensis	8–65	Toscano-Gadea (2015)
Bothriurus buecherli	7–34	Toscano-Gadea (2005)
Bothriurus rochai	20–25	*This study*
Bothriurus rochensis	18–61	Outeda-Jorge et al. (2009) and Toscano-Gadea (2005)
Urophonius brachycentrus	21–46	reviewed in Dupre (2020)
Urophonius iheringii	6–36	Toscano-Gadea (2005)
Buthidae		
Centruroides anchorellus	6–44	reviewed in Dupre (2020)
Centruroides arctimanus	14	Lourenço (2007)
Centruroides balsasensis	15–36	reviewed in Dupre (2020)
Centruroides baracoae	6–30	reviewed in Dupre (2020)
Centruroides barbudensis	22–25	Lourenço (2007)
Centruroides bicolor	6–47	Lezcano et al. (2022) and Triana and Bonilla (2023)
Centruroides edwardsii	60–72	Triana and Bonilla (2023)
Centruroides gracilis	26–91	Lourenço (2007)
Centruroides granosus	25–84	Lezcano et al. (2022)
Centruroides griseus	35	reviewed in Dupre (2020)
Centruroides guanensis	3–67	reviewed in Dupre (2020)
Centruroides hentzi	5–30	reviewed in Dupre (2020)
Centruroides hoffmanni	12–30	reviewed in Dupre (2020)
Centruroides insulanus	6–105	Lourenço (2007)
Centruroides limbatus	12–46	Lezcano et al. (2022) and Triana and Bonilla (2023)
Centruroides margaritatus	26–70	Lourenço (2007)
Centruroides meisei	15–30	reviewed in Dupre (2020)
Centruroides nigrescens	50	reviewed in Dupre (2020)
Centruroides nigropunctatus	13–16	reviewed in Dupre (2020)
Centruroides nitidus	9–46	reviewed in Dupre (2020)
Centruroides noxius	16	reviewed in Dupre (2020)
Centruroides ornatus	6–21	reviewed in Dupre (2020)
Centruroides pococki	27	Lourenço (2007)
Centruroides poncei	17–31	reviewed in Dupre (2020)

(continued)

Table 3.1 (continued)

Scorpion species	Litter size	References
Centruroides robertoi	9–24	reviewed in Dupre (2020)
Centruroides spectatus	13–15	reviewed in Dupre (2020)
Centruroides tecomanus	11–36	reviewed in Dupre (2020)
Heteroctenus garridoi	14–27	Lourenço (2007)
Heteroctenus junceus	6–72	reviewed in Dupre (2020)
Heteroctenus princeps	15–30	Lourenço (2007)
Jaguajir agamemnon	30–50	*This study*
Jaguajir pintoi	10–40	reviewed in Dupre (2020)
Jaguajir rochae	28–49	Lourenço (2007)
Microtityus consuelo	4–6	Lourenço (2007)
Microtityus fundorai	7	Lourenço (2007)
Microtityus joseantonioi	6–7	Lourenço (2007)
Microtityus minimus	1–5	reviewed in Dupre (2020)
Microtityus rickyi	3–6	Lourenço (2007)
Physoctonus amazonicus	19	reviewed in Dupre (2020)
Physoctonus debilis	2	Outeda-Jorge et al. (2009)
Rhopalurus laticauda	13–15	Lourenço (2007)
Tityopsis inexpectata	4–16	reviewed in Dupre (2020)
Tityopsis canizaresorum	6–17	reviewed in Dupre (2020)
Tityus asthenes	12–30	reviewed in Dupre (2020)
Tityus bahiensis	7–82	Lourenço (2007) and Outeda-Jorge et al. (2009)
Tityus bastosi	10–12	Lourenço (2007)
Tityus brazilae	8–12	reviewed in Dupre (2020)
Tityus cerroazul	12–28	Lezcano et al. (2022)
Tityus championi	7–32	Lezcano et al. (2022) and Triana and Bonilla (2023)
Tityus clathratus	8–18	Outeda-Jorge et al. (2009)
Tityus columbianus	13	Lourenço (2007)
Tityus confluens	6–9	Seiter (2012)
Tityus costatus	21–25	Outeda-Jorge et al. (2009)
Tityus dedoslargos	14–28	Triana and Bonilla (2023)
Tityus ecuadorensis	5–25	reviewed in Dupre (2020)
Tityus elii	10–12	Rouaud et al. (2000)
Tityus falconensis	11–24	Lourenço (2007)
Tityus fasciolatus	1–26	Lourenço (2007)
Tityus festae	12–31	Lezcano et al. (2022)
Tityus forcipula	12	reviewed in Dupre (2020)
Tityus fuehrmanni	13–16	Lourenço (2007)
Tityus insignis	17–20	Lourenço (2007)
Tityus jaimei	6–36	Lezcano et al. (2022) and Triana and Bonilla (2023)
Tityus kaderkai	12	reviewed in Dupre (2020)
Tityus kuryi	4–16	Outeda-Jorge et al. (2009)
Tityus lokiae	10	Lourenço (2007)

(continued)

Table 3.1 (continued)

Scorpion species	Litter size	References
Tityus magnimanus	7–42	reviewed in Dupre (2020)
Tityus mattogrossensis	8–12	Lourenço (2007) and Outeda-Jorge et al. (2009)
Tityus melanostictus	13–16	Lourenço (2007)
Tityus metuendus	25–35	Lourenço (2007)
Tityus neblina	1–15	reviewed in Dupre (2020)
Tityus neglectus	15–20	*This study*
Tityus obscurus	11–33	Lourenço (2007) and Outeda-Jorge et al. (2009)
Tityus ocelote	7–12	Seiter (2012)
Tityus potameis	14	*This study*
Tityus pusillus	8–12	Albuquerque and Lira (2016)
Tityus serrulatus	2–25	Lourenço (2007) and Outeda-Jorge et al. (2009)
Tityus silvestris	5–14	Outeda-Jorge et al. (2009)
Tityus smithii	15	Dupre (2020)
Tityus stigmurus	6–18	Outeda-Jorge et al. (2009)
Tityus strandi	12	Lourenço (2007)
Tityus tamayoi	14	reviewed in Dupre (2020)
Tityus tayrona	4–22	Lezcano et al. (2022)
Tityus trinitatis	28	Lourenço (2007)
Tityus trivittatus	1–26	reviewed in Dupre (2020)
Tityus uruguyaensis	6–24	Toscano-Gadea (2005)
Tityus ythieri	13–25	Lourenço (2007)
Tityus zulianus	6–24	reviewed in Dupre (2020)
Troglorhopaluruslacrau	27–30	reviewed in Dupre (2020)
Troglorhopalurustranslucidus	43	reviewed in Dupre (2020)
Zabius fuscus	12–35	Lourenço (2007)
Caraboctonidae		
Hadruroides charcasus	40–86	reviewed in Dupre (2020)
Chactidae		
Broteochactas delicatus	5	reviewed in Dupre (2020)
Brotheas sp.	8–21	Outeda-Jorge et al. (2009)
Brotheas gervaisii	8–26	reviewed in Dupre (2020)
Brotheas granulatus	8–25	reviewed in Dupre (2020)
Chactas exsul	6–16	Lezcano et al. (2022)
Chactas lepturus	14–15	Lourenço et al. (2003)
Chactas reticulatus	15–16	Lourenço et al. (2003)
Neochactas orinocensis	9	reviewed in Dupre (2020)
Teuthraustes atramentarius	20	reviewed in Dupre (2020)
Diplocentridae		
Cazierus gundlachii	7–29	reviewed in Dupre (2020)
Didymocentrus krausi	32–40	Triana and Bonilla (2023)
Didymocentrus lesueuri	26–70	reviewed in Dupre (2020)
Didymocentrus trinitarius	13–46	reviewed in Dupre (2020)

(continued)

Table 3.1 (continued)

Scorpion species	Litter size	References
Diplocentrus gerstchi	25	reviewed in Dupre (2020)
Diplocentrus tehuacanus	6	reviewed in Dupre (2020)
Diplocentrus whitei	33–38	reviewed in Dupre (2020)
Heteronebo bermudezi	11	reviewed in Dupre (2020)
Euscorpiidae		
Megacormus gertschi	19–75	reviewed in Dupre (2020)
Megacormus granosus	19–75	reviewed in Dupre (2020)
Megacormus seductus	2–32	*This study*
Hormuridae		
Opisthacanthus cayaporum	15–25	Outeda-Jorge et al. (2009)
Opisthacanthus elatus	9–15	Lezcano et al. (2022)
Vaejovidae		
Chihuahuanus bilineatus	17–26	reviewed in Dupre (2020)
Chihuahuanus coahuilae	20–41	reviewed in Dupre (2020)
Chihuahuanus crassimanus	41	reviewed in Dupre (2020)
Chihuahuanus russelli	81	reviewed in Dupre (2020)
Maaykuyak waueri	14	reviewed in Dupre (2020)
Mesomexovis caxcan	35	*This study*
Mesomexovis punctatus	27	reviewed in Dupre (2020)
Mesomexovis variegatus	27	reviewed in Dupre (2020)
Paravaejovis spinigerus	15–69	reviewed in Dupre (2020)
Paruroctonus gracilior	14–35	reviewed in Dupre (2020)

The phase during which the offspring remain with the mother (Fig. 3.2) varies according to the species. At the end of this period, the offspring disperse and begin an independent life, but already with a greater capacity for survival thanks to the initial protection phase they received. Generally, in scorpions, the number of offspring per brood tends to be inversely proportional to the size of the offspring and directly proportional to the size of the mother (Francke 1981; Lourenço 2007), although there are exceptions such as species of the genus *Ananteris*, which have on average 9–34 offspring (Lourenço and Cuellar 1999; Lourenço 2007; Outeda-Jorge et al. 2009; Lira et al. 2021).

Viviparity and parental care in scorpions can be seen as important evolutionary adaptations that allow for a higher reproductive success rate, especially in harsh environments (Polis and Sissom 1990; Shaffer and Formanowicz 1996; Lourenço 2018). The additional protection provided by viviparity and maternal care increases the chances of offspring survival, compensating for the low number of descendants compared to oviparous species, which generally produce a larger number of eggs but without intensive parental care (Polis and Sissom 1990; Warburg 2012; Lourenço 2018; Dionisio-da-Silva et al. 2023). This reproductive strategy appears to be closely linked to the longevity of scorpions, which can vary from 4 to 25 years,

Fig. 3.2 Female scorpions with offspring on their back. (**a**) *Brotheas amazonicus*, (**b**) *Tityus pusillus*, (**c**) *Physoctonus debilis*, (**d**) Vaejovidae sp., (**e**) *Megacormus orizaba,* and (**f**) *Tityus melici.* (Photos by Jonas Martins (**a**), Gabriel Celante (**b–c**), Alejandro Valdez-Mondragón (**d**), Edmundo González-Santillán (**e**), and Gracielle Braga-Pereira (**f**))

depending on the species (Polis and Sissom 1990; Warburg 2011a). In environments where competition for resources is high and environmental conditions are extreme, viviparity and parental care are crucial factors that allow scorpions to maintain stable populations and withstand climatic and ecological variations (Warburg 2011b; Lourenço 2018).

Scorpion reproduction is strongly linked to environmental factors such as temperature and food availability, which influence both reproductive success and offspring development rates. For example, the reproductive period is associated with times of greater resource availability (e.g., Cala-Riquelme and Colombo 2011; Quijano-Ravell and Ponce-Saavedra 2015; Lira et al. 2018b). During this period, females display higher receptivity to male advances, and males increase their foraging activity (e.g., Araújo et al. 2010; Dionisio-da-Silva et al. 2018; Moreira 2021). However, contrary patterns exist where some species reproduce in the opposite season (Dionisio-da-Silva et al. 2018; Lira et al. 2018b). This mechanism has been reported in small-sized species such as *Ananteris mauryi* and *Physoctonus debilis* (Dionisio-da-Silva et al. 2018, Lira et al. 2018b). According to these authors, this shift in reproductive activity compared to other scorpion species in the same assemblage may represent a survival strategy to avoid intraguild predation. Given that body size is a decisive factor in predator-prey interactions among these arachnids, this behavior could reduce predation risk (see Chap. 13).

3.9 Parthenogenesis

Organisms that reproduce asexually generate offspring without the presence of a member of the opposite sex, allowing for greater chances of colonizing disturbed environments as well as rapid proliferation in new habitats (Cuellar 1977; Lourenço 2008). Furthermore, asexual reproduction, when it has the same fertility rate as sexual reproduction, is considered more efficient since it entails lower costs (Kawatsu 2013). One such cost is the production of males, which involves little investment in offspring (Maynard-Smith 1971). However, asexual reproduction also brings evolutionary consequences, such as reproductive isolation from sexual forms and the emergence of lineages with low genetic diversity, which can lead to their extinction (Simon et al. 2003).

Parthenogenesis, a form of reproduction in which a single gamete originates from a new individual without the need for fertilization, has evolved many times independently in the animal kingdom (Simon et al. 2003). This form of reproduction has been observed, so far, in at least 11 species of scorpions; of these, eight are neotropical: *T. serrulatus*, *T. stigmurus*, *T. colombianus* Thorell 1876, *T. carrilloi*, *T. metuendus* Pocock 1897, *T. uruguayensis* Borelli 1901, *T. confluens*, and *Centruroides gracilis* (Francke 2008; Lourenço 2008; Seiter 2012). The most reliable criterion for confirming parthenogenetic species is based on the observation of offspring generation by females born in the laboratory and isolated from the presence of males, this event being observed for more than one generation (Francke 2008). Besides scorpion species that produced offspring in the laboratory from virgin females, parthenogenesis has also been suggested in other species due to the proportion of females being higher than the proportion of males (Francke 2008).

The parthenogenetic form of reproduction in scorpions was first recorded by Matthiesen (1962), who observed that female *T. serrulatus* in captivity produced offspring even when isolated from males. For many years, parthenogenesis was considered the only reproductive mode of this species, as there were no records of males of the yellow scorpion. However, Souza et al. (2009) described a male *T. serrulatus* found in the municipality of Januária, Minas Gerais, representing the first record of a sexually reproducing population of this species. Suesdek-Rocha et al. (2006) suggested that parthenogenesis was induced by symbiotic microorganisms of the genus *Wolbachia* after identifying this microorganism in a population of *T. serrulatus* located in the state of São Paulo. However, Braga-Pereira et al. (2019), after studying several populations of the yellow scorpion distributed in different regions of Brazil, demonstrated that these microorganisms are not present, indicating that asexual reproduction in *T. serrulatus* is not induced by *Wolbachia*. Finally, Braga-Pereira and Santos (2021) demonstrated that in a sexually reproducing population of *T. serrulatus*, parthenogenesis occurs facultatively. Facultative parthenogenesis is also observed in other scorpion species that reproduce in this way, such as *T. columbianus*, *T. stigmurus*, and *T. carrilloi* (Francke 2007; Ross 2010; Foerster et al. 2022).

Parthenogenetic scorpions can be found in altered environments such as urban centers (e.g., Lourenço 2015; Ojanguren-Affilastro et al. 2021; Foerster et al. 2022). For example, sexually reproducing populations of the scorpion *T. stigmurus* are common in areas of native vegetation, while parthenogenetic populations are found in urban areas (Santos et al. 2014; Foerster et al. 2022). According to Lourenço (2015), parthenogenetic species are density-independent and have population regulation. This mechanism would facilitate the colonization of urban environments.

3.10 Conclusion

The reproduction of Neotropical scorpions reveals a complex interplay of behavior, physiology, and evolutionary adaptations that ensure reproductive success in diverse and often challenging environments. From ritualized courtship to reproductive strategies such as viviparity and parental care, these arachnids exhibit refined mechanisms to maximize offspring survival. The coexistence of sexual reproduction and facultative parthenogenesis in multiple species demonstrates remarkable evolutionary flexibility, allowing population persistence under conditions of low density or environmental adversity. The behavioral diversity during mating phases, along with the use of chemical signals for sexual recognition and mate choice, indicates that sexual selection and cryptic female choice shape the genetic dynamics of these populations. Postnatal maternal care is another critical factor providing protection and physical support to vulnerable young during early development stages.

Despite recent advances, several key areas warrant further investigation:

- Chemical composition and variation of sex pheromones: A deeper understanding of the chemical compounds involved in sexual communication could elucidate mechanisms of reproductive isolation and speciation.
- Genetic and epigenetic mechanisms underlying parthenogenesis: Clarifying the factors that determine facultative versus obligate asexual reproduction and their long-term evolutionary consequences.
- Environmental influences on reproductive strategies: Studying how climate change, urbanization, and habitat modification impact reproductive behaviors and population viability.
- Population dynamics in natural and disturbed habitats: Evaluating how the interplay of viviparity, parental care, and parthenogenesis modulates population stability amid predation, competition, and environmental disturbances.
- Post-copulatory behavioral interactions and sexual selection: Exploring processes during and after copulation, including male and female behavioral influences on fertilization success and reproductive outcomes.

Acknowledgments We would like to thank the Dirección General de Asuntos del Personal Académico (DGAPA) for the postdoctoral fellowship from the Universidad Nacional Autónoma de México. We would also like to thank the doctors Jonas Martins, Gabriel Celante, Alejandro Valdez-Mondragón, and Edmundo González-Santillán for kindly providing the photos used in the chapter.

References

Albuquerque CMR, Lira AFA (2016) Insights into reproductive strategies of *Tityus* (*Archaeotityus*) *pusillus* Pocock, 1893 (Scorpiones, Buthidae). C R Biol 339(5–6):179–184. https://doi.org/10.1016/j.crvi.2016.03.003

Alexander AJ (1956) Mating in scorpions. Nature 178:867–868. https://doi.org/10.1038/178867b0

Alexander AJ (1959) Courtship and mating in the Buthid scorpions. Proc Zool Soc Lond 133:146–169. https://doi.org/10.1111/j.1469-7998.1959.tb05557.x

Araújo CS, Candido DM, Araújo HF, Dias SC, Vasconcellos A (2010) Seasonal variations in scorpion activities (Arachnida: Scorpiones) in an area of Caatinga vegetation in northeastern Brazil. Zoologia 27:372–376. https://doi.org/10.1590/S1984-46702010000300008

Barbosa-da-Silva HR, Pontes WJT, Lira AFA, Navarro DMAF, Salomão RP, Maia ACD (2025) The role of intraspecific mechanical and chemical signaling for mate and sexual recognition in male *Tityus pusillus* (Scorpiones, Buthidae). Zoology 168:126235. https://doi.org/10.1016/j.zool.2024.126235

Benton TG (1991) Reproduction and parental care in the scorpion, *Euscorpius flavicaudis*. Behaviour 117:20–28

Braga-Pereira GF (2021) Partenogênese facultativa e comportamento de acasalamento no escorpião amarelo (Scorpiones: Buthidae: *Tityus serrulatus*). Tese de doutorado. Universidade Federal de Minas Gerais

Braga-Pereira GF, Santos AJ (2021) Asexual reproduction in a sexual population of the Brazilian yellow scorpion (*Tityus serrulatus*, Buthidae) as evidence of facultative parthenogenesis. J Arachnol 49(2):185–190. https://doi.org/10.1636/JoA-S-20-001

Braga-Pereira GF, Vidigal THDA, Santos FR, Santos AJ (2019) Is the parthenogenesis of the yellow scorpion (*Tityus serrulatus*) promoted by endosymbiont bacteria (*Wolbachia* sp.)? The. J Arachnol 47(2):284–289. https://doi.org/10.1636/JoA-S-18-090

Briceño RD, Bonilla F (2009) Substrate vibrations in the scorpion *Centruroides margaritatus* (Scorpiones: Buthidae) during courtship. Rev Biol Trop 57:267–274

Brongniart C, Gaubert P (1891) Fonction de l'organe pectiniforme des scorpions. Comptes Rendus de l'Académie des Sciences 113:1062–1063

Brownell P, Polis GA (2001) Scorpion biology and research. Oxford University Press, Oxford

Bücherl W (1956) Escorpiões e escorpionismo no Brasil. V- Observações sobre o aparelho reprodutor masculino e o acasalamento de *Tityus trivittatus* e *Tityus bahiensis*. Mem Inst Butantan 27:121–155

Cala-Riquelme F, Colombo M (2011) Ecology of the scorpion, *Microtityus jaumei* in Sierra de Canasta. Cuba J Insect Sci 11(1):1–10. https://doi.org/10.1673/031.011.8601

Chantall-Rocha S, Japyassú HF (2017) Diffuse resistance courtship in the scorpion *Rhopalurus rochai* (Scorpiones: Buthidae). Behav Process 135:45–55. https://doi.org/10.1016/j.beproc.2016.11.017

Cuellar O (1977) Animal parthenogenesis: a new evolutionary-ecological model is needed. Science 197:837–843. https://doi.org/10.1126/science.887925

Cunningham EJA, Birkhead TR (1998) Sex roles and sexual selection. Anim Behav 56(6):1311–1321. https://doi.org/10.1006/anbe.1998.0953

Dionisio-da-Silva W, Lira AFA, Albuquerque CMR (2018) Distinct edge effects and reproductive periods of sympatric litter-dwelling scorpions (Arachnida: Scorpiones) in a Brazilian Atlantic forest. Zoology 129:17–24. https://doi.org/10.1016/j.zool.2018.06.001

Dionisio-da-Silva W, Albuquerque CMR, Lira AFA (2023) Reproductive investment in the Brazilian scorpion *Tityus pusillus* Pocock, 1893 (Scorpiones: Buthidae): do larger females produce better offspring? Invertebr Zool 20(1):90–96

Dupre G (1992) Observations de parturitions séquentielles chez *Didymocentrus lesueurii* (Gervais, 1844) (Scorpionida, Diplocentridae). Arachnides 98:1–16

Dupre G (2020) Nouvelle synthese sur la reproduction des scorpions. Arachnides

Eberhard WE (1996) Female control: sexual selection by cryptic female choice, Monographs in behavior and ecology. Princeton University Press, Princeton

Edward DA, Chapman T (2011) The evolution and significance of male mate choice. Trends Ecol Evol 26(12):647–654. https://doi.org/10.1016/j.tree.2011.07.012

Fabre JH (1923) The life of the scorpion. Dodd, Mead and Company, New York

Foerster SÍA, Dionisio-da-Silva W, Santos AB, Albuquerque CMR, Lira AFA (2021) Notes on the courtship behavior of the parthenogenetic scorpion *Tityus stigmurus*. Acta Arachnol 70(2):83–86. https://doi.org/10.2476/asjaa.70.83

Foerster SÍA, Dionisio-da-Silva W, Santos AB, Albuquerque CMR, Lira AFA (2022) New records of the not-so-rare males of the parthenogenetic scorpion *Tityus stigmurus* (Thorell, 1876) (Scorpiones: Buthidae). J Arachnol 50(1):27–29. https://doi.org/10.1636/JoA-S-21-009

Francke OF (1979) Observations on the reproductive biology and life history of *Megacormus gertschi* Diaz (Scorpiones: Chactidae; Megacorminae). J Arachnol 7(3):223–230

Francke OF (1981) Birth behavior and life history of *Diplocentrus spitzeri* Stahnke (Scorpiones: Diplocentridae). Southwest Nat 25(4):517–523

Francke OF (1982) Parturition in scorpions (Arachnida, Scorpiones). Revue Arachnol 4:27–37

Francke OF (2007) A critical review of reports of parthenogenesis in scorpions (Arachnida). Rev Ibér Aracnol. 16:93–104

Francke OF (2008) A critical review of reports of parthenogenesis in scorpions (Arachnida). Rev Ibér Aracnol 16:93–104

Gaffin DD, Brownell PH (1992) Evidence of chemical signaling in the sand scorpion, *Paruroctonus mesaensis* (Scorpionida: Vaejovida). Ethology 91:59–69. https://doi.org/10.1111/j.1439-0310.1992.tb00850.x

Greenfield MD (2002) Signalers and receivers. Oxford University Press, Oxford

Grimaldi D, Engel MS (2005) Evolution of the insects. Cambridge University Press, Cambridge

Guo Y, Li S, Lu S, Wang X, Cao Z, Wu Y (2024) Special survival strategy of first-instar scorpions revealed by synchronous molting behavior from social facilitation of maternal care and reciprocal aggregation. Insects 15(9):726. https://doi.org/10.3390/insects15090726

Han CS, Brooks RC, Dingemanse NJ (2020) Condition-dependent mutual mate preference and intersexual genetic correlations for mating activity. Am Nat 195(6):997–1008. https://doi.org/10.1086/708497

Jennions MD, Petrie M (2000) Why do females mate multiply? A review of the genetic benefits. Biol Rev 75(1):21–64. https://doi.org/10.1017/S0006323199005423

Johansson BG, Jones TM (2007) The role of chemical communication in mate choice. Biol Rev 82:265–289. https://doi.org/10.1111/j.1469-185X.2007.00009.x

Kawatsu K (2013) Sexual conflict over the maintenance of sex: effects of sexually antagonistic coevolution for reproductive isolation of parthenogenesis. PLoS One 8:e58141. https://doi.org/10.1371/journal.pone.0058141

Lezcano JJ, de Armas LF, Castillo LY, Murgas IL, Miranda RJ (2022) Some aspects of the reproductive biology of Panamanian scorpions (Scorpiones). Rev Ibér Aracnol 40:47–56

Lira AFA, Pordeus LM, Rego FNNA, Iannuzzi K, Albuquerque CMR (2018a) Sexual dimorphism and reproductive behavior in the Brazilian scorpion *Tityus pusillus* (Scorpiones, Buthidae). Invertebr Biol 137(3):221–230. https://doi.org/10.1111/ivb.12221

Lira AFA, DeSouza AM, Albuquerque CMR (2018b) Environmental variation and seasonal changes as determinants of the spatial distribution of scorpions (Arachnida: Scorpiones) in Neotropical forests. Can J Zool 96(9):963–972. https://doi.org/10.1139/cjz-2017-0251

Lira AFA, Araújo JC, Dionisio-da-Silva W, Albuquerque CMR (2021) Life-history traits of the Brazilian litter-dwelling scorpion: post-embryonic development and reproductive behaviour *in Ananteris mauryi* Lourenço, 1982 (Scorpiones: Buthidae). J Nat Hist 55(21–22):1323–1334. https://doi.org/10.1080/00222933.2021.1925766

Lourenço WR (2000) Reproduction in scorpions, with special reference to parthenogenesis. Eur Arachnol 2002:71–85

Lourenço WR (2007) Litter size in micro-buthoid scorpions (Chelicerata, Scorpiones). Bol Soc Entomol Aragon 40:473–477

Lourenço WR (2008) Parthenogenesis in scorpions: some history—new data. J Venom Anim Toxins Incl Trop Dis 14:19–44. https://doi.org/10.1590/S1678-91992008000100003

Lourenço WR (2015) What do we know about some of the most conspicuous scorpion species of the genus *Tityus*? A historical approach. J Venom Anim Toxins Incl Trop Dis 21:1–12. https://doi.org/10.1186/s40409-015-0016-9

Lourenço WR (2018) Scorpions and life-history strategies: from evolutionary dynamics toward the scorpionism problem. J Venom Anim Toxins Incl Trop Dis 24:1–12. https://doi.org/10.1186/s40409-018-0160-0

Lourenço WR, Cuellar O (1999) A new all-female scorpion and the first probable case of arrhenotoky in scorpions. J Arachnol 27(1):149–153

Lourenço WR, Andrzejewski V, Cloudsley-Thompson JL (2003) The life history of *Chactas reticulatus* Kraepelin, 1912 (Scorpiones, Chactidae), with a comparative analysis of the reproductive traits og three scorpion lineages in relation to habitat. Zool Anz 242(1):63–74. https://doi.org/10.1078/0044-5231-00087

Maccary A (1810) Mémoire Sur Le Scorpion Qui Se Trouve Sur La Montagne De Cette, Département De l'Hérault, Son venin, et l'usage qu'on pourrait en faire en Médecine. Gabon, Paris

Mahsberg D (2010) Brood care and family cohesion in the tropical scorpion *Pandinus imperator* (Koch) (Scorpiones: Scorpionidae). Universität Würzburg

Matthiesen FA (1962) Parthenogenesis in scorpions. Evolution 16:255–256

Mattoni CI, Peretti AV (2004) The giant and complex genital plug of the *asper* group of *Bothriurus* (Scorpiones, Bothriuridae): morphology and comparison with other genital plugs in scorpions. Zool Anz 243(1–2):75–84. https://doi.org/10.1016/j.jcz.2004.07.002

Maynard-Smith J (1971) The origin and maintenance of sex. Aldine–Atherton, Chicago

Møller A, Jennions M (2001) How important are direct fitness benefits of sexual selection? Naturwissenschaften 88(10):401–415. https://doi.org/10.1007/s001140100255

Moreira MOM (2021) Avaliação da aceitação por fêmeas às investidas reprodutivas dos machos de *Tityus pusillus* Pocock, 1893. Trabalho de Conclusão de Curso em Ciências Biológicas. Universidade de Pernambuco

Moreira MOM, Moura GJB, Lira AFA (2024) Sexual receptivity of females on Neotropical scorpion *Tityus pusillus* Pocock, 1893 in response to male reproductive advances. Biologia 79(7):2119–2124. https://doi.org/10.1007/s11756-024-01691-9

Ojanguren-Affilastro AA, Kochalka J, Guerrero-Orellana D, Garcete-Barrett B, de Roodt AR, Borges A, Ceccarelli FS (2021) Redefinición de la identidad y posición filogenética de *Tityus trivittatus* Kraepelin 1898, y descripción de *Tityus carrilloi* n. sp. (Scorpiones; Buthidae), la especie de mayor importancia médica del sur de Sudamérica. Rev Mus Argent Cienc Nat 23(1):27–55. https://doi.org/10.22179/revmacn.23.714

Olivero PA, González A, Mattoni CI, Peretti AV (2015) Chemical caressess: geographical variation of male sexual signals in a Neotropical scorpion. Behaviour 152(12–13):1745–1763. https://doi.org/10.1163/1568539X-00003302

Olivero PA, Mattoni CI, Peretti AV (2017) Differences in mating behavior between two allopatric populations of a Neotropical scorpion. Zoology 123:71–78. https://doi.org/10.1016/j.zool.2017.06.009

Olivero PA, Vrech DE, Oviedo-Diego MA, Mattoni CI, Peretti AV (2019) Courtship performance as function of body condition in an 'ancient' form of sperm transfer. Anim Biol 69(1):33–46. https://doi.org/10.1163/15707563-00001041

Outeda-Jorge S (2010) Corte e aspectos da biologia reprodutiva do escorpião brasileiro *Tityus bahiensis* (Scorpiones: Buthidae) Dissertação de Mestrado, Universidade de São Paulo

Outeda-Jorge S, Mello T, Pinto-da-Rocha R (2009) Litter size, effects of maternal body size, and date of birth in South American scorpions (Arachnida: Scorpiones). Zoologia 26:45–53. https://doi.org/10.1590/S1984-46702009000100008

Oviedo-Diego MA, Mattoni CI, Vrech DE, Michalik P, Peretti AV (2020) The morphology of mating plugs and its formation in scorpions: implications for intersexual participation. J Morphol 281(6):620–635. https://doi.org/10.1002/jmor.21125

Oviedo-Diego MA, Costa-Schmidt LE, Mattoni CI, Peretti AV (2021) Interaction between sexual communication functions leads to reproductive interference in two syntopic scorpion species. Anim Behav 181:83–93. https://doi.org/10.1016/j.anbehav.2021.08.029

Oviedo-Diego MA, Mattoni CI, Palen-Pietri R, Olivero PA, Peretti AV (2025) Communication via female resistance: sexual behavioural modulation and mutual mate choice determinants in a scorpion. Anim Behav 221:122787. https://doi.org/10.1016/j.anbehav.2023.12.005

Parker GA, Birkhead TR (2013) Polyandry: the history of a revolution. Philo Trans R Soc B Biol Sci 368(1613):20120335. https://doi.org/10.1098/rstb.2012.0335

Peretti AV (1991) Comportamiento de apareamiento de *Zabius fuscus* (Thorell) (Scorpiones, Buthidae). Bol Soc Biol Concepc 62:123–146

Peretti AV (1993) Estudio de la biología reproductiva en escorpiones argentinos (Arachnida, Scorpiones) (morfología funcional y comportamiento). Ph.D. thesis, Universidad Nacional de Córdoba, Argentina

Peretti AV (1994) Comportamiento de relación madre-cria de *Tityus trivittatus* Kraepelin, 1898 (Scorpiones, Buthidae). Bol Soc Biol Concepc 65:9–21

Peretti AV (1995) Análisis de la etapa inicial del cortejo de *Bothriurus bonariensis* (C.L. Koch) (Scorpiones, Bothriuridae) y su relación con el reconocimiento sexual. Revue Arachnologique 11:35–45

Peretti AV, Acosta LE (1999) Comparative analysis of mating in scorpions: the post-transfer stage in selected Argentinian bothriurids (Chelicerata, Scorpiones, Bothriuridae). Zool Anz 237(1998/99):259–265

Peretti AV, Carrera P (2005) Female control of mating sequences in the mountain scorpion *Zabius fuscus*: males do not use coercion as a response to unreceptive females. Anim Behav 69:453–462. https://doi.org/10.1016/j.anbehav.2004.04.019

Polis GA, Farley RD (1979) Behavior and ecology of mating in the cannibalistic scorpion, *Paruroctonus mesaensis* Stahnke (Scorpionida: Vaejovidae). J Arachnol 7(1):33–46

Polis GA, Sissom WD (1990) Life history. In: Polis GA (ed) The biology of scorpions. Stanford University Press, Stanford, pp 81–111

Pordeus LM, Lira AFA, Albuquerque CMR (2019) Male courtship behavior is triggered by female chemical cues in the scorpion *Tityus pusillus* (Scorpiones: Buthidae). Can J Zool 97:1122–1125. https://doi.org/10.1139/cjz-2019-0020

Quijano-Ravell AF, Ponce-Saavedra J (2015) Estructura poblacional de *Centruroides ornatus* (Scorpiones: Buthidae) en la cuenca de Cuitzeo, Michoacán, México. Rev Ibér Aracnol 27:35–44

Ross LK (2009a) Notes and observations on courtship and mating in *Tityus* (*Atreus*) *magnimanus* Pocock, 1897 (Scorpiones: Buthidae). J Venom Anim Toxins Incl Trop Dis 15:43–53. https://doi.org/10.1590/S1678-91992009000100005

Ross LK (2009b) Frequency of spermatophore production and regeneration in the males of *Tityus* (*Atreus*) *magnimanus* Pocock, 1897 (Scorpiones, Buthidae). J Venom Anim Toxins Incl Trop Dis 15:157–162. https://doi.org/10.1590/S1678-91992009000100014

Ross LK (2010) Confirmation of parthenogenesis in the medically significant, synanthropic scorpion *Tityus stigmurus* (Thorell, 1876). Rev Ibér Aracnol 18(1):115–121

Sánchez-Quirós C, Arévalo E, Barrantes G (2012) Static allometry and sexual size dimorphism in *Centruroides margaritatus* (Scorpiones: Buthidae). J Arachnol 40(3):338–344. https://doi.org/10.1636/B12-14.1

Santos MDS, Porto TJ, Lira-da-Silva R, Brazil TK (2014) Description of the male of *Tityus kuryi* Lourenço, 1997 and notes about males of *Tityus stigmurus* (Thorell, 1877) and *Tityus serrulatus* Lutz & Mello, 1922 (Scorpiones, Buthidae). ZooKeys 435:49–61. https://doi.org/10.3897/zookeys.435.6694

Seiter M (2012) Developmental stages and reproductive biology in *Tityus confluens* Borelli, 1899 and *Tityus ocelote* (Francke & Stockwell, 1987) (Scorpiones, Buthidae). Rev Ibér Aracnol 21:113–118

Shaffer LR, Formanowicz DR (1996) A cost of viviparity and parental care in scorpions: reduced sprint speed and behavioural compensation. Anim Behav 51(5):1017–1024. https://doi.org/10.1006/anbe.1996.0104

Simon J-C, Delmotte F, Rispe C, Crease T (2003) Phylogenetic relationships between parthenogens and their sexual relatives: the possible routes to parthenogenesis in animals. Biol J Linn Soc 79:151–163. https://doi.org/10.1046/j.1095-8312.2003.00175.x

Simone Y, Meijden A (2021) Armed stem to stinger: a review of the ecological roles of scorpion weapons. J Venom Anim Toxins Incl Trop Dis 27:e20210002. https://doi.org/10.1590/1678-9199-JVATITD-2021-0002

Souza CAR, Candido DM, Lucas SM, Brescovit AD (2009) On the *Tityus stigmurus* complex (Scorpiones, Buthidae). Zootaxa 1987:1–38. https://doi.org/10.11646/zootaxa.1987.1.1

Steiger S, Meier T, Müller JK (2012) Fitness costs associated with chemical signaling. Commun Integr Biol 5(1):57–60. https://doi.org/10.4161/cib.17988

Steinmetz SB, Bost KC, Gaffin DD (2004) Response of male *Centruroides vittatus* (Scorpiones: Buthidae) to aerial and substrate-borne chemical signals. Euscorpius 12:1–6

Suesdek-Rocha L, Bertani R, Silva-Junior PI, Selivon D (2006) The first record for Wolbachia in a scorpion: the parthenogenetic yellow scorpion *Tityus serrulatus* (Scorpiones, Buthidae). Rev Ibér Aracnol 14:183–184

Tabata J, Narai Y, Sawamura N, Hiradate S, Sugie H (2012) A new class of mealybug pheromones: a hemiterpene ester in the sex pheromone of *Crisicoccus matsumotoi*. Naturwissenschaften 99:567–574

Tallarovic SK, Melville JM, Brownell PH (2000) Courtship and mating in the giant hairy desert scorpion, *Hadrurus arizonensis* (Scorpionida, Iuridae). J Insect Behav 13:827–838. https://doi.org/10.1023/A:1007858400402

Tang V (2023) Non-aggressive competition between males of *Srilankametrus yaleensis* (Kovařík et al., 2019) (Scorpionidae), and other types of agonistic behavior observed in scorpions. Euscorpius 2023(368):1–17

Taylor MS, Cosper CR, Gaffin DD (2012) Behavioral evidence of pheromonal signaling in desert grassland scorpions *Paruroctonus utahensis*. J Arachnol 40:240–244. https://doi.org/10.1636/Hi11-75.1

Thomas ML (2011) Detection of female mating status using chemical signals and cues. Biol Rev 86(1):1–13. https://doi.org/10.1111/j.1469-185X.2010.00130.x

Tomkins JL, Radwan J, Kotiaho JS, Tregenza T (2004) Genic capture and resolving the lek paradox. Trends Ecol Evol 19(6):323–328. https://doi.org/10.1016/j.tree.2004.03.029

Toscano-Gadea CA (2005) Confirmation of parthenogenesis in *Tityus trivittatus* Kraepelin 1898 (Scorpiones, Buthidae). J Arachnol 33(3):866–869. https://doi.org/10.1636/S03-21.1

Toscano-Gadea CA (2010) Sexual behavior of *Bothriurus buecherli* (Scorpiones: Bothriuridae) and comparison with the *B. prospicuus* group. The. J Arachnol 38(2):360–363. https://doi.org/10.1636/Hi09-31.1

Trabalon M, Bagnères AG (2010) Contact recognition pheromones in spiders and scorpions. In: Blomquist GJ, Bagnères R-G (eds) Insect hydrocarbons: biology, biochemistry, and chemical ecology. Cambridge University Press, Cambridge, pp 344–374

Triana F, Bonilla F (2023) Litter size of some Costa Rican scorpions (Scorpiones: Buthidae, Diplocentridae). Rev Ibér Aracnol 42:277–279

Ugolini A, Carmignani I, Vannini M (1986) Mother-young relationship in *Euscorpius*: adaptive value of the larval permanence on the mother's Back (Scorpiones, Chactidae). J Arachnol 14(1):43–46

Vrech DE, Oviedo-Diego MA, Olivero PA, Peretti AV (2024) End of the season blues: senescence and reproductive trade-offs in male scorpions. Insects 15(12):916. https://doi.org/10.3390/insects15120916

Warburg MR (2011a) Growth and longevity of *Nebo hierichonticus* in the laboratory; a long-term study (Scorpiones, Diplocentridae). Arachnology 15(5):168–172. https://doi.org/10.13156/arac.2010.15.5.168

Warburg MR (2011b) Scorpion reproductive strategies, allocation and potential; a partial review. Eur J Entomol 108(2):173–181

Warburg MR (2012) Pre-and post-parturial aspects of scorpion reproduction: a review. Eur J Entomol 109(2):139–146

Wyatt TD (2009) Fifty years of pheromones. Nature 457:262–263. https://doi.org/10.1038/457262a

Wyman JT, Wright-Ueda J, Agnew Q, Castellano I, Simone Y (2025) First report of arm-span competition in buthid scorpions: male-male contest in *Tityus* cf. *rosenbergi* Pocock, 1898. The. J Arachnol 52(3):210–213. https://doi.org/10.1636/JoA-S-23-009

Zolessi LD (1956) Observaciones sobre el comportamiento sexual de *Bothriurus bonariensis* (Koch) (Scorpiones, Bothriuridae). Bol Fac Agron Montevideo Uruguay 35:1–10

Kovařík F (2006) Nový druh štíra Ananteris terueli sp. n. v teráriu. Akva Tera Fórum, 10:60–63

Rouaud C, Huber D, Lourenço WR (2000) Life history of Caribetityus elli (Armas & Marcano Fondeur, 1992) from the Dominican Republic (Scorpiones, Buthidae). European Arachnology, 2000:87–90

Toscano-Gadea CA (2015) Biología reproductiva, gestación y cantidad de crías de seis especies de escorpiones de Uruguay (Scorpiones: Bothriuridae,Buthidae, Euscorpiidae). Bol Soc Zool Uruguay 24(2):130–145

Chapter 4
Luminescence and Other Optical Characteristics of Scorpion's Exoskeleton

Daniel López-Cabrera, Gabriel Ramos-Ortiz, Montserrat G. Rivera-Ortiz, Edmundo González-Santillán, and Rafael Espinosa-Luna

Abstract The exoskeleton of scorpions exhibits the intriguing physical characteristic of blue-greenish light emission under ultraviolet illumination. Despite the knowledge accumulated since 1954 regarding this phenomenon, the role of this fluorescence in the biology of scorpions remains unknown. The determination of whether this phenomenon is an inherent property of scorpions or merely a byproduct of their evolutionary adaptation to their environment remains to be elucidated. In this chapter, we examine the optical characteristics of the exoskeleton, encompassing phenomena such as fluorescence and color. Such characteristics were obtained through spectroscopic techniques in conjunction with red, green, blue (RGB) color space analysis of photographs taken of scorpions in their substrate under different illumination conditions that simulate sunlight or moonlight. The objective of this chapter is to provide a comprehensive understanding of the optical characteristics of the scorpion's exoskeleton, including several Neotropical species, and its correlations with scorpion biology.

Keywords Fluorescence · Exoskeleton · RGB color base

4.1 Introduction

The physical origin of fluorescence in scorpions' exoskeletons is currently well understood. This phenomenon can be attributed to the optical excitation of distinct molecules present within the cuticle, designated as "fluorophores," which exhibit

---

D. López-Cabrera (✉) · G. Ramos-Ortiz (✉) · M. G. Rivera-Ortiz · R. Espinosa-Luna
Centro de Investigaciones en Óptica A. C, León, Mexico
e-mail: dlopez@cio.mx; garamoso@cio.mx; montserratr@cio.mx

E. González-Santillán
Colección Nacional de Arácnidos, Departamento de Zoología, Instituto de Biología, Universidad Nacional Autónoma de México, Coyoacán, Mexico City, Mexico
e-mail: edmundo.gonzalez@ib.unam.mx

A. F. de Araujo Lira, E. González-Santillán (eds.), *The Biology of Neotropical Scorpions*, https://doi.org/10.1007/978-3-032-23194-9_4

remarkable efficacy in emitting light. In natural conditions, the excitation of these fluorophores can be produced by the ultraviolet (UV) content in sunlight or moonlight.

Despite this knowledge, however, there is a paucity of studies in the literature that offer a detailed physical characterization of the fluorescence emitted by the scorpion's exoskeleton. The impetus for presenting this work stems from the identified gap in the extant literature. In this study, we present a systematic characterization of the fluorescence properties of the exoskeleton across a diverse sample of scorpions representing various species and environments within Mexico, including several Neotropical species. Additionally, the investigation encompasses supplementary salient optical characteristics of the exoskeleton such as color reflection. It is evident that these two properties hold significance in the context of the vision of living organisms. Further, they facilitate enhanced comprehension of the adaptive traits exhibited by scorpions in relation to their ecological environment.

The chapter is organized as follows: Sections 4.2 - 4.5 describe the physical principles of fluorescence and summarizes previous studies on this phenomenon in scorpions. In these aforementioned sections, the fundamentals of red, green, blue color (RGB) base analysis will be introduced. The impetus for the development of this analysis technique stemmed from the necessity of processing images of scorpions captured within a replica of their natural habitat. Furthermore, we introduce the concept of the scorpion's exoskeleton functioning as an optical surface to aid in the understanding of its physical properties. Section 4.6 is dedicated to the presentation of experimental results concerning the fluorescence and RGB-based characterization of the exoskeleton.

4.2 Principles of Fluorescence

Luminescence is defined as a physical process in which a body emits light at room temperature conditions. Depending on the underlying mechanism that produces it, luminescence can be categorized into different types, including photoluminescence, chemiluminescence, electroluminescence, and sonoluminescence. Photoluminescence (PL) is a phenomenon in which a body absorbs photons and subsequently re-emits photons of lower energy. The photon energy in an electromagnetic wave, such as light, is determined by the wavelength (λ) through the relation $E = hc/\lambda$, where h and c are the Planck constant and the speed of light in vacuum, respectively. In the Jablonski diagram, the incident photon promotes one electron of the fluorophore from the ground state (S_0) to an electronic excited state (S_1). The PL occurs when the excited molecule decays spontaneously from the lowest vibronic sublevel of S_1 to any of the vibronic sublevels of S_0. If the lifetime τ of the excited state spans over less than 10^{-8} s, then the radiative process is classified as fluorescence. The efficiency of fluorescence of a substance is quantified by a parameter called quantum yield (η), defined as the ratio between the number of emitted photons and the number of absorbed photons.

4.3 Fluorescence in Living Organisms

Fluorescence emission is a cryptic phenomenon exhibited by various biological entities, including arachnids, insects, birds, and fishes (Luen 2019). The phenomenon in question has been the subject of investigation in the context of its association with hunting, camouflage, and sexual behavior (Wilts et al. 2013; Brandt and Masta 2017; Phillips 1959; Espinosa-Luna et al. 2010).

For instance, in the case of arachnids, certain studies have indicated that fluorescence may serve as a means of confounding predators, attracting prey, or facilitating intraspecific communication (Andrews et al. 2007). In certain instances, the emission of fluorescence exhibits variation between sexes. In the case of the crab spider *M. vatia*, the intensity of fluorescence undergoes changes throughout the different life stages due to variations in foraging ecology and behavior (Brandt and Masta 2017). In contrast, in other species, the fluorescence emission is not uniform throughout the body. This is the case of the species *E. umbratica* which emits light exclusively from the palps, and this emission is influenced by factors such as age and feeding history (Painting et al. 2017). Moreover, the phenomenon of fluorescence has been observed in various chelicerates, including the orders Opiliones and Solifugae (Nieves and Santiago-Blay 2019), Xiphosura and Eurypterida (Rubin et al. 2017).

4.4 Fluorescence as Ubiquitous Optical Property in Scorpion's Exoskeleton

When exposed to UV radiation, the exoskeleton of scorpions typically emits blue-green fluorescence (Sissom et al. 1990), though there are some exceptions to this rule (Lourenco 2012). This fluorescence comprises photons within the wavelength range from 420 to 650 nm, approximately. The maximum excitation of fluorescence is around 400 nm.

To date, three fluorophores have been identified in the scorpion cuticle: β-carboline (Stachel et al. 1999), 4-methyl-7-hydroxycoumarin (Frost et al. 2001), and macrocyclic diphthalate ester (Yoshimoto et al. 2020). These fluorophores are deposited within the exoskeleton during the sclerotization process (Stachel et al. 1999). It has been established that the fluorescence intensity is positively correlated with the epicuticle thickness (Chen et al. 2015). This phenomenon elucidates the reason why scorpions exhibit no fluorescence immediately after ecdysis and only begin to exhibit it after approximately 48 h (Stahnke 1972).

The biological function of fluorescence in scorpions remains to be elucidated, if it exists at all. Several investigations have been conducted with the objective of elucidating the question (Lawrence 1954; Welch et al. 2012). Scorpions possess lateral ocelli that provide sensitive dichromatic vision in green—color at which the exoskeleton's fluorescence has the emission peak—and UV light, making them among the most sensitive photoreceptors in arthropods (Fleissner 1977). Scorpions also possess median ocelli, which are predominantly sensitive to green light (Machan 1968). Considering these findings, a hypothesis was formulated proposing that the exoskeleton may function as a light-detection system. This hypothesis emerged from observations that scorpions tend to reduce their locomotion in the presence of green and UV light (Blass and Gaffin 2008). The phenomenon of fluorescence in scorpions has been proposed as a potential mechanism for intraspecific communication, possibly aiding in the differentiation between males and females or in the recognition of individuals from other species (Jordan 2010; Kloock 2008). Additionally, it has been proposed that the intensity of moonlight detected by scorpions may serve as a signal for assessing predation risk and food availability (Skutelsky 1996).

4.5 Fluorescence under Natural Illumination Conditions

Several factors must be considered when evaluating whether a scorpion's fluorescence can be perceived or distinguished by any biological entity, including scorpions themselves. Special attention should be given to the physical conditions of the scorpion's habitat. For instance, under daylight conditions, the excitation of fluorophores in the exoskeleton occurs solely due to the UV component of sunlight. As illustrated in Fig. 4.1, the spectral composition of sunlight demonstrates that the UV content is substantial, exhibiting an irradiance of approximately 0.5 W/m^2 nm at 400 nm. This UV irradiance is absorbed to varying strengths across scorpion species, contingent upon the pigmentation of their exoskeletons. Dark-colored exoskeletons will absorb more UV radiation than lighter ones. Upon excitation, the exoskeleton exhibits its characteristic blue-greenish fluorescence with certain quantum yield (η). Here it is important to note that under daylight conditions, the blue-greenish irradiance (at 500 nm) incident on the substrate of scorpions' habitat can be as high as 1.3 W/m^2 nm (see Fig. 4.1). Considering the scorpion's exoskeleton as an optical surface, it becomes evident that the emitted blue-green fluorescence merges with the light reflected by the exoskeleton. Consequently, the visual detection of fluorescence in natural habitats would be possible if any of its properties—namely, intensity, color, or polarization—are distinguishable from the exoskeleton's reflection effects.

Under nocturnal conditions, the fluorophores present in the exoskeleton might also be excited, with maximum excitation occurring on full moonlight nights. The spectrum of nocturnal light exhibits a resemblance to that of daylight. However, the irradiance of nocturnal light is approximately 400,000 times weaker compared to sunlight (Krisciunas and Schaefer 1991).

A considerable number of living organisms are capable of perceiving light. Many organisms even have chromatic vision. The color of an object is determined by the spectral content of light reflected or emitted by the object. One approach to studying the scorpion's exoskeleton involves analyzing its characteristic color and its

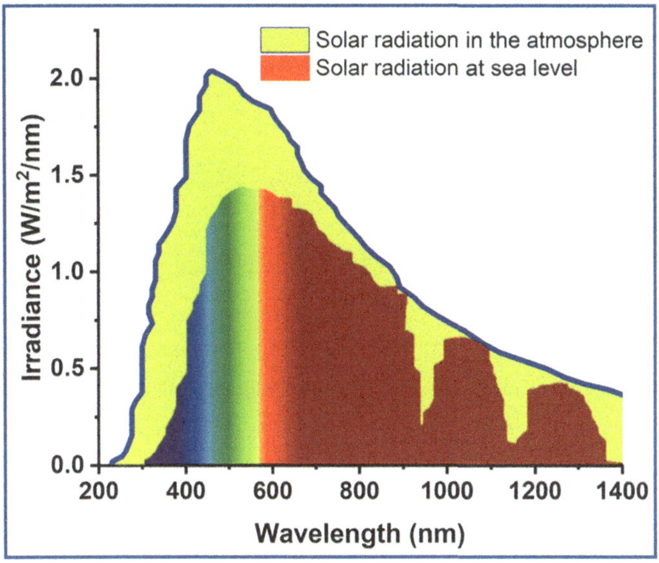

Fig. 4.1 Spectrum of solar radiation. The solar radiation is reproduced in this work by a solar simulator to induce fluorescence in scorpion's exoskeleton. (Data of solar irradiance is found in Newport (n.d.))

chromatic correlation with objects in its habitat. To provide an illustration of this color analysis, it is appropriate to delve into the subjective perception of color through the lens of human vision. The human retina contains two types of photoreceptors: rods and cones. Cones are responsible for color vision and have three types of pigments with peak sensitivity at short (S), medium (M), and long (L) wavelengths—approximately 440 nm (blue), 540 nm (green), and 560 nm (red), respectively. It is noteworthy that avian vision is characterized by a four-color channel configuration, exhibiting sensitivity to both the ultraviolet (UV) and red wavelength ranges, which extend beyond the sensitivity of human vision (Iriel and Lagorio 2010). Conversely, the visual spectrum of bees is constrained to the UV, blue and green wavelengths. In all cases, the combination of the fundamental colors and the corresponding sensitivity curves for the photoreception enables specific chromatic perception of light.

Scorpions are known to exhibit adaptive behaviors in response to their environment. One such example is the coloration of their exoskeletons, which has been observed to be closely correlated with the substrate they inhabit (Lourenco and Cloudsley-Thompson 1996). The phenomenon has been previously documented in some Neotropical species, including *Centruroides* (Stahnke and Calos 1977) and *Thorellius* (González-Santillán and Prendini 2013), as well as several species of vaejovids and hadrurids. In these species, the exoskeleton exhibits a color change in response to the soil's characteristics, ranging from dark hues in dark soils to lighter tones in light soils (Williams 1970; Williams 1987). Similarly, the fluorescent

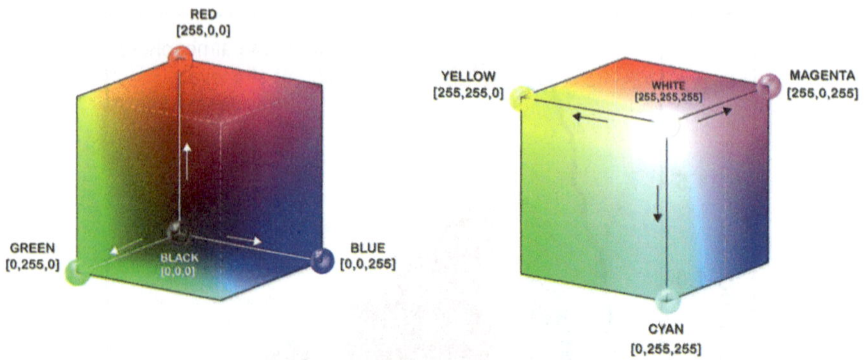

Fig. 4.2 Examples of the range of colors generated from different combinations of RGB color coordinates

characteristics of the scorpion's exoskeleton should be determined by the habitat, in case such a phenomenon possesses a biological function.

To gain insight into the optical properties of the scorpion's exoskeleton, specifically the color resulting from the reflection of environmental light or fluorescence effects, the RGB color space will be used in subsequent sections to standardize the detection of color in a quantitative manner. While this color space is predicated on human vision, the information obtained is valuable for characterizing the optical properties of the exoskeleton. The objective of this study is to provide characterizations that may correlate with biological roles.

Within the RGB color space, the representation of any color is defined by three coordinates (red, green, blue), with each coordinate ranging from 0 to 255. Therefore, the coordinates (0, 0, 0) denote the absence of color, or black, while (255, 255, 255) represent white. Consequently, the RGB color base can reproduce 16 million distinct colors. Figure 4.2 provides a schematic representation of the RGB color space. Consequently, as is the case with any image reproduced by RGB color space, the image of the scorpion's exoskeleton will be composed of the combination of R, G, and B color channels.

4.6 Characteristics of Fluorescence and Reflectance Across Various Species of Scorpions

4.6.1 Fluorescence Strength and Emission Wavelengths in Different Scorpion Species

Research has demonstrated that specific characteristics of the exoskeleton's fluorescence can serve as indicators of scorpion species. As indicated by the fluorescence spectrum, considerable variation across species is observed, contingent on the concentration and relative proportion of fluorophores. The extant literature documents a range of spectral peaks attributable to various species, including some Neotropical scorpions. For instance, the wavelength of 518 nm corresponds to *C. limpidus* (Ullrich 2016), 490 nm to *C. vittatus* (Stachel et al. 1999), 501 nm to *P. becki* (Kloock 2009), and 490 nm to *O. martensii* (Tokumoto et al. 2018). In these works, spectroscopic analyses were performed using different excitation wavelengths. In contrast, when two species were examined under the same experimental condition and the same excitation wavelength, the emission peak practically coincided, as observed in the case of *V. confusus* (500 nm) and *P. shulovi* (498 nm) (Kloock 2008). Recent spectroscopic investigations have revealed that the emission spectra of seven species from diverse habitats, and exoskeleton colors ranging from light to dark, remain largely consistent, with peaks at approximately 483 ± 3 nm (Lopez et al. 2020).

Figure 4.3a presents the emission spectra recorded for living specimens of the following species: *Diplocentrus whitei*, *Diplocentrus tehuacanus*, *Mesomexovis* sp1, *Vaejovis intermedius*, *Paruroctonus gracilior_1*, *Paruroctonus gracilior_2*, and *Centruroides infamatus*. Evidently, the fluorescence emissions are spectrally indistinguishable, exhibiting only a variation in intensity. The weakest and strongest fluorescence emissions correspond to *D. withei* and *P. gracilior_1*, respectively. The predominant trend indicates that the emissions intensity escalates as the exoskeleton hue becomes more pallid.

The fluorophores of *C. infamatus* were extracted using the method reported by Wankhede (Wankhede 2004) and were exposed to different wavelengths to test the hypothesis that fluorescence became excitation dependent. Figure 4.3b shows the fluorescence spectra obtained from the extraction of fluorophores in ethanol. The wavelength of maximum emission varied from 380 to 491 nm, as the UV excitation wavelength was tuned from 305 to 410 nm. This excitation dependence on fluorescence is consistent with the presence of multiple fluorophores in the exoskeleton. However, the nearly continuous tuning of fluorescence with excitation remains

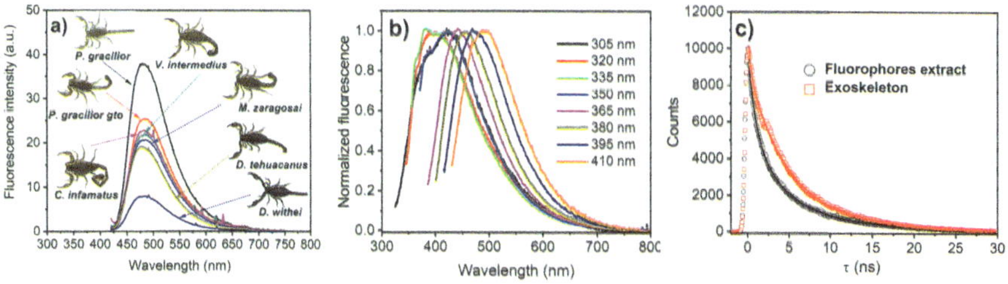

Fig. 4.3 (**a**) Fluorescence spectra from species with exoskeletons ranging from light to dark colors; (**b**) fluorescence spectra from fluorophores extracted from the exoskeleton of a *C. infamatus* under different excitation wavelengths; (**c**) temporal fluorescence decay measured in exoskeleton and fluorophores extracted from the exoskeleton of a *C. infamatus*. (Data of (**a**) and (**c**) recollected under 400 and 405 nm of excitation, respectively)

puzzling: despite the fact that only three fluorophores have been identified in the exoskeleton to date, Fig. 4.3b shows seven distinct emission peaks.

This observation appears to contradict a general spectroscopic principle, known as Kasha's rule (Demchenko et al. 2017), which states that any fluorophore emits the same fluorescence spectrum regardless of excitation. According to this established principle, the mixture of three chromophores in solution should yield no more than three emission peaks. Therefore, the observation of more than three emission peaks by tuning excitation suggests the presence of additional, unidentified fluorophores in the exoskeleton. Moreover, most synthetic or natural fluorophores exhibit luminescence in the liquid phase but undergo complete quenching in the solid phase—such as within the exoskeleton. This phenomenon may also contribute to the observed discrepancies between the behavior of the extracted fluorophores in solution and those embedded within the exoskeleton structure.

The fluorescence decay dynamics measured directly from the exoskeleton of a *C. infamatus* and from the fluorophores in liquid phase are reported in Fig. 4.3c. This fluorescence decay was obtained using a time-correlated single photon counting (TCSPC) module in a spectrofluorometer (FS5, Edinburgh Instruments) with pulsed laser excitation at 405 nm. The average excited-state lifetime τ from the exoskeleton and from the solution of extracted fluorophores resulted in 4.7 ns and 3.6 ns, respectively. Unexpectedly, the fluorescence decay rate was slower in the solid phase. Usually, fluorescence decay exhibited by common fluorophores is faster in the solid than in the liquid phase. This is because in the solid phase fluorophores are tightly packed, favoring strong molecular interactions which efficiently promote rapid non-radiative decay pathways reducing the lifetime of the excited state. To explain the faster decay of excited state in solution, it should be observed that the fluorophores were extracted in a polar protic solvent (ethanol) so that protonation can occur. On the other hand, it is not discarded that the fluorophores are packed in the exoskeleton in a manner that favors the stabilization of the excited state.

The quantum yield of fluorescence (η) was also determined in both the exoskeleton of a scorpion and the solution of extracted fluorophores. For the determination of η, the absolute method was employed with an integrating sphere module in our spectrofluorometer. Due to the difficulties involved in measuring η in a living specimen, the measurement was instead made in a non-living specimen. Thus, the chelae and telson of a *C. infamatus* exhibited $\eta \sim 5.3\%$. This value appears to be somewhat low considering the intense fluorescence observed in scorpions. However, it is important to note that a proportion of the UV photons may also be absorbed by non-fluorescent molecules, such as melanin. Consequently, the actual value of η due exclusively to fluorophores embedded within the exoskeleton can be significantly higher. To the best of our knowledge, this is the first assessment of η in the exoskeleton of a scorpion. In the ethanol solution, the value of η for all extracted molecules was determined to be 7.9%.

4.6.2 Heterogeneous Fluorescence Within the Exoskeleton

Recent research has revealed that the fluorescence intensity emitted through the exoskeleton of scorpions exhibits heterogeneity, i.e., segments such as the metasoma and chelae emit higher fluorescence intensity compared to the mesosoma (Lopez et al. 2020). As an illustration of this optical property in living scorpions, Fig. 4.4 presents a digital photograph from a specimen of *C. infamatus* under UV light illumination. Upon initial observation, the photograph demonstrates fluorescence emanating from all segments of the exoskeleton. The inhomogeneity in intensity is readily apparent during digital image processing. In this process, the three RGB color layers are initially converted into a single layer of grayscale values. Subsequently, to enhance visual perception, grayscale values are represented graphically with false colors. In this scale, red corresponds to the highest fluorescence intensity, which is predominantly observed in the metasoma and chelae. In contrast, the mesosoma segments, depicted in yellow-orange hues, emit significantly less fluorescence. The analysis was expanded to encompass 24 species of live specimens, all of which exhibited heterogeneous fluorescence patterns. This finding was statistically supported by a repeated variables ANOVA study [$F(2,69) = 7.9$, $p < 0.05$], along with its respective post-hoc tests, confirming significantly different fluorescence intensities between metasoma and chelae compared to mesosoma. A second statistical analysis was conducted, the results of which indicated that there was no significant difference in fluorescence intensity between the female and male subjects of the same species.

Furthermore, heterogeneous fluorescence was observed in the exuviae shed by second, third, and fifth instars of *Oliverius martensii* scorpions (Liu et al. 2024). Scorpions that have undergone ecdysis exhibit a lack of fluorescence, with this

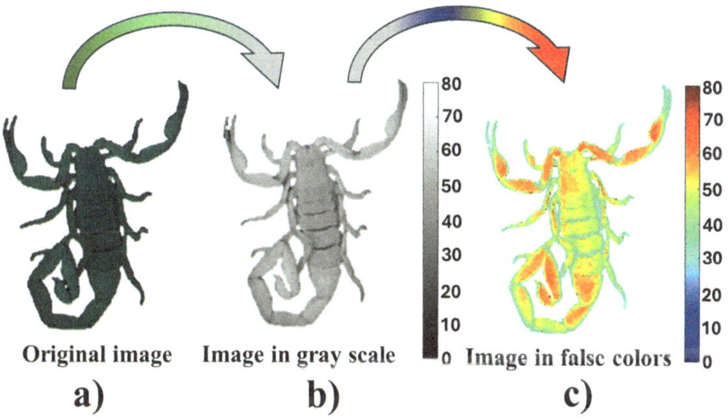

Fig. 4.4 Photographs obtained for specimen *Centruroides infamatus*: (**a**) under UV illumination; (**b**) image converted to gray scale; (**c**) image plotted in with false colors for each pixel

phenomenon becoming apparent only after a 48-h period. The study reports that the fluorescence emission exhibited heterogeneity among different body sections, including the chelae, mesosoma, metasoma, and telson, with the chelae and telson exhibiting higher levels of intensity. In addition, the recovery of fluorescence was found to be accelerated in these segments. An emission spectrum indistinguishable from that of other segments was identified, with a peak at 475 nm. This phenomenon can be explained by the fact that the synthesis of fluorescent substances by the digestive glands begins before ecdysis to ensure the formation of the new exoskeleton within the next 24 h. This process demonstrates a clear preference for areas of greater hardness, such as the chelae, metasoma, and telson, which play a critical role in survival.

4.6.3 RGB Image Analysis of Scorpions in a Substrate

A more pragmatic and perspicacious approach entails the study of scorpions under conditions that replicate their natural habitats and with illumination that emulates sunlight or moonlight. In the following sections, the optical properties of living scorpions within a microhabitat that has been engineered to replicate the substrate of their natural environment are presented. To conduct these studies, a method of RGB analysis was applied to photographs taken of scorpions under these controlled conditions. The method is applied to *C. infamatus* as an example of a Neotropical species.

In summary, a solar simulator (Oriel® Class AA) was utilized to illuminate a microhabitat ($20 \times 40 \times 15$ cm^3) that had been recreated within a plastic container. The microhabitat contained soil, stones, and tree bark, all of which were sourced from the same area in which the specimens were collected. During these experiments, a camera with an f/2.2 aperture, a 118° field of view, and an 8-megapixel capacity was utilized. To obtain well-lit photographs that were not underexposed or overexposed, the system was calibrated every experimental session with the 99% reflective target (BaSO4). The camera's shutter time, ISO, and focus parameters were manually optimized with the reflective target until balanced pixel values close to 255 for each RGB color channel were achieved. It is important to note that meticulous efforts were undertaken to guarantee linearity in the registered intensities across a substantial dynamic range. Following the identification of the optimal settings, the specimens were situated within the designated microhabitat for photographic documentation. In each of the processed photographs presented here, the specific section of interest was obtained using a segmentation routine in Matlab®. For instance, this routine was employed to obtain the RGB histograms of the reflective target, scorpion's segments, and sections of soil, stone, or tree bark.

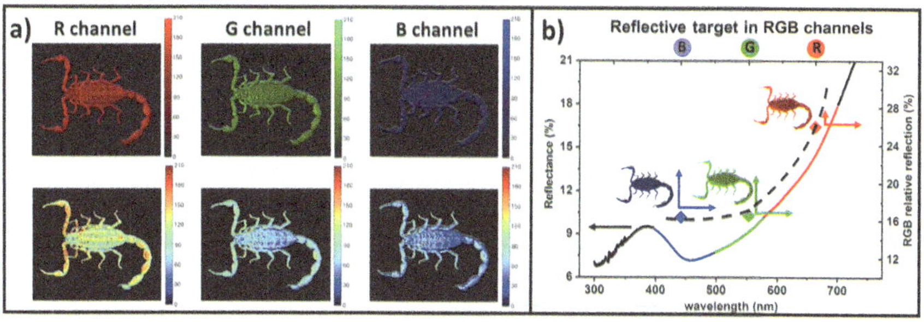

Fig. 4.5 (**a**) Upper row: decomposition in RGB color layers of the image obtained from a *C. infamatus* scorpion illuminated with white light equivalent to one sun. Lower row: each color layer plotted in scale of false colors with relative intensities in the range 0–210; (**b**) reflectance spectrum measured in the exoskeleton of the *C. infamatus* compared with the relative reflection calculated for each RGB channels in a photograph of the specimen

4.6.4 Visualization of Scorpion's Exoskeleton in RGB Channels

Figure 4.5a displays RGB images of the exoskeleton of a *C. infamatus,* photographed under white light illumination spectrally equivalent to one sun's irradiance. It was observed that the color contribution in the blue (B) channel was more subdued than for the red (R) and green (G) channels. To enhance visual interpretation, the lower row of Fig. 4.5a displays each individual RGB channel mapped using a false-color scale. The false-color scale is applied uniformly across all three channels. The results indicate that the B channel hue is nearly homogeneous across the exoskeleton. In contrast, the R and G channels exhibit substantial hue variation, particularly evident in the pedipalps and metasoma. The present findings suggest the existence of localized variations in reflectance due to cuticle pigmentation.

The values of each pixel in the photograph of Fig. 4.5a correspond to the amount of light reflected by the exoskeleton in each color channel. As will be discussed in the subsequent sections, the reflectance observed in these channels can match closely the characteristic reflectance of the substrate. By averaging the huc values for each color channel and comparing them with the pixel values registered from the BaSO4 reference target, the relative reflectance was calculated, showing a value of 16% for B and G which increases up to 26% for the R channel.

The relative reflectance values derived through RGB analysis demonstrate a substantial degree of concurrence with the spectral reflectance (see Fig. 4.5b) previously documented for *C. infamatus* (Lopez et al. 2020). The results of this study are also consistent with the findings reported by Ullrich (2016) for the *C. limpidus* specimen, which exhibited a minimum reflectance at 453 nm.

4.6.5 RGB Correlations of Scorpion's Exoskeleton with Substrate

Figure 4.6a presents a photograph of *C. infamatus* captured in the microhabitat under white light illumination equivalent to one sun. RGB histograms were generated for select sections of interest in the photograph, based on each pixel value (V_{Pi}), which range from 0 to 255 for the three RGB channels. To account for the overall color contribution from all pixels, a parameter of intensity distribution, herein denoted as I_{Pi}, was introduced. This parameter is defined as the product of the pixel value V_{Pi} and its corresponding frequency f_i in the RGB histogram. This approach effectively ponders each pixel value by its occurrence in the image. For the sake of expediency, the intensity distribution I_{Pi} is regarded as a weighted parameter as follows:

$$I_{Pi} = \frac{V_{Pi} f_i}{\sum_{i=1}^{255} V_{Pi} f_i} \tag{4.1}$$

Notice that the areas under the I_{Pi}distributions are normalized, thereby enabling a direct comparison of the relative contributions of each RGB channel in percentage terms.

As illustrated in Fig. 4.6b–d, the I_{Pi}distributions derived from the image of *C. infamatus* reveal interesting information. Although the I_{Pi} curves exhibit

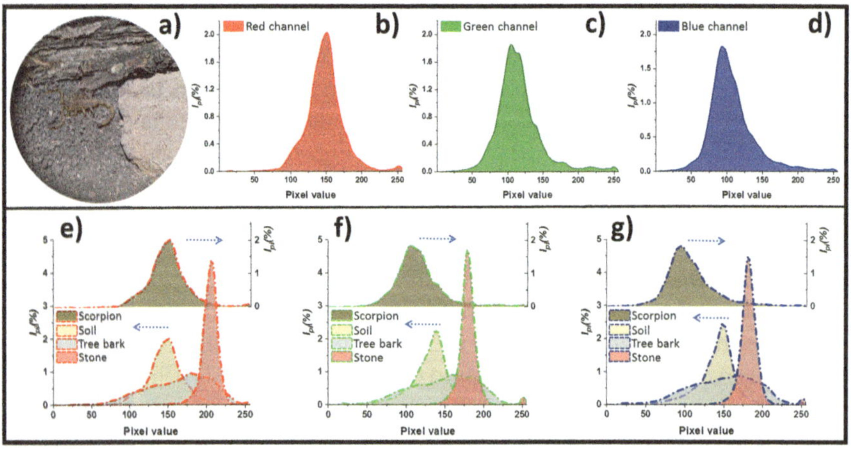

Fig. 4.6 (**a**) Image of a *Centruroides infamatus* photographed in a laboratory-generated microhabitat under one sun of illumination. Histograms for the I_{Pi} value for the whole exoskeleton at the (**b**) red, (**c**) green, and (**d**) blue channels. RGB color histograms of soil, tree bark, and stone in the substrate (lower plots) against the *C. infamatus* exoskeleton's histograms (upper plots) for the (**e**) red, (**f**) green, and (**g**) blue channels

comparable dispersion of V_{Pi} values across the three channels, the I_{Pi}distribution for the R channel is centered around a V_{Pi} value of approximately 150. In contrast, the distributions for the green and blue channels are centered around 110 and 100, respectively. The data unequivocally demonstrated that the scorpion's exoskeleton exhibited the highest reflectance in the red channel and the lowest in the blue channel.

To incorporate the scorpion habitat into the study, cross sections with equal areas to the entire scorpion's exoskeleton were generated for the soil, stone, and tree bark shown in Fig. 4.6a. Each image of equal cross-section so obtained was separated into its respective RGB color layers. The distributions of intensity I_{Pi} were then plotted in the lower panels of Fig. 4.6e–g and compared with the corresponding distributions for the exoskeleton plotted in upper panels of Fig. 4.6e–g.

The I_{Pi}distribution from the scorpion's exoskeleton exhibited a resemblance to the soil substrate's in the red channel, indicating that the reflected light patterns are analogous in both (Fig. 4.6e–g). The resemblance is evident in the I_{Pi} curves, which are centered at V_{Pi} ~ 150. A comparatively negligible disparity in percentage of reflectance is measured at the peak of I_{Pi}, which is 4.5% in the exoskeleton and 2% in soil. In the same red channel, the matching between exoskeleton and stone is significantly reduced, and to a lesser extent with bark. In the green and blue channels, however, the I_{Pi}distribution of the exoskeleton differs significantly from those of soil, stone, and bark which are centered at larger values of V_{Pi}. This observation indicates that the exoskeleton exhibits a darker coloration in green and blue hues relative to the substrate. In summary, the findings indicate that, for this specific substrate, the scorpion exhibits superior camouflage on red light sources. The contrast between the scorpion's exoskeleton and its surroundings in each color channel is more clearly visualized in Fig. 4.7. To facilitate the observation of this contrast, each RGB image in Fig. 4.7a–c is displayed in a scale of false colors in Fig. 4.7d–f.

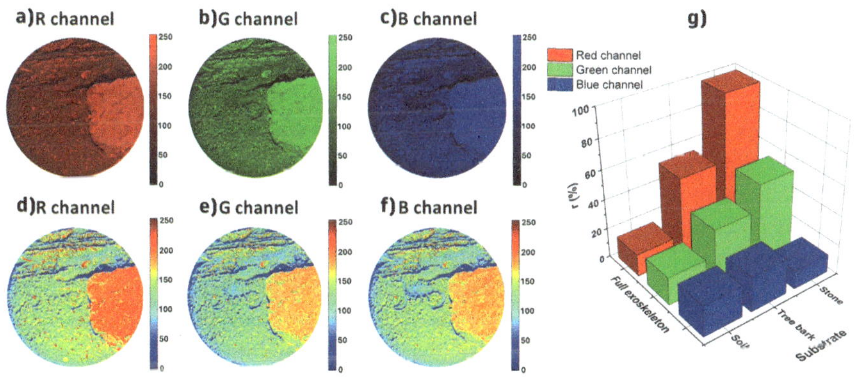

Fig. 4.7 Upper row. RGB color layers: (**a**) red, (**b**) green, and (**c**) blue of the *C. infamatus* specimen photographed on its substrate under illumination equivalent to one sun. Lower row. RGB color layers: (**d**) red, (**e**) green, and (**f**) blue graphed in false colors to observe easily similarity or variability in the color tones within one single color layer. (**g**) Correlation percentages between the scorpion exoskeleton and the soil, stone, and tree bark

To ascertain the quantitative similarity between the I_{Pi} distributions obtained from the scorpion's exoskeleton and those from soil, stone, and tree bark on each RGB channel, the statistical correlation parameter (r) determined by Eq. 4.2 was employed.

$$r = \frac{\sum_{i=1}^{255}\left(A_i - \underline{A}\right)\left(B_i - \underline{B}\right)}{\sqrt{\left(\sum_{i=1}^{255}\left(A_i - \underline{A}\right)^2\right)\left(\sum_{i=1}^{255}\left(B_i - \underline{B}\right)^2\right)}} \tag{4.2}$$

In this equation, A_i *and* B_i represent the I_{Pi} intensity distribution that is being compared. Concurrently, $\underline{A}$ *and* $\underline{B}$ signify the average values (Wackerly et al. 2007). As a means of complementing the results that were qualitatively discussed above, Fig. 4.7g presents the calculated statistical correlations. The statistical similarity between the scorpion's exoskeleton and the soil substrate in the red channel is noteworthy, with an r value of approximately 98%. However, this similarity decreases to 54% for bark and 15% for stone, suggesting distinct differences in material properties among these substrates. This tendency is replicated for the green channel; however, the correlation percentage is lower. The trend is disrupted for the blue channel, as r for the soil substrate (~12%) is lower than for bark (~23%) and stone (~20%).

Photographic documentation was also obtained in the microhabitat under conditions of illumination equivalent to full moonlight. At this low level of illumination, the highest correlations between the exoskeleton of *C. infamatus* and the soil, tree bark, and stone occurred again in the red channel, with percentage correlation values of 96.0, 68.7, and 20.6%, respectively. In the green channel, the r values for soil, tree bark, and stone were 50.2%, 50.6%, and 25.8%, respectively.

From an ecological perspective, *C. infamatus* is considered an "errant species" (Prendini 2001), also known as a "bark scorpion." This species is typically active at night and has been observed moving in both open and vegetated areas. In consideration of the ecomorphological classification of this species, *C. infamatus* has evolved certain morphological and behavioral adaptations that facilitate its camouflage within its natural environment. The results presented here align with such adaptations, as scorpions have evolved strategies to mimic the substrate in which they live, developing cryptic coloration (Lourenco and Cloudsley-Thompson 1996). As discussed in the lines above in the context of a specific substrate, *C. infamatus* exhibits a coloration pattern that closely resembles that of soil, particularly in the red channel, and to a lesser extent in the green channel.

Finally, under conditions of illumination equivalent to that of a full moon, exoskeleton's fluorescence was successfully detected. It is noteworthy that the most extant literature on the subject reports the exoskeleton's fluorescence under arbitrary UV conditions. In contrast, the present study reports the fluorescence induced by illuminating the specimen with the UV content matching that of moonlight (Fig. 4.8). However, it must be noted that under these conditions of illumination, the intensity of fluorescence was found to be weaker than the intensity of reflected light

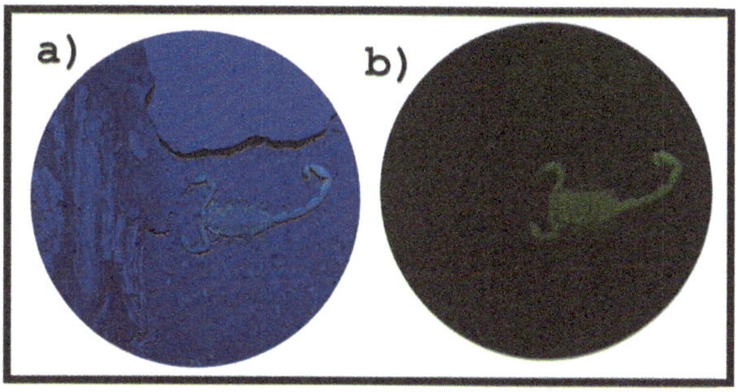

Fig. 4.8 (**a**) *Centruroides infamatus* photographed with the UV content in full moonlight. This UV illumination was generated by passing white light equivalent to full moonlight through a UV filter with a cut-off wavelength of 420 nm. (**b**) Photograph in the green channel; a green filter was placed in front of the camera to transmit wavelengths larger than 480 nm

at the green channel. To obtain this photograph, the white illumination was passed through a low-pass filter with a cutoff wavelength of 420 nm, allowing the excitation of the exoskeleton's fluorophores within the wavelength range of approximately 350–420 nm. The figure also displays the image in the green channel, where the exoskeleton's fluorescence is clearly discernible.

4.7 Conclusions

By means of digital image processing, it was determined that the fluorescence emission is different along the scorpion's exoskeleton; this heterogeneous fluorescence emission is a consequence of differentiated sclerotization processes in the exoskeleton. The intensity of fluorescence also varies across species according to color tonality of exoskeleton: the intensity increases as exoskeleton color becomes lighter, although the spectral features of the emission are practically the same in all cases. In turn, the color tone of exoskeleton may be related to the substrate of its habitat and its activity on the surface, all as an evolutionary function for survival. For instance, by using *C. infamatus* as a model of semipsammophilous species, it was demonstrated through RGB color space analysis, that the exoskeleton reflectance tends to match that of the soil of scorpion's habitat, mostly in the red channel at which the highest sun light (or moonlight) reflectance occurs, and in less extend in green channel at which the fluorescence takes place. This result is in line with the behavior of this species that spends much of its time outside its gallery wandering across soil in search of food or mating. These optical studies might provide insights into the exoskeleton characteristics that result from the evolutionary strategies allowing higher survival of scorpions.

Acknowledgments For this work, the postdoctoral position of D. López-Cabrera was sponsored by SECIHTI. The authors thank Lucero Alvarez for drawing assistance.

References

Andrews K, Reed SM, Masta SE (2007) Spiders fluoresce variably across many taxa. Biol Lett 3:265–267. https://doi.org/10.1098/rsbl.2007.0016

Blass CGR, Gaffin DD (2008) Light wavelength biases of scorpions. Anim Behav 76(2):365–373. https://doi.org/10.1016/j.anbehav.2007.12.022

Brandt EE, Masta ES (2017) Females are the brighter sex: differences in external fluorescence across sexes and life stages of a crab spider. PLoS One 15(5):12. https://doi.org/10.1371/journal.pone.0175667

Chen JY, Ju-Chiu P, Chung-Lee C (2015) Fluorescence and multilayer structure of the scorpion cuticle. Proc SPIE 9629:Article 96290P. https://doi.org/10.2307/1296207

Demchenko PA, Tomin IV, Chou P-T (2017) Breaking the Kasha rule for more efficient photochemistry. Chem Rev 117(21):13353–13381. https://doi.org/10.1021/acs.chemrev.7b00110

Espinosa-Luna R, Atondo-Rubio G, Velarde-Escobar O (2010) Métrica de despolarización escalar Q(M) como criterio para identificar sistemas retardadores o desfasadores puros. Rev Mex Fís 56(5):406–410

Fleissner GJ (1977) Scorpion lateral eyes: extremely sensitive receptors of Zeitgeber. J Comp Physiol 118(1):101–118. https://doi.org/10.1007/BF00612340

Frost L, Butler D, O'Dell B, Fet V (2001) A coumarin as a fluorescent compound in scorpion cuticle. In: Selden PA, Fet V (eds) Scorpions. British Arachnological Society, Buckinghamshire, pp 365–368

González-Santillán E, Prendini L (2013) Redefinition and generic revision of the North American vaejovid scorpion subfamily Syntropinae Kraepelin, 1905, with descriptions of six new genera. Bull Am Mus Nat Hist 382:1–71. https://doi.org/10.1206/830.1

Iriel A, Lagorio MG (2010) Implications of reflectance and fluorescence of *Rhododendron indicum* flowers in biosignaling. Photochem Photobiol Sci 9:342–348. https://doi.org/10.1039/b9pp00104b

Jordan MA (2010) Can scorpions detect fluorescing scorpions? Thesis, University of Oklahoma, The Department of Zoology, Oklahoma

Kloock CT (2008) A comparison of fluorescence in two sympatric scorpion species. J Photochem Photobiol B Biol 91(2–3):132–136. https://doi.org/10.1016/j.jphotobiol.2008.02.008

Kloock C (2009) Reducing scorpion fluorescence via prolonged exposure to ultraviolet light. J Arachnol 37(3):368–370. https://doi.org/10.1636/Sh08-87.1

Krisciunas K, Schaefer B (1991) A model of the brightness of moonlight. Publ Astron Soc Pac 103(667):1033–1039. https://doi.org/10.1086/132921

Lawrence R (1954) Fluorescence in Arthropoda. (170, Ed.). J Entomol Soc South Afr 17(2):167

Liu Y, Li S, Li Y, Wang X, Zhang H, Wu W, Cao Z, Huang K, Wu Y (2024) Differential fluorescence features and recovery speeds of different scorpion exoskeleton parts during the molting process. Spectrochim Acta A Mol Biomol Spectrosc 316:1386–1425. https://doi.org/10.1016/j.saa.2024.124309

Lopez D, Ramos G, González-Santillán E, Espinosa R (2020) Characterization of the fluorescence intensity and color tonality in the exoskeleton of scorpions. J Photochem Photobiol B Biol 209. https://doi.org/10.1016/j.jphotobiol.2020.111945

Lourenco WR (2012) Fluorescence in scorpions under UV light; can chaerilids be a possible exception? C R Biol 335(12):731–734. https://doi.org/10.1016/j.crvi.2012.11.001

Lourenco WR, Cloudsley-Thompson JL (1996) The evolutionary significance of colour, colour patterns and fluorescence in scorpions. Rev Suisse Zool 2:449–458

Luen MJ (2019) Biofluorescence in terrestrial animals, with emphasis on fireflies: a review and field observation. In: Suzuki H, Bioluminescence—analytical applications and basic biology. https://doi.org/10.5772/intechopen.86029

Machan L (1968) Spectral sensitivity of scorpion eyes and the possible role of shielding pigment effect. J Exp Biol 49(1):95–105. https://doi.org/10.1242/jeb.49.1.95

Newport(n.d.).Retrievedfromhttps://www.newport.com/n/solar-simulator-spectral-irradiance-data

Nieves LEN, Santiago-Blay JA (2019) UV-induced fluorescence of two puerto rican endemic arthropods: the harvestman, *Yunquenus portoricanus* and the millipede, *Spirobolellus richmondi*. Life Excit Biol 7(1):22–30. https://doi.org/10.9784/LEB7(1)Laboy.01

Painting JC, Chang C-C, Fen-Seah J, Li D (2017) Condition dependence of female-specific UV-induced fluorescence in a jumping spider. Anim Behav 127:233–241. https://doi.org/10.1016/j.anbehav.2017.03.022

Phillips LS (1959) Fluorescence in the colors of certian Lepidoptera observed. J Lepidopterists' Soc 13:73

Prendini L (2001) Substratum specialization and speciation in southern African scorpions: the effect hypothesis revisited. In: Selden PA, Fet V (eds) Scorpions 2001. British Arachnological Society, Burnham Beeches, Bucks, pp 113–138

Rubin M, Lamsdell JC, Prendini L, Hopkins MJ (2017) Exocuticular hyaline layer of sea scorpions and horseshoe crabs suggests cuticular fluorescence is plesiomorphic in chelicerates. J Zool 303(4):245–253. https://doi.org/10.1111/jzo.12493

Sissom W, Polis GA, Watt D (1990) Field and laboratory methods. In: Polis GA (ed) The biology of scorpions. Stanford University Press, Stanford, pp 215–221

Skutelsky O (1996) Predation risk and state-dependent foraging in scorpions: effects of moonlight on foraging in the scorpion *Buthus occitanus*. Anim Behav 52(1):49–57. https://doi.org/10.1006/anbe.1996.0151

Stachel JS, Stockwell AS, van Vranken DL (1999) The fluorescence of scorpions and cataractogenesis. Chem Biol 6(8):531–539. https://doi.org/10.1016/s1074-5521(99)80085-4

Stahnke LH (1972) UV light, a useful field tool. Bioscience 22(10):604–607. https://doi.org/10.2307/1296207

Stahnke LH, Calos M (1977) A key to the species of the genus *Centruroides* Marx (Scorpionida: Buthidae). Entomol News 88(5):111–120

Tokumoto H, Shimomura H, Hakamatsuka T, Ozeki Y, Goda Y (2018) Fluorescence coupled with macro and microscopic examinations of morphological phenotype give key characteristics for identification of crude drugs derived from scorpions. Biol Pharm Bull 41(4):510–523. https://doi.org/10.1248/bpb.b17-00801

Ullrich BD (2016) Optical properties of a scorpion (*Centruroides limpidus*). Phys Scr 91(4). https://doi.org/10.1088/0031-8949/91/4/045802

Wackerly DD, Mendenhall W, Scheaffer LR (2007) Mathematical statistics with applications. Thomson, Belmont

Wankhede R (2004) Extraction, isolation, identification and distribution of soluble fluorescent compounds from the cuticle of the scorpion (*Hadrurus arizonensis*). Master thesis

Welch LV, van Hooijdonk E, Intrater N, Vigneron JP (2012) Fluorescence in insects. The Nature of Light: light in nature. https://doi.org/10.1117/12.929547

Williams SC (1970) Coexistence of desert scorpions by differential habitat preference. Pan Pac Entomol 46(4):254–267

Williams SC (1987) Scorpion bionomics. Annu Rev Entomol 32:275–295. https://doi.org/10.1146/annurev.en.32.010187.001423

Wilts DB, Michielsen K, De Raedt H, Stavenga GD (2013) Sparkling feather reflections of a bird of paradise explained by finite difference time domain modeling. PNAS 111(12):4363–4368. https://doi.org/10.1073/pnas.1323611111

Yoshimoto Y, Tanaka M, Miyashita M, Abdel-Wahab M, Megaly AMA, Nakagawa Y, Miyagawa H (2020) A fluorescent compound from the exuviae of the scorpion, Liocheles australasiae. J Nat Prod 83(2):542–546. https://doi.org/10.1021/acs.jnatprod.9b00972

Chapter 5
Diversity, Taxonomy, and Distribution of Neotropical Scorpions

Edmundo González-Santillán, José Antonio Ochoa, and Andrés Alejandro Ojanguren-Afilastro

Abstract We provide an overview of the diversity of the 12 scorpion families and 91 genera present in the neotropical region: Ananteridae (Pocock, The fauna of British India, including Ceylon and Burma. Arachnida. Taylor and Francis, London, 1900), Bothriuridae Simon, 1880, Buthidae C.L.Koch, 1837, Caraboctonidae (Kraepelin, Zool Jahrb Abt Syst 22:321–364, 1905), Chactidae (Pocock, Ann Mag Nat Hist 6:303–330, 1893), Diplocentridae Karsch, 1880, Euscorpiidae Laurie, 1896, Hadruridae (Stahnke, J Arachnol 1:107–141, 1974), Hormuridae Laurie, 1896, Troglotayosicidae Lourenco 1998, Typhlochactidae Mitchell, 1971, Vaejovidae, Thorell 1876. We focused only on the neotropical representatives of each family. We provide a brief characterization of each family, data about their phylogenetic relationships and distribution, and we also provide data about the genera of each family, number of species, distribution, and habitat. When possible, we provide a brief generic key. We do not include data and characterization of families Troglotayiscidae and Typlochactidae because these data are presented in the chapter about Troglobius neotropical scorpions.

Keywords Scorpiones · Neotropics · Diversity · Taxonomy · Distribution · Characterization

E. González-Santillán (✉)
Colección Nacional de Arácnidos, Departamento de Zoología, Instituto de Biología, Universidad Nacional Autónoma de México, Coyoacán, Mexico City, Mexico
e-mail: edmundo.gonzalez@ib.unam.mx

J. A. Ochoa
Facultad de Ciencias Biológicas, Universidad Nacional de San Antonio Abad del Cusco, Wanchaq, Cusco, Peru
e-mail: jose.ochoa@unsaac.edu.pe

A. A. Ojanguren-Afilastro
División Aracnología, Museo Argentino de Ciencias Naturales "Bernardino Rivadavia", Buenos Aires, Argentina
e-mail: ojanguren@macn.gov.ar

A. F. de Araujo Lira, E. González-Santillán (eds.), *The Biology of Neotropical Scorpions*, https://doi.org/10.1007/978-3-032-23194-9_5

5.1 Introduction

Notwithstanding the fact that Neotropical scorpions comprise a substantial percentage of the world's scorpion fauna, the diversity, taxonomy, and distribution of these arachnids have never been the subject of a comprehensive synthesis. One of the pillars of the taxonomy of extant scorpions rests in the work of Kraepelin (1899), a seminal contribution that stands as the first work that conducted a comprehensive survey of the global scorpion fauna. In particular for the Neotropics, there were some historical epitomic works, instrumental in establishing the foundations of modern taxonomy in the Neotropics: Mello-Leitão (1945), Francke (1978), de Armas (1988), Cekalovic (1983), González-Sponga (1984, 1996), Francke and Stockwell (1987), and Maury (1973), among many others. However, since the year 2000, a significant increase in the number of new researchers has been observed, leading to a substantial proliferation of species descriptions. Maybe the hallmark in that year was the publication of the Catalog of the Scorpions of the World (Fet et al. 2000). To illustrate the proliferation of newly discovered species from 2000 to 2025, 195 species of the Neotropical family Buthidae were described. In contrast, for 137 years, from the description of the type species of the Neotropical genus *Centruroides*, *C. exilicauda* (Wood, 1863) to the year 2000, 236 species were described. This implies that within a mere 25 years, the number of species described has nearly equaled that of the 137-year period.

In addition to technological advancements, such as the World Wide Web, genomics, bioinformatics, and other advances, the advent of the new millennium has precipitated a resurgence of interest in scorpions. This could be, at least in part, due to the high medical importance of this group in the region, being scorpions responsible for one of the highest number of accidents by poisonous animals in the region. In this chapter, we present a comprehensive compilation and analysis of the taxonomic diversity of scorpions distributed in the Neotropics. Given the extensive diversity that has been observed, a family-based approach is employed, wherein each family is briefly characterized, accompanied by a report on its salient features pertaining to natural history and biology. We have omitted two families, Troglotayosicidae and Typhlochactidae, which have been treated in Chap. 9, not only taxonomically but with an emphasis on their troglomorphic features and ecological and evolutionary significance. On a final note, we tried to use more and less the current use of morphological nomenclature, but since there is no consensus on several structures we have taken, at times, the editorial dissection of using one that we deem appropriated.

In the Neotropical region, scorpions are found in different types of terrestrial ecosystems in tropical, subtropical, and temperate zones, including Amazon rainforest, mountain forests, dry forests, savannas, deserts, Patagonian steppe, Chaco, paramos, and punas, among others. Currently, the scorpion fauna of this region comprises 12 families, 91 genera, and 1155 species (Table 5.1).

Table 5.1 Families and number of scorpion genera and species from the Neotropical region

Family	Genera	Species	Distribution in the Neotropical region
Ananteridae	1	103	Central America, South America
Bothriuridae	14	159	South America
Buthidae	15	426	Caribbean, Central America, South America, North America
Caraboctonidae	2	26	South America
Chactidae	15	201	Caribbean, Central America, South America
Diplocentridae	9	122	Caribbean, Central America, South America, North America
Euscorpiidae	3	13	Central America, North America
Hadruridae	1	2	North America
Hormuridae	1	10	Caribbean, Central America, South America
Troglotayosicidae	1	7	South America
Typhlochactidae	4	11	North America
Vaejovidae	25	75	North America
Total	91	1155	

5.2 Family Ananteridae Pocock, 1900

The family Ananteridae comprises small scorpions with a Gondwanian distribution, occurring in Central and South America, Africa, Madagascar, and Asia. Initially, the family was classified as a subfamily of Buthidae by Pocock (1900), with the type genus designated as *Ananteris*. This criterion was also followed by Kraepelin (1905) and by most subsequent authors. The concept of Ananterinae is largely congruent with the *Ananteris* genus group, as recently delineated by Fet et al. (2005) and subsequently adopted by other contemporary authors. This family was finally recognized as distinct by Ythier (2024), yet further analysis is necessary to ascertain its precise familial or sub-familial classification.

To date, seven extant genera have been included in this family, but only two are present in the Neotropics: *Ananteris* Thorell, 1891 (Fig. 5.1a) and *Microananteris* Lourenço, 2003. Nonetheless, the validity of *Microananteris* is a subject of debate. We therefore opt to align with the perspectives of Botero-Trujillo and Noriega (2011), regarding *Microananteris* as a synonym of *Ananteris*.

In the Neotropical region, the distribution of this family extends from Costa Rica to Argentina, with southern Chacoan woodlands identified as the southernmost distribution record (Ojanguren-Affilastro 2005). In this biogeographical realm, the species occupies tropical and subtropical environments, predominantly woodlands, but is also present in open savanna environments. While it is more diversified in lowlands, some species have been observed to reach elevations of up to 2300 m above sea level, as evidenced by the mountain of Pico da Neblina in northern Brazil (Ythier 2024).

Fig. 5.1 Anateridae, habitus in life. (**a**) *Ananteris balzani* Thorell, 1891, ♂, Argentinean Chaco. (**b**) *Ananteris balzani*, from Argentinean Chaco with autotomized metasomal segments IV and V

A distinctive and noteworthy characteristic of *Ananteris* is its capacity for autotomizing metasomal segments II–V (Fig. 5.1b). To date, it is the sole scorpion genus documented to demonstrate this defensive mechanism (Lira et al. 2014; Mattoni et al. 2015). The presence of cleavage planes between segments facilitates relatively facile autotomy, characterized by a diminished loss of hemolymph and expedited, efficient healing. In the aftermath of autotomy, scorpions are incapable of regenerating the severed segments; however, they have been observed to survive for several months.

5.2.1 Phylogenetic Relationships

Ananteridae forms part of Parvorder Buthida and Superfamily Buthoidea. It is closely related to the family Buthidae.

5.2.2 Characterization

Smallest species: *Ananteris intermedia* Lourenço, 2012 (9.3 mm).

Largest species: *Ananteris ehrlichi* Lourenço, 1994 (36 mm).

Chelicerae: Fixed finger with a median tooth, a basal tooth, a subdistal tooth, a distal tooth, and an internal tooth; movable finger with a basal internal tooth, one or two basal external teeth, one median internal tooth, one medial external tooth, a subdistal tooth, and an external distal tooth.

Carapace: Anterior margin straight or with a shallow notch; median eyes large, ocular tubercle clearly in the anterior half of the carapace (Fig. 5.2a); with three lateral ocelli; lateral ocelli pattern type 4A (Loria and Prendini 2014).

Coxoexternal region: sternum type I, much longer than wide, subtriangular. Pectens poorly developed, without fulcra, or with very poorly developed and barely visible fulcra (Fig. 5.2b). Pectinal tooth count 10–25.

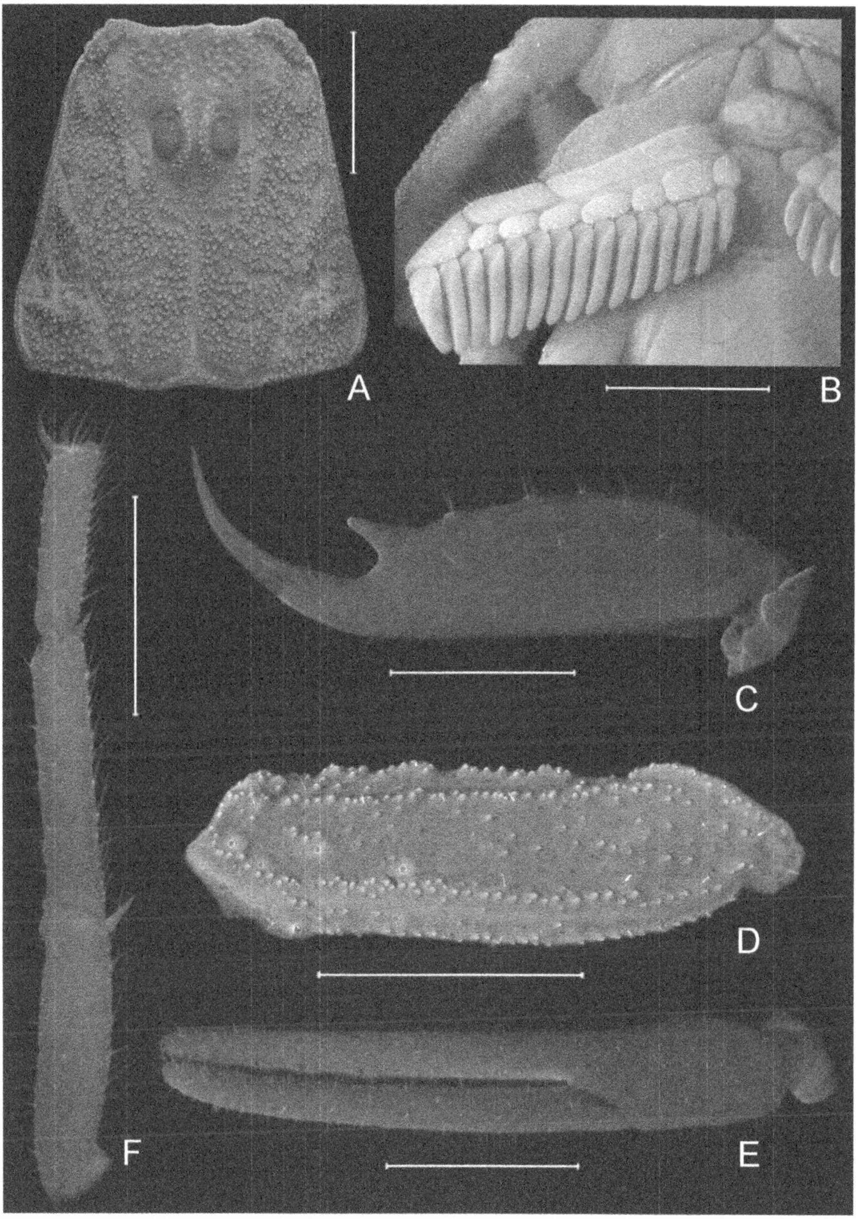

Fig. 5.2 *Ananteris* sp. (**a**) Prosoma; (**b**) Right pecten, ventral aspect; (**c**) Telson, lateral aspect; (**d**) Right pedipalp femur, dorsal aspect; (**e**) Left pedipalp chela, external aspect; (**f**) Tibia, basitarsus, and telotarsus, leg IV. Scale bars: 1 mm

Pedipalp: smooth and acarinate; with a small manus and long fingers (Fig. 5.2e). Chela fingers usually with six or seven rows of internal granules. Chelae without conspicuous differences between genders.

Trichobothrial pattern Type A (sensu Vachon 1973), configuration β, with trichobothrium *d3* posterior to the line formed by *d1* and *d4* (Fig. 5.2d).

Legs: telotarsi with abundant ventral setae. With prolateral and retrolateral pedal spurs, legs III and IV with a single tibial spur (Fig. 5.2f).

Metasoma: VL and VSM carinae usually present in segments I–VI; Metasomal segment V with VL, and VM carinae straight and extending the entire length of the segment. With cleavage planes between metasomal segments that allow autotomy.

Telson: Vesicle slender, extremely elongated in males, shorter in females, with a VM carina and a well-developed subtriangular subaculear apophysis (Fig. 5.2c). Aculeus is poorly curved, less than half the length of the vesicle.

Hemispermatophore: flageliform, "no fold" bauplan (Monod et al. 2017). Pars reflexa or flagellum is usually very coiled, with snail-like appearance, but in some species it can be poorly developed and poorly curved (Botero-Trujillo and Flórez 2011); capsular region, or lobe region, with two or three lobes.

5.2.3 Habitat and Distribution of the Genera Within Ananteridae

Ananteris Thorell, 1891. From 0 m up to 2300 m asl. Argentina, Bolivia, Brazil, Colombia, Costa Rica, Ecuador, French Guyana, Guyana, Panama, Paraguay, Peru, Surinam, and Venezuela. In tropical and subtropical environments, in woodlands and savannas. 103 spp.

5.3 Family Bothriuridae Simon, 1880

The family Bothriuridae comprises small to medium-sized burrowing scorpions (Fig. 5.3) distributed in South America, Australia, Tasmania, and Southern Africa; a record from the Himalayas, in the Indian subcontinent, resulted to be a mislabeled specimen.

In South America, this family occurs from southern Patagonia to the Amazonian Forest. It occupies several different environments being more common and diverse in all southern South American arid regions, such as Patagonian steppe, Chaco, Caatinga, Monte, and the Coastal Pacific desert, among others; it is also highly diversified in high altitude habitats of the Andes and other mountain ranges, almost reaching 5000 m asl. Several species occupy mesic habitats as Pampean grasslands. To a lesser extent it also occurs in different South American forests, such as the humid Patagonian Valdivian woods, the Andean Yungas, the Atlantic forests, and the Amazonian forests, among others.

Fig. 5.3 Bothriuridae, habitus in life. (**a**) *Bothriurus bertae* Abalos, 1955, Argentinean Paranaense forest. (**b**) *Centromachetes* sp., Southern Chile. (**c**) *Brachistosternus ehrenbergii* (Gervais, 1841), ♀, Lima, Peru. (**d**) *Orobothriurus parvus* Maury, 1976, ♂, Junín, Peru. (**e**) *Timogenes* sp., ♂, Santa Cruz, Bolivia. (**f**) *Pachakutej* sp. Cusco, Peru

This family is remarkable since many of its species are highly adapted to extreme cold conditions. Many species of the genus *Urophonius* develop their surface activity in winter, contrary to all remaining known species from temperate environments which are active in spring-summer. Many species occur at very high altitudes as *Orobothriurus huascaran* Ochoa, Ojanguren-Affilastro, Mattoni and Prendini, 2011 reaching 4910 m asl and sharing with the Tibetan buthid *Hottentotta leetzi* Ythier, 2024 the highest altitudinal record in the order (4920 m asl). It also bears the highest latitudinal record for the order with *Urophonius granulatus* Pocock 1898 reaching 52° of southern latitude in extreme southern Patagonia.

It currently comprises 17 genera, 14 of which are present in South America: *Bothriurus* Peters 1861, *Brachistosternus* Pocock, 1893, *Brazilobothriurus* Lourenço and Monod, 2000, *Centromachetes* Lonnberg, 1897, *Mauryius* Ojanguren-Affilastro and Mattoni, 2017, *Orobothriurus* Maury, 1976, *Pachakutej* Ochoa, 2004, *Phoniocercus* Pocock, 1893, *Rumikiru* Ojanguren-Affilastro, Mattoni, Ochoa and Prendini, 2012, *Tehuankea* Cekalovic, 1973, *Thestylus* Simon, 1880, *Timogenes* Simon, 1880, *Urophonius* Pocock, 1893, and *Vachonia* Abalos, 1954.

It was initially recognized as a family by Peters in 1861 as Teligonini Peters, 1861, with type genus *Telogonus* C.L. Koch 1837 (= *Thestylus* Simonn 1880), which was a junior synonym of *Telogonus* Hubner, 1816 (Lepidoptera) (Lowe and Fet 1998). It was later recognized as a family by Peters (1880) with type genus *Bothriurus*, to also include *Timogenes*, *Cercophonius*, and *Thestylus*. It has been divided into two subfamilies by Maury (1971): Bothriurinae Simon, 1880 and Brachistosterninae Maury, 1973; however, this division has not been widely accepted and most recent phylogenetic studies on the family clearly do not support it.

5.3.1 Phylogenetic Relationships

It forms part of parvorder Iurida, in which it separates very early, being relatively more closely related with superfamily Iurioidea than with the rest of the families of the parvorder.

5.3.2 Characterization

Smallest species: *Pachakutej juchuicha* Ochoa, 2004 (13.4 mm).

Largest species: *Timogenes elegans* (Mello-Leitão, 1931) (125 mm).

Chelicerae: Movable finger with one or two subdistal teeth, without a prominent basal tooth in the internal surface. Serrula absent.

Carapace: Anterior margin usually slightly convex, but it can be straight, with a median bulge or even with a deep median notch; ocular tubercle usually located in the anterior half of the carapace, usually raised, but in some genera it can be shallow and placed in a depression; lateral ocelli pattern type 3A, PDMi smaller than others ocelli (Loria and Prendini 2014).

Coxoexternal region: Sternum type II, much wider than long, usually more than twice as wide as long, but in some genera it can be sub-pentagonal (Fig. 5.4h, i) (Soleglad and Fet 2003b). Pectens are usually well developed, with one or two intermediate rows of lobulliform lamellae (Fig. 5.4j, k). Pectinal tooth count 8–58. *Brachistosternus multidentatus* Maury, 1984 bears the highest record in the number of pectinal teeth in the order with a male with 58 teeth in each pecten.

Pedipalp: Chela most usually smooth and acarinate; in most species only a VM carina is visible. Chela fingers usually have a single internal median denticle row,

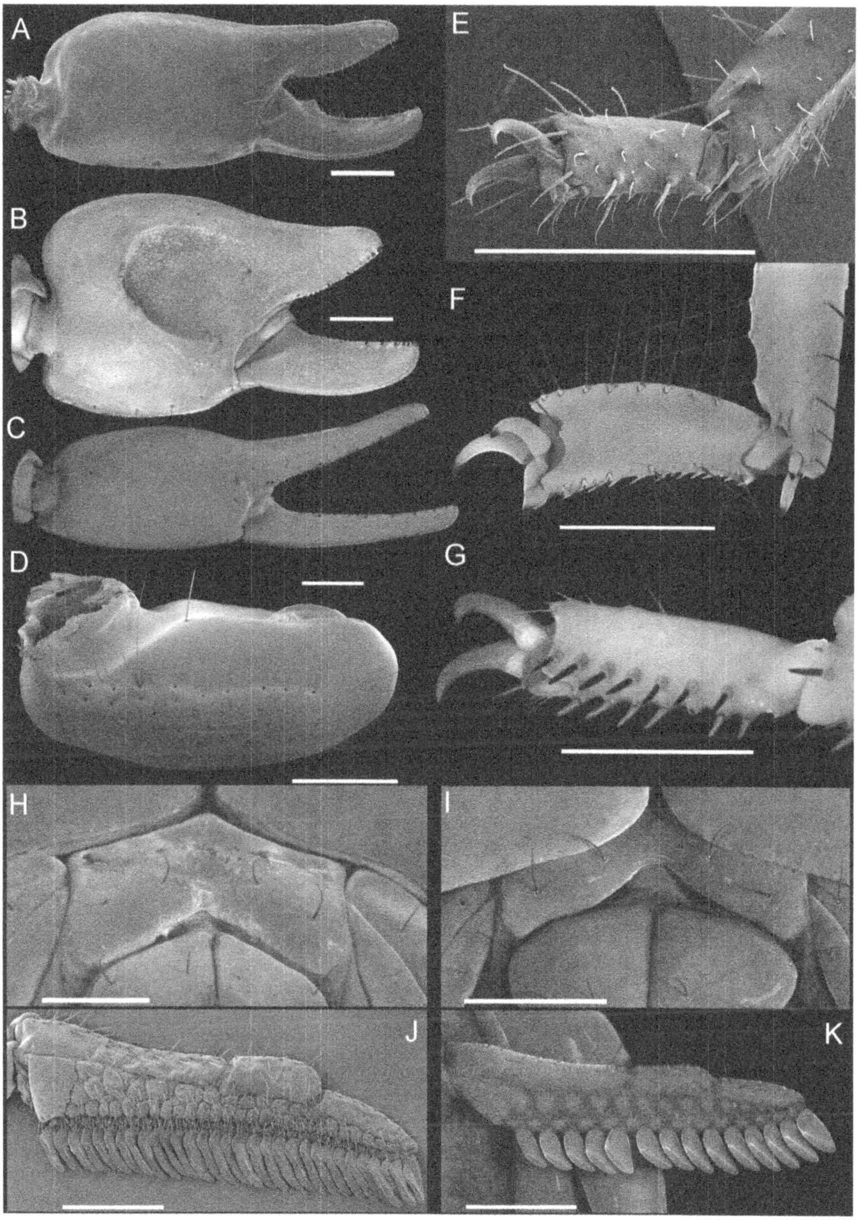

Fig. 5.4 (**a**) *Rumikiru*, left pedipalp chela, internal aspect; (**b**) *Timogenes*, male, left pedipalp chela, internal aspect; (**c**) *Urophonius*, male, left pedipalp chela, internal aspect; **d** *Vachonia*, right pedipalp patella, ventral aspect; (**e**) *Phoniocercus*, telotarsus I, internal aspect; (**f**) *Brachistosternus*, telotarsus III, internal aspect; (**g**) *Thestylus*, telotarsus III, internal aspect; (**h**) *Tehuankea*, sternum; (i) *Rumikiru*, sternum; (**j**) *Brachistosternus*, left pecten; (**k**) *Rumikiru*, left pectin. Scale bars: 1 mm

and four to eight pairs of accessory denticles (internal and external respect to the median row). Some genera can bear two or three complete median denticle rows, and in some genera these double rows can be only partial, and reduced to the median or basal margin of finger. Males usually with an apophysis or an excavation in the internal surface of pedipalp manus near the base of movable finger (Fig. 5.4b, c). In some genera in males there is also a separate group of granules in the base of the fixed finger.

Trichobothrial pattern neobothriotaxic major Type C, with one or more accessory trichobothria in *V* series; femur with 3 trichobothria (*d, i, e*); patella, with 19–51 trichobothria (2 *d, i,* 3 *et, est,* 2 *em,* 2 *esb,* 5 *eb,* 3 to 32 *V*) (Fig. 5.4d), *esb1* can be petit; chela with 26–53 trichobothria (*Dt, Db,* 5 *Et, Est, Esb,* 3 *Eb, dt, dst, dsb, db, et, est, esb, eb, ib, it,* 4–31 *V*), *Et4* can be petit (sensu Vachon 1973). *Vachonia martinezi* Abalos, 1954 bears the highest record in the order in the number of trichobothria with 107 in each pedipalp.

Legs: Leg, telotarsus with ventromedian row of spinules and paired ventrolateral spines or setae. Usually with prolateral and retrolateral pedal spurs, but in some general the retrolateral pedal spur can be absent (Fig. 5.4e–g).

Hemispermatophore: lamelliform "four-fold" bauplan (Monod et al. 2017), highly variable between genera (Fig. 5.5g–k); with well-developed distal lamina, and a lobular region that can be very complex in some genera.

Metasoma: VL and VSM carinae usually present in segments I–III, more developed in females; Metasomal segment V with VL, VSM, and VM carinae, more developed in males. VSM, and even VM carinae can be absent in some genera (Fig. 5.5a–c). Males of some genera with paired dorsal glands, also called androvestigia (sensu Cekalovic 1973), in metasomal segment V (Fig. 5.5d).

Telson: Vesicle slightly concave to convex dorsally, males of some genera can bear one or two dorsal glands. Aculeus medium-sized to large, varying among genera (Fig. 5.5e, f). Venom gland Type II (sensu Pavlovsky 1913).

5.3.3 Key to the Identification of Valid Genera of the Family Bothriuridae

An exception to the valid genera is that of *Brazilobothriurus.* Even if not formally synonymized yet, in this key we will consider *Brazilobothriurus* as a synonym of *Bothriurus* since both genera are not objectively separated using morphology, and the original diagnosis of *Brazilobothriurus* falls inside the definition of *Bothriurus.*

1. Median basal denticle of pedipalp chela movable finger conspicuously enlarged, about five times bigger than remaining denticle (Fig. 5.4a)................................ .... *Rumikiru*
   – Median basal denticles of pedipalp chela movable finger of size similar to remaining denticles or only slightly bigger (Fig. 5.4b, c) 2

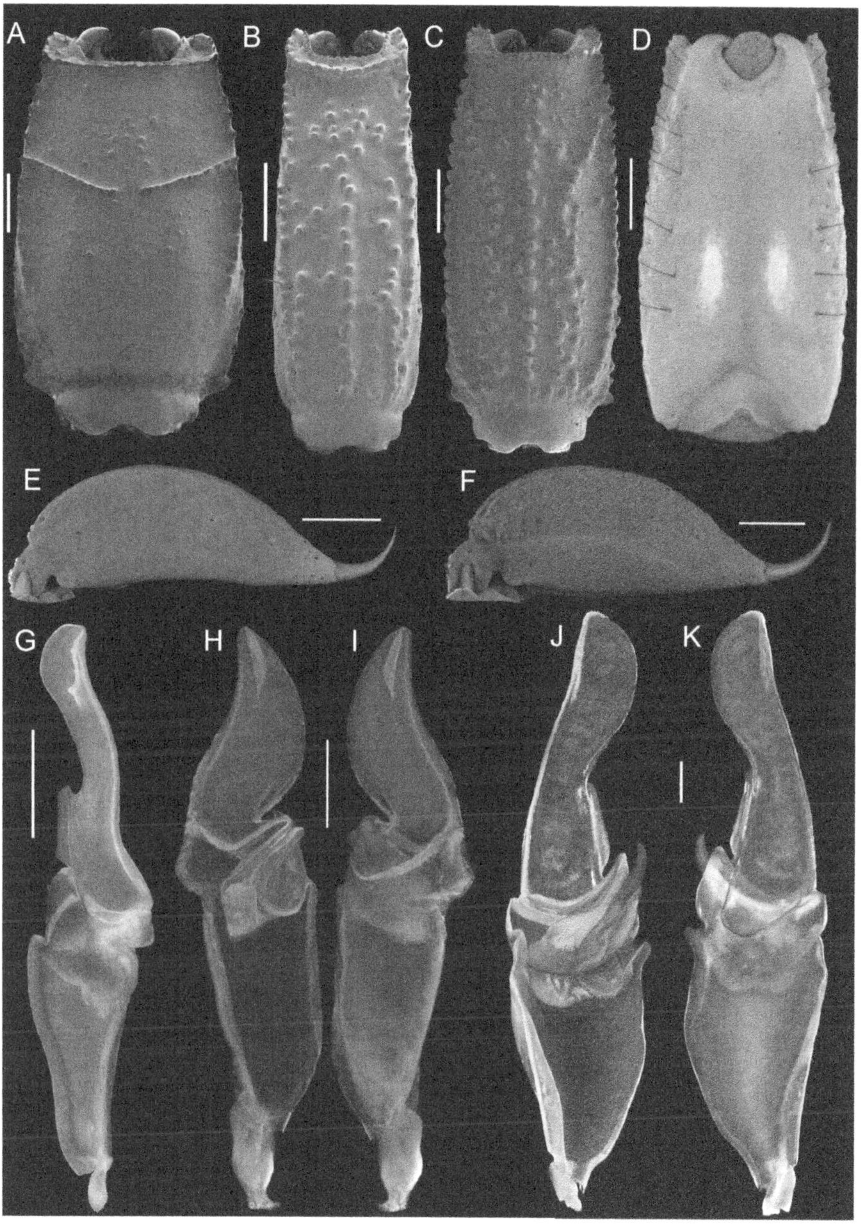

Fig. 5.5 (**a**) *Timogenes*, metasomal segment V, ventral aspect; **b** *Orobothriurus*, metasomal segment V, ventral aspect; (**c**) *Bothriurus*, metasomal segment V, ventral aspect; (**d**) *Brachistosternus*, male, metasomal segment V, dorsal aspect; (**e**) *Urophonius*, male, telson, lateral aspect; (**f**) *Urophonius*, female, telson, lateral aspect; (**g**) *Mauryius*, left hemispermatophore, ectal aspect; (**h**) *Urophonius*, left hemispermatophore, ental aspect; (**i**) *Urophonius*, left hemispermatophore, ectal aspect; (**j**) *Brachistosternus*, left hemispermatophore, ental aspect; (**k**) *Brachistosternus*, left hemispermatophore, ectal aspect. Scale bars: 1 mm

2. Pedipalp patella with 29 or more ventral trichobothria (Fig. 5.4d) .... *Vachonia*
   –Pedipalppatellawith7orfewerventraltrichobothria... 3
3. Tarsomere II of legs with ventrolateral setae (Fig. 5.4f). Pectines with two or three intermediate rows of lobuliform lamellae (Fig. 5.4j) .... *Brachistosternus*
   – Tarsomere II of legs with ventrolateral spines or setiform spines (Fig. 5.4e, g). Pectines with a single intermediate row of lamellae (Fig. 5.4k)......................... 4
4. Ventral surface of metasomal segment V with a well-defined transverse carina (Fig. 5.5a). Internal surface of adult male chela with a semicircular excavation (Fig. 5.4b) .... *Timogenes*
   – Ventral surface of metasomal segment V without a well-defined transverse carina (in some species there is a semicircular carina but never an actual transverse carina) (Fig. 5.5b, c). Internal surface of adult male chela with an apophysis or smooth (Fig. 5.4c).................. 5
5. Tarsomeres of legs III and IV with prolateral pedal spurs, retrolateral pedal spur absent or strongly reduced (Fig. 5.4e, g)... 6
   – Tarsomeres of legs III and IV with prolateral and retrolateral pedal spurs (Fig. 5.4f)....... 7
6. Tarsomere II with ventrolateral setiform spines (Fig. 5.4e). VSM carinae of metasomal segments I and II not converging in posterior margins of segments. Pedipalp chela of males with an internal apophysis.. *Phoniocercus*
   – Tarsomere II with ventrolateral spines (Fig. 5.4f). VSM carinae of metasomal segments I and II converging in posterior margins of segments. Pedipalp chela of males smooth, without an internal apophysis.. *Thestylus*
7. Ventrolateral spinal formula of tarsomere II of legs III–IV: 4-4 to 7-7. 8
   – Ventrolateral spinal formula of tarsomere II of legs III–IV: 3-3 or 3-4. 9
8. Internal surface of pedipalp fingers with a complete double or triple median row of denticles and 3–5 internal accessory denticles, but without conspicuous external accessory denticles. Endemic to central and south-central Chile. .... *Centromachetes*
   – Internal surface of pedipalp fingers with a single median row of denticles that in some species can be double in its median area, and with 4–6 conspicuous pairsofinternalandexternalaccessorydenticles.. *Urophonius*
9. Sternum slightly compressed anteroposteriorly, subpentagonal (Fig. 5.4h)................ 10
   – Sternum strongly compressed anteroposteriorly, usually being more than twice as wide as long, and even reduced to two separate slits (Fig. 5.4i)............. 11

10. Internal dentate margin of pedipalp fingers with two median denticle rows and four to six pairs of accessory denticles............................*Tehuankea*
– Internal dentate margin of pedipalp fingers with a single denticle row and four to eight pairs of accessory denticles.. *Mauryius*
11. VSM carinae of metasomal segment V subparallel to VL carinae throughout their course and placed very close to them, joining VL carinae in anterior and posterior thirds of segment (Fig. 5.5. B)... *Orobothriurus*
– VSM carinae of metasomal segment V not subparallel to VL carinae, or in some species partially subparallel only anteriorly. Never joining VL carinae in anterior third of segment (Fig. 5.5 .C)... 12
12. External surface of pedipalp chela prismatic, with external secondary, digital, and dorsal secondary carinae represented as an elevation of tegument. M1 macroseta of pedipalp femur placed closer to *e* trichobothria than to *d* trichobothria .... *Pachakutej*
– External surface of pedipalp chela smooth, without traces of carinae. M1 macroseta of pedipalp femur placed closer to *d* trichobothria than to *e* trichobothria .... *Bothriurus* (and *Brazilobothriurus*)

5.3.4 Habitat and Distribution of the Genera Within Bothriuridae

Bothriurus Peters 1861. From Sea level up to 4000 m. Argentina, Bolivia, Brazil, Chile, Paraguay, and Uruguay. In a wide range of environments, from arid steppes of southern Argentinean Patagonia to humid Amazonian forest. It also occurs in Pampean grasslands, Monte, Chaco, Caatinga, Atlantic forest, Amazonian forest, Pacific Coastal Desert, and High Andes, among other environments. 50 spp.

Brachistosternus Pocock, 1893. In all arid to semiarid areas from southern Argentina, Bolivia, Brazil, Chile, and Peru up to southern Ecuador. From sea level, to higher altitudes of the Andes, up to 4500 m. In the Patagonian steppe, Monte, Chaco, Caatinga, Pacific Coastal desert, Atacama and Sechura deserts, and High Andes, among others. 51 spp.

Brazilobothriurus Lourenço and Monod, 2000. Described for Pantanal environments of Matto Grosso do Sul State in Central Brazil. 1 sp.

Centromachetes Lonnberg, 1897. In central and south-central Chilean woods. 3 spp.

Mauryius Ojanguren-Affilastro and Mattoni, 2017. Endemic to a small area of intermediate altitudes (2500–2700 m) of Argentinean central Precodillera mountain chain. 1 sp.

Orobothriurus Maury, 1976. Known from high altitudes of the Andes and related mountain chains of Argentina, Chile, Bolivia, and Peru; also in some areas of northern Chilean and southern Peruvian inhabits Lomas biotopes below 1000 m and *Prosopis* forest near the Coastal Pacific deserts. 15 spp.

Pachakutej Ochoa, 2004. In south-central Peruvian inter-Andean valleys. 6 spp.

Phoniocercus Pocock, 1893. In southern Patagonian humid woods, mostly in Chile, with few records in Argentina. 2 spp.

Rumikiru Ojanguren-Affilastro, Mattoni, Ochoa and Prendini, 2012. In the Rocky slopes of the central Chilean Pacific coastal desert. 2 spp.

Tehuankea Cekalovic, 1973. Endemic to a small area of temperate forests of south-central Chile. 1 sp.

Thestylus Simon, 1880. In the rainforests of Mata Atlantica of southeastern Brazil. 3 spp.

Timogenes Simon, 1880. In the Arid to semiarid lowland areas of Argentina, Bolivia, and Paraguay. In northern Patagonia, Monte and Chaco environments. 6 spp.

Urophonius Pocock, 1893. In mesic to arid environments of southern South America. From southern Argentina, Chile, and Uruguay up to southern Brazil. From sea level up to 2200 m asl. 17 spp.

Vachonia Abalos, 1954. In arid lowland areas of central Argentina, most usually associated with psammophilous environments. 1 sp.

5.4 Family Buthidae C.L. Koch, 1837

Buthids comprise the vast majority of medically relevant species in the world. The venom of these organisms contains small peptides that bind to sodium and potassium ion-gated transmembrane proteins. The inoculation of these peptides via stinging has been demonstrated to disrupt membrane potential in neural and skeletal vertebrate tissue (Chap. 15). This disruption has the potential to result in fatal outcomes for humans. Dated phylogenomic analyses placed the main diversification of Buthidae in the Late Mesozoic, which overlaps diversification times for mammals known to be scorpion predators. These analyses suggest that target scorpions and their predators may have been involved in an arm race for several millions of years (Santibáñez-López et al. 2022). Most medically important species of southern South America are sinanthropic (they are adapted to human environments), and are also parthenogenetic (i.e., females can reproduce asexually, without being fertilized); therefore, scorpionism in southern South America is an increasing urban problem, contrary to most regions of the world (Barrios-Montivero et al. 2025). This is not exclusive to South America. *Centruroides balsasensis* Ponce-Saavedra and Francke, 2004, a medically important species, has been recorded in Aguascalientes and Zacatecas, north of the Balsas Depression province within the Neotropics. The type locality is in Michoacán, approximately 350 km away from the Neotropics. Furthermore, available data suggest that this species may be parthenogenetic (Chávez-Samayoa et al. 2024).

Although in the Old World have been adapted to xeric habitats, the New World buthid scorpions have been adapted to tropical and subtropical habitats. This could be related to the relatively recent dispersal of American buthids from Africa approximately 50 million years ago, alongside other animal groups (Ojanguren-Affilastro et al. 2017). Neotropical buthids are characteristically wandering scorpions (Figs. 5.6 and 5.7), unlike lithophilous, pelophilous, or lapidicolous ecomorphotypes, buthids are cryptic scorpions with no evident adaptations to a substrate. However, the *Thorelli* species group of *Centruroides* has been regarded as arboreal (Goodman et al. 2021a, b).

The Buthidae family comprises 15 genera and 432 species in the Americas, one of which is an introduced species. All of these species are found in the Neotropics, except for certain *Centruroides* species that have spread to desert and semidesert areas in northern Mexico and southern USA. Neotropical buthids account for 14% of the genera and 33% of the species worldwide. In the Americas, all species belong to the *Tityus* genus group, as defined by Fet et al. (2005), which means that they may represent a monophyletic group (Santibáñez-López et al. 2022; Štundlová et al. 2022). Three genera dominate Neotropical buthids: *Centruroides* Marx, 1890 (104 species), *Microtityus* Kjellesvig-Waering, 1966 (41 species), and *Tityus* C. L. Koch,

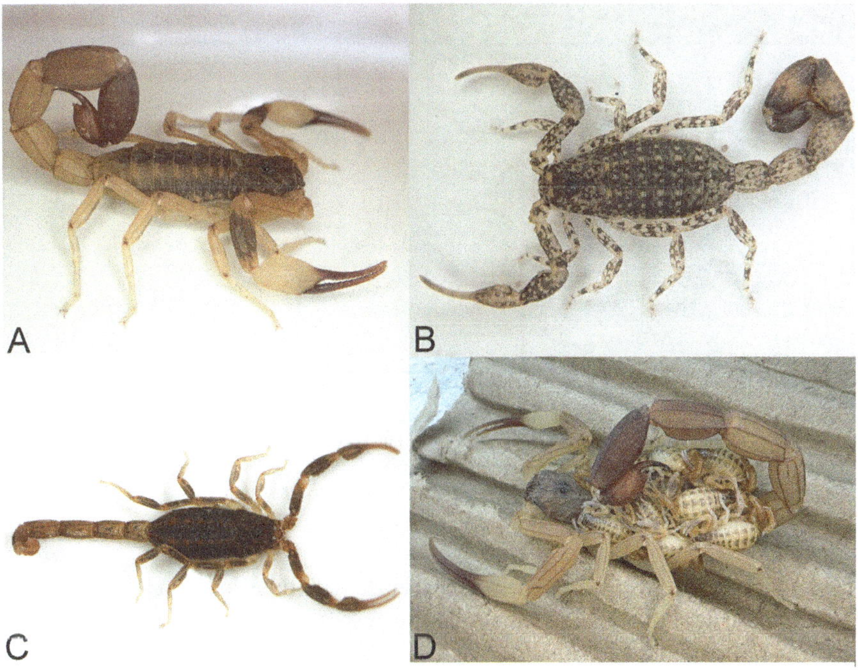

Fig. 5.6 Buthidae, habitus in life. (**a**) *Tityus carrilloi* Ojanguren-Affilastro, 2021, ♂, northern Argentinean Chacoan woods. (**b**) *Tityus paraguayensis* Kraepelin, 1895, ♀, northern Argentinean Chacoan woods. (**c**) *Tityus uruguayensis* Borelli, 1901, central argentinean grasslands. (**d**) *Tityus carrilloi* Ojanguren-Affilastro, 2021, parthenogenetic ♀ with her offspring, from an Argentinean synanthropic population

Fig. 5.7 Buthidae, habitus in life. (**a**) *Tityus cisandinus* Lourenço and Ythier, 2017, ♂, Napo, Ecuador. (**b**) *Tityus filodendron* González-Sponga, 1981, Amazonas, Venezuela. (**c**) *Tityus* sp., ♀, Cusco, Peru. (**d**) *Centruroides margaritatus* (Gervais, 1841), ♀, Loja, Ecuador. (**e**) *Mesotityus vondangeli* González-Sponga, 1981, ♂, Aragua, Venezuela. (**f**) *Rhopalurus ochoai* Esposito, Yamaguti, Souza, Pinto da Rocha and Prendini, 2017, ♂, Zulia, Venezuela

1836 (230 species). Noteworthy, several species with the genera *Centruroides* and *Tityus* produce highly toxic venom and are responsible for scorpionism in their respective distribution areas (Chap. 16). As is the case with numerous other scorpion families around the world, we are only beginning to comprehend the true diversity present within our respective countries. Given that several species are described on an annual basis, the available data has reached a critical mass for a taxonomic synthesis. It is imperative that we undertake such a synthesis to enhance our understanding of the epidemiological aspect of these scorpions.

5.4.1 Phylogenetic Relationships

Buthidae is a family belonging to the Parvor Buthidae and it is sister to the clade comprising Chaerilidae and Pseudochactidae, at the base of the Three of Life of Scorpions.

5.4.2 Characterization

Smallest species: *Femtobuthus shutuae* Lowe, 2010 (13 mm).

Largest species: *Parabuthus transvaalicus* Purcell, 1899 (140 mm).

Chelicerae: Movable finger dorsal edge with two basal denticles, internal distal and external distal denticles opposable. Fixed finger with a ventral medial denticle (Chávez-Samayoa et al. 2024).

Carapace: Anterior margin usually with a pronounced median notch. Tegument densely granular with several carinae. Ocular tubercle usually in the median part of the carapace, or in the anterior half. With 2–5 lateral eyes, with different type patterns (2A, 3A, 3B, 4A, 4C, and 5).

Trichobothrial pattern: Orthobothriotaxic pattern type A, femur with 11 trichobothria, α or β configurations; patella with 13 trichobothria; chela with 15 trichobothria. In some genera and species many trichobothria are missing.

Coxoexternal region: sternum type I, longer than wide, usually subtriangular. Pectens in general are well developed, with fulcra. Pectinal tooth count usually ranges from 10 to more than 30.

Pedipalp: All segments usually with well-developed carinae. Internal margin of fingers in *Tityus* group with several subparallel denticle rows, that may or may not be accompanied by accessory or supernumerary granules. In several species of different genera the pedipalp chela of males can be stouter than in females and with a lobe and a notch.

Legs: *Tityus* group without tibial spurs; with prolateral and retrolateral pedal spurs. Ventral surface of telotarsi with scattered setae.

Metasoma: Highly variable between genera and species, even inside the *Tityus* group. Usually with well-developed carinae that can bear some granulation, as in the VL carinae of metasomal segment V of *Zabius birabeni* Mello-Leitao 1938, or even a serrula as in DL carinae of metasomal segment IV of *Tityus serrulatus* Lutz and Mello, 1922 (Fig. 5.8g). Metasomal segments IV and V can be extremely broad as in males of *Rhopalurus ochoai* Esposito et al., 2017, extremely high as in males of *Tityus paraguayensis* Kraepelin, 1895 or conspicuously elongated as in males of *Tityus argentinus* Borelli, 1899.

Telson: Vesicle rounded or elongated, many species with a subaculear tubercle (Fig. 5.8c), with different shape and development. Aculei are usually similar in size or slightly shorter than the vesicle.

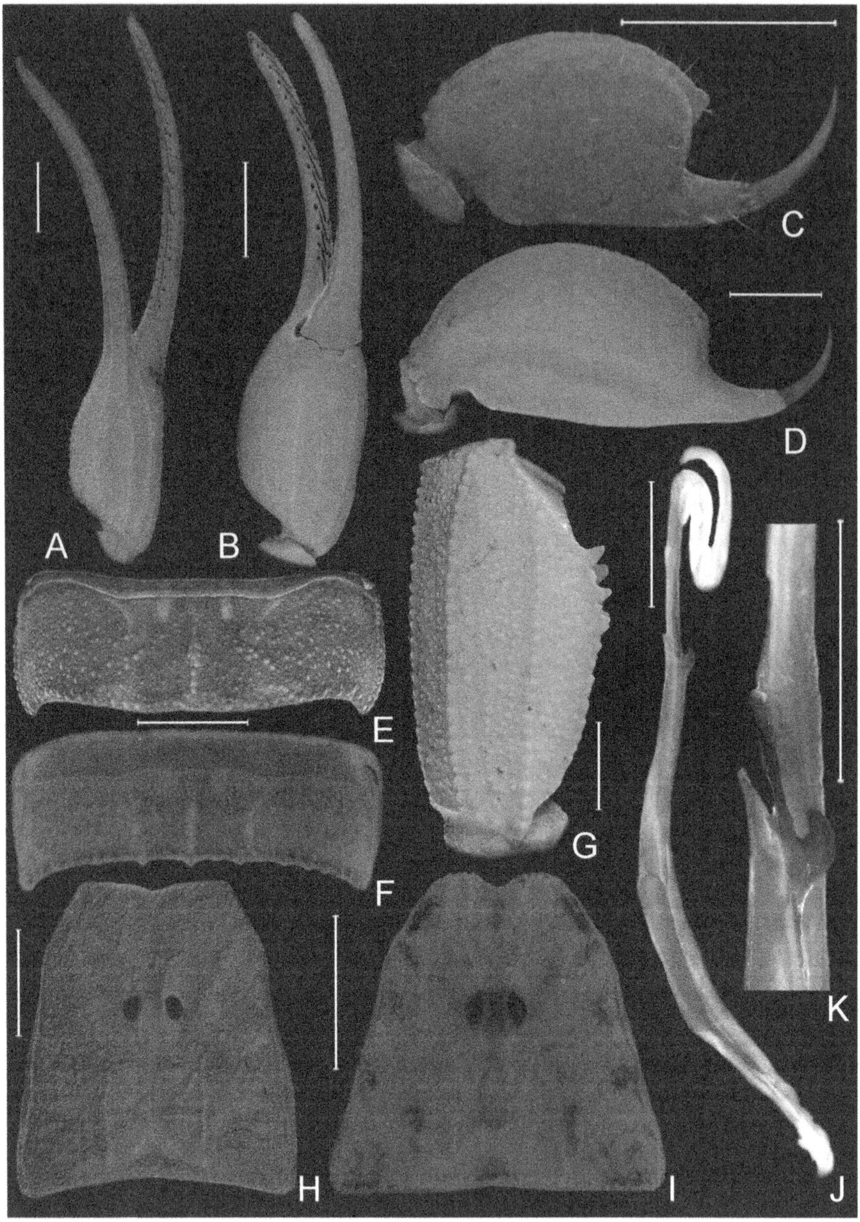

Fig. 5.8 (**a**) *Centruroides*, right pedipalp chela, dorsal aspect; **b** *Tityus*, left pedipalp chela, ventral aspect; (**c**) *Alayotityus*, telson, lateral aspect; (**d**) *Zabius*, telson, lateral aspect; (**e**) *Tityus*, tergite IV, lateral aspect; (**f**) *Alayotityus*, Tergite IV, lateral aspect; (**g**) *Tityus serrulatus*, metasomal segment IV, lateral aspect; (**h**) *Tityus*, prosoma, dorsal aspect; (**i**) *Microtityus*, prosoma, dorsal aspect; (**j**) *Tityus carrilloi*, left hemispermatophore, internal aspect; (**k**) *Tityus carrilloi*, left hemispermatophore, lobe region. Scale bars: 1 mm

Hemispermatophore: flageliform, "no fold" bauplan (Monod et al. 2017). In the *Tityus* group the pars reflexa or flagellum is usually poorly coiled (Fig. 5.8j); capsular region, or lobe region, with two to four lobes (Fig. 5.8k).

5.4.3 Key of Neotropical Genera of Family Buthidae

This key is based mostly on Esposito et al. (2017), Kovařík (2009), Teruel and Kovařík (2012), and Francke et al. (2014).

1. Trichobothrial pattern type A, β configuration (Introduced in several parts of the neotropics).. *Isometrus*
   – Trichobothrial pattern type A, α configuration..................................2
2. Pedipalp chela fingers with accessory (supernumerary) denticles, sub-parallel to the median denticle rows (Fig. 5.8a)...3
   – Pedipalp chela fingers without accessory (supernumerary) denticles, lateral to the median denticle rows (Fig. 5.8b).. 9
3. Leg I, prolateral pedal spur simple (occurs in north-eastern Brazil)............ ....*Physoctonus*
   – Leg I, prolateral pedal spur bifurcated......................4
4. Metasoma of male extremely elongated and narrow, longer than the sum of prosoma and mesosoma; pedipalp patella, prodorsal and proventral carinae adjacent (occurs in southern North America, Central America, and northern South America).................*Centruroides*
   – Metasoma of male not elongated, similar in length to the sum of prosoma and mesosoma. Pedipalp patella, prodorsal and proventral carinae clearly separated......................5
5. Pectinal peg sensilla elongated and acuminated (mostly from hypogean environments of northern Brazil)..............................*Troglorhopalurus*
   – Pectinal peg sensilla peg shaped, short, and rhomboid 6
6. Telson laterally compressed, vesicle width about half its height; mesosoma width similar to carapace width; pectinal plate with two lateral depressions; spiracles not elongated, less than three times wider than long (occurs in central eastern Brazil)......... ....*Ischnotelson*
   – Telson rounded, vesicle width similar to its height; mesosoma wider than carapace; pectinal plate with one median depression or without depressions; spiracles elongated, more than five times wider than long..............7
7. Sternite III, surface with large, acuminate and evenly distributed granules; proventral carina of pedipalp chela manus present; anterior margin of pectinal plate without a furrow (central and northern Brazil)................................*Jaguajir*

– Sternite III, surface finely granular or with small, irregularly distributed granules; proventral carina of pedipalp chela manus absent; anterior margin of pectinal plate with a furrow 8

8. Telson with a subaculear tubercle; carapace, central lateral and posteromedian carinae fused; pectinal plate without depressions (occurs in northern South America)..........................*Rhopalurus*

   – Telson without a subaculear tubercle; carapace, central lateral and posteromedian carinae separated; pectinal plate with single median depression (occurs in the Greater Antilles)..*Heteroctenus*

9. Anterior margin of carapace very narrow, carapace rhomboidal in shape (Fig. 5.8i) (northern South America and the Antilles) ....*Microtityus*

   –Anterior margin of carapace slightly narrower than posterior margin (Fig. 5.8h) 9

10. Tergites I–VI with three longitudinal carinae (Fig. 5.8f).........................11

   –Tergites I–VI with one longitudinal carinae (Fig. 5.8e)...................................12

11. Telson with a conspicuous subaculear tubercle (Fig. 5.8c) (occurs in Cuba).....*Alayotityus*

   – Telson without a conspicuous subaculear tubercle, only with a blunt bulge (Fig. 5.8d) (occurs in south eastern South America).........................*Zabius*

12. Trichobothrial pattern orthobothriotaxic type A; femur with 11 trichobothria, α configuration; patella with 13 trichobothria; chela with 15 trichobothria...13

   – Trichobothrial pattern with decreasing neobothriotaxy type A, ...14

13. Spiracles subcircular; Trichobotrium *d5* of femur situated at the same level of *e1*, (occurs in Cuba) ....*Tityopsis*

   – Spiracles elongated or slitlike; Trichobothrium *d5* of femur basal to *e1* (occurs in South America, southern Central America, and part of the Antilles) ....*Tityus*

14. Base color yellowish, with a marked reticular dark pigment pattern. Subaculaer tubercle of telson with clearly curved anterodorsal margin, and with two conspicuous dorsal denticles (occurs in northern South America, in Venezuela)...... ....*Mesotityus*

   – Base color yellowish, it can present some faintly pigmented areas. Subaculear tubercle of telson flat, crest-like, its width same as that of base of aculeus, ending in a small finger-like projection (occurs in south western Mexico)........ ....*Chaneke*

5.4.4 Habitat and Distribution of the Genera Within Neotropical Buthidae

Isometrus Ehrenberg, 1828. 1 introduced species, *Isometrus maculatus* (DeGeer, 1778). Pantropical; in the neotropics with several records in northern South America and the Antilles.

Tityus Group

Alayotityus Armas, 1973. Forests of eastern Cuba, some species with one troglobiont species. 8 spp.

Centruroides Marx, 1890. In diverse environments of southern North America, Central America, and northern South America. 107 spp. Inhabitat from sea level in tropical deciduous forest to more than 2000 m in montane habitats dominated by pine-oak forest.

Chaneke Francke, Teruel and Santibanez-Lopez, 2014. Tropical deciduous forest of south-western Mexico. 4 spp.

Heteroctenus Pocock, 1893. Greater Antilles. 10 spp.

Ischnotelson Esposito, Yamaguti, Souza, Pinto da Rocha and Prendini, 2017. Caatinga and Cerrado, Brazil. 2 spp.

Jaguajir Esposito, Yamaguti, Souza, Pinto da Rocha and Prendini, 2017. Caatinga, Cerrado, and Amazon forest, Brazil. 3 spp.

Mesotityus Gonzalez-Sponga, 1981. Tropical forests of Aragua state, Venezuela. 1 sp.

Microtityus Kjellesvig-Waering, 1966. Scrubland and forests of northern South America, and the Antilles. 41 spp.

Physoctonus Mello-Leitao, 1934. Caatinga and Amazon forest, Brazil. 3 spp.

Rhopalurus Thorell, 1876. Northern South America. 3 spp.

Tityopsis Armas, 1974. Forests of western Cuba. 7 spp.

Tityus C. L. Koch, 1836. Tropical and subtropical environments of South America, southern Central America, and part of the Antilles; in forests, savannahs, grasslands, and shrublands, from sea level up to 4000 m asl. 230 spp.

Troglorhopalurus Lourenço, Baptista and Giupponi, 2004. Mostly in hypogean environments (caves) of northeastern Brazil. 3 spp.

Zabius Thorell, 1893. Semiarid forests and scrublands of South America. 3 spp.

5.5 Family Caraboctonidae Kraepelin, 1905

The family Caraboctonidae comprises small to medium-sized scorpions distributed in the Andes and coastal Pacific desert of South America, currently comprising two genera: *Caraboctonus* Pocock, 1893 (Fig. 5.9) and *Hadruroides* Pocock, 1893, and

Fig. 5.9 Caraboctonidae, habitus in life. (**a**) *Carboctonus keyserlingi* Pocock, 1893, ♂, Central Chile. (**b**) *Hadruroides carinatus* Pocock, 1900, ♂, Cajamarca, Peru. (**c**) *Hadruroides vichayitos* Ochoa and Prendini, 2010, ♂, Piura, Peru. (**d**) *Hadruroides charcasus* (Karsch, 1879), ♂, Lambayeque, Peru

26 species. It was initially recognized as a subfamily of Vaejovidae (Kraepelin 1905) and later transferred to the family Iuridae Thorell, 1876 by Francke and Soleglad (1981) who reorganized the generic composition, and included some genera that were formerly placed in the Chactidae and Vaejovidae. The subfamily Caraboctoninae, as recognized by Francke and Soleglad, includes two tribes: Hadrurini for the North American genus *Hadrurus* and the tribe Caraboctonini for *Hadruroides* and *Caraboctonus*. This taxonomic status remained without changes until the *Catalog of the Scorpions of the World* (Sissom and Fet (2000). The formerly Caraboctoninae was elevated to the family rank in 2003 by Soleglad and Fet, including *Hadruroides*, *Caraboctonus*, and *Hadrurus* Thorell, 1876, in two subfamilies: Caraboctoninae and Hadrurinae Stahnke 1974, in which later was included the genus *Hoffmannihadrurus* Fet and Soleglad, 2004. Later in 2020, the subfamily Hadrurinae was removed from Caraboctonidae and elevated to family status by Santibáñez-López et al. (2020).

5.5.1 Phylogenetic Relationship

Caraboctonidae appears to be most closely related to the family Superstitioniidae, and both are included in the Superfamily Caraboctonoidea Kraepelin 1905 (Santibáñez-López et al. 2019, 2020). These families are similar in the trichobothrial pattern (type C), tarsal armature (with a medium row of spinule clusters), sternum shape (type II), ocular pattern (type 4C), and hemispermatophore (one-fold bauplan) (Santibáñez-López et al. 2020).

5.5.2 Characterization

Smallest species: *Hadruroides leopardus* Pocock, 1900 (30 mm).

largest species: *Haduroides charcasus* (Karsch, 1879) (80 mm).

Chelicerae: Movable finger: two subdistal teeth and one prominent basal tooth in the internal surface; dorsal distal tooth shorter than ventral distal tooth. Serrula absent.

Carapace: Anterior margin slightly convex, median eyes raised and located in the anterior half of the carapace (Fig. 5.10a), lateral ocelli pattern type 4C, ADMi smaller than other ocelli (Loria and Prendini 2014), in *Hadruroides* ADMi is slightly perceptible.

Coxoexternal region: Sternum subpentagonal with Y-shaped sulcus, type II wider than long (Soleglad and Fet 2003b). Pectinal tooth count 9–23.

Pedipalp: Chela smooth acarinate in most species except *H. charcasus* and *H. pachamama*, in which granular carinae are present on internal surface; Chela movable finger, chela movable finger, median denticle row with six or seven median subrows of denticles (Fig. 5.10e), and variable number of accessory denticles in *Hadruroides* (absent in *Caraboctonus*).

Trichobothrial pattern: Neobothriotaxic major Type C (Fig. 5.10c–e) (Vachon 1973): femur with three trichobothria (*i*, *e*, *d*); patella with 20: one internal (*i*), two dorsal ($d_{1,2}$), three ventral (v_{1-3}), v_3 situated on external surface, and 14 external trichobothria (et_{1-3}, *est*, em_{1-3}, $esb_{1,2}$, eb_{1-5}); chela with 26 trichobothria: 14 on manus (Et_{1-5}, *Est*, *Esb*, Eb_{1-3}, V_{1-4}) and 12 on fixed finger (*Dt*, *Db*, *et*, *est*, *esb*, *eb*, *dt*, *dst*, *dsb*, *db*, *it*, *ib*), *it* and *ib* situated on the base of the finger, *Db* and *Dt* in the middle of the fixed finger (Vachon 1973).

Legs: Leg telotarsus with ventromedian row of spinule clusters (setaceous tufts) (Fig. 5.10b), unguicular spine reduced (Fet et al. 2004). Tibial spurs absent, prolateral and retrolateral pedal spurs present.

Metasoma: DL carinae complete on segments I–IV (Fig. 5.10g), ML carinae complete on I–III, LIM carinae complete on segment I; metasomal segment V with complete, granular VL and VM carinae in *Hadruroides*, VL incomplete and VM carinae absent in *Caraboctunus* (Ochoa and Prendini 2010).

Telson: Slightly concave dorsally, with short aculeus (Fig. 5.10g). Venom gland Type II (Pavlovsky 1913).

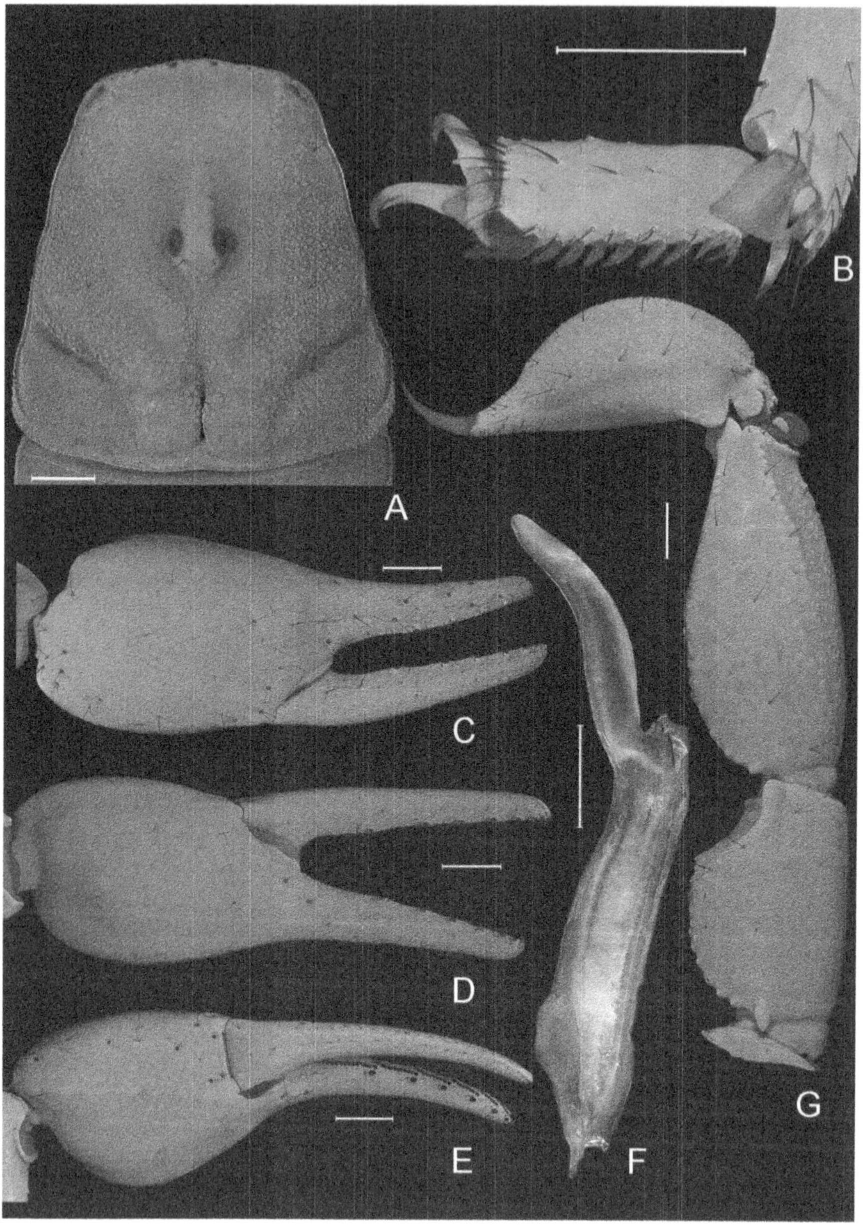

Fig. 5.10 *Caraboctonus keyserlingi*. (**a**) Carapax, male; (**b**) telotarsus III, internal margin; (**c**) Male pedipalp chela, internal margin; (**d**) Male pedipalp chela, external margin; (**e**) Female pedipalp chela, ventral margin; (**f**) Left hemispermatophore, external margin; (**g**) Metasomal segments IV and V, and telson, male, lateral aspect. Scale bars: 1 mm

Hemispermatophore: Lamelliform (Fig. 5.10f) "one-fold" bauplan (Monod et al. 2017), truncal flexure absent, interno basal reflection of the sperm duct absent.

5.5.3 *Key to the Identification of Valid Genera of the Family Caraboctonidae*

1.- Base color dark brown or black. Metasomal segment V with VL carinae restricted to the posterior half of the segment and VM carinae absent. Pedipalp chela finger dentition without internal and external accessory denticles. Hemispermatophore with a digital-shaped process, without crest, free lobe slightly developed.. *Caraboctonus*

- Base color yellowish, with a brown dark pattern. Metasomal segment V with VL and VM carinae present and complete or restricted to the posterior two-thirds of the segment. Pedipalp chela finger dentition with internal and external accessory denticles present, flanking the median denticles row. Hemispermatophore with an elongated crest accompanied externally by medial slit, free lobe more developed.. *Hadruroides*

5.5.4 *Habitat and Distribution of the Genera Within Neotropical Caraboctonidae*

Caraboctonus Pocock, 1893. Central Chile. 1 sp.

Hadruroides Pocock, 1893. Northern Chile, Ecuador, Perú, including several offshore islands and the Galapagos. The habitat includes dry forests, Coastal Pacific desert, and inter Andean valleys. 25 spp.

5.6 Family Chactidae Pocock, 1893

The family Chactidae was created by Pocock (1893) as a subfamily "Chactini" of Iuridae to include some Neotropical genera as *Brotheas* C.L. Koch, 1837, *Broteochactas* Pocock, 1893, *Chactas* Gervais, 1844, *Hadrurochactas* Pocock, 1893, and *Teuthraustes* Simon, 1878, including also the European genus *Euscorpius* Thorell, 1876. During subsequent years, many genera were included and removed from the family, including *Alacran* Francke, 1982, *Calchas* Birula, 1899, *Chactopsis* Kraepelin, 1912, *Chaerilus* Simon, 1877, *Megacormus* Karsch, 1881, *Plesiochactas* Pocock, 1900, *Superstitionia* Stahnke, 1940, *Troglotayosicus* Lourenço, 1981, and *Typhlochactas* Mitchell, 1968 among others (Kraepelin 1899, 1905; Birula 1917;

Stahnke 1940; Mitchell 1968; Francke and Soleglad 1981; Lourenço 1981; Francke 1982). Many authors noted problems with the delimitation of Chactidae and Vaejovidae; two pairs of lateral eyes were traditionally considered diagnostic for Chactidae and three for Vaejovidae, but this character is variable at generic and even specific level in scorpions (Loria and Prendini 2014).

By 1990, Chactidae included 15 genera and four subfamilies (Sissom 1990): Chactinae; Euscorpioninae; Megacorminae; Superstitioninae. Stockwell (1992) subsequently elevated Euscorpiinae and Superstitioninae to family level, and transferred Megacorminae to Euscorpiidae, reducing Chactidae to eleven genera. Lourenço (1998) disagreed with the placement of *Troglotayosicus* and *Belisarius* Simon, 1879 to the Superstitionidae, created the new family Trogotayosicidae in order to accommodate both genera. Sissom (2000) in the Catalog of Scorpions of the World (Fet et al. 2000) listed eleven genera without subfamily divisions: *Broteochactas*, *Brotheas*, *Chactas*, *Chactopsis*, *Guyanochactas* Lourenço, 1998, *Hadrurochactas*, *Nullibrotheas* Williams, 1974, *Teuthraustes,* and tree genera created by González-Sponga, *Cayooca* González-Sponga, 1996, *Taurepania* González-Sponga, 1978, and *Vachoniochactas* González-Sponga, 1978. Previously, the genus *Auyantepuia* González-Sponga, 1978 was synonymized with *Broteochactas* (Francke and Boss 1986).

Major taxonomic changes were introduced by Soleglad and Sissom (2001) and Soleglad and Fet (2003a), including the transfer of Chactopsis from Chactidae to Euscorpiidae; the restoration of the subfamily divisions (Chactinae, Brotheinae, and Uroctoninae), and the establishment of new taxa: new tribe (Nullibrotheini), a new subtribe (Neochactina) and a new genus (*Neochactas* Soleglad and Fet, 2003b); additionally, they transferred the genus *Belisarius* from Troglotayosicidae to Chactidae and synonymized the genera *Cayooca*, *Guyanochactas* and *Taurepania* with *Broteochactas*. The phylogenetic analysis presented by Soleglad and Fet (2003b) was severely criticized by Prendini and Wheeler (2005), who rejected all changes to the classification proposed by these authors, and restored the generic and suprageneric classification published prior to 2001. Despite this, Fet and Soleglad (2005) reinstated the classification rejected by Prendini and Wheeler (2005). The suprageneric classification proposed by Soleglad and Fet (2003b), in which Chactidae is divided into three subfamilies (Chactinae, Brotheinae, Uroctoninae) and four tribes (Belisariini, Chactini, Nullibroteini, Broteini), is not widely accepted.

Currently, there is no consensus on the generic composition of Chactidae; some authors do not recognize the synonymy of the polemic genera *Auyantepuia* and *Guyanochactas* (Lourenço and Qi 2007; Ythier 2015, 2018a, b; Lourenço and Ythier 2011; Lourenço 2018; Ochoa and Rojas-Runjaic 2019). Meanwhile, the synonymy of *Cayooca* and *Taurepania* with *Broteochactas* seems to be more widely accepted by most of the authors (e.g., Rojas-Runjaic and De Sousa 2009); some other authors considered them valid (Ochoa and Rojas-Runjaic 2019).

Finally, three new genera were added to the family: *Chactopsoides* Ochoa et al., 2013, *Megachactops* Ochoa et al., 2013, and *Spinochactas* Lourenco, 2016 (Ochoa et al. 2013; Lourenço 2016); while the genus *Uroctonus* Thorell, 1876 is considered as *incertae sedis* (Santibáñez-López et al. 2019). The affinity of *Chactopsis* with other genera of Neotropical Chactidae is also confirmed (Ochoa et al. 2013), as was found previously by Stockwell (1989) and Flórez and Mattoni (2007).

Most chactid genera, including those with the greatest species diversity, are exclusively Neotropical and inhabit the tropical rainforests of northwestern and central South America; additionally, it includes a small monotypic North American genus, *Nullibrotheas*.

5.6.1 Phylogenetic Relationship

Family Chactidae forms part of superfamily Chactoidea and appears to be most closely related to Scorpiopidae and Euscorpiidae (Sharma et al. 2015); however, the relationships with other chactoid families remain to be rigorously tested and the internal phylogenetic relationships among the chactid genera.

5.6.2 Characterization

Smallest species: *Spinochactas mitaraka* Lourenço, 2016 (12.9 mm).

Largest species: *Brotheas libinallyi* González-Sponga, 1978 (92.2 mm).

Chelicerae: Movable finger with two subdistal; dorsal distal tooth smaller than ventral distal tooth. Fixed finger usually with median and basal teeth fused into a bicuspid. Serrula present.

Carapace: Anterior margin almost straight or with moderate to shallow median notch or weak median projection median eyes raised and located in the anterior half of the carapace, lateral ocelli pattern variable: types 2A, 3C, 4C; most genera with type 4C (Loria and Prendini 2014).

Coxoexternal region: sternum subpentagonal with Y-shaped sulcus and two lateral lobes, type II wider than long (Soleglad and Fet 2003b). Pectinal tooth count 4–12.

Pedipalp: Chela carinate with different degrees of development, usually with D, SD, DMA, DI, E, IM, VE, and VI carinae (Fig. 5.12a–c). Male pedipalp chela without evident sexual dimorphism in most of the genera (Fig. 5.11), usually slightly more incrassate in females, except in the genus *Chactas* where the chela is longer than the female (Fig. 5.11e). Chela movable finger with median denticle row continuous in some genera (Fig. 5.12b) (e.g., *Chactopsis*, *Megachactops,* and *Chactopsoides*) or discontinuous (e.g., *Chactas*, *Teuthraustes*) with five to seven median subrows of denticles; variable number of internal and external denticles; some genera with numerous accessory denticles (Fig. 5.12b) (e.g., *Chactopsis*).

Trichobothrial pattern: Neobothriotaxic major Type C (Vachon 1973), the number and position are variable, usually the femur with three trichobothria, patella with 25–34 (Fig. 5.12d–e): one internal, two dorsal, 3–8 ventral and 17–26 external trichobothria; chela with 26 trichobothria (Fig. 5.12a–c) (Vachon 1973;

Fig. 5.11 Chactidae, habitus in life. (**a**) *Neochactas efreni* (González-Sponga, 1978), ♀, Bolivar, Venezuela. (**b**) *Teuthraustes whymperi* (Pocock, 1893), ♀, Napo, Ecuador; (**c**) *Chactopsoides anduzei* (González-Sponga, 1982), ♂, Amazonas, Venezuela. (**d**) *Brotheas libinallyi* González sponga, 1978, ♂, Bolivar, Venezuela. (e) *Chactas* sp., ♂, Cundinamarca, Colombia. (f) *Taurepania manisapanensis* González-Sponga, 1992, ♀, Bolivar, Venezuela

González-Sponga 1978, 1996; Soleglad and Sissom 2001; Soleglad and Fet 2003b; Ochoa et al. 2013).

Legs: Leg telotarsus setose with ventromedian row of spinule, ungues well developed. Tibial spurs absent, prolateral and retrolateral pedal spurs present.

Metasoma: Carinae variable, usually with DL, ML, VL, VSM carinae distinct in metasomal segments I–IV; DL and VL complete in all segments in most of the genera; VM carina variably developed on segment V (Ochoa et al. 2010).

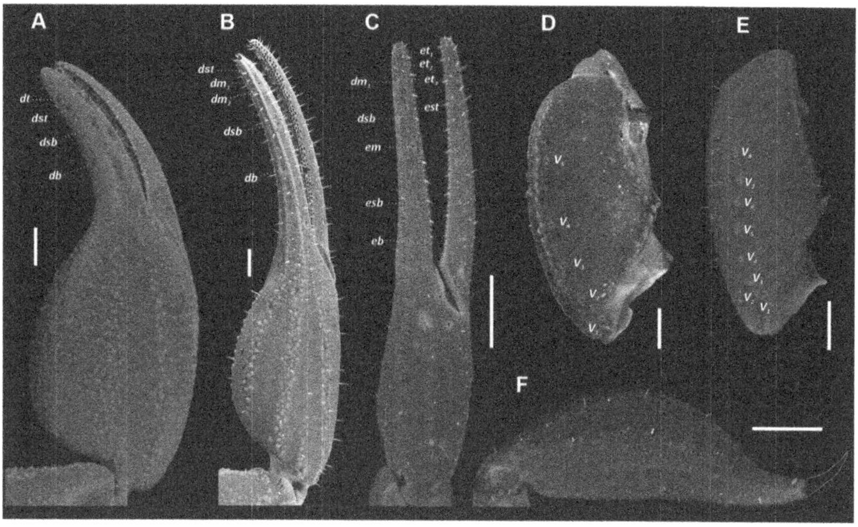

Fig. 5.12 **a, d, k**), (**a**); (**d**); (**k**). (**b, c, e, f, j**), (**B**); (**c**) d; (**e**) (**f**); (**j**). (**g**). (**h**). (**i**). (**l**) *s*. Chactidae. (a) *Guyanochactas* sp., dextral pedipalp chela, dorsal aspect; (b) *Megachactops kuemoi*, dextral pedipalp chela, dorsal aspect; (c) *Chactopsis sujirima*, dextral pedipalp chela, external aspect; (d) *Teuthraustes camposi*, pedipalp patella, ventral aspect; (e) *Cayooca venezuelensis*, pedipalp patella, ventral aspect; (f) *Chactopsis sujirima*, telson lateral aspect.

Telson: slightly elongated and convex dorsally, with short aculeus (Fig. 5.12f). Venom gland Type II (Pavlovsky 1913).

Hemispermatophore: lamelliform "two-fold" bauplan (Monod et al. 2017), truncal flexure present, lobe region (capsule) complex different degrees of development; interno basal reflection of the sperm duct present.

5.6.3 *Key to the Identification of Valid Genera of the Family Chactidae*

1. Fingers of the pedipalp manus with median denticle row continuous, complete, flanked by 9–10 internal and external denticles, and numerous accessory denticles arranged in rows (Fig. 5.12a–b). Fixed-fingers with 16 trichothoria (Fig. 5.12b, c) 13
   - Fingers of the pedipalp manus with median denticles row discontinuous, comprising 5–7 primary subrows slightly oblique. Fixed-fingers with 10–12 trichobothria (Fig. 5.12a) 2
2. - Pedipalp patella with 4–6 ventral (Fig. 5.12d) and 17 external trichobothria 3

- Pedipalp patella with 7–8 ventral (Fig. 5.12e) and 22 or more external trichobothria (Fig. 5.12f) 5

3. - Telson with 3 spinoid subaculear tubercles situated at the base of aculeus ... *Vachoniochactas*
   - Telson without spinoid subaculear tubercles (Fig. 5.12f) 4
4. Pedipalp chela fixed-finger with basal external denticle greatly enlarged, which corresponds with a well-developed sub-basal notch in movable finger. Chela manus swollen in both sexes, without evident sexual dimorphism, fingers shorter than manus length (Fig. 5.12a) ....*Teuthraustes*
   - Pedipalp chela fixed-finger with basal external denticle not greatly enlarged. Male pedipalp chela longer than female chela (Fig. 5.11e) .... *Chactas*
5. - Pedipalp patella with 7 trichobothria 6
   - Pedipalp patella with 8 ventral trichobothria (Fig. 5.12e).................. *Cayooca*
6. - Telson with subaculear tubercles .... *Hadrurochactas*
   - Telson without subaculear tubercles 7
7. - Sternites with relatively large elongated spiracles; leg tarsus with two ventral submedian rows of subspiniform setae .... *Brotheas*
   - Sternites with small rounded or elliptic spiracles; leg tarsus with two ventral submedian rows of thinner setae 8
8. - Sternites with elliptic spiracles .... *Guyanochactas*
   - Sternites with rounded spiracles 9
9. - Metasomal segments III and IV with VSM carinae present and discontinuous .... *Taurepania*
   - VSM carinae obsolete or absent on metasonal segments III and IV 10
10. - Pedipalp chela trichobothria *eb* and *et* situated on proximal two-thirds of the fixed-finger. Chela *Db* trichobothria situated proximally on manus .... *Neochactas*
    - Pedipalp chela trichobothria *eb* and *et* situated on distal two-thirds of the fixed-finger *Db* situated medially on chela manus 11
11. - Pedipalp patella with 22 external trichobothria .... *Spinochactas*
    - Pedipalp patella with 24 external trichobothria 12
12. - Metasomal carinae DL on segments III and IV with significantly larger posterior granule .... *Auyantepuia* - Metasomal carinae DL on segments III and IV with the same size granules .... *Broteochactas*
13. - DM carinae of pedipalp patella present, DPP prominent; patellar trichobothria *v6* and *v7* aligned with *v1*–*v5* and situated close to VE carina; metasomal seg-

ment IV, VSM carinae incomplete, VM carina present medially; hemispermatophore apex short and subtriangular, lobe region well developed, occupying third of length, dorsal apophysis absent ...... *Megachactops*

- DM carinae of pedipalp patella absent, DPP weakly developed; patellar trichobothria *v6* and *v7* not aligned with *v1–v5* and situated submedially; metasomal segment IV, VSM carinae complete, VM carina absent; hemispermatophore apex elongated, lobe region reduced to one-fifth of length, dorsal apophysis present 14

14. - Pedipalp chela trichobothrium *ist* absent, *isb* present, *Et1* situated proximal to *Et2*; hemispermatophore slender, flagellum slightly undulated, not curved, ental fold absent, lobe region with single lobe, ental lobe absent, median lobe reduced . . .. *Chactopsoides*

– Pedipalp chela trichobothrium *ist* absent, *isb* present, *Et1* and *Et2* situated in same axis; hemispermatophore broad, flagellum curved or straight, ental fold present, pronounced on ental margin, lobe region with two lobes, ental lobe slightly sclerotized .... *Chactopsis*

5.6.4 Habitat and Distribution of the Genera Within Neotropical Chactidae

Auyantepuia González-Sponga, 1978. This genus is currently synonymized with *Broteochactas* (Soleglad and Fet 2005). However, Ythier (2018a, b) listed 13 species from Brazil, French Guiana, Suriname, and Venezuela, some of these previously included in *Broteochactas* and *Neochactas*. After Ochoa and Rojas-Runjaic (2019), *Auyantepuia* comprises a single described species, *A. scorzai* Dagert, 1957, known only from Guayaraca, located on the middle slopes of the Auya'n-tepui (1014 m asl).

Broteochactas Pocock, 1893. The distribution of this genus (sensu stricto) includes the coastal mountain range of Venezuela and the islands Trinidad and Tobago. Additionally, it could also include the Amazon and Guianan forests from Brazil and Guyana. 10 spp.

Brotheas C.L. Koch, 1837. Amazon and Guianan forests from Bolivia, Brazil, Colombia, French Guiana, Guyana, Suriname, and Venezuela. 31 spp.

Cayooca González-Sponga, 1996. Endemic to the base of the tepui La Neblina from Amazonas State in Venezuela. 1 sp. Currently synonymized with *Broteochactas*.

Chactas Gervais, 1844. Most species have an Andean distribution pattern in Venezuela, Colombia, Ecuador, and Peru. They are also present on Trinidad Island and in some areas of the Amazon rainforest from Brazil, Ecuador, and Peru. In Central America, it is present in Costa Rica and Panama. 51 spp.

Chactopsis Kraepelin, 1912. Amazon rainforest from Brazil and Peru, and Guiana region of Venezuela. 8 spp.

Chactopsoides Ochoa et al., 2013. Tropical rainforest in the basins of the Orinoco and Rio Negro rivers from Brazil (state of Amazonas), Colombia, and Venezuela. 4 spp.

Guyanochactas Lourenço, 1998. Amazon and Guianan forests from Brazil and French Guiana. 5 spp.

Hadrurochactas Pocock, 1893. Amazon and Guianan forests from Brazil, French Guiana, Suriname, Guyana, and Venezuela. This genus is also distributed in Brejo formation in the State of Ceará, Brazil. 9 spp.

Megachactops Ochoa et al., 2013. Endemic of the tropical rainforest of the state of Amazonas in Venezuela. 3 spp.

Neochactas Soleglad and Fet, 2003b. Amazon and Guianan forests, and savannas from Brazil, Colombia, French Guiana, Guyana, Trinidad and Tobago, Venezuela. 38 spp.

Spinochactas Lourenço, 2016. Tropical rainforests from French Guiana. 2 spp.

Taurepania González-Sponga, 1978. It is the only known scorpion genus that inhabits the top of the tepuis in the Pantepui up to 2810 m asl. Some other species occur in middle tepuian slopes (1000–1300 m) in Bolivar state of Venezuela and Roraima state of Brazil. 5 spp. Currently synonymized with *Broteochactas*.

Teuthraustes Simon, 1878. Andean mountain forests from southern Colombia, Ecuador, and northern Peru up to 3300 m asl. The distribution includes the Guiana Shield of Venezuela and Northern Brazil, and some localities in amazon forest in Brazil. 28 spp.

Vachoniochactas González-Sponga, 1978. Guiana Shield from Brazil and Venezuela, four species are located above 1200 m related to the Pantepui area. 5 spp.

5.7 Family Diplocentridae Karsch, 1880

Since its inception (Pocock 1894), the Diplocentridae family has retained its status, except during a period of turbulent changes in the Order Scorpiones classification (Soleglad and Fet 2003b). This issue has since been rectified. The family Diplocentridae comprises borrowing scorpions (Fig. 5.13) that are characterized by their incrasset pedipalp chelae and tarsi that are armed with spiniform setae, which are used for digging. Cavernicolous and endogean species are present in the Neotropical realm (Chap. 9), as are two lithophilic species. The Neotropical fauna consists of 122 species belonging to nine genera, accounting for 87% of the 141 global species belonging to ten genera. This demonstrates the overwhelming richness of the Neotropics. The family's most striking biogeographic trait is the presence of *Heteronebo* and *Nebo* species in the Middle East and Arabian Peninsula. *Heteronebo* has 16 species in the Antillean Islands and three in the Arabian Peninsula. This pattern has never been satisfactorily explained and remains a challenge for scientists.

Fig. 5.13 Diplocentridae habitus in life. (**a**) *Diplocentrus zacatecanus* Hoffmann, 1932, ♂, Guanajuato, Mexico. (**b**) *Tarsoporosus* sp. ♀, Zulia, Venezuela

5.7.1 Phylogenetic Relationship

The family Diplocentridae has not been thoroughly assessed. However, with poor sampling, it has consistently been retrieved within the superfamily Scorpionoidea as the sister group to the families Scorpionidae and Hormuridae (Sharma et al. 2015; Santibáñez-López et al. 2022).

5.7.2 Characterization

Smallest species: *Bioculus parvulus* Martín-Frías, 2004 (23 mm).

Largest species: *Kolotl magnus* (Beutelspacher and Lopez-Forment, 1991) (101 mm).

Chelicerae: Manus elongated. Fixed finger with the median and basal teeth fused into a bicusp; movable finger with one subdistal and one basal tooth on the external margin; the distal external tooth is smaller than the distal internal tooth and is usually opposable to it; the internal margin is smooth.

Carapace: Surface usually flat with a conspicuous emargination, creating a bilobulated anterior margin (Fig. 5.14a). Ocular tubercle low, median ocelli present in most species except in four troglobite species. Lateral ocelli type 3A and 2B (Loria and Prendini 2014).

Coxoexternal region: Sternum subpentagonal (Fig. 5.14b), type II wider than long (Soleglad and Fet 2003b).

Pedipalp: Pedipalp chela stout, with short to medium-sized fingers (Fig. 5.14c). Internal margins of fingers with a principal row of denticles flanked by external and internal accessory denticles partially merged with the principal row.

Trichobothrial pattern: Trichobothrial pattern Type C, Orthobothriotaxic or neobothriotaxic major.

Legs: Tibial spurs absent. Prolateral pedal spurs present or absent. The tarsi with two rows of subventral spines (Fig. 5.14d).

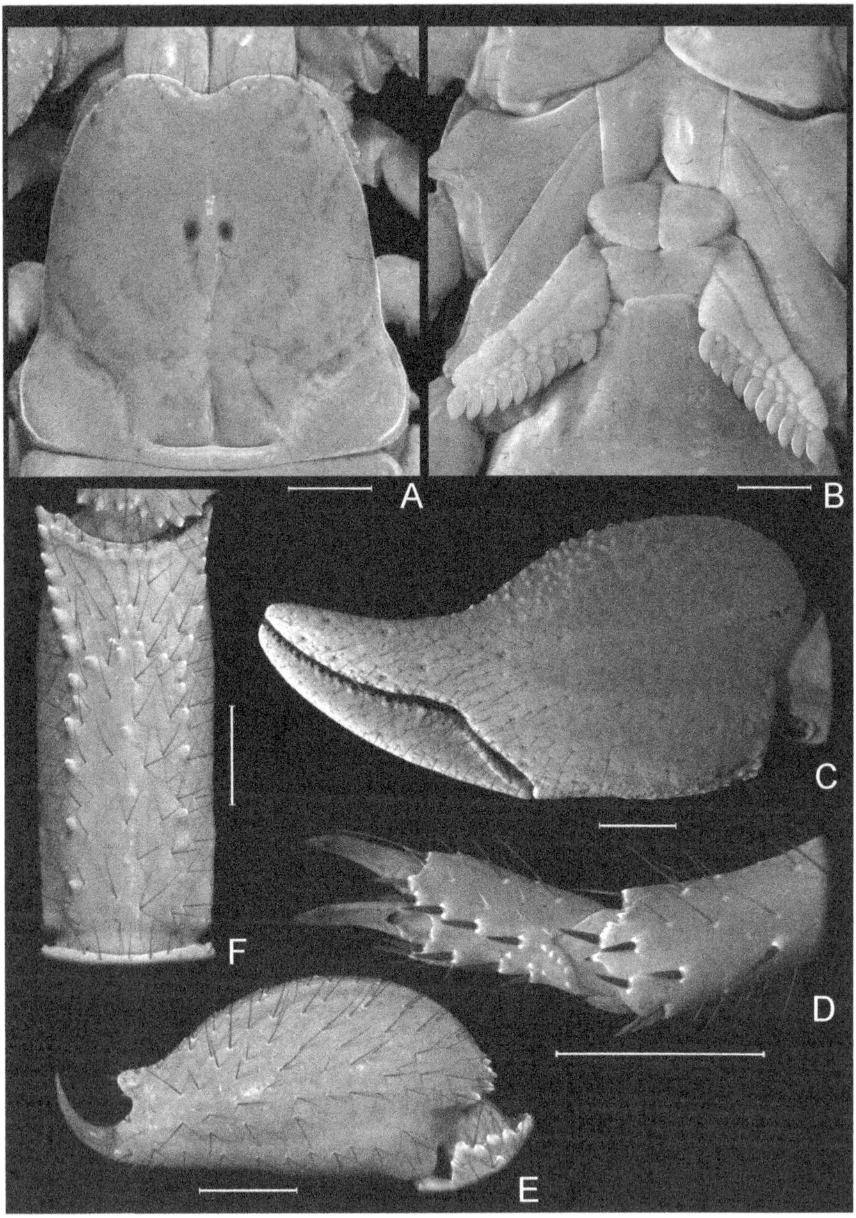

Fig. 5.14 *Didimocentrus*. (**a**) Prosoma, dorsal aspect; (**b**) Sternum, genital operculum and pectens; (**c**) Left pedipalp chela, external aspect; (**d**) Basitarsus and telotarsus IV, ventral aspect; (**e**) Telson, lateral aspect; (**f**) Metasomal segment V, ventral aspect. Scale bars: 1 mm

Metasoma: Metasomal segment V with or without a conspicuous median transversal carina, depending on the genus (Fig. 5.14f).

Telson: With a subacular tubercle blunt and covered with minor macrosetae (Fig. 5.14e). Aculeus shorter than vesicle.

Hemispermatophore: Lamelliform, four-fold *bauplan* (Monod et al. 2017).

5.7.3 Key to the Identification Neotropical Genera of the Family Diplocentridae

Partially based on Polis (1990) and Teruel and Kovařík (2012).

1. Trichobothrial pattern neobothriotaxic major, with accessory trichobothria on the retrolateral and ventral surfaces of the patella (western Mexico)..*Kolotl*
   - Trichobothrial pattern Orthobothriotaxic..2
2. Legs without pedal spurs..3
   - Legs with prolateral pedal spurs..4
3. Median eyes small, similar in size to lateral eyes; metasomal segment V without ventral transversal carina (endemic to eastern Cuba)..................*Cryptoiclus*
   - Median eyes well developed, larger than lateral eyes; metasomal segment V with well-developed ventral transversal carina (Lesser Antilles)...............*Oiclus*
4. Metasomal segment V without well-developed ventral transversal carina (Antilles) ....*Heteronebo*
   - Metasomal segment V with well-developed ventral transversal carina............5
5. Telotarsi with abundant ventral and lateroventral pores (Colombia and Venezuela)...*Tarsoporosus*
   - Telotarsi without ventral and lateroventral pores..................................6
6. Ventral median carina of pedipalp chela oblique (Antilles, Central America, and North America) ....*Didmocentrus*
   - Ventral median carina of pedipalp chela parallel to its longitudinal axis 7
7. Pedipalp femur always deeper than wide; cheliceral movable finger shorter than cheliceral manus (Antilles) ....*Cazierius*
   - Pedipalp femur usually wider than deep; cheliceral movable finger never shorter than cheliceral manus (North America and Central America)..................8
8. Two pairs of lateral eyes (North America)...................................... ....*Bioculus*
   - Three pairs of lateral eyes (Northern and Central America)........*Diplocentrus*

5.7.4 *Habitat and Distribution of the Genera Within Neotropical Diplocentridae*

Bioculus Stahnke, 1968. North America, Baja California, and Mexico. 5 spp. This taxa inhabits desertic areas with sandy to clayey soils, where it constructs burrows.

Cazierius Francke, 1978. Antilles, 13 spp. Species within this genus can be found from tropical forest to grasslands in low altitudes, to tropical deciduous forest and xeric coastal landscapes.

Cryptoiclus Teruel and Kovařík (2012). Endemic to eastern Cuba. 1 species. This is an endogean species, inhabiting deep leaf litter close to coastal humid forest.

Didymocentrus Kraepelin, 1905. Antilles, Central America, and North America. 12 spp. This species inhibits coastal xeric and semideciduous forest, although savannas and grasslands are also included.

Diplocentrus Peters, 1861. Northern and Central America. 65 spp. The species of this genus inhabits most kinds of forest in México, from xeric to tropical forest. It has also adapted to deep soils in pine-oak forest in montane landscapes.

Heteronebo Pocock, 1899. Antilles. 19 spp. These species are found in coastal xeric and semideciduous forest.

Kolotl Santibanez-Lopez, Francke and Prendini, 2014. Mexico. 2 spp. These two species are lithophilous, exclusively found close to karstic or granitic outcrops with cracks and flakes.

Oiclus Simon, 1880. Lesser Antilles. 8 spp. These species are found also in coastal xeric and semideciduous forest.

Tarsoporosus Francke, 1978. Northern South America, Colombia, and Venezuela. 6 spp. These species are found in coastal tropical forests.

5.8 Family Euscorpiidae Laurie, 1896

Scorpions belonging to the family Euscorpiidae are typically hydrophilic and lithophilic, utilizing crevices formed by rocks and analogous microhabitats. However, recent fieldwork has indicated that certain species may also exhibit arboreal behavior, which is a behavior that is typically exhibited by animals that live in trees. A limited number of troglobionts and endogean species have been documented in Mexico. The family Euscorpiidae in the Neotropics comprises three genera and 13 species, a stark contrast to the Old World, where only three genera are observed but 104 species have been documented. A revisionary study was conducted in which cladistic analyses were performed (Soleglad and Sissom 2001). The results of this study indicated that *Troglocormus* Francke, 1981, which was originally classified in the subfamily Megacorminae, should be transferred to the subfamily Scorpiopinae (Scorpiopsinae). Subsequent to this, the taxonomic placement has never undergone rigorous testing to substantiate the position of *Troglocormus*. Nonetheless, it was hypothesized that the analyses conducted by Soleglad and Sissom require reevaluation, as many of their original proposed placement of taxa has been modified on the grounds of numerical analysis. Presently, the classification of the Catalog of Scorpions of the World (Fet et al. 2000) is to be followed, wherein *Troglocormus* is placed within the family Euscorpiidae. This classification may persist until new analyses are presented.

5.8.1 Phylogenetic Relationship

Recent filogenomic analyses have revealed that Euscorpiidae belongs to the Parvorder Iurida, Superfamily Chactoidea, and is sister to Chactidae (in part) and Scorpiopsidae, although the sampling has been insufficient to confidently confirm its position (Santibáñez-López et al. 2019, 2020, 2023).

5.8.2 Characterization

Smallest species: *Megacormus granosus* (Gervais, 1844) 14 mm.

Largest species: *Plesiochactas mitchelli* Soleglad 1976 53.3 mm.

Chelicerae: Fix finger with basa, medial finger separated, unlike other families it does not form a bicuspid single-based denticle. Movable finger with two denticles on the dorsal surface and one to seven denticles on the ventral edge; basal, medial, and subdistal denticles well-developed, distal denticles similar in size and opposable.

Carapace: The shape of the carapace varies; *Megacormus* and *Plesichactas* tend to be trapezoid-shaped shorter on the proximal margin with a feeble median notch. *Troglocormus* species tend to be more rectangular with a deep median notch, lacking median ocular tubercle.

Coxoexternal region: sternum subpentagonal with distal deep rounded sulcus, type II wider than long (Soleglad and Fet 2003b). Pectens with or without medial lamella and fulcra. Pectinal tooth count 3–10.

Pedipalp: Patella with patellar spurs (prolateral processes) uneven, dorsal patellar spur larger than ventral patellar spur, except for *Troglocormus* with spurs similarly developed. Pedipalp manus with nine distinctive carinae from granular to costate. Ventral and dorsal carianae virtually parallel in a perspective facing the distal end of the fingers (Soleglad and Sissom 2001).

Trichobothrial pattern: Neobothriotaxic major Type C (Vachon 1973): Major trichobothorial counts vary within species from two to five trichobothria. Femur trichobothrium *d* positioned on the midpoint of the dorsal surface. Patellar trichobothrial series *eb*, *em*, *est*, and *eb* with accessory trichobothria. Pedipalp chelae trichobothrial series V_1-V_4 restricted to the medial distal area of the pedipalp. Trichobothria V_4 positioned on the external surface. Trichobothrium Et_4 not configured as petite but the same size as the rest of the series (González-Santillán et al. 2017).

Legs: Telotrasi with ventral median row of spinules flanched by rows of four lateral long setae.

Metasoma: Metasomal median carinae on segments I–IV present, granular; ventral submedian carinae absent. Except in *Troglocormus* species, in which ventral submedian are present and ventral medial absent and are costated instead of granular.

Telson: Ventral surface granular and aculeus relatively long, except in *Troglocormus* species, which is smooth and relatively short.

Hemispermatophore: Lamelliform, capsular area conforming to the "two-fold" *bauplan* (Monod et al. 2017). Crown-like structure, ventral, and dorsal trough distal margins with variable number of denticles. Physema comprises a spicule-coated membrane that forms the sperm duct. *Troglocormus* differs in the development of the crown-like structure and the distal ventral and dorsal troughs, which are shorter and tapering instead of more sclerotized and elongated.

5.8.3 Key to the Identification of Valid Genera of the Family Euscorpiidae

1. Metasomal segments I–IV ventral median carina present, pedipalp patella prolateral processes (spur after Soleglad and Sissom 2001) dorsal larger than ventral, median tubercle present with conspicuous median oce lli...2
   - Metasomal segments I–IV ventral median carina absent, pedipalp patella prolateral processes (spur after Soleglad and Sissom 2001) dorsal and ventral equally developed, median tubercle absent without median ocelli.. *Troglocormus*
2. Pedipalp chela fixed finger with rows of median denticles flaked by retro and prolateral rows of granules, patella retrolateral surface *et* series with five trichobotria.............. *Plesiochactas*
   - Pedipalp chela fixed finger with rows of median denticles flaked by retrolateral rows of granules, prolateral rows absent, patella retrolateral surface *et* series with three trichobotria.... *Megacormus*

5.8.4 Habitat and Distribution of the Genera Within Neotropical Euscorpiidae

Megacormus Karsch, 1881 (8 spp) (Fig. 5.15) is endemic to Mexico, mostly occupying the ridges of the Sierra Madre Oriental and part of Veracruzean province. The species are common in humid subtropical to montane forest although some species are found in tropical forest at low altitudes.

Pleasiochactas Pocock, 1900 (3 spp.) inhabits the Veracruzean and Chiapas highlands provinces in Mexico and Guatemala. However, it inhabits tropical forests at relatively high altitudes (1000–1900 m. asl) and recent field work (Lira and Gonzále-Santillán *in prep.*). indicated that it may be of arboreal habitus.

Fig. 5.15 Euscopiidae habitus in life. (**a**) *Megacormus orizaba* Teruel, Kovarik, Lowe and Stahlavsky, 2023 ♂, Veracruz, Mexico. **b** *Megacormus* sp. ♀ Puebla, Mexico

Troglocormus Francke, 1981 (2 spp.) is endemic to Mexico, exclusively found in caves of the Northern ranges of the Sierra Madre Oriental. Both known species exhibit strong adaptations to live inside troglomorphic habitats.

5.9 Family Hadruridae Stahnke, 1974

After Thorell's description of the genus *Hadrurus* Thorell, 1876, the taxonomic placement has been in constant debate until recently. Firstly, Pocock (1893) created a catch-all group with taxa from the Old and New world, the subfamily Iurini (Iurinae), as part of the family Vaejovidae. It was Stahnke (1974) who created the subfamily Hadrurinae, wherein placed the genus *Hadrurus* and *Anoroctonus*. Afterwards, Francke and Soleglad (1981) redefined the composition of the family Iuridae and restricted *Hadrurus* within the tribe Hadrurini Fracke and Soleglad belonging to the subfamily Caraboctoninae and 1981. Subsequently, Fet and Soleglad (2008) proposed the separation of Iuridae and Caraboctonidae wherein refurbished to the family Hadruridae with the genus *Hadrurus* and *Hoffmannihadrurus*. It was until recently that Hadruridae regained its hierarchy within the superfamily Hadruroidea with only two genera recognized earlier after a comprehensive phylogenomic analysis (Santibáñez-López et al. 2020).

5.9.1 Phylogenetic Relationship

The family Hadruridae has been retrieved in two positions: sister to the genus *Uroctonus* or Sister to (Scorpionoidea + Vaejovoidea) (Santibáñez-López et al. 2020). However, the position of Hadruridae as the sister group to Scorpionoidea and Vaejovidae has been gaining support in subsequent analyses (Santibáñez-López et al. 2022, 2023).

5.9.2 Characterization

Smallest and largest species in Neotropics: The body size of these scorpions is similar in the two species inhabiting the Neotropics, ranging from 68 to 100 mm.

Chelicerae: Movable finger with two subdistal teeth; dorsal distal tooth smaller than ventral distal tooth; prominent ventral basal present on the proximal half of the finger. Fixed finger with median and basal teeth fused into a bicusp. Serrula absent.

Carapace: Anterior margin convex, with anterior paired lobes on anterior half of the carapace, ocular tubercle situated medially. Lateral ocelli pattern type 4C (Loria and Prendini 2014).

Coxoexternal region: Sternum subpentagonal with proximal deep, forked, rounded sulcus, type II wider than long (Soleglad and Fet 2003b). Pectinal tooth count 22–44. Males with reduced, not visible genital operculum.

Pedipalp: Patella with patellar spurs (prolateral processes) weakly developed to vestigial. Retrolateral carinae on palp manus absent. Pedipalp chela fingers with 9–10 oblique rows of principal denticles.

Trichobothrial pattern: Neobothriotaxic major Type C (Vachon 1973): Major trichobothrial counts vary within species from two to five trichobothria. Patella with less than 50 trichobothria on the prolateral surface and circa 30 on ventral surface. Pedipalp chela manus with two dorsal trichobothria, and less than 12 on ventral aspect.

Legs: Pedal spurs pectinate; tarsus with single ventral row of spinules, flanked by macrosetae on the four walking legs; Uniquium strongly developed.

Metasoma: Most carinae finely granular, moniliform; lateral carinae partially present on metasomal segment IV.

Telson: Incrasset, surface covered with macrosetae. Some species with paired dorso-subaculear glands in mature males. Aculeus as long as vesicle.

Hemispermatophore: Lamelliform, laminal short and widened on base, tapering; capsular area conforming to the "two-fold" bauplan (Monod et al. 2017), with weak folding, producing a relatively short sperm duct.

5.9.3 Habitat and Distribution of the Genera Within Neotropical Hadruridae

The only genus, *Hoffmannihadrurus* (Fig. 5.16) in the family Hadruridae distributed in the Neotropics inhabits the Balsas Basin province in the states of Morelos, Guerrero, Puebla, and Oaxaca, Mexico. They are found in semideserts of low altitude intermontane valleys, characterized by xeric and deciduous tropical forest. Only two species have been described and well documented.

Fig. 5.16 Hadruridae, habitus in life. (**a**) ♀. (**b**) ♂, *Hoffmannihadrurus aztecus* Pocock, 1902 from Puebla, Mexico

5.10 Family Hormuridae Laurie, 1896

The family Hormuridae comprises medium-sized to large scorpions distributed in Africa, South and Southeast Asia, Australia, and Neotropics representing a typical Eurogondwanaland distribution (Sissom 1990; Fet et al. 2000; Soleglad and Fet 2003b; Monod and Prendini 2015). Some species of this family are among the largest scorpions in the world (> 20 cm, e.g., *Hadogenes*). Currently comprising ten genera: *Cheloctonus* Pocock, 1892, *Chiromachetes* Pocock, 1899, *Chiromachus* Pocock, 1893, *Hadogenes* Kraepelin, 1894, *Hormiops* Fage, 1933, *Hormurus* Thorell, 1876, *Iomachus* Pocock ,1893, *Liocheles* Sundevall, 1833, *Opisthacanthus* Peters, 1861, and *Palaeocheloctonus* Lourenço, 1996. Only *Opisthacanthus* is distributed in the Neotropics (Fig. 5.17).

Many of those genera were assigned to the subfamilies Ischnuridae, Heteroscorpioninae, and Hemiscorpiinae, in the family Scorpionidae. Stockwell (1989) recognized Ischnuridae as a family level comprising the three mentioned subfamilies. Lourenço (1996) elevated Heteroscorpionidae Kraepelin, 1905 to family rank, based on Stockwell's (1989) phylogeny, while Hemiscorpiidae Pocock, 1893 was elevated to family rank by Prendini (2000).

The name Ischnuridae was created by Simon (1879), and was accepted for a long time; however, Fet and Bechly (2001) noticed the homonymy with Ischnurinae Fraser, 1957 (Insecta, Odonata) and proposed the name Liochelidae as a substitute, which was accepted by ICZN (2003). Many taxonomic changes were proposed by Soleglad and Fet (2003b), but posteriorly rejected by Prendini and Wheeler (2005). Finally, Monod and Prendini (2015), based on phylogenetic and biogeographical arguments, elevated Hormurini Laurie, 1896 (considered as subfamily: type genus: *Hormurus* Thorell, 1876) to the family rank, considering as a senior synonym of Liochelidae Fet and Bechly 2001.

Fig. 5.17 Hormuridae, habitus in life. (**a**) *Opisthacanthus weyrauchi* Mello-Leitão, 1948, Cajamarca, Peru. (**b**) *Opisthacanthus elatus* (Gervais, 1844), Santander, Colombia

5.10.1 Phylogenetic Relationship

Hormuridae forms part of superfamily Scorpionoidea and appears to be most closely related with the families Hemiscorpiidae and Heteroscorpionidae (Prendini 2000; Soleglad and Fet 2003b). However, Sharma et al. (2015) consider that some Hormuridae could be related to Scorpionidae.

5.10.2 Characterization

Smallest species in Neotropics: *Opisthacanthus weyrauchi* Mello-Leitão, 1948 (47.3 mm).

largest species in Neotropics: *Opisthacanthus elatus* (Gervais, 1844) (90.7 mm).

Chelicerae: Movable finger with one subdistal tooth; dorsal distal tooth smaller than ventral distal tooth. Fixed finger with median and basal teeth fused into a bicusp. Serrula absent.

Carapace: Anterior margin with deep median notch (Fig. 5.18a). Median eyes raised and located in the center of the carapace, lateral ocelli pattern variable, most genera with type 3A (Loria and Prendini 2014).

Coxoexternal region: Sternum subpentagonal with Y-shaped sulcus and two lateral lobes (Fig. 5.18b), type II wider than long (Soleglad and Fet 2003b). Pectinal tooth count 5–13.

Pedipalp: Patella with anterior process well developed in Neotropical species. Chela carinate, usually with dorsal marginal, external secondary, ventrointernal, and ventromedian carinae well developed formed by strong granules, other carinae obsolete or vestigial. Chela movable finger, with two median denticle rows, often becoming fused at the base (Fig. 5.18d).

Trichobothrial pattern: Orthobothriotaxic Type C in most of the genera, except *Hadogenes* and some *Opisthacanthus* and *Liocheles*, which present Neobothriotaxic

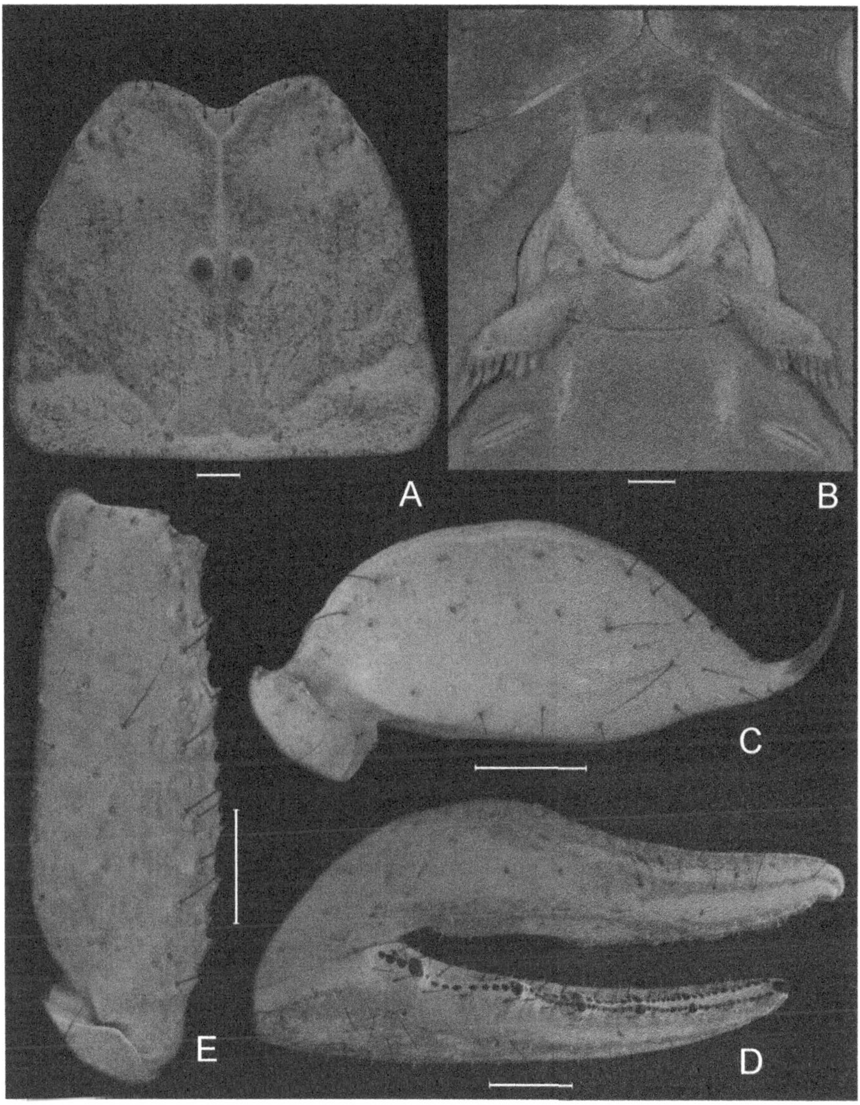

Fig. 5.18 (**a**) *Opisthacanthus elatus*. Prosoma, dorsal aspect; (**b**) *Opisthacanthus cayaporum*, Sternum, genital operculum and pectens; (**c**) *Opisthacanthus elatus*, telson, lateral aspect; (**d**) *Opisthacanthus elatus*, right pedipalp movable finger, internal aspect; **e** *Opisthacanthus elatus*, Metasomal segment V, lateral aspect. Scale bars: 1 mm

major Type C (Vachon 1973; Prendini 2000). Neotropics species: femur with three trichobothria (*i*, *e*, *d*); patella with 20: one internal (*i*), two dorsal ($d_{1,2}$), three ventral ($v_{1\text{-}3}$), and 13–14 external trichobothria ($et_{1\text{-}3}$, *est*, $em_{1\text{-}2}$, $esb_{1\text{-}3}$, $eb_{1\text{-}5}$), some species with three *esb* trichobothria (Lourenço, 2017); chela with 26 trichobothria: 16 on manus (*Dt*, *Db*, Et_{1-5}, *Est*, *Esb*, Eb_{1-3}, V_{1-4}) and 10 on fixed finger (*et*, *est*, *esb*, *eb*, *dt*, *dst*, *dsb*, *db*, *it*, *ib*), *it*, *ib*, *esb* and *eb* situated on the base of the finger (Vachon 1973).

Legs: Telotarsus with paired ventrosubmedian rows of spinules (Prendini 2000). Tibial spurs and retrolateral pedal spurs absent, prolateral pedal spurs present (Sissom 1990).

Metasoma: Relatively short, compared with total length. Neotropical species with metasomal segments I–V with most carinae often absent or vestigial (Fig. 5.18e).

Telson: Slightly convex dorsally, with short aculeus (Fig. 5.18c). Venom gland Type I (Pavlovsky 1913).

Hemispermatophore: Lamelliform "four-fold" bauplan (Monod et al. 2017), truncal flexure present, lobe region (capsule), very complex different degrees of development; interno basal reflection of the sperm duct present.

5.10.3 Habitat and Distribution of the Genera Within Neotropical Hormuridae

Opisthacanthus Peters, 1861. This genus is present in the Central America and Caribbean regions from Cuba, Panama, Costa Rica (Cocos Island), and Hispaniola. In South America, it is distributed in northern Brazil, Colombia, French Guiana, Peru, Suriname, and Venezuela. Its distribution includes the Amazon rainforest, savannas, and dry Andean forests. 10 spp.

5.11 Family Troglotayosicidae Lourenço 1998

This family comprises small-sized troglomorphic scorpions, which are included in the monotypic genus *Troglotayosicus* Lourenço 1981. The members of this genus inhabit the leaf litter and caves of tropical rainforests from Ecuador and Colombia (see Chap. 9).

5.12 Family Typhlochactidae

This family is endemic to Mexico and includes four genera: *Alacran* Francke, 1982, *Sotanochactas* Francke, 1986, *Stygochactas* Vignoli and Prendini, 2009, and *Typhlochactas* Mitchell, 1971. All species are troglomorphic and inhabit hypogean and endogean habitats (see Chap. 9).

5.13 Family Vaejovidae Thorell, 1876

The original composition of the Vaejovidae family included the genera *Vaejovis* C. L. Koch and *Hadrurus* Thorell, 1876. Its taxonomic scope later increased to include two subfamilies: the Iurini and the Vejovini (Lauri 1896). Kraepelin later recognized five subfamilies: Syntropinae, Vaejovidae, Caraboctoninae, Iurinae, and Scorpiopsinae. These subfamilies included scorpions distributed worldwide. Francke and Soleglad (1981) removed the Iurinae subfamily and elevated it to family rank and transferred the members of Caraboctoninae into it. Sissom (1990) considered Vaejovidae and Chactidae to be problematic taxa and was unable to diagnose or define them due to the persistence of classical character systems in the literature that are highly convergent. Stockwell (1992) removed Scorpiopsidae from Vaejovidae and transferred *Nullibrotheas* Williams, 1974 to Chactidae. Stockwell then restricted Vaejovidae to North America, a classification with which most authorities agree today. The family currently comprises the following four subfamilies: Smeringurinae (Soleglad and Fet 2008), Syntropinae (Kraepelin 1905), Vaejovinae (Thorell, 1876), and Stahnkeninae (Soleglad and Fet 2008). However, the last two subfamilies have not been tested for monophyly (Soleglad and Fet 2008; González-Santillán and Prendini 2013, 2015; Santibáñez-López et al. 2019). While some authors have recognized species groups (Haradon 1984a, b, 1985; Soleglad and Fet 2005), this has received little attention in recent literature. Others have been elevated to the generic level. The family Vaejovidae encompasses 25 genera and 248 species. Eight of these genera and 75 species inhabit the Neotropics. Most arid-adapted scorpions have been documented in the deserts of northern Mexico and the southwestern United States. In contrast, endogean, hypogean, and hydrophilic scorpions are predominantly found in the Neotropics.

Currently, the family Vaejovidae is a Nearctic scorpion group that has dispersed into the Neotropical realm. The vaejovid species have demonstrated a high degree of habitat specialization across North America, with some species exhibiting ecomorphological adaptations that include lithophilic, troglobitic, humicolous, and psammophilous (Fig. 5.19) characteristics (González-Santillán and Prendini 2013). Vaejovids exhibit a remarkable capacity for adaptability, thriving in environments characterized by high humidity and exhibiting a dominant presence in arid habitats. The Neotropical species under consideration are distributed across the following biogeographical provinces: the Sierra Madre Oriental, the Sierra Madre Occidental, and the Pacific Lowlands provinces. The family has extended south to the mountains of Guatemala. It is noteworthy that these organisms are not present along the littorals of the Gulf of Mexico and are absent from the Yucatan Peninsula. Only two subfamilies inhabit the Neotropics, Syntropinae and Vaejovinae; Sthankeinae and Smeringurinae are exclusively found in the Neartic real.

Fig. 5.19 Selected representative of the family Vaejovidae. (**a**) *Kochius punctipalpi* (Wood, 1863) ♂ from Baja California Sur. (**b**) *Vizcaino viscainenesis* Gonzalez-Santillan and Prendini, 2013 ♂ from Baja California Sur. (**c**) *Konetontli chamelaensis* ♂ from Jalisco (Williams, 1986). (**d**) *Chihuahuanus glabrimanus* (Sissom and Hendrixson, 2005) ♀ from Zacatecas. (**e**) *Paruroctonus utahensis* (Williams, 1968) ♂ from Chihuahua. (**f**) *Thorellius cristimanus* (Pocock, 1898) ♂ from Colima. Photo credits A and B Warren Savary

5.13.1 Phylogenetic Relationship

The most recent phylogenetic analysis indicates that Vaejovoidea is the sister group of Scorpionoidea, thereby invalidating the historical relationship with the superfamily Chactoideae (Santibáñez-López et al. 2019). The relationship between Vaejovidea and Scorpionoidea as sister groups has been consistently retrieved in many instances (Sharma et al. 2015; Santibáñez-López et al. 2020, 2022) and is further supported by the increased of complexity observed in the number of folds within the capsular area, which increases from three in Vaejovidea to four in Scorpionoidea (Monod et al. 2017). Further investigation is needed to determine the limits and composition of the subfamilies; in fact, they may need to be elevated as families.

5.13.2 Characterization

Smallest species: *Konetontli chamelaensis* (Williams, 1986) (11–13 mm).

Largest species: *Syntropis williamsi* Soleglad, Lowe and Fet, 2007 (83–95 mm).

Chelicerae: Movable finger with internal and external distal tooth located in the same plane, not opposed. Two smaller subdistal and one basal tooth. Median teeth are larger than subdistal and basal, but smaller than external distal. Fix finger with a basal bicuspid denticle and one median smaller teeth. Serrula from vestigial to well-developed.

Carapace: Surface with a variety of ornamentations, ranging from smooth and glabrous in *Balsateres cisnerosi* (Saavedra and Sissom, 2004) to heavily granular as in all species within *Mesomexovis* González-Santillán and Prendini 2013 (Fig. 5.20a). Anterior margin is frequently straight with a median, weak notch of protuberance. Lateral ocelli of type 3A (Loria and Prendini 2014).

Coxoexternal region: Sternum type 2 (Soleglad and Fet 2003b), subequilateral pentagonal, wider than long, posterior with evident emargination and convexed lateral lobes. Pectinal teeth count: 9–40.

Pedipalp: Carinae with ample variation in ornamentation from absent to well-developed granular crenular or smooth, even obsolete, expressed as distinct pigmentation stripes (Fig. 5.20b). Pedipalp chelae may be slender, elongated to stout, incrasset with or without carinae. Chela finger dentate margins armed with median denticles aligned in a row interrupted by larger retrolateral denticles and flanked by subpaired prolateral denticles.

Trichobothrial pattern: Orthotricobotaxia type C (Vachon 1973). Pedipalp chela fixed finger trichobothrim *eb* positioned near the dorsal surface of the finger, *esb* near the dentate margin, *ib* and *it* located on the finger, either basal or distal, never on the manus. Patella trichobothrium v_3 is located on the retrolateral surface (Fig. 5.20f). Femur trichobothrium *d* and *i* are located contiguous to the dorsal retrolateral carina (Fig. 5.20e).

Legs: Pedal spurs present, tibial spur absent. Basitarsus with the presence of setal combs (Fig. 5.20d) in psammophilous species. Tarsus with single median row of spinules on ventral surface, terminating distally with bifurcating rows of three to six of spinules.

Metasoma: Carinae variable from obsolete to granular; dorsal submedian carina on segments I–IV posterior margin with a spiniform projection pointing backwards, in most species; dorsal lateral carinae of segment IV posterior margin projecting beyond the dorsal intercarinal surface.

Telson: Laterobasal aculear serration on the base of telson aculeus present in the early stage of development, most adults lose it. Dorsal gland expressed as a hyaline area or a depression of variable size.

Hemispermatophore: Lamelliform (Fig. 5.20c), with capsula conforming to the "three-fold" bauplan (Monod et al. 2017). The Sperm duct and the subex connected (Lira and González-Santillán 2025). Mating plug present, either soft or sclerotized.

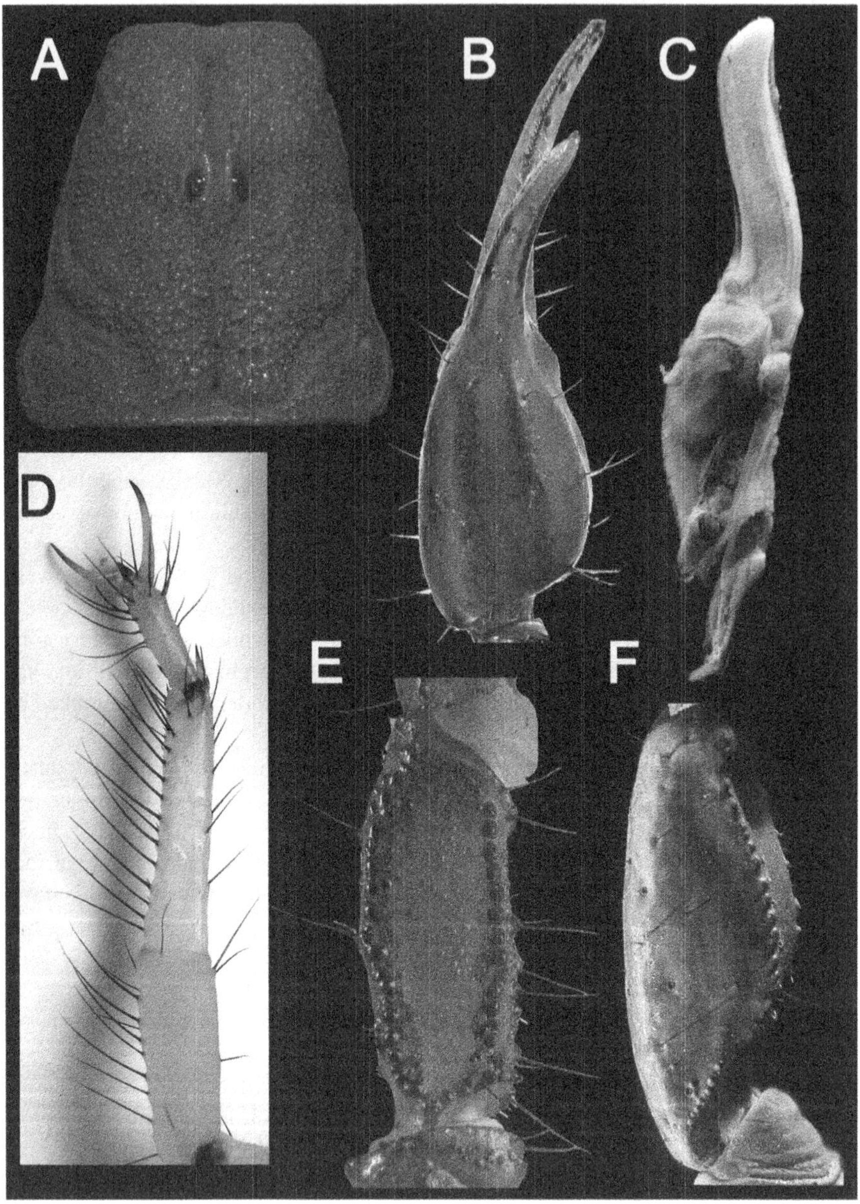

Fig. 5.20 (**a–c**, **e**, **f**) *Mesomexovis* sp. (**d**) *Smeringurus mesaensis*. (**a**) Carapace, (**b**) Pedipal chela dorsal view, (**c**) Hemispermatophore contra-lateral view. (**d**) Tibia, basitasus, and telotarsus dorsal view. (**e**) Pedipalp femur dorsal view. (**f**) Pedipalp patella ventral view

5.13.3 Key to the Identification of Valid Subfamilies of Vaejovidae

1. Metasomal segments I–III posterior end of lateral median carinae rounded, dorsal and retrodorsal macrosetae of telotarsal legs I–IV are arranged in a sublinear or linear row forming setal combs (distributed in the Neartic)..... Smeringurinae (included genera: *Smeringurus* Haradon, 1983; *Paruroctonus* Werner, 1934; *Vejovoidus* Stahnke, 1974).
   - Metasomal segments I–III posterior end of lateral median carinae spiniform, dorsal and retrodorsal macrosetae of telotarsal legs I–IV are arranged in two independent divergent rows not forming setal combs..........2
2. Female genital operculum sclerites separated on posterior one to two-fifths, trichobothria *ib* and *it* positioned basal on the fixed finger .... Vaejovidae (included genera: *Catalinia* Soleglad, Ayrey, Graham and Fet, 2017; *Franckeus* Soleglad and Fet, 2005; *Graemeloweus* Soleglad, Fet, Graham and Ayrey, 2016; *Kovarikia* Soleglad, Fet and Graham, 2014; *Pseudouroctonus* Stahnke, 1974; *Uroctonites* Williams and Savary, 199; *Vaejovis* C.L. Koch, 1836)
   -Female genital operculum sclerites connected their entire length, trichobothria *ib* and *it* positioned medial on fixed finger...3
3. Hemispermatophore with the sclerotized hemimating plug bearing the margin of distal barb armed with spines, the first to third pectinal teeth are unmodified............................ Syntropinae (included genera: *Balsateres* Gonzalez-Santillan and Prendini, 2013; *Chihuahuanus* Gonzalez-Santillan and Prendini, 2013; *Kochius* Soleglad and Fet, 2008; *Konetontli* Gonzalez-Santillan and Prendini, 2013; *Kuarapu* Francke and Ponce-Saavedra, 2010; *Maaykuyak* Gonzalez-Santillan and Prendini, 2013; *Mesomexovis* Gonzalez-Santillan and Prendini, 2013; *Paravaejovis* Williams, 1980; *Syntropis* Kraepelin, 1900; *Thorellius* Soleglad and Fet, 2008; *Vizcaino* Gonzalez-Santillan and Prendini, 2013).
   - Hemispermatophore of the sclerotized hemimating plug bearing the margin of distal barb without spines, smooth; the first to third pectinal teeth are enlarged, ovoid, and devoid of sensilla .... Stanhkeinae (included genera: *Gertschius* Graham and Soleglad, 2007; *Serradigitus* Stahnke 1974; *Stahnkeus* Soleglad and Fet, 2006; *Wernerius* Soleglad and Fet, 2008)

References

Armas LF (1988) The Greater Antillean scorpions (Arachnida, Scorpiones) [Abstract]. In Abstracts of the XIV international congress of arachnology and 22nd annual meeting of the American Arachnological society (p. 50)

Barrios-Montivero A, Martínez PA, Ojanguren-Affilastro, AA (2025) The ongoing expansion of the medically important scorpions in Southern South America. EcoHealth Advance online publication. https://doi.org/10.1007/s10393-025-01722-0

Birula AA (1917) *Arachnoidea Arthrogastra Caucasica. Pars I. Scorpiones*. Zapiski Kavkazskogo Muzeya. [In Russian]. (English translation: Byalynitskii-Birulya, A. A. (1964). *Arthrogastric Arachnids of Caucasia. 1. Scorpions*. Israel Program for Scientific Translations)

Botero-Trujillo R, Flórez E (2011) A revisionary approach of Colombian *Ananteris* (Scorpiones, Buthidae): two new species, a new synonymy, and notes on the value of trichobothria and hemispermatophore for the taxonomy of the group. Zootaxa 2904(1):1–44. https://doi.org/10.11646/zootaxa.2904.1.1

Botero-Trujillo R, Noriega JA (2011) On the identity of *Microananteris* with a discussion on pectinal morphology, and description of a new *Ananteris* from Brazil (Scorpiones, Buthidae). Zootaxa 2747(1):37–52. https://doi.org/10.11646/zootaxa.2747.1.3

Cekalovic KT (1973) Nuevo carácter sexual secundario en los machos de *Brachistosternus* (Scorpiones, Bothriuridae). Bol Soc Biol Concepc 46:99–102

Cekalovic KT (1983) Catalogo de los escorpiones de Chile (Chelicerata, Scorpiones). Bol Soc Biol Concepc 54:43–70

Chávez-Samayoa F, Escoto-Moreno JA, González-Santillán E (2024) Richness analysis and completeness of the scorpion fauna of Aguascalientes, Mexico with an identification key to species. Rev Mex Biodivers 95:1–46. https://doi.org/10.22201/ib.20078706e.2024.95.5380

Esposito LA, Yamaguti HY, Souza CA, Pinto-da-Rocha R, Prendini L (2017) Systematic revision of the neotropical club-tailed scorpions, *Physoctonus*, *Rhopalurus*, and *Troglorhopalurus*, resurrection of *Heteroctenus*, and description of two new genera and three new species (Buthidae: Rhopalurusinae). Bull Am Mus Nat Hist 415:1–134

Fet V, Bechly G (2001) Case 3120a. Liochelidae, fam. nov. (Scorpiones): Proposed introduction as a substitute name for Ischnuridae Simon, 1879. Bulletin of Zoological Nomenclature 58(4):280–281

Fet V, Soleglad ME (2005) Contributions to scorpion systematics. I. On recent changes in high-level taxonomy. Euscorpius 2005(31):1–13

Fet V, Sissom WD, Lowe G, Braunwalder ME (2000) Catalog of the scorpions of the world (1758–1998). New York Entomological Society, New York

Fet V, Soleglad ME, Neff DPA, Stathi I (2004) Tarsal armature in the superfamily Iuroidea (Scorpiones: Iurida). Rev Ibér Aracnol 10:17–40

Fet V, Soleglad ME, Lowe G (2005) A new trichobothrial character for the high-level systematics of Buthoidea (Scorpiones: Buthida). Euscorpius 2005b(23):1–40

Fet V (2008) Cladistic analysis of superfamily Iuroidea, with emphasis on subfamily Hadrurinae (Scorpiones: Iurida). Boletín de la Sociedad Entomológica Aragonesa 43:255–281

Flórez E, Mattoni CI (2007) Los escorpiones del género *Chactopsis*: ¿euscórpiidos o cháctidos? Darwiniana 45(Suppl):83–95

Francke OF, Teruel R, Santibáñez-López CE (2014) A new genus and a new species of scorpions (Scorpiones: Buthidae) from southeastern Mexico. The Journal of Arachnology 42:220–232

Francke OF (1978) Systematic revision of diplocentrid scorpions (Diplocentridae) from Circum-Caribbean lands. Texas Tech University Press, Lubbock

Francke OF (1982) Studies of the scorpion subfamilies Superstitioninae and Typhlochactinae, with description of a new genus (Scorpiones, Chactidae). Bull Assoc Mex Cave Stud 8:51–61

Francke OF, Boss J (1986) Chactidae (Scorpiones) from Trinidad and Tobago. J Arachnol 14(1):15–28

Francke OF, Soleglad ME (1981) The family Iuridae Thorell (Arachnida, Scorpiones). J Arachnol 9:233–258

Francke OF, Stockwell SA (1987) Scorpions (Arachnida) from Costa Rica. Texas Tech University Press, Lubbock

González-Santillán E, Prendini L (2013) Redefinition and generic revision of the North American vaejovid scorpion subfamily Syntropinae Kraepelin, 1905, with descriptions of six new genera. Bull Am Mus Nat Hist 382:1–71

González-Santillán E, Prendini L (2015) Phylogeny of the North American vaejovid scorpion subfamily Syntropinae Kraepelin, 1905, based on morphology, mitochondrial and nuclear DNA. Cladistics 31:341–405

González-Santillán E, González-Ruíz JM, Escobedo-Morales LA (2017) A new species of *Megacormus* (Scorpiones, Euscorpiidae) from an oak-pine forest in Guanajuato, México with an identification key to the species in the genus. Zootaxa 4299(2):221–237. https://doi.org/10.11646/zootaxa.4299.2.4

González-Sponga MA (1978) Escorpiofauna de la Región Oriental del Estado Bolívar, en Venezuela. Roto-Impresos C.A

González-Sponga MA (1984) Escorpiones de Venezuela. Cuadernos Lagoven

González-Sponga MA (1996) Guía para identificar escorpiones de Venezuela. Cuadernos Lagoven

Goodman AM, Prendini L, Esposito LA (2021a) Systematics of the arboreal Neotropical 'thorellii' clade of *Centruroides* bark scorpions (Buthidae) and the efficacy of mini-barcodes for museum specimens. Diversity 13(9):441. https://doi.org/10.3390/d13090441

Goodman AM, Prendini L, Francke OF, Esposito LA (2021b) Systematic revision of the arboreal Neotropical "Thorellii" clade of *Centruroides* Marx, 1890, bark scorpions (Buthidae C.L. Koch, 1837) with descriptions of six new species. Bull Am Mus Nat Hist 452(1):1–92. https://doi.org/10.1206/0003-0090.452.1.1

Haradon RM (1984a) New and redefined species belonging to the *Paruroctonus baergi* group (Scorpiones, Vaejovidae). J Arachnol 12:205–221

Haradon RM (1984b) New and redefined species belonging to the *Paruroctonus borregoensis* group (Scorpiones, Vaejovidae). J Arachnol 12:317–340

Haradon RM (1985) New groups and species belonging to the nominate subgenus *Paruroctonus* (Scorpiones, Vaejovidae). J Arachnol 13:19–42

Kovařík F (2009) Illustrated catalog of scorpions. Part I. Introductory remarks; keys to families and genera; subfamily Scorpioninae with keys to Heterometrus and Pandinus species. Clairon Production, Prague

Kraepelin K (1899) Scorpiones und Pedipalpi. In: Dahl F (ed) Das Tierreich. 8. Lieferung. R. Friedländer und Sohn, pp 1–265

Kraepelin K (1905) Die geographische Verbreitung der Skorpione. Zool Jahrb Abt Syst 22(3):321–364

Lira AFA, González-Santillán E (2025) The scorpions of the Estación de Biología Chamela, Jalisco, Mexico with the description of a new species of *Mesomexovis* (Scorpiones, Vaejovidae) and an identification key. ZooKeys 1243:241–267. https://doi.org/10.3897/zookeys.1243.125789

Lira AFA, Sarinho NMS, DeSouza AM, Albuquerque CMR (2014) Metasomal autotomy in *Ananteris mauryi* Lourenço, 1982 (Scorpiones: Buthidae). J Insect Behav 27:279–282. https://doi.org/10.1007/s10905-013 9431-3

Loria SF, Prendini L (2014) Homology of the lateral eyes of Scorpiones: a six-ocellus model. PLoS One 9(12):e112913. https://doi.org/10.1371/journal.pone.0112913

Lourenço WR (1981) Scorpions cavernicoles de l'Équateur: *Tityus demangei* n. sp. et *Ananteris ashmolei* n. sp. (Buthidae); *Troglotayosicus vachoni* n. gen., n. sp. (Chactidae), scorpion troglobie. Bull Mus Natl Hist Nat, 4e sér 3:635–662

Lourenço WR (1996) Faune de Madagascar. 87. Scorpiones (Chelicerata, Scorpiones). Muséum national d'Histoire naturelle

Lourenço WR (1998) Panbiogéographie, les distributions disjointes et le concept de famille relictuelle chez les scorpions. Biogeographica 74(3):133–144

Lourenço WR (2016) Scorpions from the Mitaraka Massif in French Guiana: description of one new genus and species (Scorpiones: Chactidae). C R Biol 339(3–4):141–146. https://doi.org/10.1016/j.crvi.2016.02.003

Lourenço WR (2018) The scorpions from the Mitaraka Massif in French Guiana (Scorpiones: Buthidae, Chactidae). Zoosystema 40(14):367–374. https://doi.org/10.5252/zoosystema2018v40a14

Lourenço WR, Qi J-X (2007) Additions à la faune des scorpions de l'Etat du Amapá, Brésil (Chelicerata, Scorpiones). Rev Suisse Zool 114(1):3–12. https://doi.org/10.5962/bhl.part.80384

Lourenço WR, Ythier E (2011) A new species of *Guyanochactas* Lourenço, 1998 (Scorpiones, Chactidae) from French Guiana. Bol Soc Entomol Aragonesa 48:203–206

Lowe G, Fet V (1998) Bothriuridae. In: Fet V, Sissom WD, Lowe G, Braunwalder ME (eds) Catalog of the scorpions of the world (1758–1998). New York Entomological Society, New York, pp 17–53

Mattoni CI, García-Hernández S, Botero-Trujillo R, Ochoa JA, Ojanguren-Affilastro AA, Pinto-da-Rocha R, Prendini L (2015) Scorpion sheds 'tail' to escape: consequences and implications of autotomy in scorpions (Buthidae: *Ananteris*). PLoS One 10(1):e0116639. https://doi.org/10.1371/journal.pone.0116639

Maury EA (1971) Essai d'une classification des sous-familles de scorpions Bothriuridae. In Proceedings of the 5th international congress of arachnology (pp. 29–36). Brno

Maury EA (1973) Essai d'une classification des sous familles de scorpions Bothriuridae. In Proceedings of the 5th International Congress of Arachnology, Brno. 1971pp 29–36

Mello-Leitão C (1945) Escorpiões sul-americanos. Arq Mus Nac 40:7–468

Mitchell RW (1968) *Typhlochactas*, a new genus of eyeless cave scorpion from Mexico (Scorpionida, Chactidae). Ann Spéléol 23(4):753–777

Monod L, Prendini L (2015) Evidence for Eurogondwana: the roles of dispersal, extinction and vicariance in the evolution and biogeography of Indo-Pacific Hormuridae (Scorpiones: Scorpionoidea). Cladistics 31(1):71–111. https://doi.org/10.1111/cla.12067

Monod L, Cauwet L, González-Santillán E, Huber S (2017) The male sexual apparatus in the order Scorpiones (Arachnida): a comparative study of functional morphology as a tool to define hypotheses of homology. Front Zool 14(1):51. https://doi.org/10.1186/s12983-017-0231-z

Ochoa JA, Prendini L (2010) The genus *Hadruroides* Pocock, 1893 (Scorpiones: Iuridae), in Peru: new records and descriptions of six new species. Am Mus Novit 3687:1–56

Ochoa JA, Rojas-Runjaic FJM (2019) Scorpions. In: Guerrero RR, Chapman AA (eds) Biodiversity of Pantepui: the pristine "lost world" of the Neotropical Guiana Highlands. Academic, London, pp 223–246

Ochoa JA, Rojas-Runjaic FJM, Pinto-da-Rocha R, Prendini L (2013) Systematic revision of the neotropical genus *Chactopsis* Kraepelin, 1912 (Chactoidea: Chactidae), with descriptions of two new genera and four new species. Bull Am Mus Nat Hist 378:1–121

Ochoa JA, Botero-Trujillo R, Prendini L (2010) On the troglomorphic scorpion Troglotayosicus humiculum (Scorpiones, Troglotayosicidae), with first description of the adults. American Museum Novitates 3691:1–19

Ojanguren-Affilastro AA (2005) Estudio monográfico de los escorpiones de la República Argentina. Rev Ibér Aracnol 11:75–241

Ojanguren-Affilastro AA, Adilardi RS, Mattoni CI, Ramírez MJ, Ceccarelli FS (2017) Dated phylogenetic studies of the southernmost American buthids (Scorpiones; Buthidae). Mol Phylogenet Evol 110:39–49. https://doi.org/10.1016/j.ympev.2017.03.009

Pavlovsky EN (1913) Scorpiotomische Mitteilungen. I, Ein Beitrag zur Morphologie der Giftdrusen der Skorpione. Z Wiss Zool 105:157–177

Pocock RI (1893) Notes on the classification of scorpions, followed by some observations on synonymy, with descriptions of new genera and species. Ann Mag Nat Hist 6(12):303–330

Pocock RI (1894) Scorpions and their geographical distribution. Nat Sci 4(24):353–364

Pocock RI (1900) The fauna of British India, including Ceylon and Burma. Arachnida. Taylor and Francis, London

Polis GA (1990) The biology of scorpions. Stanford University Press, Stanford

Prendini L (2000) Phylogeny and classification of the superfamily Scorpionoidea Latreille 1802 (Chelicerata, Scorpiones): an exemplar approach. Cladistics 16(1):1–78. https://doi.org/10.1111/j.1096-0031.2000.tb00348.x

Prendini L, Wheeler WC (2005) Scorpion higher phylogeny and classification, taxonomic anarchy, and standards for peer review in online publishing. Cladistics 21(5):446–494. https://doi.org/10.1111/j.1096-0031.2005.00073.x

Rojas-Runjaic FJM, de Sousa L (2009) Catálogo de los escorpiones de Venezuela (Arachnida: Scorpiones). Boletín de la Sociedad Entomológica Aragonesa 40:281–307

Santibáñez-López CE, González-Santillán E, Monod L, Sharma PP (2019) Phylogenomics facilitates stable scorpion systematics: reassessing the relationships of Vaejovidae and a new higher-level classification of Scorpiones (Arachnida). Mol Phylogenet Evol 135:22–30. https://doi.org/10.1016/j.ympev.2019.02.021

Santibáñez-López CE, Ojanguren-Affilastro AA, Sharma PP (2020) Another one bites the dust: taxonomic sampling of a key genus in phylogenomic datasets reveals more non-monophyletic groups in traditional scorpion classification. Invertebr Syst 34(2):133–143. https://doi.org/10.1071/IS19033

Santibáñez-López CE, Aharon S, Ballesteros JA, Gainett G, Baker CM, González-Santillán E, Harvey MS, Hassan MK, Abu Almaaty AH, Aldeyarbi SM, Monod L, Ojanguren-Affilastro AA, Pinto-da-Rocha R, Zvik Y, Gavish-Regev E, Sharma PP (2022) Phylogenomics of scorpions reveal contemporaneous diversification of scorpion mammalian predators and mammal-active sodium channel toxins. Syst Biol 71(6):1281–1289. https://doi.org/10.1093/sysbio/syac021

Santibáñez-López CE, Ojanguren-Affilastro AA, Graham MR, Sharma PP (2023) Congruence between ultraconserved element-based matrices and phylotranscriptomic datasets in the scorpion tree of life. Cladistics 39(6):533–547. https://doi.org/10.1111/cla.12551

Sharma PP, Fernández R, Esposito LA, González-Santillán E, Monod L (2015) Phylogenomic resolution of scorpions reveals multilevel discordance with morphological phylogenetic signal. Proc R Soc B Biol Sci 282(1804):20142953. https://doi.org/10.1098/rspb.2014.2953

Simon E (1879) 3e Ordre. Scorpiones. In: Les Arachnides de France. Tome septième. Roret, Paris, pp 79–115

Sissom WD (1990) Systematics, biogeography, and paleontology. In: Polis GA (ed) The biology of scorpions. Stanford University Press, Stanford, pp 64–160

Sissom WD (2000) Family Chactidae Pocock, 1893. In: Fet V, Sissom WD, Lowe G, Braunwalder ME (eds) Catalog of the scorpions of the world (1758–1998). New York Entomological Society, New York, pp 287–322

Soleglad ME, Fet V (2003a) High-level systematics and phylogeny of the extant scorpions (Scorpiones: Orthosterni). Euscorpius 2003a(11):1–56

Soleglad ME, Fet V (2003b) The scorpion sternum: structure and phylogeny (Scorpiones: Orthosterni). Euscorpius 2003b(5):1–34. https://doi.org/10.18590/euscorpius.2003.vol2003.iss5.1

Soleglad ME, Fet V (2005) The genus Auyantepuia González-Sponga, 1978 is confirmed as a synonym of Broteochactas Pocock, 1893 (Scorpiones: Chactidae). Euscorpius 29:1–14

Soleglad ME, Fet V (2008) Contributions to scorpion systematics. III. Subfamilies Smeringurinae and Syntropinae (Scorpiones: Vaejovidae). Euscorpius 71:1–115

Soleglad ME, Sissom WD (2001) Phylogeny of the family Euscorpiidae Laurie, 1896: a major revision. In: Fet V, Selden PA (eds) Scorpions 2001. In memoriam Gary A. Polis. British Arachnological Society, Beeches, Bucks, pp 25–111

Stahnke HL (1940) The scorpions of Arizona. Iowa State Coll J Sci 15(1):101–103

Stahnke HL (1974) Revision and keys to the higher categories of Vejovidae. J Arachnol 1(2):107–141

Stockwell SA (1989) Revision of the phylogeny and higher classification of scorpions (Chelicerata) [Unpublished doctoral dissertation]. University of California, Berkeley

Stockwell SA (1992) Systematic observations on North American Scorpionida with a checklist of the families and genera. J Med Entomol 29(3):407–422. https://doi.org/10.1093/jmedent/29.3.407
Štundlová J, Šťáhlavský F, Opatova V, Stundl J, Kovařík F, Dolejš P, Šmíd J (2022) Molecular data do not support the traditional morphology-based groupings in the scorpion family Buthidae (Arachnida: Scorpiones). Mol Phylogenet Evol 173:107511. https://doi.org/10.1016/j.ympev.2022.107511
Teruel R, Kovařík F (2012) Scorpions of Cuba. Clairon Production, Prague
Vachon M (1973) Étude des caractères utilisés pour classer les familles et les genres de scorpions (Arachnides). 1. La trichobothriotaxie en arachnologie. Sigles trichobothriaux et types de trichobothriotaxie chez les scorpions. Bull Mus Natl Hist Nat (3e série) 140:857–958
Ythier E (2015) A new species of *Auyantepuia* González-Sponga, 1978 (Scorpiones, Chactidae) from French Guiana. ZooKeys 539:97–109. https://doi.org/10.3897/zookeys.539.6664
Ythier E (2018a) A synopsis of the scorpion fauna of French Guiana, with description of four new species. ZooKeys 764:27–90. https://doi.org/10.3897/zookeys.764.25108
Ythier E (2018b) A new species of *Auyantepuia* González-Sponga, 1978 (Scorpiones, Chactidae) from Brazil. Arachnida Riv Ital Aracnol 4(20):13–22
Ythier E (2024) A new high-altitude scorpion species of the genus *Ananteris* Thorell, 1891 (Scorpiones: Ananteridae) from the Pico da Neblina, Brazil. Faunitaxys 12(19):1–9

Chapter 6
Advances in the Phylogenomic Relationships of Neotropical Scorpions

Carlos E. Santibáñez-López and Prashant P. Sharma

Abstract The field of scorpion systematics has achieved a relative stability in its higher-level relationships, largely due to the advent of phylogenomics. However, significant gaps persist, particularly at lower taxonomic levels where relationships are often still based solely on morphology. In this chapter, we revisit the historical context of scorpion systematics, synthesizing foundational contributions and defining the outstanding challenges, especially the systematic placement of potentially pivotal unsampled Neotropical lineages. In addition, to investigate the utility of classic morphological characters in a modern phylogenomic framework, we performed an ancestral state reconstruction for trichobothrial patterns—a character set originally established by Max Vachon over 50 years ago—as well as new character systems (hemispermatophore types; lateral ocelli architecture) using the most recent and comprehensive phylogenomic topology available. Since our work focuses on the diverse scorpions of the Neotropical realm, we define this group by conducting a biogeographical area reconstruction, contrasting the competing models presented by Holt et al. (Science 339(6115):74–78, 2013) and Morrone (Zootaxa 3782(1):1–110, 2014) to provide a robust biogeographical context. Lastly, we articulate the most pressing challenges in Neotropical scorpion systematics, emphasizing the critical need for targeted field efforts and the sampling of key taxa and regions. We emphasize that surmounting challenges in scorpion phylogeny requires collaborative efforts and exchanges of expertise to fill long-standing knowledge gaps.

Keywords Systematics · Phylogeny · Ancestral state reconstruction · Biogeography

C. E. Santibáñez-López (✉)
Department of Biology, Western Connecticut State University, Danbury, CT, USA
e-mail: santibanezlopezc@wcsu.edu

P. P. Sharma
Department of Integrative Biology, University of Wisconsin-Madison, Madison, WI, USA
e-mail: prashant.sharma@wisc.edu

A. F. de Araujo Lira, E. González-Santillán (eds.), *The Biology of Neotropical Scorpions*, https://doi.org/10.1007/978-3-032-23194-9_6

6.1 Introduction

Scorpions, an iconic group of chelicerates, constitute an ancient lineage with more than 2900 described species distributed worldwide (Fig. 6.1). Beyond their renowned venom potency, they offer a captivating window into the evolutionary history of arthropods in general. Traditionally, the systematics of the order Scorpiones C.L. Koch, 1837 relied on morphological cladistic analyses, which often yielded

Fig. 6.1 Scorpion representatives from the genera distributed in the Neotropical region. (**a**) *Tityus smithi* (female); (**b**) *Tityus obscurus* (male); (**c**) *Brotheas granulatus* (male); (**d**) *Centruroides limbatus* (male); (**e**) *Hoffmannihadrurus aztecus* (male); and (**f**) *Vaejovis zapoteca* (female). (Photographs by C. Santibáñez-López (**a**, **d**–**f**), and Gonzalo Giribet (**b**, **c**))

conflicting interpretations and dubious relationships (Stockwell 1989; Soleglad and Fet 2003; Fet and Soleglad 2005; Prendini and Wheeler 2005).

Early cladistic studies (the unpublished work by Stockwell 1989) recognized four superfamilies (Buthoidea Simon, 1879, Chactoidea Pocock, 1893, Scorpionoidea Peters, 1862, and Vaejovoidea Thorell, 1876) and 14 families (see Fig. 6.2a, Table 6.1). Subsequent revisions, such as the one by Sissom (1990) in the seminal book *Biology of Scorpions*, rejected the superfamilial classification and

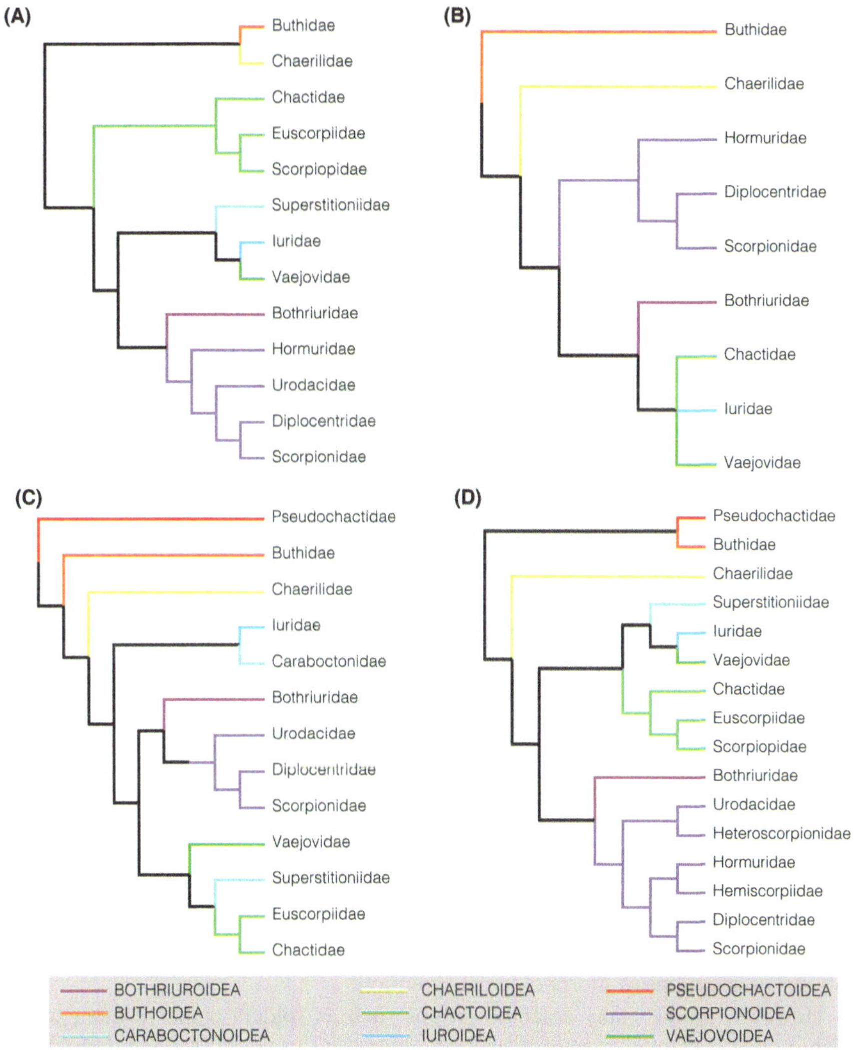

Fig. 6.2 Early scorpion morphological phylogenetic hypotheses. (**a**) Stockwell (1989). (**b**) Sissom (1990). (**c**) Soleglad and Fet (2003). (**d**) Coddington et al. (2004)

Table 6.1 High-level classification of scorpions proposed by Stockwell (1989)

Superfamily	Family	Subfamily
Buthoidea	Buthidae	
	Chaerilidae	
Chactoidea	Chactidae	
	Euscorpiidae	
	Scorpiopsidae	
Vaejovoidea	Vaejovidae	
	Iuridae	Iurinae
		Caraboctoniae
		Hadrurinae
	Superstitioniidae	Superstitioninae
		Troglotayosinae
		Belisariinae
		Typhlochactinae
	Vaejovidae	Vaejovinae
		Syntropinae
Scorpionoidea	Bothriuridae	Lisposominae
		Bothriurinae
	Ischnuridae	Ischnurinae
		Hemiscorpiinae
		Heteroscorpioninae
	Urodacidae	
	Scorpionidae	
	Diplocentridae	Nebinae
		Diplocentrinae

only considered nine scorpion families as valid (Fig. 6.2b). A decade later, Prendini (2000) revised the internal classification of superfamily Scorpionoidea. Soleglad and Fet (2003) proposed a more complex classification with four parvorders and six superfamilies (Fig. 6.2c; Table 6.2). Coddington et al. (2004) presented a different topology but it was not based on any formal cladistic analysis (Fig. 6.2d). Next, Prendini and Wheeler (2005) challenged this classification based on Soleglad and Fet's (2003) character interpretation and re-classified scorpions into 18 families though no phylogeny was presented (Table 6.3). All these classifications were based mainly on interpretations of morphological traits, such as trichobothrial patterns, sternum shape, and the anatomy of the hemispermatophore (the male reproductive structure in scorpions, Sissom 1990; Soleglad and Fet 2003; Prendini and Wheeler 2005; Coddington et al. 2004; Monod et al. 2017). The advent of molecular phylogenetics in the past 30 years offered a new source of data to test many of these traditional conceptions of morphological evolution.

However, unlike many arachnid groups such as spiders or harvestmen whose higher-level systematics were intensively addressed in the 2000s and early 2010s with Sanger sequencing, scorpion systematics initially suffered from a paucity of molecular datasets that tested relationships independent of reference to

Table 6.2 High-level classification of scorpions proposed by Soleglad and Fet (2003). Comments provided in this table clarify the systematic position of taxa classified in new high-level taxonomic groups (see Table 6.5)

Parvorder	Superfamily	Family	Subfamily	Comments
Buthida	Buthoidea	Buthidae		
Chaerilida	Chaeriloidea	Chaeriliidae		
Iurida	Chactoidea	Chactidae	Chactinae	
			Brotheinae	*Belisarius* is included here
			Uroctoninae	*Anuroctonus* is included here
		Euscorpiidae	Euscorpiinae	
			Megacorminae	
			Scorpiopinae	
		Superstitioniidae	Superstitioninae	*Troglotayosicus* is included here
			Typhlochactinae	
		Vaejovidae		
	Iuroidea	Iuridae		
		Caraboctonidae	Caraboctoninae	
			Hadrurinae	
	Scorpionoidea	Bothriuridae		
		Liochelidae	Liochelinae	
			Hemiscorpiinae	
		Scorpionidae	Scorpioninae	
			Diplocentrinae	
		Urodacidae	Urodacinae	
			Heteroscorpioninae	
Pseudochactida	Pseudochactoidea	Pseudochactidae		

morphological characters. Early molecular works prioritized the higher-level relationships of significant groups, such as Buthidae (Fet et al. 2003). Intensive taxonomic sampling combined with Sanger data is limited to a few studies, such as (a) within New World Bothriuridae Simon, 1880 (e.g., genus *Brachistosternus* Pocock, 1893, Ojanguren-Affilastro et al. 2016); (b) Buthidae C.L. Koch, 1837 (e.g., Fet et al. 2003, subfamily Centruroidinae Kraus, 1955, Esposito et al. 2018; family-level Štundlová et al. 2022); (c) Diplocentridae Karsch, 1880 (e.g., genus *Diplocentrus* Peters, 1861, Santibáñez-López et al. 2014a); (d) Iuridae Thorell, 1876 (e.g., genus *Iurus* Thorell, 1876, Parmakelis et al. 2022); (e) Pseudochactidae Gromov, 1998 (Loria et al. 2022); (f) Scorpionidae Latreille, 1802 (Prendini et al. 2003; subfamily Heterometrinae Simon, 1879, Loria and Prendini 2020, 2021); and (g) Vaejovidae Thorell, 1876 (e.g., the subfamily Syntropinae Kraepelin, 1905, González-Santillán and Prendini 2015). Other Sanger datasets have been presented at international meetings, but never published (Prendini and Wheeler 2004; Mattoni et al. 2010). Surprisingly, a subset of recent phylogenetic efforts continues to

Table 6.3 High-level classification of scorpions proposed by Prendini and Wheeler (2005)

Family	Subfamily
Bothriuridae	
Buthidae	
Chactidae	
Chaerilidae	
Diplocentridae	Diplocentrinae
	Nebinae
Euscorpiidae	
Hemiscorpiidae	
Heteroscorpionidae	
Iuridae	Caraboctoninae
	Hadrurinae
	Iurinae
Liochelidae	
Microcharmidae	
Pseudochactidae	
Scorpionidae	
Scorpiopidae	
Superstitioniidae	Superstitioniinae
	Typhlochactinae
Troglotayosicidae	Belisariinae
	Troglotayosicinae
Urodacidae	
Vaejovidae	

prioritize phylogenetic inference based only on morphological data (e.g., Moreno-González et al. 2024).

The continued reliance on morphological data in scorpion systematics is surprising, given that molecular data have overturned an array of putative relationships that were once strongly held by arthropod biologists, but are now understood to be the result of evolutionary convergence. Examples of erstwhile morphological hypotheses that were overturned in this way by molecular data include the subdivision of the animal tree of life as a function of coelomic condition (replaced by Ecdyzosoa, Aguinaldo et al. 1997); Articulata, or annelids as the sister group to arthropods (replaced by Panarthropoda, Nielsen et al. 1997); Atelocerata, or the sister group relationship of insects and myriapods (replaced by Pancrustacea, Zrzavý et al. 1998); and Isoptera as a distinct order of insects (termites were later shown by molecular data be derived eusocial cockroaches, Inward et al. 2007). Throughout the tree of life, nodes spanning a range of phylogenetic depths were questioned and falsified when molecular data entered the arena and became a widely available, scalable, and practical tool of inference.

Scorpions proved no exception, save for the fact that the first higher-level scorpion phylogeny to sample all superfamilies leapfrogged over Sanger data. This phylogenomic analysis, based on RNAseq data for 25 exemplars, revealed that several

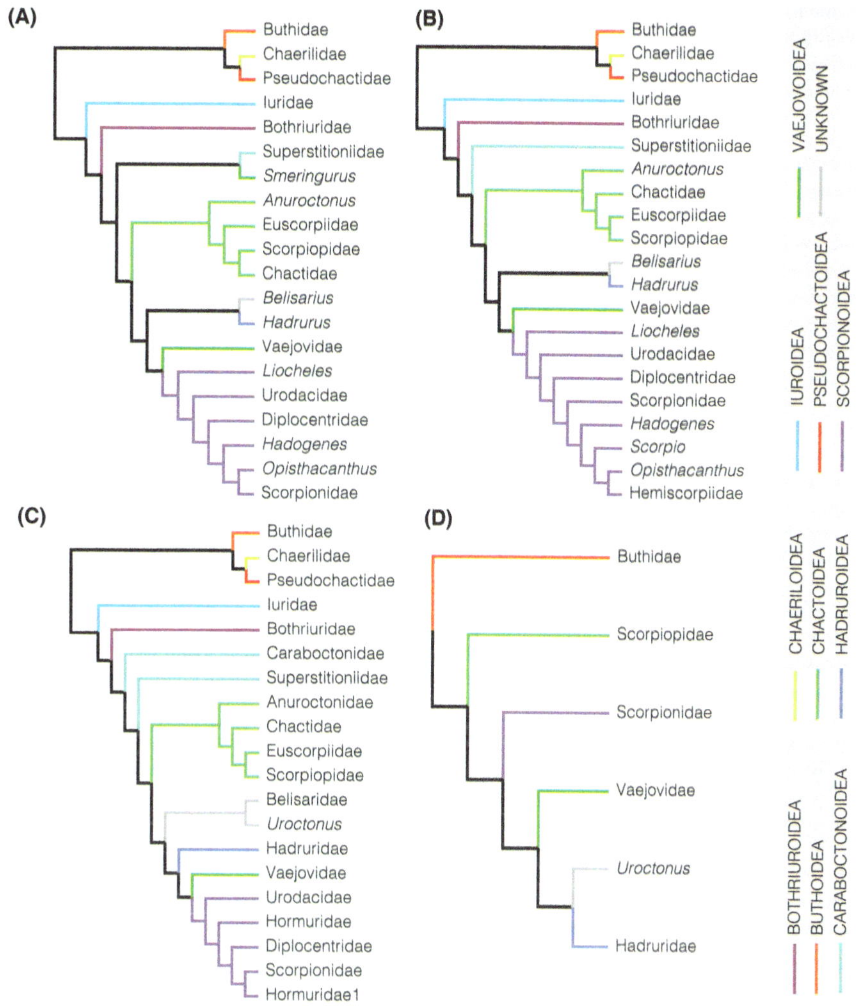

Fig. 6.3 Scorpion phylogenomic analyses. (**a**) Sharma et al. (2015). (**b**) Santibáñez-López et al. (2018). (**c**) Santibáñez-López et al. (2023). (**d**) Graham et al. (2024). Topologies (**a**)–(**c**) were recovered from the analyses of RNAseq. Topology (**d**) was recovered from the analysis of mitogenomes

key morphological traits were in discordance with some scorpion groups (Sharma et al. 2015; Fig. 6.3a). For example, Chaerilidae Pocock, 1893, the putative sister group to the remaining Iurida, was recovered as closely related to Buthidae and Pseudochactidae, a relationship that was suggested earlier but not well supported (Prendini et al. 2006). Pseudochactidae, a putative "living fossil" long considered the sister group to the remaining scorpions, were shown to be nested within the tree near buthids as well. Sharma et al. (2015) then amended scorpion classification and reinstated the parvorder and superfamilial levels (Table 6.4).

Table 6.4 High-level classification of scorpions proposed by Sharma et al. (2015)

Parvorder	Superfamily	Family
Buthida	Buthoidea	Buthidae
	Chaeriloidea	Chaerilidae
	Pseudochactoidea	Pseudochactidae
Iurida	Bothriuroidea	Bothriuridae
	Chactoidea[a]	Caraboctonidae
		Chactidae[a]
		Euscorpiidae
		Scorpiopidae
		Superstitioniidae
		Troglotayosicidae
		Typhlochactidae?
		Vaejovidae[a]
	Iuroidea	Iuridae
	Scorpionoidea	Diplocentridae
		Hemiscorpiidae?
		Heteroscorpionidae?
		Hormuridae[a]
		Scorpionidae[a]
		Urodacidae

[a]Indicates superfamilies and families with dubious monophyly. ? Indicates taxa with unknown phylogenetic position based on molecular data

In the 9 years since that work, phylogenomic analyses have revealed more non-monophyletic groups, suggesting broader discrepancies with the traditional morphological classification mentioned above (Graham et al. 2024; Santibáñez-López et al. 2018, 2019, 2020, 2022, 2023; Fig. 6.3b–d). Based on these analyses, the current scorpion classification (Table 6.5) comprises 10 superfamilies in two parvorders, with some having better resolution (e.g., Buthoidea and Vaejovoidea), and others still with limited sampling (e.g., Chactoidea, Scorpionoidea). The next challenges in scorpion systematics then are to uncover the phylogenetic position of the missing taxa (i.e., Akravidae Levy, 2007, Heteroscorpionidae Kraepelin, 1905, Rugodentidae Bastawade et al. 2005, Troglotayosicidae Lourenco, 1998, and Typhlochactidae Mitchell, 1971), increase exemplars from families Caraboctonidae Kraepelin, 1905, Chactidae Pocock, 1893, Hormuridae Laurie, 1896, and Scorpiopidae Kraepelin, 1905 to reassess the monophyly of current superfamilies, their composition and phylogenetic relationships, and revisit Buthidae systematics (currently, this family is rendered paraphyletic due to the position of the newly raised to family level Ananteridae; Ythier 2024).

Morphological evidence suggests a close relationship between the Malagasy family Heteroscorpionidae and genus *Urodacus* Peters, 1861, within the superfamily Scorpionoidea (Prendini 2000). Similarly, *Troglotayosicus* Lourenco, 1981 and *Belisarius* Simon, 1879 have been proposed as closely related, though their exact phylogenetic relationship remains uncertain (Prendini and Wheeler 2005). While

Table 6.5 Revised higher level classification of extant scorpions

Order Scorpiones Koch 1837
Suborder Neoscorpionina Thorell and Lindström 1885
Infraorder Orthosterni Pocock 1911
Parvorder Buthida Soleglad and Fet 2003
Superfamily Buthoidea Koch 1837
Family Buthidae Koch 1837
Superfamily Chaeriloidea Pocock 1893
Family Chaerilidae Pocock 1893
Superfamily Pseudochactoidea Gromov 1998
Family Pseudochactidae Gromov 1998
Parvorder Iurida Soleglad and Fet 2003
Superfamily Bothriuroidea Simon 1880
Family Bothriuroidea Simon 1880
Superfamily Caraboctonoidea Kraepelin 1905
Family Caraboctonidae Kraepelin 1905
Family Superstitioniidae Stanhke 1940
Superfamily Chactoidea Pocock 1893
Family Anuroctonidae Santibáñez-López et al. 2023
Family *Chactidae Pocock 1893
Family Euscorpiidae Laurie 1896
Family Scorpiopidae Kraepelin 1905
Superfamily Iuroidea Thorell 1876
Family Iuridae Thorell 1876
Superfamily Hadruroidea Stahnke 1974
Family Hadruridae Stahnke 1974
Superfamily Scorpionoidea Latreille 1802
Family Diplocentridae Karsch 1880
Family Hemiscorpiidae Pocock 1893
Family *Hormuridae Laurie 1896
Family *Scorpionidae Latreille 1802
Family Urodacidae Pocock 1893
Superfamily Vaejovoidea Thorell 1876
Family Vaejovidae Thorell 1876
Incertae sedis
Family? *Belisariidae Lourenço 1998
Family? Heteroscorpionidae Kraepelin 1905
Family? Troglotayosicidae Loureno 1998
Family? Typhlochactidae Mitchell 1971
Family? Rugodentidae Bastawade et al. 2005

Modified from Santibáñez-López et al. (2023)
Taxa of questionable monophyly are indicated with asterisks. Taxa with unknown phylogenetic position based on phylogenomic data are indicated with question marks

Tropea and Onnis (2020) transferred *Belisarius* to its own family, a robust phylogenetic framework is needed to definitively resolve the placement of both *Troglotayosicus* and *Belisarius*. Further, a better understanding of the position of these two taxa may shed light on the evolutionary history of *Uroctonus* Thorell, 1876 (Fig. 6.4).

The family Akravidae is currently considered extinct (Fet et al. 2011). However, morphological similarities between *Akrav israchanani* Levy, 2007 and two species

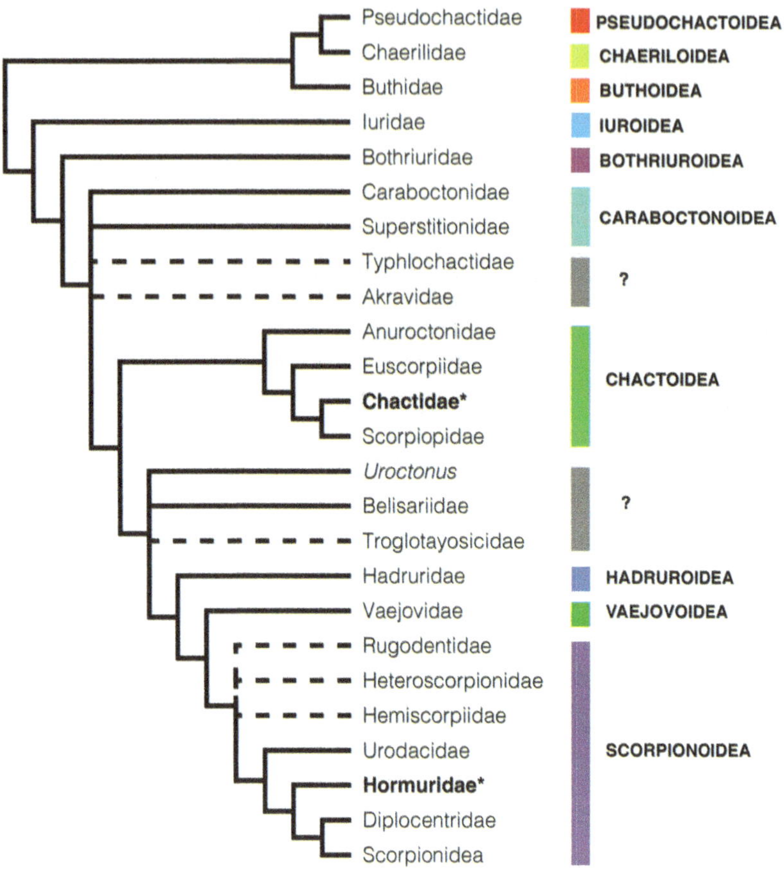

Fig. 6.4 Putative phylogenetic relationships of key scorpion families. Dotted lines indicate hypothesized positions based on current morphological data. Bold text denotes taxa with questionable monophyly

of *Typhlochactas* Mitchell, 1971 (*T. mitchelli* Sissom, 1988, and *T. sylvestris* Mitchell and Peck, 1977) have led to the inference of a close relationship between these taxa. While these similarities may be attributed to convergent evolution due to their troglomorphic habitat, a phylogenomic analysis is necessary to support or reject this hypothesis. Finally, the placement of family Typhlochactidae as sister group to Superstitioniidae Stahnke, 1940 (e.g., Prendini et al. 2010; Santibáñez-López et al. 2014b) is putatively supported by morphological evidence (Fig. 6.4). However, the propensity of evolution in caves to drive similar suites of phenotypic traits underscores the imperative to test this hypothesis with a dataset that is independent of anatomical characters.

6.2 The Role of Scorpion Morphology in the Era of Genomics

Since the publication of the first phylogenomic analysis, our understanding of scorpion relationships has been revolutionized. Nevertheless, a reappraisal of traditional morphological characters has been largely overlooked. As mentioned before, the use of the shape of the sternum, trichobothria patterns, and shape of hemispermatophore have been considered the most useful morphological characters to delimit the higher-level scorpion categories. To explore this potential role, we conducted ancestral state reconstruction of trichobothrial patterns (Vachon 1973), the lateral ocelli patterns (Loria and Prendini 2014), and the folds in the hemispermatophore (Monod et al. 2017) scoring these characters on the taxa and topology recovered by Santibáñez-López et al. (2023), using the function ace form the package *ape* (Paradis and Schliep 2019) and the Maximum Likelihood method [*ace(matrix,tree, method = "ML", type = "discrete")*].

6.2.1 Pedipalp Trichobothrial Patterns (Sensu Vachon 1973)

One of the main challenges in scorpion trichobothria is their homology hypotheses. These modified setae located on three segments of the scorpion pedipalp are traditionally used to diagnose scorpion superfamilies and families (e.g., Vachon 1973; Prendini 2000). While these trichobothria are useful to differentiate parvorders (types A, B, and D for parvorder Buthida, and type C for parvorder Iurida), and the three superfamilies within Buthida (type A for Buthidae, type B for Chaerilidae, and type D for Pseudochactidae), these patterns are highly variable between families within parvorder Iurida, without even considering the neobothriotaxic patterns in these scorpions (e.g., additional or absent trichobothria). The ancestral reconstruction of trichobothrial patterns supports the origin of each type described for these taxa. While the ancestral state of extant scorpions (infraorder Orthosterni Pocock, 1911) is equivocal, exhibiting a 50% probability of a type C pattern (Fig. 6.5a), our analysis reveals distinct patterns within the constituent parvorders. Buthida's ancestral state is unresolved, whereas Iurida's most recent common ancestor (MRCA) is strongly inferred to possess a type C pattern. Within Buthida, the inferred ancestral state for each superfamily is congruent with the established patterns for their respective taxa: Chaeriloidea (type B), Pseudochactoidea (type D), and Buthoidea (type A). To further elucidate the evolutionary history of this character system, future studies should prioritize formal analysis of neobothriotaxic patterns. We also advocate for the implementation of a placeholder methodology to generate specific homology statements for individual trichobothria (e.g., Prendini and Wheeler 2005; Prendini et al. 2010), moving beyond the current reliance on broader trichobothrial pattern analyses (e.g., Soleglad and Fet 2001).

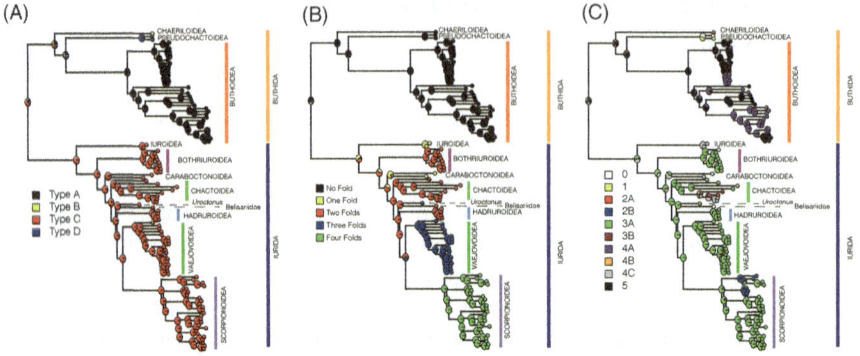

Fig. 6.5 Ancestral state reconstruction of three of the three relevant morphological systems used in scorpion systematics. (**a**) Pedipalp trichobothrial patterns (sensu Vachon 1973). (b) Hemispermatophore fold patterns (sensu Monod et al. 2017). (**c**) Lateral ocelli types (sensu Loria and Prendini 2014)

6.2.2 Hemispermatophore Fold Patterns (Sensu Monod et al. 2017)

Genital morphology is a valuable character in traditional cladistic analyses of various animal taxa, including chelicerates. In scorpions, sperm transfer occurs via spermatophores, external, partially sclerotized structures composed of two hemispermatophores. Stockwell's (1989) dissertation laid the groundwork for utilizing hemispermatophore morphology in scorpion systematics. More recently, Monod et al. (2017) established homology statements for these structures, facilitating the diagnosis of parvorders and some superfamilies. Here, we expanded the ancestral state reconstruction of the hemispermatophore folding pattern, as defined, and initially analyzed by Monod et al. (2017). Our likelihood analyses (support >60%) suggest the MRCA of extant scorpions likely possessed a hemispermatophore without folds. Notably, the MRCA of parvorder Buthida unequivocally exhibited this no-fold state (however, the hemispermatophore of the two pseudochactid species used here needs to be verified). Iurida displays the greatest diversity of folding patterns, with its MRCA inferred to have possessed a single fold, a trait retained in the MRCA of superfamily Iuroidea (Fig. 6.5b). The ancestral state at the divergence of Bothriuroidea and other superfamilies remains uncertain, although the two-fold pattern received more support. The MRCA of Caraboctonoidea showed a two-fold pattern, while the MRCA of Chactoidea, Hadruroidea, and the clade comprising *Uroctonus* and *Belisarius* (with *Uroctonus* showing three folds) exhibited a single fold. Finally, the divergence of Vaejovoidea and Scorpionoidea likely involved an MRCA with a two-fold pattern (support >60%). Subsequently, the MRCA of Vaejovoidea evolved a three-fold pattern, and Scorpionoidea, a four-fold pattern. Based on these analyses, the two-fold pattern dominates the diversity of scorpion hemispermatophores. However, establishing the phylogenetic position of the missing taxa (i.e., Troglotayosicidae, Belisariidae, Typhlochactidae) will clarify the ancestral state of this folding pattern at the MRCA of Iurida.

6.2.3 Lateral Ocelli Patterns (Sensu Loria and Prendini 2014)

The number and type of lateral ocelli in scorpions, as explored by Loria and Prendini (2014), represent another character system recently explored. Scorpions possess both median and lateral eyes, each composed of simple ocelli. While a pair of median ocelli is nearly constant (with the exception of troglobitic species), the number of lateral ocelli pairs varies from zero to five, exhibiting limited intraspecific variation. Loria and Prendini (2014) proposed six-ocellus models for extant scorpions and discussed the utility of this character set; however, a formal ancestral state reconstruction of these patterns remained absent. Utilizing our recent phylogenomic topology (Santibáñez-López et al. 2022), we attempted to reconstruct the evolutionary history of these characters, revealing that ocelli patterns generally align with higher-level scorpion systematics (parvorder level). Due to the absence of key taxa from Loria and Prendini (2014) in our phylogenomic topology, a comprehensive assessment of phylogenetic signal was not feasible. While ancestral state reconstruction did not support a specific lateral ocelli pattern for scorpions or parvorder Buthida, it recovered Type 3A as the ancestral state for the MRCA of Iurida, with strong support. This pattern was recovered as ancestral across most Iurida superfamilies (Bothriuroidea, Chactoidea, Hadruroidea, Vaejovoidea, and Scorpionoidea), but not Iuroidea, Caraboctonoidea, or certain genera within Scorpionidae, Hormuridae, and Urodacidae (Fig. 6.5c). Finally, our analysis suggests Type 5 as the ancestral state for the MRCA of Buthidae.

6.2.4 Other Character Systems

The use of quantitative characters in scorpion studies is restricted to few studies, with some of them showing these systems retain phylogenetic signal (Santibáñez-López et al. 2017), or are less informative than their discrete counterparts (González-Santillán and Prendini 2015). Here, we emphasize the use of these characters in future scorpion phylogenetics to explore their potential as diagnostic characters (i.e., synapomorphies). Additionally, exploring the geometric morphology and functional data of scorpion morphology has the potential to identify non-traditional morphological characters to support monophyletic groups (e.g., van der Meijden et al. 2012, 2013; Coelho et al. 2017; Koch et al. 2017; Santibáñez-López et al. 2017; Foerster 2025). We recommend focusing the study of these character systems expanding the taxonomic sampling to cover taxa with known atypical patterns or inhabiting extreme habitats (i.e., all troglobitic scorpions).

In contrast to morphology, molecular synapomorphies broadly support the two parvorders and most of the superfamilies currently recognized (i.e., Santibáñez-López et al. 2018). For example, calcins, a scorpion peptide with affinity to ryanodine receptors in mammals' cells, have been reported only from scorpions from the parvorder Iurida (summarized in Xiao et al. 2016), with molecular and comparative evolutionary analyses supporting their use as a putative molecular synapomorphy for this clade of scorpions (Santibáñez-López et al. 2018). Similarly, the presence of a toxin affecting potassium channels in mammals (lambda Potassium Toxins, KTx) found in the venom of scorpions from parvorder Buthida has been suggested as a putative molecular synapomorphy for the parvorder Buthida (Santibáñez-López et al. 2018).

6.3 Biogeographic Origin of Scorpion Families and Neotropical Clades

Scorpions are distributed in almost all major landmasses, except Antarctica and circumpolar regions. Assessments of scorpion richness are available for several targeted regions across the globe (e.g., Lourenço 2002; Ochoa 2005; Ojanguren-Affilastro 2005; Prendini 2004, 2005; Prendini and Bird 2008; Cao et al. 2014; Santibáñez-López et al. 2015). Few attempts have been made to study scorpion faunal composition at the level of whole biogeographic regions (e.g., Lourenço 2003; Sharma et al. 2018; Ojanguren-Affilastro et al. 2023), with most works focusing at the phylogeographic scale of scorpions mostly distributed in the Nearctic (examples listed in Table 6.6).

The Neotropical region originally comprised South America, Central America, and some areas in central Mexico (Wallace 1876). Holt et al. (2013) updated the biogeographic realms and defined 11 zoogeographic regions. Under their definition, the Neotropical realm was restricted only to South America, setting the Panamanian realm covering Central America (Panama) up to the transition volcanic area in Mexico, including the Caribbean (Fig. 6.6a). Subsequently, using cladistic biogeography, Morrone (2014) then redefined the biogeographical regionalization of the Neotropical region, and redelimited this area comprising South America, Central America reaching up to the transition zone in Central Mexico, and the Antilles (Fig. 6.6b). The discrepancy between Holt et al.'s (2013) and Morrone's (2014) delimitation of this region is important, as measures of faunal diversity change dramatically under these two different scenarios. Under Holt et al.'s (2013) regionalization, scorpion diversity comprises 739 species in 44 genera and seven families. However, under Morrone's (2014) hypothesis, the scorpion diversity increases to 11 families, 69 genera, and 1114 species (Fig. 6.6c, Table 6.7).

Table 6.6 Examples of biogeographic and/or phylogeographic studies on scorpions. High-level classification follows Table 6.5

Parvorder	Superfamily	Family	Continent	References
Buthida	Buthoidea	Buthidae	North America	Yamashita and Rhoads (2013)
		Buthidae	South America	Ojanguren-Affilastro et al. (2017) and Esposito and Prendini (2019)
		Buthidae	Africa, Asia	Gantenbein and Largiadèr (2003), Othmen et al. (2009), Graham et al. (2012), Graham et al. (2019), Cain et al. (2021) and Shi et al. (2023)
	Chaeriloidea	Chaerilidae	None	
	Pseudochactoidea	Pseudochactidae	Asia	Loria et al. (2022)
Iurida	Bothriuroidea	Bothriuridae	South America	Mattoni et al. (2012) and Ceccarelli et al. (2016, 2017)
		Bothriuridae	Australia, South America	Sharma et al. (2018) and Ojanguren-Affilastro et al. (2023)
	Caraboctonoidea	Caraboctonidae	None	
		Superstitioniidae	None	
	Chactoidea	Anuroctonidae	North America	Graham et al. (2014)
		Chactidae	North America	Graham et al. (2014)
		Euscorpiidae	Europe	Salomone et al. (2007)
		Scorpiopidae	None	
	Hadruroidea	Hadruridae	North America	Graham et al. (2013a, 2014)
	Iuroidea	Iuridae	Europe	Parmakelis et al. (2006)
	Scorpionoidea	Diplocentridae	North America	Graham et al. (2014)
		Hemiscorpiidae	None	
		Hormuridae	Asia, Oceania	Monod (2014) and Monod and Prendini (2015)
		Rugodentidae	None	
		Scorpionidae	Africa, Asia	Loria and Prendini (2020), Visser et al. (2020) and Khammassi et al. (2024)

(continued)

Table 6.6 (continued)

Parvorder	Superfamily	Family	Continent	References
		Urodacidae	None	
	Vaejovoidea	Vaejovidae	North America	Bryson Jr et al. (2013a, b, 2014), Graham et al. (2013b, 2014, 2017, 2021) and Miller et al. (2014)

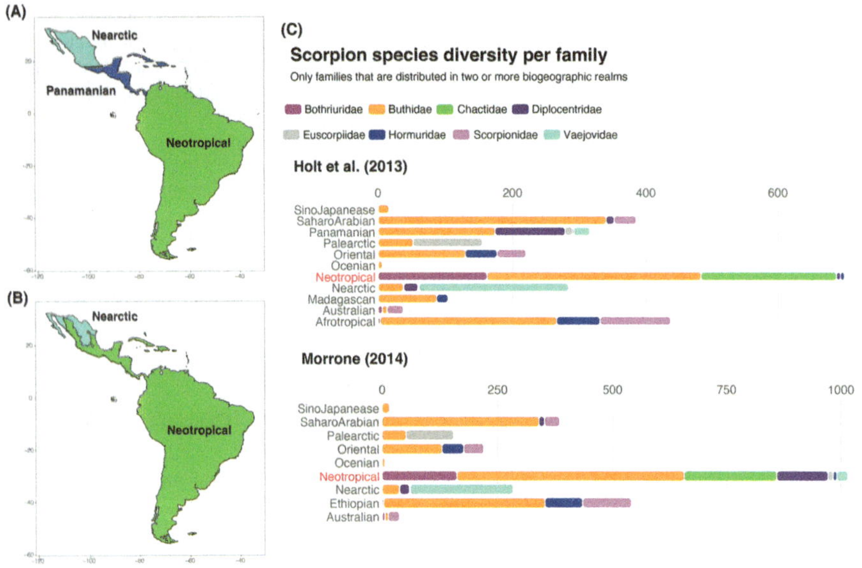

Fig. 6.6 Neotropical biogeographic regions and scorpion family diversity. (**a**) Map based on Holt et al. 2013. (**b**) Map based on Morrone 2014 using the shapefile by Löwenberg-Neto (2014). (**c**) Scorpion family diversity per biogeographic region sensu Holt et al.'s 2013 (top), and Morrone's 2014 (bottom) hypotheses

6.3.1 Neotropical Scorpion Diversity (Holt et al. 2013)

Under the definition of Holt et al. (2013), the diversity of scorpions within the Neotropics exceeds all other realms. Seven families are distributed in the Neotropics, and two of these are endemic (Caraboctonidae with two genera and 26 species; and Troglotayosicidae with one genus and seven species). The three most diverse families are (a) Buthidae (12 genera and 328 species), (b) Chactidae (13 genera and 204 species), and (c) Bothriuridae (14 genera and 162 species). Lastly, family Diplocentridae is represented by one genus (endemic) and six species, whereas Hormuridae is represented by one genus (not endemic) and six species (Fig. 6.6c, Table 6.7). The Neotropics share common scorpion families with the Afrotropical (Bothriuridae, Buthidae, and Hormuridae), the Australian (Bothriuridae and

Table 6.7 Species and genera richness per family distributed in the Neotropical region sensu Holt et al. (2013) and Morrone (2014)

	Parvorder	Superfamily	Family	# Genera	# spp
Holt et al. (2013)	Buthida	Buthoidea	Buthidae	12	328
	Iurida	Bothriuroidea	Bothriuridae	14	162
		Caraboctonoidea	Caraboctonidae	2	26
		Chactoidea	Chactidae	13	204
		Scorpionoidea	Diplocentridae	1	6
			Hormuridae	1	6
		Insertae sedis	Troglotayosicidae	1	7
Morrone (2014)	Buthida	Buthoidea	Buthidae	16	495
	Iurida	Bothriuroidea	Bothriuridae	14	162
		Caraboctonoidea	Caraboctonidae	2	26
		Chactoidea	Chactidae	13	204
			Euscorpiidae	3	11
		Hadruroidea	Hadruridae	1	2
		Scorpionoidea	Diplocentridae	8	116
			Hormuridae	1	8
		Vaejovoidea	Vaejovidae	6	72
		Insertae sedis	Troglotayosicidae	1	7
			Typhlochactidae	4	11

Buthidae), the Malagasy (Buthidae, and Hormuridae), the Nearctic (Buthidae, Chactidae, and Diplocentridae), the Oriental (Buthidae), the Panamanian (Buthidae and Diplocentridae), and the Saharo-Arabian (Buthidae and Diplocentridae) regions.

6.3.2 *Neotropical Scorpion Diversity (Morrone 2014)*

Under the definition of Morrone (2014), most of the diversity of the Nearctic and the Panamanian regions under the Holt et al. (2013) definition becomes subsumed into the Neotropics. Thus, 11 families are distributed in the Neotropics, of which three families are endemic (the same two listed above Caraboctonidae and Troglotayosicidae, plus Typhlochactidae represented by four genera and 11 species). Buthidae is the most diverse in the region, represented by 16 genera and 495 species, followed by Chactidae (represented by 13 genera and 204 species), and Bothriuridae (14 genera and 162 species). Diplocentridae is represented in the Neotropics by eight genera and 116 species, followed by Vaejovidae represented by six genera and 72 species, and Euscorpiidae with three genera and 11 species (Fig. 6.6c, Table 6.7). Lastly, the two least diverse families are Hormuridae (one genus and eight species) and Hadruridae (one genus and two species). The Neotropics as defined by Morrone shares families with more regions than in the previous scenario (Holt et al. 2013). Families shared with the other regions include: (a) Australian (Bothriuridae and Buthidae), (b) the Ethiopian (Ananteridae, Bothriuridae, Buthidae, and Hormuridae), (c) the Nearctic (Buthidae, Chactidae, Diplocentridae, Hadruridae, Vaejovidae), (d) the Oriental (Ananteridae and Buthidae), and (e) the Palearctic (Buthidae and Euscorpiidae) regions.

6.3.3 Ancestral Area Biogeographic Reconstruction

To demonstrate how the delimitation of the Neotropics impacts biogeographic reconstruction, we performed ancestral area reconstruction using the Holt et al. (2013) definition, followed by the definition of Morrone (2014). We ran BioGeoBEARS (Matzke 2013) using the tree topology from Santibáñez-López et al. (2022). We constructed the distribution matrix for the species from that study by designating each terminal taxon to the biogeographic regions they inhabit for the two Neotropical scenarios: (Holt et al. 2013; Morrone 2014). We implemented the DEC model with a maximum number of areas occupied set to two.

The resulting ancestral area analyses are shown in Fig. 6.7. Both biogeographic scenarios suggest that within the family Buthidae, the "*Tityus* group" (subfamily Centruroidinae sensu Esposito et al. 2018) originated in the Neotropics (~100 Mya). Based on this scenario, genus *Centruroides* might have originated in South America and migrated North into the Nearctic realm later in agreement with the previous hypothesis (Esposito and Prendini 2019). In contrast, the origin of the largely

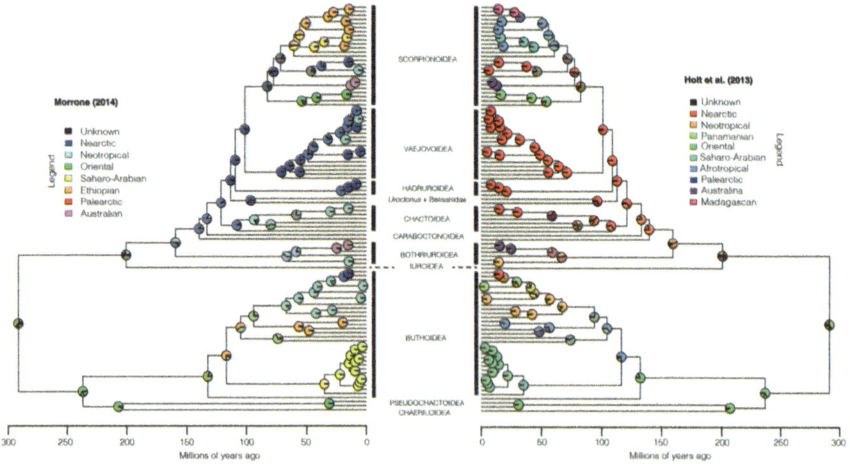

Fig. 6.7 Historical biogeography of scorpion superfamilies. Ancestral areas estimated by BioGeoBEARS using DEC mapped on MRCA nodes for each terminal taxa from Santibáñez-López et al.'s (2022) topology. Each terminal was assigned to the region they inhabit following Morrone's (2014; left) and Holt et al.'s (2013; left) biogeographic hypotheses

Neotropical family Chactidae (represented only by two species here) had its origin in an area that might have been close to the Nearctic. However, Chactidae was recovered as paraphyletic, as *Brotheas* (a Neotropical chactid) was recovered as sister group to genus *Scorpiops* (East Asian scorpion), with *Nullibrotheas* (a Nearctic chactid) recovered as sister group to this clade. This relationship is congruent with previous proposals (e.g., Soleglad and Fet 2003) that Chactidae comprises two subfamilies, Brotheinae and Chactinae, both of which were separated by morphological characters (Soleglad and Fet 2003; new characters also suggested by Loria and Prendini 2014; Monod et al. 2017). Due to the sparse sampling of Neotropical chactids, the origin of this family remains unclear. Next, the origin of Bothriuridae has been widely studied before and suggested to have a Gondwanan origin, with considerable diversity in South America (Sharma et al. 2018; Ojanguren-Affilastro et al. 2023). Since there is only one species of Caraboctonidae sampled (*Caraboctous keyserlingi*), the origin of this family remains unclear, as this taxon was recovered sister to *Superstitionia donensis* (a Nearctic scorpion). Lastly, no members of Troglotayosicidae or Typhlochactidae have yet been included in molecular datasets, and the origin of these families cannot be assessed at present.

6.4 Challenges in the Neotropical Scorpion Systematics: Where Do We Put Our Chips?

Phylogenomic sampling of scorpion species has prioritized the representation of higher-level taxa (e.g., superfamilies and families), to the detriment of intensive species sampling conducive to biogeographic analysis (e.g., Graham et al. 2014; Esposito and Prendini 2019). The main goal for future phylogenetic efforts is to focus on intensively sampling species-rich Neotropical clades. Below, we highlight urgent needs for closing gaps in modern understanding of Neotropical scorpion diversity.

6.4.1 Buthidae C.L. Koch, 1837

The highly diverse family Buthidae, which comprises almost half of described scorpion diversity, requires formal revision of internal generic relationships, as suggested by two recent phylogenetic analyses (i.e., Santibáñez-López et al. 2022; Štundlová et al. 2022). Previous attempts to organize this diversity have resulted in informal species groups (Fet et al. 2005): (a) *Buthus* group (Paleoarctic and the Saharo-Arabian region); (b) *Isometrus* group (Africa and Southeast Asia); (c) *Charmus* group (Indian subcontinent and adjacent islands); (d) *Uroplectes* group (Africa); (e) *Tityus* group (New World); and (f) *Ananteris* group (Madagascar, Southeast Asia, and South America). The last of these has recently been elevated to family status (Ananteridae; Ythier 2024), but is treated as part of Buthidae here, because the deeply nested placement of Ananteridae renders Buthidae as paraphyletic. A reassessment of these groups is crucial, particularly through expanded sampling within the *Isometrus*, *Charmus*, *Uroplectes*, and *Ananteris* groups. For Neotropical buthids, a key next step—regardless of whether these groups are ultimately recognized as subfamilies or families—is to resolve phylogenetic relationships at the genus level. While existing phylogenomic analyses are informative, the inclusion of additional genera, such as *Zabius*, Caribbean genera like *Tityopsis* and *Alayotityus*, and the enigmatic Mexican genus *Chaneke*, will shed light on the biogeographic origins of this clade.

Critically, generic revisions of the highly diverse Neotropical genera *Centruroides* and *Tityus* are still needed. While *Centruroides* has been the subject of several molecular phylogenetic studies (e.g., Esposito et al. 2018; Esposito and Prendini 2019), *Tityus*, the most diverse Neotropical buthid genus, requires extensive tests of monophyly (i.e., Moreno-Gonzalez 2021; Moreno-González et al. 2021). We expect that a rigorous phylogenetic analysis will likely result in the division of these two genera into more genera.

6.4.2 *Bothriuridae Simon, 1880*

This family exhibits the highest diversity within the Neotropical region and has been the subject of extensive research (e.g., Mattoni and Peretti 2004; Mattoni et al. 2012; Ojanguren-Affilastro and Kovaŕík 2013; Ojanguren-Affilastro et al. 2023). However, a comprehensive biogeographic reconstruction of this superfamily is hampered by the lack of molecular data (Sanger and genomic) for much of the South American fauna, as well as African genera (i.e., *Brandbergia* Prendini, 2003 and *Lisposoma* Lawrence, 1928). Future research should prioritize sampling these taxa and generating genome-scale datasets, ideally through ultraconserved element (UCE) sequencing of fresh tissue or, where necessary, preserved museum specimens.

6.4.3 *Caraboctonidae Kraepelin, 1905*

This family has been redefined recently based on phylogenomic analysis (Santibáñez-López et al. 2019). It comprised genera *Caraboctonus* Pocock, 1893, *Hadruroides* Pocock, 1893, *Hadrurus* Thorell, 1876, and *Hoffmannihadrurus* Fet and Soleglad, 2004, with the latter two transferred to their own family (Hadruridae Stahnke 1974; Santibáñez-López et al. 2019). Unfortunately, only *Caraboctonus keyserlingi* Pocock, 1893 is represented in the phylogenomic reconstructions. A goal of future sampling efforts must be to include exemplars of the genus *Hadruroides*.

6.4.4 Chactidae Pocock, 1893

This family remains one of the most understudied groups in scorpions. Of the 13 genera and 205 species, only two genera (*Brotheas granulatus* Simon, 1877 and *Nullibrotheas allenii* (Wood, 1863)) have been included in phylogenomic analysis. This limited number of taxa has brought the monophyly of the family into question. To address this gap in scorpion systematics, it is imperative to include additional species from key genera, particularly genus *Chactas* Gervais, 1844 (the type genus for the family), *Brotheas* C.L. Koch, 1837, *Teuthraustes* Simon, 1878 and *Neochactas* Soleglad and Fet 2003. These genera should be favored over others based on their inclusion in subfamilies Brotheinae and Chactinae as established by Soleglad and Fet (2003).

6.4.5 Family Diplocentridae Karsch, 1880

The scorpion family Diplocentridae exhibits a striking disjunct distribution. Its greatest diversity is found in North America and the Caribbean, with some genera also occurring in Central and South America. By contrast, the genus *Nebo* Simon, 1878 is found in the Middle East, specifically the Arabian Peninsula and Israel. While the monophyly of Diplocentridae is well-supported by both Sanger-based phylogenies (Santibáñez-López et al. 2014a, 2017) and by phylogenomic datasets (e.g., Santibáñez-López et al. 2023), limited sampling of Caribbean genera, particularly *Heteronebo* Pocock, 1899, hinders a robust assessment of generic monophyly and relationships within the family.

6.4.6 Family Hormuridae Laurie, 1896

While the monophyly of the family was challenged in the last two phylogenomic analyses (Santibáñez-López et al. 2022, 2023), this family is represented in the Neotropical region only by genus *Opisthacanthus* Peters, 1861 and up to eight species. The monophyly of this genus is also refuted in these analyses, as the two South African exemplars clustered together, but the sole representative from Madagascar (*O. madagascariensis* Kraepelin, 1894) was recovered as part of a clade with Old World taxa such as *Chiromachus ochropus* (C. L. Koch, 1837) (Seychelles) and *Palaeocheloctonus pauliani* Lourenço, 1996 (Madagascar). Outside of the Neotropics, *Opistacanthus* has a disjunct distribution, with 16 additional species in South Africa and 14 species in Madagascar. We anticipate that this family will undergo substantial systematic revision; the Neotropical species within *Opisthacanthus* are almost certainly in need of transfer to newly defined genera.

6.4.7 Troglotayosicidae Lourenço, 1998 and Typhlochactidae Mitchell, 1971

These two families are key missing taxa in all phylogenomic analyses, and their sampling is direly needed. Their sampling would aid in solving the phylogenetic position of Superstitioniidae, Belisariidae, and genus *Uroctonus* Thorell, 1876, in addition to testing the hypothesis that troglomorphism is an evolutionary dead-end with a non-morphological dataset (e.g., Prendini et al. 2010; Santibáñez-López et al. 2014b; Moreno-González et al. 2024).

6.5 Concluding Remarks

Since the publication of the first phylogenomic study of scorpions 10 years ago, genome-scale phylogenies have dramatically advanced our understanding of scorpion systematics. While RNAseq has been a prominent tool, a recent study demonstrated the robust congruence between phylogenies derived from RNAseq and Ultraconserved Elements (UCEs). This finding opens a significant avenue for future research by enabling the use of museum specimens to address the current taxonomic gaps presented here. Prioritizing UCE sequencing over other technologies (i.e., Sanger sequencing) over the next decade, in tandem with mining available natural history collections from around the world, may prove instrumental to closing the last gaps surrounding higher-level relationships, clarifying phylogenetic positions of enigmatic families, and reconstructing historical biogeographic patterns and morphological evolution.

Looking ahead to the next decade, our focus should shift toward intensive taxonomic sampling at the genus and species levels. This will facilitate in-depth studies of: (a) historical biogeographic, especially those explaining disjunct distributions in families like Diplocentridae and Hormuridae; (b) phylogeographic patterns to understand how past climate fluctuations have shaped scorpion distribution; (c) diversification processes that have given rise to asymmetrically diverse groups, such as Buthidae; and (d) the reevaluation and establishment of diagnostic characters for higher-level taxa. Concurrently, we must delve into chromosome-level genomes to unveil: (a) the genetic basis of morphological adaptations to extreme environments (caves, deserts, mountains); and (b) the mechanisms of venom synthesis, including epigenetic regulation in response to variable environments that are prone to climate change (e.g., arid and subarid regions of the world). We thus foresee dramatic improvements in our limited understanding of scorpion evolution by 2036.

References

Aguinaldo AMA, Turbeville JM, Linford LS, Rivera MC, Garey JR, Raff RA, Lake JA (1997) Evidence for a clade of nematodes, arthropods and other moulting animals. Nature 387(6632):489–493. https://doi.org/10.1038/387489a0

Bryson RW Jr, Savary WE, Prendini L (2013a) Biogeography of scorpions in the *Pseudouroctonus minimus* complex (Vaejovidae) from south-western North America: Implications of ecological specialization for pre-Quaternary diversification. J Biogeogr 40(10):1850–1860. https://doi.org/10.1111/jbi.12134

Bryson RW Jr, Riddle BR, Graham MR, Smith BT, Prendini L (2013b) As old as the hills: montane scorpions in southwestern North America reveal ancient associations between biotic diversification and landscape history. PLoS One 8(1):e52822. https://doi.org/10.1371/journal.pone.0052822

Bryson RW Jr, Prendini L, Savary WE, Pearman PB (2014) Caves as microrefugia: Pleistocene phylogeography of the troglophilic North American scorpion *Pseudouroctonus reddelli*. BMC Evol Biol 14:1–17. https://doi.org/10.1186/1471-2148-14-9

Cain S, Loria SF, Ben-Shlomo R, Prendini L, Gefen E (2021) Dated phylogeny and ancestral range estimation of sand scorpions (Buthidae: *Buthacus*) reveal Early Miocene divergence across land bridges connecting Africa and Asia. Mol Phylogenet Evol 164:107212. https://doi.org/10.1016/j.ympev.2021.107212

Cao Z, Di Z, Wu Y, Li W (2014) Overview of scorpion species from China and their toxins. Toxins 6(3):796–815. https://doi.org/10.3390/toxins6030796

Ceccarelli FS, Ojanguren-Affilastro AA, Ramírez MJ, Ochoa JA, Mattoni CI, Prendini L (2016) Andean uplift drives diversification of the bothriurid scorpion genus *Brachistosternus*. J Biogeogr 43(10):1942–1954. https://doi.org/10.1111/jbi.12760

Ceccarelli FS, Pizarro-Araya J, Ojanguren-Affilastro AA (2017) Phylogeography and population structure of two *Brachistosternus* species (Scorpiones: Bothriuridae) from the Chilean coastal desert–the perils of coastal living. Biol J Linn Soc 120(1):75–89. https://doi.org/10.1111/bij.12877

Coddington JA, Giribet G, Harvey MS, Prendini L, Walter DE (2004) Arachnida. In: Cracraft J, Donoghue PCJ (eds) Assembling the tree of life. Oxford University Press, New York, pp 296–318

Coelho P, Kaliontzopoulou A, Rasko M, van der Meijden A (2017) A 'striking'relationship: scorpion defensive behaviour and its relation to morphology and performance. Funct Ecol 31(7):1390–1404. https://doi.org/10.1111/1365-2435.12855

Nielsen C (1997) The phylogenetic position of the Arthropoda. In: Fortey, R.A. & Thomas, R.H. (eds.) Arthropod Relationships, Systematics Association Special Volume Series 55. Chapman & Hall, London. 1995:11–22

Esposito LA, Prendini L (2019) Island ancestors and new world biogeography: a case study from the scorpions (Buthidae: Centruroidinae). Sci Rep 9(1):3500. https://doi.org/10.1038/s41598-018-33754-8

Esposito L, Yamaguti H, Pinto-da-Rocha R, Prendini L (2018) Plucking with the plectrum: phylogeny of the New World buthid scorpion subfamily Centruroidinae Kraus, 1955 (Scorpiones: Buthidae) reveals evolution of three pecten-sternite stridulation organs. Arthropod Syst Phylogeny 76:87–122. https://doi.org/10.3897/asp.76.e31942

Fet V, Soleglad ME (2005) Contributions to scorpion systematics. I. On recent changes in high-level taxonomy. Euscorpius 31:1–13

Fet V, Gantenbein B, Gromov AV, Lowe G, Lourenço WR (2003) The first molecular phylogeny of Buthidae (Scorpiones). Euscorpius 4:1–10

Fet V, Soleglad ME, Lowe G (2005) A new trichobothrial character for the high-level systematics of Buthoidea (Scorpiones: Buthida). Euscorpius 23:1–40

Fet V, Soleglad ME, Zonstein SL (2011) The genus *Akrav* Levy, 2007 (Scorpiones: Akravidae) revisited. Euscorpius 134:1–49

Foerster SÍA (2025) Body size prediction in scorpions: a phylogenetic comparative examination of linear measurements of individual body parts. PeerJ 13:e18621. https://doi.org/10.7717/peerj.18621

Gantenbein B, Largiadèr CR (2003) The phylogeographic importance of the Strait of Gibraltar as a gene flow barrier in terrestrial arthropods: a case study with the scorpion *Buthus occitanus* as model organism. Mol Phylogenet Evol 28(1):119–130. https://doi.org/10.1016/S1055-7903(03)00031-9

González-Santillán E, Prendini L (2015) Phylogeny of the North American vaejovid scorpion subfamily Syntropinae Kraepelin, 1905, based on morphology, mitochondrial and nuclear DNA. Cladistics 31(4):341–405. https://doi.org/10.1111/cla.12091

Graham MR, Oláh-Hemmings V, Fet V (2012) Phylogeography of codistributed dune scorpions identifies the Amu Darya River as a long-standing component of Central Asian biogeography: (Scorpiones: Buthidae). Zool Middle East 55(1):95–110. https://doi.org/10.1080/09397140.2012.10648924

Graham MR, Jaeger JR, Prendini L, Riddle BR (2013a) Phylogeography of Beck's desert scorpion, *Paruroctonus becki*, reveals Pliocene diversification in the eastern California Shear Zone and postglacial expansion in the Great Basin Desert. Mol Phylogenet Evol 69(3):502–513. https://doi.org/10.1016/j.ympev.2013.07.028

Graham MR, Jaeger JR, Prendini L, Riddle BR (2013b) Phylogeography of the Arizona hairy scorpion (*Hadrurus arizonensis*) supports a model of biotic assembly in the Mojave Desert and adds a new Pleistocene refugium. J Biogeogr 40(7):1298–1312. https://doi.org/10.1111/jbi.12079

Graham MR, Bryson RW Jr, Riddle BR (2014) Late Pleistocene to Holocene distributional stasis in scorpions along the Baja California peninsula. Biol J Linn Soc 111(2):450–461. https://doi.org/10.1111/bij.12216

Graham MR, Wood DA, Henault JA, Valois ZJ, Cushing PE (2017) Ancient lakes, Pleistocene climates and river avulsions structure the phylogeography of a large but little-known rock scorpion from the Mojave and Sonoran deserts. Biol J Linn Soc 122(1):133–146. https://doi.org/10.1093/biolinnean/blx058

Graham MR, Myers EA, Kaiser RC, Fet V (2019) Cryptic species and co-diversification in sand scorpions from the Karakum and Kyzylkum deserts of Central Asia. Zool Scr 48(6):801–812. https://doi.org/10.1111/zsc.12381

Graham MR, Garcia EL, Hendrixson BE, Sampognaro AM, Cushing PE (2021) Pliocene origins, Pleistocene refugia, and postglacial range expansions in southern devil scorpions (Vaejovidae: *Vaejovis carolinianus*). Org Divers Evol 21:575–590. https://doi.org/10.1007/s13127-021-00505-z

Graham MR, Santibáñez-López CE, Zehnpfennig JR, Tillman DS, Murdoch B (2024) Serendipitous discovery of desert hairy scorpion mitogenomes as bycatch in venom data via nanopore sequencing. Arthropoda 2(2):119–129. https://doi.org/10.3390/arthropoda2020009

Holt BG, Lessard JP, Borregaard MK, Fritz SA, Araújo MB, Dimitrov D, Fabre PH, Graham CH, Graves GR, Jønsson KA, Nogués-Bravo D, Wang Z, Whittaker RJ, Fjeldså J, Rahbek C (2013) An update of Wallace's zoogeographic regions of the world. Science 339(6115):74–78. https://doi.org/10.1126/science.1228282

Inward D, Beccaloni G, Eggleton P (2007) Death of an order: a comprehensive molecular phylogenetic study confirms that termites are eusocial cockroaches. Biol Lett 3(3):331–335. https://doi.org/10.1098/rsbl.2007.0102

Khammassi M, Nouira S, Badry A, Sadine SE, Harris DJ (2024) Phylogeography and evolutionary history of the burrowing scorpion genus *Scorpio* Linnaeus, 1758 (Scorpiones: Scorpionidae) in the Mediterranean Basin. Org Divers Evol 24(3):335–352. https://doi.org/10.1007/s13127-024-00649-8

Koch NM, Ceccarelli FS, Ojanguren-Affilastro AA, Ramirez MJ (2017) Discrete and morphometric traits reveal contrasting patterns and processes in the macroevolutionary history of a clade of scorpions. J Evol Biol 30(4):814–825. https://doi.org/10.1111/jeb.13050

Loria SF, Prendini L (2014) Homology of the lateral eyes of Scorpiones: a six-ocellus model. PLoS One 9(12):e112913. https://doi.org/10.1371/journal.pone.0112913
Loria SF, Prendini L (2020) Out of India, thrice: diversification of Asian forest scorpions reveals three colonizations of Southeast Asia. Sci Rep 10(1):22301. https://doi.org/10.1038/s41598-020-78183-8
Loria SF, Prendini L (2021) Burrowing into the forest: phylogeny of the Asian forest scorpions (Scorpionidae: Heterometrinae) and the evolution of ecomorphotypes. Cladistics 37(2):109–161
Loria SF, Ehrenthal VL, Nguyen AD, Prendini L (2022) Climate relicts: Asian scorpion family Pseudochactidae survived Miocene aridification in caves of the Annamite Mountains. Insect Syst Divers 6(6):3. https://doi.org/10.1093/isd/ixac028
Lourenço WR (2002) Scorpions of Brazil. Les editions de l'If, Paris, p 308
Lourenço WR (2003) Scorpion biogeography: a review. Una perspectiva Latinoamericana de la biogeografía. Las Prensas de Ciencia, Facultad de Ciencias, UNAM, México, pp 227–240
Löwenberg-Neto P (2014) Neotropical region: a shapefile of Morrone's (2014) biogeographical regionalisation. Zootaxa 3802:300. https://doi.org/10.11646/zootaxa.3802.2.12
Mattoni CI, Peretti AV (2004) The giant and complex genital plug of the asper group of *Bothriurus* (Scorpiones, Bothriuridae): morphology and comparison with other genital plugs in scorpions. Zool Anz 243(1–2):75–84. https://doi.org/10.1016/j.jcz.2004.07.002
Mattoni CI, Ochoa JA, Ojanguren-Affilastro AA, Prendini L (2010) Towards an all-species phylogeny of the scorpion family Bothriuridae. Cladistics 26(2):216–217
Mattoni CI, Ochoa JA, Ojanguren-Affilastro AA, Prendini L (2012) *Orobothriurus* (Scorpiones: Bothriuridae) phylogeny, Andean biogeography, and the relative importance of genitalic and somatic characters. Zool Scr 41(2):160–176. https://doi.org/10.1111/j.1463-6409.2011.00508.x
Matzke NJ (2013) BioGeoBEARS: BioGeography with Bayesian (and likelihood) evolutionary analysis in R Scripts. R package, version 0.2, 1:2013
Miller AL, Makowsky RA, Formanowicz DR, Prendini L, Cox CL (2014) Cryptic genetic diversity and complex phylogeography of the boreal North American scorpion, *Paruroctonus boreus* (Vaejovidae). Mol Phylogenet Evol 71:298–307. https://doi.org/10.1016/j.ympev.2013.11.005
Monod L (2014) The genus *Hormiops* Fage, 1933 (Hormuridae, Scorpiones), a palaeoendemic of the South China Sea: systematics and biogeography. C R Biol 337(10):596–608. https://doi.org/10.1016/j.crvi.2014.07.009
Monod L, Prendini L (2015) Evidence for Eurogondwana: the roles of dispersal, extinction and vicariance in the evolution and biogeography of Indo-Pacific Hormuridae (Scorpiones: Scorpionoidea). Cladistics 31(1):71–111. https://doi.org/10.1111/cla.12067
Monod L, Cauwet L, González-Santillán E, Huber S (2017) The male sexual apparatus in the order Scorpiones (Arachnida): a comparative study of functional morphology as a tool to define hypotheses of homology. Front Zool 14:1–48. https://doi.org/10.1186/s12983-017-0231-z
Moreno-Gonzalez JA (2021). Análise filogenética do grupo de espécies *Tityus clathratus* e outros grupos e subgêneros de *Tityus* (Scorpiones: Buthidae) baseada em caracteres moleculares e morfológicos. Doctoral dissertation, Universidade de São Paulo
Moreno-González JA, Pinto-da-Rocha R, Gallão JE (2021) Bringing order to a complex system: phenotypic and genotypic evidence contribute to the taxonomy of *Tityus* (Scorpiones, Buthidae) and support the description of a new species. ZooKeys 1075:33–75. https://doi.org/10.3897/zookeys.1075.67459
Moreno-González JA, Luna-Sarmiento DA, Prendini L (2024) Phylogeny of the troglomorphic scorpion genus *Troglotayosicus* (Scorpiones: Troglotayosicidae) with description of a new species from Colombia. Am Mus Novit 2024(4011):1–39. https://doi.org/10.1206/4011.1
Morrone JJ (2014) Biogeographical regionalisation of the Neotropical region. Zootaxa 3782(1):1–110. https://doi.org/10.11646/zootaxa.3782.1.1
Ochoa JA (2005) Patrones de distribución de escorpiones de la región andina en el sur peruano. Rev Peru Biol 12(1):49–68
Ojanguren-Affilastro AA (2005) Estudio monográfico de los escorpiones de la República Argentina. Rev Ibér Aracnol 11:75–241

Ojanguren-Affilastro AA, Kovarík F (2013) Bothriuridae. In: Kovarík F, Ojanguren-Affilastro AA (eds) Illustrated catalog of scorpions. Part II. Bothriuridae; Chaerilidae; Buthidae I., genera *Compsobuthus*, *Hottentotta*, *Isometrus*, *Lychas*, and *Sassanidotus*. Jakub Rolcík Publisher, Czech Republic, pp 6–130

Ojanguren-Affilastro AA, Mattoni CI, Ochoa JA, Ramírez MJ, Ceccarelli FS, Prendini L (2016) Phylogeny, species delimitation and convergence in the South American bothriurid scorpion genus *Brachistosternus* Pocock 1893: integrating morphology, nuclear and mitochondrial DNA. Mol Phylogenet Evol 94:159–170. https://doi.org/10.1016/j.ympev.2015.08.007

Ojanguren-Affilastro AA, Adilardi RS, Mattoni CI, Ramírez MJ, Ceccarelli FS (2017) Dated phylogenetic studies of the southernmost American buthids (Scorpiones; Buthidae). Mol Phylogenet Evol 110:39–49. https://doi.org/10.1016/j.ympev.2017.02.018

Ojanguren-Affilastro AA, Pizarro-Araya J, Santibáñez-López CE (2023) Old and cold: diverse phylogenomic datasets support an ancient transantarctic dispersive route on the scorpion family Bothriuridae in temperate Gondwana. Mol Phylogenet Evol 187:107886. https://doi.org/10.1016/j.ympev.2023.107886

Othmen AB, Said K, Mahamdallie SS, Testa JM, Haouas Z, Chatti N, Ready PD (2009) Phylogeography of *Androctonus* species (Scorpiones: Buthidae) in Tunisia: diagnostic characters for linking species to scorpionism. Acta Trop 112(1):77–85. https://doi.org/10.1016/j.actatropica.2009.07.001

Paradis E, Schliep K (2019) ape 5.0: an environment for modern phylogenetics and evolutionary analyses in R. Bioinformatics 35(3):526–528. https://doi.org/10.1093/bioinformatics/bty633

Parmakelis A, Stathi I, Spanos L, Louis C, Mylonas M (2006) Phylogeography of *Iurus dufoureius* (Brulle, 1832) (Scorpiones, Iuridae). J Biogeogr 33(2):251–260. https://doi.org/10.1111/j.1365-2699.2005.01386.x

Parmakelis A, Dimitriadou D, Gkigkiza E, Karampatsou L, Stathi I, Fet V, Yagmur EA, Kovařík F (2022) The evolutionary history of the relict scorpion family Iuridae of the eastern Mediterranean. Mol Phylogenet Evol 177:107622. https://doi.org/10.1016/j.ympev.2022.107622

Prendini L (2000) Phylogeny and classification of the superfamily Scorpionoidea Latreille 1802 (Chelicerata, Scorpiones): an exemplar approach. Cladistics 16(1):1–78. https://doi.org/10.1006/clad.1999.0127

Prendini L (2004) On the scorpions of Gabon and neighboring countries, with a reassessment of the synonyms attributed to *Babycurus buettneri* Karsch and a redescription of *Babycurus melanicus* Kovaík. Calif Acad Sci Mem 28:235–267

Prendini L (2005) Scorpion diversity and distribution in southern Africa: pattern and process. In: Huber BA, Sinclair BJ, Lampe K-H (eds) African biodiversity: molecules, organisms, ecosystems. Proceedings of the 5th international symposium on tropical biology, museum Alexander Koenig, Bonn. Springer Verlag, New York, pp 25–68

Prendini L, Bird TL (2008) Scorpions of the Brandberg Massif, Namibia: species richness inversely correlated with altitude. Afr Invertebr 49(2):77–107

Prendini L, Wheeler WC (2004) Assembling the scorpion tree of life: phylogeny of extant scorpions based on morphology, molecules and exemplars. Cladistics 20(6):602–603

Prendini L, Wheeler WC (2005) Scorpion higher phylogeny and classification, taxonomic anarchy, and standards for peer review in online publishing. Cladistics 21(5):446–494

Prendini L, Crowe TM, Wheeler WC (2003) Systematics and biogeography of the family Scorpionidae (Chelicerata: Scorpiones), with a discussion on phylogenetic methods. Invertebr Syst 17(2):185–259

Prendini L, Volschenk ES, Maaliki S, Gromov AV (2006) A 'living fossil'from Central Asia: the morphology of *Pseudochactas ovchinnikovi* Gromov, 1998 (Scorpiones: Pseudochactidae), with comments on its phylogenetic position. Zool Anz 45(3–4):211–248. https://doi.org/10.1016/j.jcz.2006.07.001

Prendini L, Francke OF, Vignoli V (2010) Troglomorphism, trichobothriotaxy and typhlochactid phylogeny (Scorpiones, Chactoidea): more evidence that troglobitism is not an evolutionary dead-end. Cladistics 26(2):117–142
Zrzavý J, Hypša V & Vlášková M (1998) Arthropod phylogeny: taxonomic congruence, total evidence and conditional combination approaches to morphological and molecular data sets. In Arthropod relationships (pp. 97–107). Dordrecht: Springer Netherlands
Salomone N, Vignoli V, Frati F, Bernini F (2007) Species boundaries and phylogeography of the *Euscorpius carpathicus* complex (Scorpiones: Euscorpiidae) in Italy. Mol Phylogenet Evol 43(2):502–514
Santibanez CE, Francke OF, Prendini L (2014a) Phylogeny of the North American scorpion genus *Diplocentrus* Peters, 1861 (Scorpiones: Diplocentridae) based on morphology, nuclear and mitochondrial DNA. Arthropod Systematics and Phylogeny 72:257–279
Santibáñez-López CE, Francke OF, Prendini L (2014b) Shining a light into the world's deepest caves: phylogenetic systematics of the troglobiotic scorpion genus *Alacran* Francke, 1982 (Typhlochactidae: Alacraninae). Invertebr Syst 28(6):643–664. https://doi.org/10.1071/IS14035
Santibáñez-López CE, Francke OF, Ureta C, Possani LD (2015) Scorpions from Mexico: from species diversity to venom complexity. Toxins 8(1):2. https://doi.org/10.3390/toxins8010002
Santibáñez-López CE, Kriebel R, Sharma PP (2017) *eadem figura manet*: measuring morphological convergence in diplocentrid scorpions (Arachnida: Scorpiones: Diplocentridae) under a multilocus phylogenetic framework. Invertebr Syst 31(3):233–248. https://doi.org/10.1071/IS16078
Santibáñez-López CE, Kriebel R, Ballesteros JA, Rush N, Witter Z, Williams J, Janies DA, Sharma PP (2018) Integration of phylogenomics and molecular modeling reveals lineage-specific diversification of toxins in scorpions. PeerJ 6:e5902. https://doi.org/10.7717/peerj.5902
Santibáñez-López CE, González-Santillán E, Monod L, Sharma PP (2019) Phylogenomics facilitates stable scorpion systematics: reassessing the relationships of Vaejovidae and a new higher-level classification of Scorpiones (Arachnida). Mol Phylogenet Evol 135:22–30
Santibáñez-López CE, Ojanguren-Affilastro AA, Sharma PP (2020) Another one bites the dust: taxonomic sampling of a key genus in phylogenomic datasets reveals more non-monophyletic groups in traditional scorpion classification. Invertebr Syst 34(2):133–143. https://doi.org/10.1071/IS19033
Santibáñez-López CE, Aharon S, Ballesteros JA, Gainett G, Baker CM, González-Santillán E, Harvey MS, Hassan MK, Abu Almaaty AH, Aldeyarbi SM, Monod L, Ojanguren-Affilastro AA, Pinto-da-Rocha R, Zvik Y, Gavish-Regev E, Sharma PP (2022) Phylogenomics of scorpions reveal contemporaneous diversification of scorpion mammalian predators and mammal-active sodium channel toxins. Syst Biol 71(6):1281–1289. https://doi.org/10.1093/sysbio/syac021
Santibáñez-López CE, Ojanguren-Affilastro AA, Graham MR, Sharma PP (2023) Congruence between ultraconserved element-based matrices and phylotranscriptomic datasets in the scorpion Tree of Life. Cladistics 39(6):533–547. https://doi.org/10.1111/cla.12551
Sharma PP, Fernández R, Esposito LA, González-Santillán E, Monod L (2015) Phylogenomic resolution of scorpions reveals multilevel discordance with morphological phylogenetic signal. Proc R Soc B Biol Sci 282(1804):20142953. https://doi.org/10.1098/rspb.2014.2953
Sharma PP, Baker CM, Cosgrove JG, Johnson JE, Oberski JT, Raven RJ, Harvey MS, Boyer SL, Giribet G (2018) A revised dated phylogeny of scorpions: phylogenomic support for ancient divergence of the temperate Gondwanan family Bothriuridae. Mol Phylogenet Evol 122:37–45. https://doi.org/10.1016/j.ympev.2018.01.003
Shi CM, Zhang XS, Liu L, Ji YJ, Zhang DX (2023) Phylogeography of the desert scorpion illuminates a route out of Central Asia. Curr Zool 69(4):442–455. https://doi.org/10.1093/cz/zoac061
Sissom WD (1990) Systematics, biogeography, and paleontology. In: Polis GA (ed) The biology of scorpions. Stanford University Press, Stanford, pp 64–160
Soleglad ME, Fet V (2001) Evolution of scorpion orthobothriotaxy: a cladistic approach. Euscorpius 1:1–40

Soleglad ME, Fet V (2003) High-level systematics and phylogeny of the extant scorpions (Scorpiones: Orthosterni). Euscorpius 3:1–210

Stockwell SA (1989) Revision of the phylogeny and higher classification of scorpions (Chelicerata). PhD thesis. Berkeley

Štundlová J, Šťáhlavský F, Opatova V, Stundl J, Kovařík F, Dolejš P, Šmíd J (2022) Molecular data do not support the traditional morphology-based groupings in the scorpion family Buthidae (Arachnida: Scorpiones). Mol Phylogenet Evol 173:107511. https://doi.org/10.1016/j.ympev.2022.107511

Tropea G, Onnis C (2020) A remarkable discovery of a new scorpion genus and species from Sardinia (Scorpiones: Chactoidea: Belisariidae). Arachn Riv Aracnol Italiana 26:3–25

Vachon M (1973) Étude des caractères utilisés pour classer les familles et les genres de scorpions (Arachnides). 1. La trichobothriotaxie en arachnologie. Sigles trichobothriaux et types de trichobothriotaxie chez les scorpions. Bull Mus Natl Hist Nat 3:857–958

van der Meijden A, Kleinteich T, Coelho P (2012) Packing a pinch: functional implications of chela shapes in scorpions using finite element analysis. J Anat 220(5):423–434. https://doi.org/10.1111/j.1469-7580.2012.01485.x

van der Meijden A, Coelho PL, Sousa P, Herrel A (2013) Choose your weapon: defensive behavior is associated with morphology and performance in scorpions. PLoS One 8(11):e78955. https://doi.org/10.1371/journal.pone.0078955

Visser JH, Geerts S, van Vuuren BJ (2020) Phylogeographic patterns in a semi-lithophilous burrowing scorpion, *Opistophthalmus pallipes*, from South Africa. Zool Sci 38(1):36–44

Wallace AR (1876) The geographical distribution of animals. Macmillan, London, p 1110

Xiao L, Gurrola GB, Zhang J, Valdivia CR, SanMartin M, Zamudio FZ, Zhang L, Possani LD, Valdivia HH (2016) Structure–function relationships of peptides forming the calcin family of ryanodine receptor ligands. J Gen Physiol 147(5):375–394

Yamashita T, Rhoads DD (2013) Species delimitation and morphological divergence in the scorpion *Centruroides vittatus* (Say, 1821): insights from phylogeography. PLoS One 8(7):e68282. https://doi.org/10.1371/journal.pone.0068282

Ythier E (2024) A new high-altitude scorpion species of the genus *Ananteris* Thorell, 1891 (Scorpiones: Ananteridae) from the Pico da Neblina. Brazil Faunitaxys 12(19):1–9

Chapter 7
Fossil Scorpions, with a Focus on the Neotropical Species

Jason A. Dunlop

Abstract Fossil scorpions go back as far as the Silurian, where a key question remains whether the oldest examples lived in water or on land. Fossils can be preserved in several ways including shales, nodules, macerated cuticles, or as inclusions in amber and copal. A challenge for the study of older fossils is their higher systematics, with several poorly defined groups raised on questionable characters. The Neotropics represent an important source of data, particularly for younger species. The oldest, and only, Palaeozoic record from this region comes from the Permian of Brazil and remains unplaced within Orthosternina. Two species originate from the Cretaceous Crato Formation, also in Brazil, and belong to Chactidae and either Hemiscorpiidae or an extinct family, Protoischnuridae. From the Miocene, both Chiapas (Mexican) and Dominican amber yield fossils assignable to modern genera of Buthidae. Finally, Colombian Copal, which may be Pleistocene or even Holocene, yields a record of Chactidae.

Keywords Arachnids · Evolution · Paleontology

7.1 Introduction

Scorpions (Arachnida: Scorpiones) have an extensive fossil record, with more than 150 extinct species going back at least 433 million years to the middle of the Silurian period. Fossil scorpions can be found during all subsequent periods of Earth history, albeit somewhat concentrated into particular windows of space and time. For example, the Late Carboniferous Coal Measures of Europe and North America, the mid-Cretaceous Burmese amber of Myanmar, and the Palaeogene Baltic amber of northern Europe have all yielded several species offering valuable insights into the

J. A. Dunlop (✉)
Museum für Naturkunde, Leibniz Institute for Evolution and Biodiversity Science, Berlin, Germany
e-mail: jason.dunlop@mfn.berlin

A. F. de Araujo Lira, E. González-Santillán (eds.), *The Biology of Neotropical Scorpions*, https://doi.org/10.1007/978-3-032-23194-9_7

diversity and distribution of scorpions in the distant past. Other localities have produced only one or two species, but also contribute to the wider picture of scorpion evolution. A regularly updated summary of the known fossil species can be found in Dunlop et al. (2023) and references therein.

The aim of this chapter is to offer (1) a general overview of fossil scorpions, their significance, and some of the challenges of working with fossil material and (2) a more detailed account of the Neotropical records. This region of the world has yielded one fossil scorpion species from the Permian of Canoinhas in southern Brazil, two species from the Cretaceous Crato Formation of northeastern Brazil, either four or six species from the Neogene (Miocene) Dominican Republic amber (two species may be junior synonyms), three species from the probably slightly older Chiapas amber of Mexico, and one either Pleistocene or Holocene species from Colombian copal.

7.2 Preservation and Study Methods

Many fossil scorpions are preserved as compression fossils in shales (Fig. 7.1a). These are usually studied using traditional methods of photography, plus an accompanying interpretative drawing. Selden and Penney (2017) discussed how to image fossil spiders, and most of these techniques would be applicable to scorpions too. Sometimes a shale splits into a part and counterpart which, ideally, reveals the dorsal and ventral surfaces of the specimen, respectively. What may happen in compressed material is that dorsal and ventral features became superimposed, which can make interpretation of the fossil challenging. Depending on the type of shale and the degree of surface relief, it may be helpful to photograph the specimen using low-angle lighting, or immersing the specimen in alcohol to enhance the contrast between the fossil and its surrounding matrix. Most of the older (i.e., Palaeozoic) species are preserved as compression fossils, and both dead animals and/or their moulted exoskeletons could potentially have been preserved. McCoy and Brandt (2009) suggested criteria for determining whether fossil scorpions are likely to be moults or mortalities. For example, a straight body is probably from a dead scorpion (i.e., a mortality) while a curved body is more suggestive of a moulted exuvium.

Some Carboniferous Coal Measures fossils are preserved in ironstone concretions (or nodules). Here, the original animal is no longer present, but has left a three-dimensional space in the rock matrix which reveals its external anatomy (Fig. 7.1b). Such material can be drawn and photographed like a shale fossil, but may also be amenable to computed microtomography (Legg et al. 2012). In this technique the space in the matrix is scanned and turned into a three-dimensional positive image which can be rotated or otherwise manipulated on the computer, and which may reveal details of the limbs which were hidden deeper in the matrix. Under some circumstances, fossil scorpions preserved in shales can also be studied using microtomography (Kühl et al. 2012; Carvalho et al. 2023), so long as the specimen retains a degree of three-dimensionality.

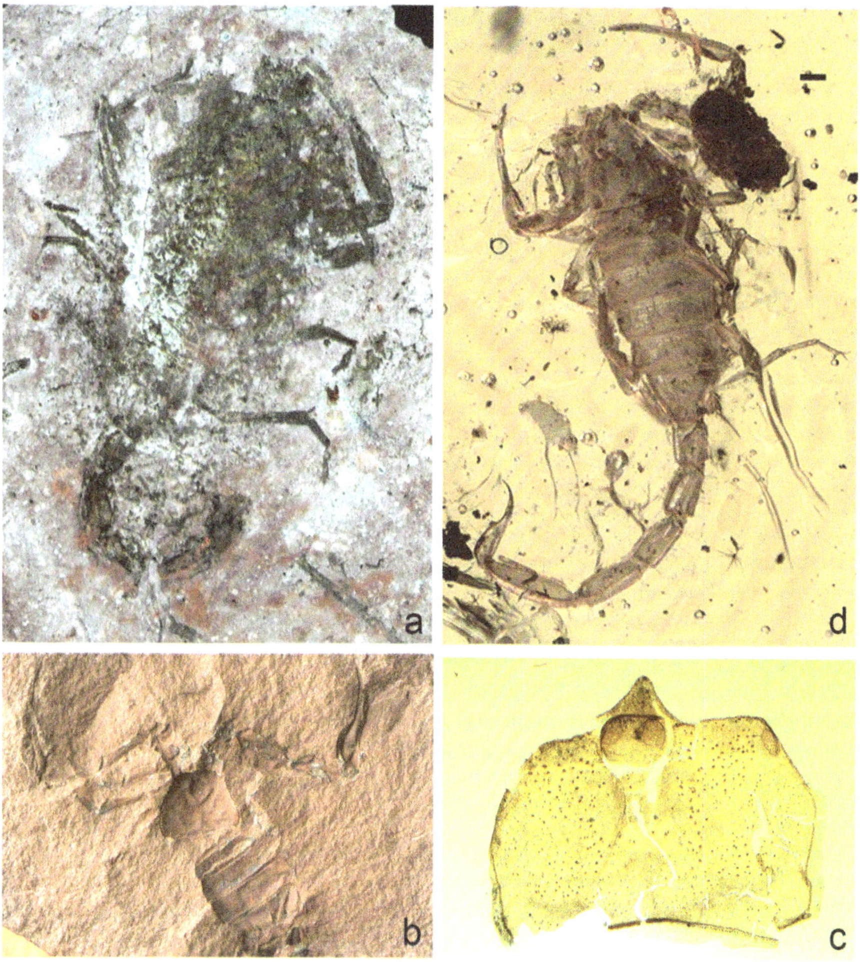

Fig. 7.1 Selected fossil scorpions showing different modes of preservation. (**a**) *Opsieobuthus tungeri*, a shale-preserved fossil from the Permian of Germany. (**b**) *Comsoscorpius buthiformis* in an ironstone concretion from the Late Carboniferous of England. (**c**) *Mesophonus perornatus*, acid-macerated carapace cuticle from the Triassic of England. (**d**) *Betaburmesebuthus bellus*, an inclusion in Cretaceous Burmese amber. (Image (**a**) courtesy of the Chemnitz Museum (CC BY-NC-SA@Museum für Naturkunde Chemnitz), (**b** and **c**) courtesy of Carsten Kamenz, (**d**) courtesy of Wilson Lourenço via ZooKeys and a Wikimedia Commons Creative Commons (CC-BY-4.0) license

Another way of studying fossil scorpions is through macerated cuticles (e.g., Wills 1947; Jeram 1994a; Shear et al. 1996). Rock containing the fossil material is dissolved in a strong acid, leaving behind any plant or animal cuticles which were inside. These pieces of cuticle can be mounted on microscope slides (Fig. 7.1c), and then drawn and photographed as above. The advantage here is that preservation can be excellent, revealing details of cuticle ornamentation or taxonomically informative structures such as the positions of trichobothria. The disadvantage is that the material is often fragmentary, complete specimens are very rare (but see Jeram 1994a), and this makes it difficult to compare such fossils taxonomically with more traditional fossils in shales. If there are no distinctive, scorpion-specific features (e.g., a pectine or a sting) it may be difficult to determine if the material really is from a scorpion, as opposed to another arachnid or arthropod. It should be added that Bartram et al. (1987) argued that the scorpion exocuticle is unusually resilient, and is thus perhaps preserved more frequently than that of other arachnids.

The final major source of data is amber. Inclusions preserved here can again be either dead animals or moulted exoskeletons. The oldest scorpions in amber are Cretaceous in age (Fig. 7.1d), and several Palaeogene and Neogene amber deposits have also yielded scorpions (reviewed by Lourenço 2023). Copal should also be mentioned here. This subfossil resin has not yet fully hardened into amber and can range in age from only decades to about two and half million years. For a modern definition of the difference between copal and amber see Solórzano-Kraemer et al. (2020). A few scorpions have been found in copal, including one Neotropical record (see below). Photographs of fossils in resin are best made by submerging the specimen in water or glycerine to reduce surface refraction. Inclusions are three-dimensional, but may lie in awkward orientations such that a problem in the past was obtaining sharp images of the entire specimen. Modern image-stacking software has largely overcome this problem and allows parts of a specimen which lie deeper in the matrix to be brought into focus too. Fossils in resins can also be studied using computer tomography, but this technique has not yet been applied to any of the amber scorpions. Suggested protocols for conserving and imaging specimens in amber can be found in, e.g., Sadowski et al. (2021).

A note of caution is the fact that amber and copal, being tree resin, tend to preferentially trap animals which live on or near tree bark (Solórzano-Kraemer et al. 2018). It may not have sampled all the ecological niches in a given habitat and may not reflect the true diversity of the original ecosystem. Buthid scorpions are adept at climbing and often associated with tree bark habitats (e.g., Schawaller 1979a; Lourenço and Pézier 2002). This may partly explain why buthids, or buthid-like scorpions, are the most frequently encountered groups in amber. Other scorpion families may have been similarly abundant at these localities, but rarely came into contact with tree resin. Also large animals, and some extant scorpions, which are quite large, are less likely to be preserved in amber as they would probably have been strong enough to free themselves from the sticky resin. In this context, caution should be exercised when large, and unusually well-preserved scorpions, allegedly in amber, are offered for sale commercially. The possibility that they are modern scorpions encased in plastic resin, or even modern scorpions inserted into genuine amber, needs to be excluded. For further discussion of amber forgeries, and ways to detect them, see Grimaldi et al. (1994) or Eriksson and Poinar (2015).

7.3 Fossils as Calibration Points

Perhaps one of the most important potential roles for fossil scorpions is as calibration points for molecular phylogenies (e.g., Wolfe et al. 2016; Sharma et al. 2018; Howard et al. 2019). As noted above, the oldest scorpions go back to the Silurian. Anatomically modern scorpions can be found in the Late Carboniferous. Fossils potentially assignable to modern families, or at least belonging to living clades, have been proposed from the Mesozoic onwards, and fossils assignable to living Neotropical genera are known from the Miocene. However, when working with fossil scorpions first several challenges in their study and interpretation should be considered; see also comments in Dunlop and Garwood (2024).

7.4 Aquatic or Terrestrial?

A recurring question in scorpion palaeontology is whether the oldest fossils were aquatic or terrestrial animals. This topic has been debated since the late nineteenth century, with Howard et al. (2019) offering a recent review. At one extreme, the monograph of Kjellesvig-Waering (1986) consolidated information on all Palaeozoic and Mesozoic scorpions known at that time, beginning with the presumption that almost all of these fossils were aquatic animals who had gills concealed beneath abdominal plates on the underside of the body. While this position is no longer tenable, for example, Jeram (1990) identified book lungs in a Carboniferous scorpion, questions still remain about the habitat of the Silurian and Devonian species. Complicating the issue is the presence of at least two Devonian species from Germany in the genus *Waeringoscorpio* with what appear to be external respiratory organs protruding from the lateral sides of the body in about the position where we would expect book lungs to occur (Størmer 1970; Poschmann et al. 2008). The latter authors suggested that these scorpion structures are similar to the tracheal gills of secondarily aquatic insects, and speculated that these could be secondarily aquatic scorpion which evolved from terrestrial ancestors.

Other early scorpions show a diversity of leg morphologies (Fig. 7.2), ranging from short, pointed, almost crab-like limbs in the Silurian genus *Palaeophonus*, through to longer, more anatomically modern legs with a short tarsus which could perhaps have been placed flat on the ground as in modern species. The implication here is that fossil scorpions may reveal a transition from aquatic animals to forms better adapted for locomotion in a terrestrial environment (e.g., Waddington et al. 2015). Further discussion of the physiological changes which would have been required when moving from water onto land, with particular reference to scorpions, can be found in Selden and Jeram (1989).

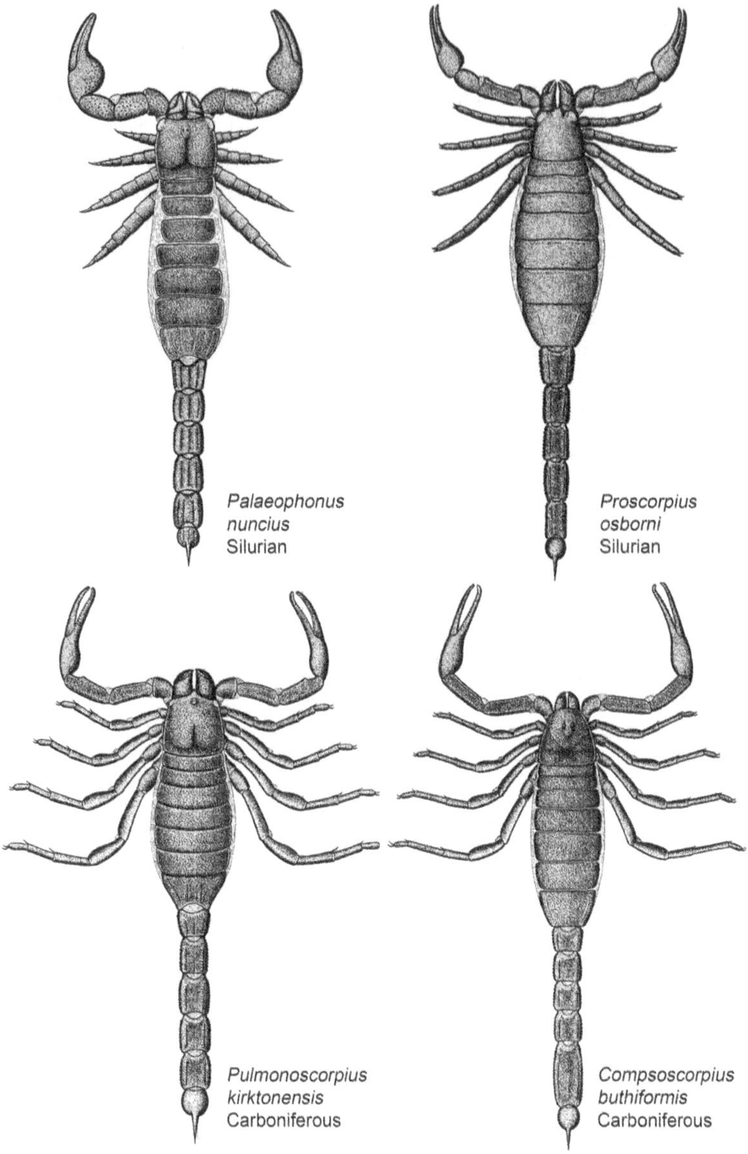

Fig. 7.2 Reconstructions of some of the better known Palaeozoic scorpions (Adapted from Dunlop and Garwood (2024, fig. 9)). *Paleophonus nuncius* from the Silurian of Sweden is a palaeophonid. *Proscorpius osborni* from the Silurian of the USA is a proscorpiid. *Pulmonoscorpius kirktonensis* from the Early Carboniferous of Scotland is a mesoscorpion. *Comsoscorpius buthiformis* from the Late Carboniferous of England belongs to the anatomically modern orthostern scorpions. Note the trend toward longer, thinner legs, and both the reduction of the lateral eyes and the posterior migration of the median eyes on the carapace. Drawings not to scale

7.5 Fossils and Phylogeny

The aquatic versus terrestrial debate is especially relevant for the wider question of where scorpions belong in the arachnid family tree. Related to this, are arachnids monophyletic and did they evolve from a single common terrestrial ancestor, or did several arachnid lineages make the transition from water onto land independently? For further discussion on this topic see Garwood and Dunlop (2023) or Sharma and Gavish-Regev (2025), and references therein. If the oldest scorpions were aquatic, it would imply that there were *at least* two terrestrialisation events across arachnids as a whole. Another group which should be mentioned here are the extinct (Ordovician to Permian) eurypterids. Commonly referred to as sea scorpions, some eurypterids have a distinctly scorpion-like appearance, leading several authors such as Braddy et al. (1999) to suggest that eurypterids could be the sister-group of scorpions or even ancestral to them. Cladistic studies of arachnid relationships (e.g., Shultz 2007; Garwood and Dunlop 2014) did not find strong support for a close phylogenetic relationship between scorpions and sea scorpions.

A prevalent theme in arachnid phylogeny is the suggestion that scorpions are the sister-group of all other arachnids. The long fossil record of the scorpions may have influenced the idea that they were the first group to branch off from the arachnid evolutionary tree. As an example, the highly influential study of Weygoldt and Paulus (1979) recognised a fundamental split into scorpions and non-scorpion arachnids. An alternative hypothesis (Shultz 2007) recovered scorpions closest to harvestmen, forming a group named Stomothecata. Fossil scorpions are relevant to this debate, because the main diagnostic character supporting Shultz's hypothesis was a unique preoral chamber (the stomotheca) formed by projections of the anterior leg coxae. These structures are evidently present in living scorpions, with their well-developed coxapophyses, but several Palaeozoic scorpions differ in the anatomy of this region (Dunlop et al. 2008; Poschmann et al. 2008; Haug et al. 2019) and lack these coxal projections. This implies that a key character of Stomothecata evolved *within* the scorpions, and should not have been present in a putative common ancestor of scorpions and harvestmen.

Living scorpions respire through four pairs of book lungs, and in recent years a consensus has emerged that scorpions are probably related to the other lung-bearing arachnids, i.e., spiders, whip spiders, whip scorpions, schizomids, and a couple of extinct orders. Sharma et al. (2014) proposed the name Arachnopulmonata for this entire group and several lines of evidence, including molecular phylogeny, whole gene duplications, and several anatomical features, now support this hypothesis. Ontano et al. (2021) went further to argue that pseudoscorpions belong in this group too, specifically as the sister-group of scorpions for which the name Panscorpiones was proposed. The significance of fossil data here is in offering a minimum age for these hypotheses. In other words, the oldest unequivocal scorpion, *Dolichophonus loudonensis*, from the mid Silurian Pentland Hills of Scotland implies that both Arachnopulmonata and Panscorpiones must be at least 433–438 million years old.

7.6 Large or Small?

There are no minute fossil scorpions. Most fossils discovered so far have body lengths in the centimetre range (e.g., Kjellesvig-Waering 1986) and in this sense they are similar in size to living scorpion species. If scorpions are most closely related to pseudoscorpions (see above), then the fossil record offers no indication that both groups started as tiny, millimetre-long animals, with scorpions later evolving to the sizes that we know today. Interestingly, at least some Palaeozoic scorpions were demonstrably larger than the largest living species. There is, however, a problem with the very largest putative fossil scorpions. Body lengths of up to a metre have been estimated for two species from the Devonian of England (Kjellesvig-Waering 1972, 1986). However, both are based on fragmentary arthropod material, which does not preserve any diagnostic characters of scorpions such as pectines or a curved sting. One giant scorpion is only based on the isolated free finger of a claw. We should entertain the possibility that they are actually misidentified parts of something like a large decapod crustacean.

There are still at least two unequivocal scorpions from the Early Carboniferous of Scotland which were larger than any living species. *Gigantoscorpio willsi* is somewhat disarticulated, but preserves diagnostic scorpion features such as pectines and has an estimated body length of 30–35 cm (Størmer 1963). *Pulmonoscorpius kirktonensis* is known from a spectacular complete specimen with a length of 28 cm (Jeram 1994a), as well as fragments which suggest it could have reached body lengths approaching 70 cm. The explanation for how and why some scorpions became so large is unclear. We might speculate whether it was related to increased oxygen levels in the Carboniferous, which may also have permitted the evolution of large insects and myriapods. Alternatively, a lack of competition on land may have allowed scorpions to evolve into large predators, with tetrapods only becoming established as fully terrestrial animals toward the end of the Carboniferous.

7.7 Challenges to Their Classification

One of the main difficulties when working with fossil scorpions, at least for older fossils (Fig. 7.2), is the current state of their systematics. At the species/genus level two of the major contributors, Alexander Petrunkevitch and Erik Kjellesvig-Waering, described numerous taxa based on rather trivial differences which could be influenced by the way in which the fossils were preserved, or which might fall within the expected range of variation seen within a modern species. As an example, Legg et al. (2012) recognised nine Carboniferous scorpion species from the British Middle Coal Measures as junior synonyms of *Compsoscorpius buthiformis* (Pocock, 1911). Other scorpion faunas from similarly productive Coal Measures localities in continental Europe and North America have not been revised to this level of detail, but similar results might be expected.

Another problem is species names raised for highly incomplete material, such as an isolated pedipalp claw (Kjellesvig-Waering 1954). These are often best treated as *nomina dubia*, meaning a name of dubious application, as it is difficult to compare such fossils in a meaningful way with more complete specimens. Dunlop and Garwood (2023) addressed two examples of such problematic names, one of which may not even be a scorpion. Another series of fragmentary fossils named by Kjellesvig-Waering (1986) were interpreted as the gills of an aquatic scorpion species, but turned out to be parts of an extinct arthropleurid myriapod (Selden and Shear 1992).

The second issue, again largely affecting geologically older records, is the higher classification of fossil scorpions; see Dunlop and Garwood (2024) for a review. In brief, most current problems stem from the monograph of Kjellesvig-Waering (1986). Here, the majority of the fossil scorpions known at that time were divided into four infraorders, each defined by the condition of the ventral sclerites on the opisthosoma which, as noted above, Kjellesvig-Waering considered to be abdominal plates covering gills. These plates could be entire, or they could show increasing degrees of lobation. Kjellesvig-Waering's scheme thus suffers from being effectively a single character system, such that ostensibly similar-looking fossils ended up in different infraorders based on the supposed shape of their abdominal plates. For a recent critique see, e.g., Dunlop and Garwood (2023), who proposed synonymising two genera artificially split across two infraorders. In addition to this, Kjellesvig-Waering (1986) proposed several new fossil families and superfamilies, rendering his whole classification rather top-heavy with many higher taxa and little resolution of relationships between them.

Subsequent work (Stockwell 1989; Jeram 1994a, b, 1998) applied cladistic methods to recover relationships between fossil scorpions based on a wider range of morphological characters. This resulted in a series of rather better-defined and probably more reliable groups, usually referred to as palaeophonids, proscorpiids, and mesoscorpions as well as the anatomically modern orthostern scorpions (Figs. 7.2 and 7.3). However, there are still some discrepancies between these studies. Stockwell's scheme derives from his PhD thesis, which was never formally published, and Jeram's papers did not cover all of the known Palaeozoic and Mesozoic genera leaving several taxa in an uncertain systematic position. In addition to this, there have also been attempts to address and synonymise some of the apparently superfluous fossil family names (e.g., Dunlop et al. 2008), but in lieu of a comprehensive alternative classification many of Kjellesvig-Waering's taxonomic groups remain in the literature. A fundamental reappraisal of fossil scorpion higher systematics would be welcome.

A final challenge comes from the younger Mesozoic fossils. Several species have been assigned to extinct families which, at least based on their names, are implicitly closely related to living families. For example, Protobuthidae (Lourenço and Gall 2004) from the Triassic of France, Archaeobuthidae (Lourenço 2001) from Cretaceous Lebanese amber, and Palaeoburmesebuthidae (Lourenço 2015) from Cretaceous Burmese amber were all considered buthoids. Among iurids Protochactidae (Magnani et al. 2022) from the Triassic of Switzerland are implicitly

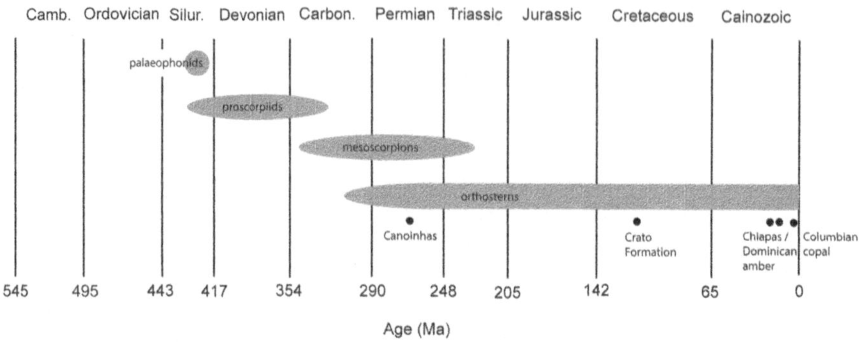

Fig. 7.3 Geological timeline showing the stratigraphic ranges of the major fossil groups as per Fig. 7.2, largely based on Jeram (1998: table 2). Note that this differs markedly from the scheme proposed by Kjellesvig-Waering (1986). All Neotropical scorpion fossils are orthosterns, and the approximate ages of the relevant localities are also indicated: Canoinhas (259–273 Ma), Crato Formation (ca. 115 Ma), Chiapas amber (ca. 20–24 Ma), Dominican amber (ca. 15–20 Ma), and Colombian copal (ca. 2.6–0 Ma). See text for details

close to Chactidae and Palaeoeuscorpiidae (Lourenço 2003) from Cretaceous French and Burmese amber should be related to Euscorpiidae. While the placement of these fossils in extinct families may be justified, cladistic analyses combining both extinct and living families to explicitly test these inferred relationships are largely lacking. Fet and Soleglad (2005: Table 2) listed several Mesozoic families as superfamily *incertae sedis*, while Baptista et al. (2006) cautioned against deriving names for extinct families and genera from modern scorpion groups unless there is convincing evidence that the fossils really are closely related. At least for amber fossils, some of the diagnostic characters used in the systematics of living scorpion species are potentially available. This is less straightforward for older (Palaeozoic) scorpions where character sets used to describe living species, such as patterns of trichobothrial distribution on the pedipalps, are often equivocal.

7.8 The Neotropical Fossils

7.8.1 Permian Record

The Canoinhas locality, a roadside outcrop in the municipality of Canoinhas in Santa Catarina State, southern Brazil, samples the Teresina Formation of the Paraná Basin. This has been dated to the Guadalupian epoch of the Permian, ca. 259–273 Ma. Details of the locality and its geological setting can be found in, e.g., Fanton et al. (2006), who also presented evidence based on the plant fossils that the original locality sampled a somewhat arid environment. Canoinhas has also yielded the oldest Neotropical fossil scorpion (Fig. 7.4), which was described by Martine et al. (2020) as *Suraju itayma*. The holotype is a partially complete specimen, about

Fig. 7.4 The oldest Neotropical scorpion, *Suraju itayma*, from the Permian of Canoinhas, Santa Catarina, Brazil. (**a**) Original specimen. (**b**) Suggested reconstruction of its appearance in life. This species belongs to the anatomically modern orthosterns based on its lung spiracles, but cannot be placed further. (Images reproduced with permission from Martine et al. (2020) and courtesy of Ariel Martine)

40 mm long, preserved in ventral view. It is housed in the Paleontology Scientific Collection of the Institute of Geosciences at the State University of Campinas (UNICAMP).

Suraju itayma has book-lung spiracles opening in the middle of the sternite which allowed Martine et al. (2020) to determine that the fossil was terrestrial and belongs to the anatomically modern Orthosternina group. However, given the state of preservation they were unable to place it further at family level. During the Permian the area which now corresponds to South America was part of the supercontinent Pangea. *Suraju itayma* indicates that scorpions were present at that time in the southern parts of Pangea. It fits a wider pattern of scorpions having been distributed across the whole supercontinent in the Permian, as evidenced by records from Russia (Fet et al. 2011), Germany (Dunlop et al. 2016), and China (Lei et al. 2020) as well as trackways and resting traces possibly left by scorpions from the southern USA (Sadler 1993; Lucas et al. 2013).

7.8.2 Cretaceous Records

Trace fossils, as opposed to body fossils, can be defined as the work of an animal and usually consist of impressions left in the sediment such as burrows, borings, and the resting trace and trackways alluded to above. Study of trace fossils is conventionally referred to as ichnology and these animal impressions are often formally named as ichnogenera and ichnospecies. Criteria for recognising fossil burrows made by scorpions, and other arachnids, were reviewed by Hembree (2023). Perhaps more commonly encountered in the fossil record are arthropod trackways, consisting of several series of footprints or scratch marks arranged in a distinct, repeating pattern; see e.g. Minter et al. (2007) for a review. Some of these are consistent with walking patterns generated by modern arachnids, such as trackways described from the Permian of Arizona by Sadler (1993) which could be directly compared to trackways made by living species.

For the Neotropics, the doctoral thesis of Fernandes (2005) mentioned some putative scorpion trackways from the Jurassic to Cretaceous (ca. 145 Ma) Botucatu Formation found at the São Bento Quarry in Araraquara, São Paulo State, Brazil. These trace fossils were also mentioned by Martine et al. (2020), while Mángano et al. (2020) formally described them as the ichnospecies *Paleohelcura araraquarensis*. Ichnospecies in the ichnogenus *Paleohelcura* have sometimes been interpreted as scorpion trackways, although a general challenge for ichnotaxonomy is the fact that different arthropods can make similar-looking trackways depending on the behaviour of the animal and the type of sediment it was walking over. A corollary of this is that the same animal may be able to produce more than one trace fossil; for example, a scorpion dragging its tail behind it in the sand will leave a different trackway to a scorpion walking with its tail raised up. Mángano et al. (2020) provided evidence that the São Bento fossils were probably produced by insects rather than arachnids, but trackways remain a potentially useful alternative for documenting the presence of groups like scorpions in the fossil record, even when the scorpions themselves have not been recorded.

7.8.3 Crato Formation

The Crato Formation from the Araripe Basin of Ceara Province, northeastern Brazil, is a world famous locality dating to the Early Cretaceous (Aptian: ca. 115 Ma). It preserves a rich fauna of plants as well as both vertebrate and invertebrate animals, with fossils often exquisitely preserved as orange-brown specimens on a pale yellow limestone matrix. In these cases, the original organic material has been replaced by the iron mineral goethite. Barling et al. (2020) discussed in detail the preservation of the insects, and their results may be applicable to arachnids too. Although Crato fossils are usually flattened, a certain degree of three-dimensionality can remain and even fine details like hairs, or hair sockets, can be preserved. With respect to the original habitat, there is evidence that this was a lake, or lake system, surrounded by a semi-arid hinterland. For example, among the arachnids, the Crato Formation preserves camel spiders which are typical for drier environments. For a recent summary of the locality and its palaeoenvironment see Ribeiro et al. (2021), and references therein. These authors suggested that the fossils were deposited in a seasonal, semi-arid, shallow lacustrine wetland, with Lake Chad in Africa as a possible modern analogue.

The first Crato scorpion fossil was described by Campos (1986) as *Araripescorpius ligabuei*. The ca. 27 mm long holotype is partially complete in dorsal view, but missing most of the legs. It comes from the Tatajuba quarry and was deposited in the Seção de Paleontologia da Divisão de Geologia e Mineralogia do Departamento Nacional da Produção Mineral, Rio de Janeiro. The original description was brief and the new genus was effectively diagnosed on a combination of general scorpion features and remarks on the fossil's preservation. Campos (1986) referred the fossil to the superfamily Scorpionoidea, but left it as family *incerate sedis*. Carvalho and Lourenço (2001) tentatively included it in their new family Protoischnuridae (see below), while Baptista et al. (2006) regarded it as a stem chactoid.

Menon (2007) later redescribed *Araripescorpius ligabuei* based on new material in the natural history museums of Stuttgart and Berlin, both in Germany. The genus was redefined as small (ca. 11–27 mm) scorpions with bulky pedipalps and a type 2 sternum, as well as other details such as an oval telson without a subaculear tooth and trichobothria in the "C" pattern. Menon (2007) discussed several characters supporting the assignment of *Araripescorpius ligabuei* to the modern American family Chactidae, drawing particular attention to a "v" trichobothrium on the pedipalp patella which is characteristic of neobothriotaxy and typical for modern South American chactids (e.g., Lira et al. 2018). Chactids are known from several countries in South America, including Brazil. As noted above, if the assignment of *Araripescorpius ligabuei* to Chactidae is correct, it would be the oldest scorpion belonging to a living family.

The second Crato scorpion is *Protoischnurus axelrodorum* described by Carvalho and Lourenço (2001). This species (Fig. 7.5) is larger, up to ca. 40 mm, and was based on a complete male holotype in both dorsal and ventral view in the Museu Nacional, Rio de Janeiro, and several paratypes in the American Museum of Natural History, New York. A specific collecting locality within the Crato Formation was not recorded. Carvalho and Lourenço (2001) placed their new species in Scorpionoidea, noting similarities to both Scorpionidae and Ischnuridae, albeit with some features implying a more plesiomorphic condition compared to their living relatives. For this reason they assigned their species to a new, extinct family, Protoischnuridae, defined on a suite of characters including a pentagonal sternum, pectines with 14 teeth and no fulcra, a type "C" trichobothria pattern again with evidence of neobothriotaxy, and dentate margins of the fixed finger of the claw made up of linear row of granules.

The correct name for Ischnuridae is now Hemiscorpiidae; see Soleglad et al. (2005) for an explanation. Baptista et al. (2006) suggested that *Protoischnurus axelrodorum* could be a stem scorpionoid. Menon (2007) redescribed the species based on new material from the Santana Museum in Brazil, the Manchester Museum, UK, and the two German museums noted above. She also noted that while Carvalho and Lourenço (2001) postulated characters by which Protischnuridae differs from living families, they neither listed nor explicitly defined them. Menon (2007) thus treated Protoischnuridae as a junior synonym of Hemiscorpiidae, documenting five putative hemiscorpiid characters in the fossils: (1) the central suture of the carapace having an inverse Y-shape, (2) trichobothrium Est placed centrally on the ventral

Fig. 7.5 One of the two scorpion species from the Cretaceous Crato Formation of Brazil. *Protoischnurus axelrodorum* belongs either to Hemiscorpiidae or an extinct family Protoischnuridae. (**a**) Well-preserved example held in the Museum für Naturkunde Berlin, also figured by Menon (2007). (**b**–**c**) Specimen described by Carvalho et al. (2023) under light microscopy (**b**) and reconstructed using computer tomography (**c**). (Images (**b**) and (**c**) courtesy of John Maisey)

surface of the pedipalp chela, (3) an internal well-developed projection of the pedipalp patella, (4) the position of the ventral carinae V2 and V3 on the pedipalp chelae, and (5) a sternum longer than wide. If this interpretation is correct it would place another extant scorpion family in the Cretaceous of Brazil, with Menon further arguing that the Crato species belongs in the subfamily Hormurinae. Following Monod and Prendini (2015) this group is now regarded as a separate family, Hormuridae. At least one hormurid genus, *Opisthocanthus*, is found today in the Neotropics (e.g., Lourenço et al. 2018), as well as in other former areas of Gondwana. See also Monod and Prendini (2015: fig. 8) for a biogeographical scenario for the origins of the family.

Scorpions from the Crato Formation were often briefly mentioned in studies of other arthropods from this locality, or in reviews of the original ecosystem. They were also included in a review of the Crato arachnids by Dunlop et al. (2007), who followed Menon's systematics. It should be noted that Lourenço et al. (2018) rejected the synonymy of Protoischnuridae with Hemiscorpiidae as "arbitrary" describing here, and in other papers (e.g., Lourenço and Velten 2021), additional Cretaceous fossils from Burmese amber as protoischnurids. Carvalho et al. (2023) redescribed *Protoischnurus axelrodorum* with the aid of computer microtomography (Fig. 7.5c); the first time this technique has been applied to a Mesozoic scorpion. Using these methods, they were able to recover much of the detailed anatomy of this species. They maintained Protoischnuridae as a valid family and suggested that it may best be interpreted as a stem hormurine.

7.8.4 Miocene Records

The amber from the Dominican Republic is probably the best studied source of terrestrial fossil arthropods from the Neotropics. This amber outcrops on the island of Hispaniola, part of the Greater Antilles, at three main sites in the north and east of the Dominican Republic. As with many ambers, its age has been debated in the literature, but a recent account (Iturralde-Vinent and MacPhee 2019) summarised the evidence for a Miocene age of ca. 15–20 Ma. The amber is thought to be resin from *Hymenaea* trees and presumably reflects a similar climate to that of the island of Hispaniola today. A rich fauna of insects, arachnids, and other invertebrates has been recorded. For a summary of the locality and its broader geological and geochemical setting see, e.g., Penney (2010) and Stach et al. (2021), and references therein.

The first fossil scorpion to be described from Dominican amber was *Centruroides beynai* Schawaller, 1979a, which was placed in a modern genus of the family Buthidae. The holotype is an almost complete and well-preserved juvenile, about a centimetre in body length, which originally came from the private collection of Erich Beyna and was subsequently deposited in the State Museum of Natural History in Stuttgart, Germany. The amber mine which produced the specimen was not recorded. Schawaller (1979a) offered a detailed description, including several characters in support of its assignment to *Centruroides*. At the same time he commented on difficulties when comparing it with its living relatives. About 60 species of *Centruroides* are known from the southern USA through Central America and the Caribbean down into northern South America (Esposito and Prendini 2019, fig. 1A). Some may be diagnosed on unreliable characters, such as colouration. At least six *Centruroides* species are found today on Hispaniola (Teruel et al. 2015).

Armas (1988) even went so far as to consider *Centruroides beynai* as a synonym of one of these living species, *Centruroides nitidus* Thorell, 1876. If this assignment is correct, it would be a rare example of a living arthropod species with a fossil record going back at least 15 million years. This does have some precedence in that Judson (1998) explicitly assigned a Dominican amber pseudoscorpion to a living species while the Dominican amber whip spider described, albeit as an extinct species, by Schawaller (1979b) is barely distinguishable from a modern Hispaniolan form. Esposito and Prendini (2019) considered the assignment of a Miocene amber scorpion to a living species unlikely, but accepted that *C. beynai* is similar to modern Hispaniolan elements and probably belongs to the Greater Antilles clade recovered in their phylogenetic analysis. In terms of ecology, Schawaller (1979a) inferred that the Dominican amber *Centruroides* species probably lived under tree bark and speculated that it may have fed on termites, two examples of which were found in the same piece of amber.

Schawaller (1982) also described the second scorpion species from Dominican amber, another buthid which he named *Tityus ambarensis*. The holotype is a complete and well-preserved juvenile, a little under a centimetre long, also deposited in the State Museum of Natural History in Stuttgart. For this species the original locality is known, namely the Bayaguana mine which is one of the two eastern sites. Again, Schawaller's (1982) description was detailed and several characters in support of its assignment to *Tityus* were documented. In the same paper Schawaller (1982) also described the isolated postabdomen of an evidently larger scorpion, but was unable to place it taxonomically.

Armas (1988) transferred Schawaller's second species to a different genus, *Microtityus*, to give the combination *Microtityus ambarensis*. It should be noted that this transfer was based on the published description, as opposed to restudy of the holotype. Santiago-Blay et al. (1990) obtained a second example of this species from the La Toca mine in the Cordillera Septentrional in the north of the Dominican Republic, which appears from the text to stem from the private collection of J. Yellen. It is another complete specimen, probably representing a second instar and one instar younger than the holotype. Unlike the holotype, the new specimen revealed the presence of small, suboval spiracles and two of the three dorsal metasomal keels. These and other characters (e.g., pedipalp dentition, pedipalp femoral trichobothria, and pectinal tooth count) enabled Santiago-Blay et al. (1990) to confirm the placement of this species in *Microtityus*. As the name implies, *Microtityus* scorpions are noted for their small size, adults usually have a body length of less than 17 mm, and for the large number of modern species both on the mainland of South and Central America and on Caribbean islands, where each species may show quite restrictive habitat preferences and thus a limited geographical distribution (e.g., Teruel 2019). The specific habitat of the amber species remains unknown.

Santiago-Blay et al. (1990) also briefly discussed whether *Microtityus* species in general colonised the Caribbean islands via vicariance or dispersal events. This is a significant question for all the Dominican amber fossil arachnids which were present on Hispaniola at least 15 million years ago. In the vicariance hypothesis, the ancestors of the amber scorpions were already present in the region which now forms the Greater Antilles (Cuba, Hispaniola, Puerto Rico, Jamaica, etc.) from a time when these islands were still connected to a continental land mass. Changes in sea level resulted in the eventual isolation of the faunas on their respective islands. Dispersal implies that scorpions reached the Greater Antilles by sea, for example, by rafting on plant material, at a time when they were already islands and separate from the mainland. Further discussion as related to scorpions can be found in Esposito and Prendini (2019). Differentiating between these hypotheses requires an understanding of the geological development of the Greater Antilles. For example, in the GAARlandia hypothesis (*Greater Antilles Aves Ridge*) there was a putative land bridge during the Eocene/Oligoene, i.e., before the deposition of the Miocene amber, which allowed terrestrial animals to migrate from South America onto what is now Hispaniola and the neighbouring islands. Further evidence in support of this can be found in Philippon et al. (2020) and references therein; see also Esposito and Prendini (2019: fig. 3).

Three more Dominican amber scorpions have been assigned to *Tityus*. The holotype of *Tityus geratus* Santiago-Blay and Poinar, 1988 is a complete specimen, possibly a second instar nymph, nearly 16 mm long, from the La Toca mine held in the private collection of Carlton Rood, Santo Domingo, Dominican Republic. The new species was diagnosed on a range of characters by Santiago-Blay and Poinar (1988), who discussed at some length the differences to Schawaller's (1982) species; which in any case was later transferred to a different genus (see above). In relation to the vicariance versus dispersal hypotheses, Santiago-Blay and Poinar (1988) suggested that *Centruroides* species tend to be hardier and more tolerant in captivity than *Tityus* species. They speculated that *Centruroides* may have been better able to expand their distribution range across Caribbean islands by rafting on things like plant debris. By contrast, in their scenario *Tityus* species are less resilient and vicariance may offer a better explanation. Thirteen living species of *Tityus* are currently known from Hispaniola (Teruel and de los Santos 2018).

The other Dominican amber *Tityus* species are more problematic. *Tityus hartkorni* Lourenço 2009 is a complete specimen, probably a juvenile male, nearly 19 mm long, with the holotype held in the private collection of Joachim Hartkorn (Germany). The mine which yielded the specimen was not recorded. In the original description, it was also placed in the subgenus *Brazilotityus*. Several characters were proposed by Lourenço (2009) to differentiate the new fossil from both living *Tityus* species and *Tityus geratus*. These relate to pectinal tooth count and pectines with a vestigial fulcra, the short spiracles, granulation of the fixed and free finger of the pedipalp, trichobothrial pattern, a smooth carapace and tergites, a pentagonal sternum, and a strong spinoid subaculear tooth on the telson. *Tityus azari* Lourenço, 2013 (Fig. 7.6a) is less complete, the end of the metasoma is missing, and the holotype is again probably a juvenile male estimated at ca. 17 mm long. The holotype is in the personal collection of Jörg Wunderlich (Germany), but should eventually be transferred to the Senckenberg Museum in Frankfurt am Main, Germany. Again, the mine which yielded the specimen was not recorded. The diagnostic characters of *Tityus azari* were similar to those of *T. hartkorni*, with an emphasis on the pectinal fulcra being absent and strong and sharp accessory spinoid granules on the fixed and movable fingers of the pedipalps.

Teruel (2017) reconsidered the Dominican amber *Tityus* fossils and suggested that their slender body shapes, and several other details, justified their placement in the "*crassimanus*" species group together with three of the Recent Hispaniolan taxa. Teruel (2017) was also critical of the species proposed by Lourenço (2009, 2017), arguing that several key observations like the reduced or absent fulcra are misinterpretations based on his reexamination of the original photographs. Teruel also questioned the assignment of one of the species to a putatively Brazilian subgenus without an accompanying biogeographic explanation, and made some general criticisms of Lourenço's approaches to the taxonomy of both modern and fossil scorpions. *Tityus hartkorni* and *T. azari* were thus regarded as junior synonyms of *T. geratus*. This proposal was vigorously rejected by Lourenço (2017), who outlined several challenges in studying amber scorpions (as compared to modern material) and criticised the reliance on photographs to justify taxonomic changes in lieu of studying the original material. Lourenço (2017) thus reinstated his species, which in turn was countered by Teruel and de los Santos (2018) who reaffirmed their opinion that there is only one fossil *Tityus* species (*T. geratus*) in Dominican amber. It is beyond the scope of the present review to resolve this issue, which would require a

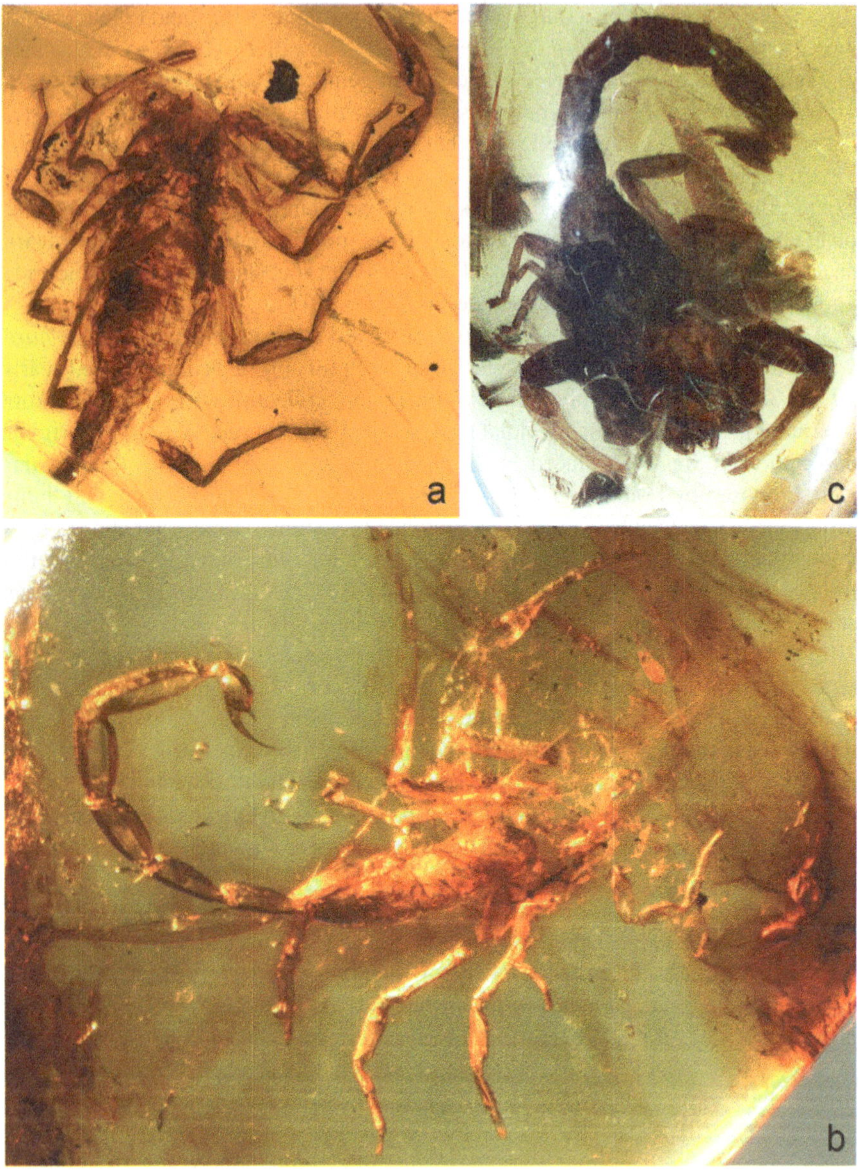

Fig. 7.6 Examples of scorpions preserved in Neotropical ambers. All belong to the family Buthidae and have been placed in living Neotropical genera. (**a**) *Tityus azari* from the Miocene Dominican Republic amber. (**b**) *Rhopalurus renelauerae*, also from Dominican Republic amber. (**c**) *Tityus apozonalli* from the Oligocene/Miocene Chiapas (Mexican) amber. (Images (**a**) and (**b**) courtesy of Wilson Lourenço, (**c**) courtesy of Francisco Riquelme)

high level of familiarity with the diagnostic characters used for modern *Tityus* species.

The most recently described Dominican amber scorpion is *Rhopalurus renelauerae* Lourenço *in* Lourenço and Velten, 2016. The holotype (Fig. 7.6b) is a complete specimen, probably the exuvium of a juvenile male, about 17 mm long, deposited in the personal collection of René Lauer, Wheaton, USA. In their description, Lourenço and Velten (2016) noted (but did not figure) the presence of stridulatory lyriform files on the internal side of the pectinal teeth as their justification for including the fossil in *Rhopalurus* rather than the probably closely related *Centruroides*. Three characters were cited by Lourenço and Velten (2016) as differentiating the amber fossil from living *Rhopalurus* species, namely spiracles ranging from oval and slit-like (albeit very short), a pentagonal sternum, and a telson with an extremely long aculeus with a subaculear tooth which is very long and spinoid. Modern *Rhopalurus* species were revised in depth by Esposito et al. (2017), although they may not have been aware of the fossil species as it was not mentioned in their monograph. An important point to note is that *Rhopalurus* has not been recorded from the modern Hispaniola fauna and is distributed today in northern South America and some of its adjacent islands; see also Esposito and Prendini (2019, fig. 1B). This raises the possibility that the amber fossil has not been assigned correctly at the genus level, as it occurs much further north than any of the modern species.

Chiapas amber, sometimes referred to more generally as Mexican amber, is another important source of fossil data for terrestrial arthropods. It originates from several mines in the Simojovel region of the Sierra Madre de Chiapas in the state of Chiapas in southern Mexico, and thus just inside the Neotropical region as it is usually defined. Chiapas amber was traditionally interpreted as being of a similar (Miocene) age to Dominican amber, although recent data suggests it may be slightly older than the Dominican deposits and close to the Oligogene/Miocene transition at ca. 20–24 Ma (Riquelme et al. 2025). Chiapas amber shares with the Dominican material its usually clear, translucent appearance and its probable origins from *Hymenaea* trees. There is good evidence that it sampled an estuarine environment, such as the presence of several semi-aquatic organisms (Serrano-Sánchez et al. 2015), and a mangrove-like surrounding vegetation has been inferred. For a general overview and the geological setting see, e.g., Solórzano-Kraemer (2010). If the Oligocene/Miocene age is correct, Chiapas amber is also of biogeographical interest for having been deposited at about the time that the Central American Seaway was starting to close (e.g., Montes et al. 2015), creating a permanent landmass between the Americas which facilitated faunal exchange. While the Isthmus of Panama was definitely closed by the Pliocene (ca. 2.8 Ma), there is evidence that it began to emerge as a connection between North and South America during the Miocene (McGirr et al. 2021) potentially offering a land, or perhaps island-hopping, route for lineages coming from the south; see also Esposito and Prendini (2019: fig. 3).

The first scorpion to be described from Chiapas amber was a juvenile, probably a fourth instar male, ca. 17 mm long. It comes from an unspecified mine near Simojovel and is held in the private collection of Dr. Rodolfo Molina. It was described by Santiago-Blay and Poinar (1993) who tentatively assigned it to *Centruroides* sp., a genus previously recorded from Dominican amber (see above). Refraining from formally describing a species, Santiago-Blay and Poinar (1993) mentioned that some characters of the fossil were equivocal and cautioned that there is limited comparative data for juveniles of living members of this genus. They did, however, note that the chelicerae resembled those of the modern Neotropical species *Centruroides gracilis*.

A named fossil belonging to this genus was added later by Lourenço (2017), *Centruroides knodeli*. The holotype is a complete but slightly desiccated juvenile female, ca. 11 mm long, held in the private collection of Herbert Knodel, Rotunda, USA. The precise mine was again not recorded. Lourenço (2017) proposed several differences between *Centruroides knodeli* and living species in this genus, as well as between the new fossil and the specimen described by Santiago-Blay and Poinar (1993). Three characters were highlighted: pectines with 20 teeth, a pentagonal sternum, and a strongly curved aculeus with a strong, but less spinoid, subaculear tooth. Lourenço (2017) again reiterated the observation that the fossil scorpions from Dominican and Chiapas amber are probably closely related to their living Neotropical relatives.

The second genus recorded from Chiapas amber is *Tityus*, another group also known from Dominican amber (see above). The Chiapas species, *Tityus knodeli* Lourenço, 2014a, is known from a complete, but slightly desiccated specimen of a juvenile female, about 12 mm, long which also originates from the Herbert Knodel private collection. Again, it comes from the Simjovel region but the precise mine which yielded the fossil was not recorded. Lourenço (2014a) was slightly more tentative in his generic assignment and discussed the fact that the fossil shares features with both *Tityus* and *Tityopsis*. The situation for *Tityopsis* is complicated by the fact that this genus is now considered endemic to Cuba. In the same year that the Chiapas fossil was described, Francke et al. (2014) transferred the putative Mexican *Tityopsis* species to a new genus: *Chaneke*. There is still an issue in that *Tityus* has not been recorded from the modern Mexican fauna. Lourenço (2014a) went further and assigned his amber fossil to the South American subgenus *Brazilotityus*.

A second species, *Tityus apozonalli* Riquelme, Villegas & González *in* Riquelme et al., 2015, was described a year later, confirming the presence of *Tityus* in the Miocene of Mexico. Unlike most of the Neotropical amber scorpions which are based on juveniles, the holotype is a mature male (Fig. 7.6c), almost 18 mm long. The original locality was recorded as Guadalupe Victoria near Simjovel. In contrast to several previously described amber scorpions from private collections, the holotype is deposited in the Museo del Ámbar de Chiapas, San Cristóbal de Las Casas, Chiapas; see also Riquelme et al. (2015) for a critique of designating holotypes from private collections where the origin of the fossil may be difficult to verify and where accessibility to future generations of researchers is not guaranteed. Riquelme et al. (2015) noted challenges in resolving phylogenetic relationships among the diverse species of *Tityus* and suggested that their new fossil is closest to living members of the subgenus *Archaeotityus* which occurs in South and Central America as well as the Caribbean. The presence of discernible fulcra at the base of the pectinal teeth in the fossil is a key character here. The authors also attempted to resolve the position of the fossil cladistically, recovering it in a basal position with respect to other members of *Tityus*. Note that in their study *Tityus* as a whole was also paraphyletic with respect to *Microtityus* and *Centruroides*.

This still leaves the question of why *Tityus* is absent in modern Mexico? The subgenus *Archaeotityus* occurs today in northern South America and the Caribbean Islands, which implies that this lineage migrated into what is now the Chiapas region from the south, rather than from the north. As noted above, there is evidence that the Central American Seaway was closing during the Miocene. Riquelme et al. (2015) discussed possible explanations for the distribution of the fossil and living taxa, including a suggestion by Lourenço (2014a) that southern Mexican *Tityus* populations were eventually replaced by more opportunistic *Centruroides* species. As Riquelme et al. (2015) correctly noted, this debate relies on a robust hypothesis for the tectonic evolution of southern Mexico. Changes to the geology of the Chiapas region since the Oligocene/Miocene almost certainly impacted on the Mexican terrestrial arthropod fauna.

As a brief aside, Menge (1869) described a putative species of *Tityus* from Baltic amber. The presence of a modern Neotropical genus in the Eocene of Europe would be surprising; a fact commented upon by several authors working both on fossil and living scorpions. The situation is not helped by the fact that Menge's type material housed in Gdańsk (= Danzig) was probably lost during the Second World War, thus a re-examination of his specimen is impossible. This issue was formally addressed by Lourenço and Weitschat (1996) who here, and in subsequent papers, showed that Baltic amber scorpions invariably belong to extinct genera which appear to have Old World affinities. Menge's fossil was presumably misidentified at the genus level and the only unequivocal fossil *Tityus* species are those from the Dominican and Chiapas ambers.

7.8.5 Quaternary Record

The youngest Neotropical scorpion is *Chactas pleistocenicus* (Chactidae) described by Lourenço and Weitschat (2005) from Colombian copal. The holotype is another specimen from the private collection of Herbert Knodel. This fairly complete specimen, about 22 mm long, is a pre-adult female and may be a moulted exuvium. The specimen comes from Penia Blanca in the Province of Santander, Colombia. It is better considered as a subfossil rather than true fossil. The species name alludes to its probably Pleistocene age, but as noted above it is possible that Colombian copal is only decades or hundreds of years old, in which case the specimen would be Holocene.

In this context, it may be worth asking whether *C. pleistocenicus* could actually be a modern species. *Chactas* is an exclusively Neotropical genus found in South and Central America, with most of the known species found in Colombia and Venezuela (e.g., González-Sponga 1996; Lourenço 1997; Flórez et al. 2010; Lira et al. 2018). The original diagnosis of *C. pleistocenicus* highlighted five ventral trichobothria on the chela, six ventral trichobothria on the patella, pectines with 8-8 teeth and no fulcra, and an especially globular vesicle for a female specimen. However, it was not clearly stated how this character combination differs from living representatives of the genus and Lourenço himself later commented on the fact that several living *Chactas* species are very similar to each other and can be difficult to differentiate morphologically (Lourenço 2014b).

Acknowledgements I thank Lauren Esposito and Lorenzo Prendini for helpful comments on Neotropical buthids. For making images of fossil scorpions available, I am grateful to Carsten Kamenz and John Maisey (photos published with permission of the American Museum of Natural History, New York), Wilson Lourenço, Ariel Martine, and Francisco Riquelme. Finally, André Lira is thanked for inviting this chapter and commenting on the typescript.

References

Armas LF (1988) Sinopsis de los escorpiones antillanos. (Editorial Cientifico-Tècnica, 1988), 102 pp

Baptista C, Santiago-Blay JA, Fet V, Soleglad ME (2006) The Cretaceous scorpion genus, *Archaeobuthus*, revisited (Scorpiones: Archaeobuthidae). Euscorpius 35:1–40

Barling N, Martill DM, Heads S (2020) A geochemical model for the preservation of insects in the Crato Formation (Lower Cretaceous) of Brazil. Cretac Res 116:104608. https://doi.org/10.1016/j.cretres.2020.104608

Bartram KM, Jeram AJ, Selden PA (1987) Arthropod cuticles in coal. J Geol Soc 144:513–517

Braddy SJ, Aldridge RJ, Gabbott SE, Theron JN (1999) Lamellate book-gills in a late Ordovician eurypterid from the Soom Shale, South Africa: support for a eurypterid–scorpion clade. Lethaia 32:72–74

Campos DRB (1986) Primeiro registo fóssil de Scorpionoidea na Chapada do Araripe (Cretáceo Inferior) Brasil. An Acad Bras Cienc 58:135–137

Carvalho MGP, Lourenço WR (2001) A new family of fossil scorpions from the Early Cretaceous of Brazil. C R Acad Sci Paris Sci Terre Planèt 332:711–716. https://doi.org/10.1016/S1251-8050(00)01501-9

Carvalho MGP, Maisey JG, Mendes ID, Carvalho IS (2023) Micro-tomographic analysis of a scorpion fossil from the Aptian Crato Formation of Northeastern Brazil. Cretac Res 147:105454. https://doi.org/10.1016/j.cretres.2022.105454

Córdova-Tabares VM, Riquelme F (2023) Annotated catalog of Arachnida in the fossil record of Mexico. Southwest Entomol 48:605–628. https://doi.org/10.3958/059.048.0312

Dunlop JA, Garwood RJ (2023) The status of two fossils assigned to the scorpion genus *Palaeophonus* and its interpretation as a senior synonym of *Allopalaeophonus*. Bull Br Arachnol Soc 19:940–943. https://doi.org/10.13156/arac.2023.19.6.940

Dunlop JA, Garwood RJ (2024) A review of fossil scorpion higher systematics. PeerJ 12:e18557. https://doi.org/10.7717/peerj.18557

Dunlop JA, Menon F, Selden PA (2007) Arachnida: spiders, scorpions and allies. In: Martill DM et al (eds) The Crato Fossil Beds of Brazil. Window into an Ancient World. Cambridge University Press, Cambridge, pp 103–132

Dunlop JA, Tetlie OE, Prendini L (2008) Reinterpretation of the Silurian scorpion *Proscorpius osborni* (Whitfield): integrating data from Palaeozoic and recent scorpions. Palaeontology 51:303–320. https://doi.org/10.1111/j.1475-4983.2007.00749.x

Dunlop JA, Legg DA, Selden PA, Fet V, Schneider JW, Rößler R (2016) Permian scorpions from the Petrified Forest of Chemnitz, Germany. BMC Evol Biol 16:72. https://doi.org/10.1186/s12862-016-0634-z

Dunlop JA, Penney D, Jekel D (2023) A summary list of fossil spiders and their relatives. In World Spider Catalog. Natural History Museum Bern, online at http://wsc.nmbe.ch, version 23.5. Accessed {date of access}

Eriksson ME, Poinar GO Jr (2015) Fake it till you make it—the uncanny art of forging amber. Geol Today 31:21–27. https://doi.org/10.1111/gto.12083

Esposito LA, Prendini L (2019) Island ancestors and new world biogeography: a case study from the scorpions (Buthidae: Centruroidinae). Sci Rep 9:3500. https://doi.org/10.1038/s41598-018-33754-8

Esposito LA, Yamaguti HY, Souza CA, Pinto-da-Rocha R, Prendini L (2017) Systematic revision of the neotropical club-tailed scorpions, *Physoctonus*, *Rhopalurus*, and *Troglorhopalurus*, revalidation of *Heteroctenus*, and descriptions of two new genera and three new species (Buthidae: Rhopalurusinae). Bull Am Mus Nat Hist 415:1–136

Fanton J, Rohn R, Ricardi-Branco F, Rösler O (2006) Canoinhas Outcrop, State of Santa Catarina—unique Record of the Permian Conifer *Krauselcladus* from the Paraná Basin. In: Winge M et al (eds) Sítios Geológicos e Paleontológicos do Brasil. Available on line 26/11/2006 at the address: http://www.unb.br/ig/sigep/sitio126/sitio126english.pdf

Fernandes MA (2005) Paleoicnologia em ambientes desérticos: análise da icnocenose de vertebrados da pedreira São Bento (Formacão Botucatu, Jurássico Superior—Cretáceo Inferior, Bacia do Paraná), Araraquara, SP. Tese de Doutorado, Universidade Federal do Rio de Janeiro (UFRJ), Rio de Janeiro

Fet V, Soleglad ME (2005) Contributions to scorpion systematics. I. On recent changes in high-level taxonomy. Euscorpius 31:1–13

Fet V, Shcherbakov DE, Soleglad ME (2011) The first record of Upper Permian and Lower Triassic scorpions from Russia (Chelicerata: Scorpiones). Euscorpius 121:1–16

Flórez E, Helbig F, Gasca-Álvarez H (2010) Registro del escorpión *Chactas keyserlingi* Pocock, 1893 (Scorpiones, Chactidae) en el dosel de un bosque altoandino de Colombia. Rev Ibér Aracnol 18:97–99

Francke OF, Teruel R, Santibáñez-López CE (2014) A new genus and a new species of scorpion (Scorpiones: Buthidae) from southeastern Mexico. J Arachnol 42:220–232. https://doi.org/10.1636/Ha13-33.1

Garwood RJ, Dunlop JA (2014) Three-dimensional reconstruction and the phylogeny of extinct chelicerate orders. PeerJ 2:e641. https://doi.org/10.7717/peerj.641

Garwood RJ, Dunlop JA (2023) Consensus and conflict in studies of chelicerate fossils and phylogeny. Arachnol Mitt 66:2–16. https://doi.org/10.30963/aramit6602

González-Sponga MA (1996) Guía para identificar Escorpiones de Venezuela. Cuadernos Lagoven. Editorial Artes, Caracas, 203 pp

Grimaldi DA, Shedrinsky A, Ross A, Baer NS (1994) Forgeries of fossils in "amber": history, identification and case studies. Curator Mus J 37:251–274. https://doi.org/10.1111/j.2151-6952.1994.tb01023.x

Haug C, Wagner P, Haug JT (2019) The evolutionary history of body organisation in the lineage towards modern scorpions. Bull Geosci 94:389–408

Hembree DI (2023) Linking burrow morphology to the behaviors of predatory soil arthropods: applications to continental ichnofossils. Palaeontol Electron 26(2):1–30. https://doi.org/10.26879/1257

Howard RJ, Edgecombe GD, Legg DA, Pisani D, Lozano-Fernandez J (2019) Exploring the evolution and terrestrialization of scorpions (Arachnida: Scorpiones) with rocks and clocks. Org Divers Evol 19:71–86. https://doi.org/10.1007/s13127-019-00390-7

Iturralde-Vinent MA, MacPhee RD (2019) Remarks on the age of Dominican amber. Palaeoentomology 2:236–240. https://doi.org/10.11646/palaeoentomology.2.3.7

Jeram AJ (1990) Book-lungs in a Lower Carboniferous scorpion. Nature 343:360–361. https://doi.org/10.1038/343360a0

Jeram AJ (1994a) Scorpions from the Viséan of East Kirkton, West Lothian, Scotland, with a revision of the infraorder Mesoscorpionina. Trans R Soc Edinb 84:283–299

Jeram AJ (1994b) Carboniferous Orthosterni and their relationship to living scorpions. Palaeontology 37:513–550

Jeram AJ (1998) Phylogeny, classification and evolution of Silurian and Devonian scorpions. In Selden PA (ed) Proceedings of the 17th European colloquium of arachnology, Edinburgh pp 17–31

Judson ML (1998) A sternophorid pseudoscorpion (Chelonethi) in Dominican amber, with remarks on the family. J Arachnol 26:419–428

Kjellesvig-Waering EN (1954) Note on a new Silurian (Downtonian) scorpion from Shropshire, England. J Paleontol 28:485–486

Kjellesvig-Waering EN (1972) *Brontoscorpio anglicus*: a gigantic Lower Paleozoic scorpion from central England. J Paleontol 46:39–42

Kjellesvig-Waering EN (1986) A restudy of the fossil Scorpionida of the world. Palaeontogr Am 55:1–287

Kühl G, Bergmann A, Dunlop JA, Garwood RJ, Rust JES (2012) Redescription and palaeobiology of *Palaeoscorpius devonicus* Lehmann, 1944 from the Lower Devonian Hunsrück slate of Germany. Palaeontology 55:775–787. https://doi.org/10.1111/j.1475-4983.2012.01152.x

Legg DA, Garwood RJ, Dunlop JA, Sutton M (2012) A taxonomic revision of orthosternous scorpions from the English Coal-Measures aided by X-ray micro-tomography (XMT). Palaeontol Electron 15(2):1–16. https://doi.org/10.26879/253

Lei X, Zhou W, Wan M, Wei H, Wang B (2020) A new scorpion from a Permian peat swamp in Inner Mongolia, China. Proc Geol Assoc 131:160–167. https://doi.org/10.1016/j.pgeola.2020.02.006

Lira AFA, Pordeus LM, de Souza MB, Guilherme E (2018) Description of the male of *Chactas braziliensis* Lourenço, Aguiar & Franklin 2005 (Scorpiones: Chactidae). Acta Arachnol 67(1):1–5. https://doi.org/10.2476/asjaa.67.1

Lourenço WR (1997) Synopsis de la faune de scorpions de Colombie, avec des considérations sur la systématique et la biogéographie des espèces. Rev Suisse Zool 104:61–94

Lourenço WR (2001) A remarkable scorpion fossil from the amber of Lebanon. Implications for the phylogeny of Buthoidea. C R Acad Sci Paris Sci Terre Planets 332:641–646. https://doi.org/10.1016/S1251-8050(01)01583-X

Lourenço WR (2003) The first scorpion fossil from the Cretaceous amber of France. New implications for the phylogeny of Chactoidea. C R Palevol 2:213–219. https://doi.org/10.1016/S1631-0683(03)00042-3

Lourenço WR (2009) A new species of *Tityus* C. L. Koch, 1836 (subgenus *Brazilotityus* Lourenço, 2006) from the Dominican amber (Scorpiones: Buthidae). Euscorpius 83:1–5

Lourenço WR (2013) A new species of *Tityus* C. L. Koch, 1836 (Scorpiones: Buthidae) from Dominican amber. Euscorpius 156:1–5

Lourenço WR (2014a) A new species of scorpion from Chiapas amber, Mexico (Scorpiones: Buthidae). Rev Ibér Aracnol 24:59–63

Lourenço WR (2014b) The second confirmed record of the scorpion genus *Chactas* Gervais, 1844 (Scorpiones, Chactidae) from Ecuador with description of a new species from the Amazonian Province of Sucumbíos. Zookeys 372:17–26. https://doi.org/10.3897/zookeys.372.6665

Lourenço WR (2015) A new subfamily, genus and species of fossil scorpions from Cretaceous Burmese amber (Scorpiones: Palaeoeuscorpiidae). In Wunderlich J (ed) Beiträge zur Araneologie 9:457–464

Lourenço WR (2017) A new species of *Centruroides* Marx, 1890 from Chiapas amber, Mexico (Scorpiones: Buthidae). Rev Ibér Aracnol 30:100–106

Lourenço WR (2018) A new remarkable scorpion genus and species from Cretaceous Burmese amber (Scorpiones: Protoischnuridae). Arachnida Riv Aracnol Italiana 18:2–14

Lourenço WR (2023) Scorpions trapped in amber: a remarkable window on their evolution over time from the Mesozoic period to present days. J Venom Anim Toxins Incl Trop Dis 29:e20230040. https://doi.org/10.1590/1678-9199-JVATITD-2023-0040
Lourenço WR, Gall J-C (2004) Fossil scorpions from the Buntsandstein (Early Triassic) of France. C R Palevol 3:369–378
Lourenço WR, Pézier A (2002) Additions to the scorpion fauna of the Manaus region (Brazil) with a description of two new species of *Tityus* from the canopy. Amazoniana 17:177–186
Lourenço WR, Velten J (2016) A new species of *Rhopalurus* Thorell, 1876 from Dominican amber (Scorpiones, Buthidae). Rev Ibér Aracnol 29:61–66
Lourenço WR, Velten J (2021) One more new genus and species of scorpion from Early Cretaceous Burmese amber (Scorpiones: Protoischnuridae). Faunitaxys 9(14):1–5
Lourenço WR, Weitschat W (1996) More than 120 years after its description, the enigmatic status of the genus of the Baltic amber scorpion *Tityus eogenus* Menge, 1869 can finally be clarified. Mitt Geol-Paläontol Inst Univ Hamburg 79:183–193
Lourenço WR, Weitschat W (2005) First sub-fossil scorpion of the genus *Chactas* Gervais from Colombian copal (Scorpiones, Chactidae). Mitt Geol-Paläontol Inst Univ Hamburg 89:179–182
Lourenço WR, Wilmé L, Waeber PO (2018) The genus *Opisthacanthus* Peters, 1861 (Scorpiones: Hormuridae), a remarkable Gondwanian group of scorpions. C R Biol 341:131–143. https://doi.org/10.1016/j.crvi.2017.12.003
Lucas SG, Lerner AJ, Voigt S (2013) Scorpionid resting trace from the Lower Permian of southern New Mexico, USA. Ichnos 20:195–201. https://doi.org/10.1080/10420940.2013.845096
Magnani F, Stockar R, Lourenço WR (2022) A new family, genus and species of fossil scorpion from the Meride Limestone (Middle Triassic) of Monte San Giorgio (Switzerland). Faunitaxys 10(24):1–7
Mángano MG, Minter NJ, dos Reis Fernandes LB, Fernandes MA (2020) A new insect trackway from the Upper Jurassic—Lower Cretaceous eolian sandstones of São Paulo State, Brazil: implications for reconstructing desert paleoecology. PeerJ 8:e8880. https://doi.org/10.7717/peerj.8880
Martine AM, Ricardi-Branco F, Beloto B, Jurigan I (2020) *Suraju itayma*: the first Paleozoic fossil scorpion in South America. J S Am Earth Sci 101:102600. https://doi.org/10.1016/j.jsames.2020.102600
McCoy VE, Brandt DS (2009) Scorpion taphonomy: criteria for distinguishing fossil scorpion molts and carcasses. J Arachnol 37:312–320
McGirr R, Seton M, Williams S (2021) Kinematic and geodynamic evolution of the isthmus of Panama region: implications for Central American Seaway closure. GSA Bull 133:867–884
Menge A (1869) Über einen Scorpion und zwei Spinnen in Bernstein. Schr naturwiss Ges Danzig, NS 2:1–9
Menon F (2007) Higher systematics of scorpions from the Crato Formation, Lower Cretaceous of Brazil. Palaeontology 50:185–195
Minter NJ, Braddy SJ, Davis RB (2007) Between a rock and a hard place: arthropod trackways and ichnotaxonomy. Lethaia 40:365–375
Monod L, Prendini L (2015) Evidence for Eurogondwana: the roles of dispersal, extinction and vicariance in the evolution and biogeography of Indo-Pacific Hormuridae (Scorpiones: Scorpionoidea). Cladistics 31:71–111. https://doi.org/10.1111/cla.12067
Montes C, Cardona A, Jaramillo C, Pardo A, Silva JC, Valencia V, Ayala C, Pérez-Angel LC, Rodriguez-Parra LA, Ramirez V, Niño H (2015) Middle Miocene closure of the Central American seaway. Science 348:226–229. https://doi.org/10.1126/science.aaa2815
Ontano AZ, Gainett G, Aharon S, Ballesteros JA, Benavides LR, Corbett KF, Gavish-Regev E, Harvey MS, Monsma S, Santibáñez-López CE, Setton EVW, Zehms JT, Zeh JA, Zeh DW, Sharma PP (2021) Taxonomic sampling and rare genomic changes overcome long-branch attraction in the phylogenetic placement of pseudoscorpions. Mol Biol Evol 38:2446–2467. https://doi.org/10.1093/molbev/msab038

Penney D (2010) Dominican amber. In: Penney D (ed) Biodiversity of fossils in amber from the major world deposits. Siri Scientific Press, Manchester, pp 22–41

Philippon M, Cornée J-J, Münch P, van Hinsbergen DJJ, BouDagherFadel M, Gailler L, Boschman LM, Quillevere F, Montheil L, Gay A, Lebrun JF, Lallemand S, Marivaux L, Antoine P-O, with the GARANTI Team (2020) Eocene intra-plate shortening responsible for the rise of a faunal pathway in the northeastern Caribbean realm. PLoS One 15:e0241000. https://doi.org/10.1371/journal.pone.0241000

Poschmann M, Dunlop JA, Kamenz C, Scholtz G (2008) The Lower Devonian scorpion *Waeringoscorpio* and the respiratory nature of its filamentous structures, with the description of a new species from the Westerwald area, Germany. Paläontol Z 82:418–436

Ribeiro AC, Ribeiro GC, Varejão FG, Battirola LD, Pessoa EM, Simões MG, Warren AG, Riccomini C, Poyato-Ariza FJ (2021) Towards an actualistic view of the Crato Konservat-Lagerstätte paleoenvironment: a new hypothesis as an Early Cretaceous (Aptian) equatorial and semi-arid wetland. Earth Sci Rev 216:103573. https://doi.org/10.1016/j.earscirev.2021.103573

Riquelme F, Villegas-Guzmán G, González-Santillán E, Córdova-Tabares V, Francke OF, Piedra-Jiménez D, Estrada-Ruiz E, Luna-Castro B (2015) New fossil scorpion from the Chiapas amber Lagerstätte. PLoS One 10(8):e0133396. https://doi.org/10.1371/journal.pone.0133396

Riquelme F, Ortega-Flores B, Estrada-Ruiz E, Córdova-Tabares V (2025) Zircon U–Pb ages of the Chiapas Amber-Lagerstätte in the uppermost Simojovel Formation, southwest Mexico. Can J Earth Sci 62:1088–1103. https://doi.org/10.1139/cjes-2024-0100

Sadler CJ (1993) Arthropod trace fossils from the Permian DeChelly sandstone, northeastern Arizona. J Paleontol 67:240–249

Sadowski EM, Schmidt AR, Seyfullah LJ, Solórzano-Kraemer MM, Neumann C, Perrichot V, Hamann C, Milke R, Nascimbene PC (2021) Conservation, preparation and imaging of diverse ambers and their inclusions. Earth Sci Rev 220:103653. https://doi.org/10.1016/j.earscirev.2021.103653

Santiago-Blay JA, Poinar j GO (1988) A fossil scorpion *Tityus geratus* new species (Scorpiones: Buthidae) from Dominican amber. Hist Biol 1:345–354

Santiago-Blay JA, Poinar GO Jr (1993) First scorpion (Buthidae: *Centruroides*) from Mexican amber (Lower Miocene to Upper Oligocene). J Arachnol 21:147–151

Santiago-Blay JA, Schawaller W, Poinar j GO (1990) A new specimen of *Microtityus ambarensis* (Scorpiones, Buthidae), fossil from Hispaniola: evidence of taxonomic status and possible biogeographic implications. J Arachnol 18:115–117

Schawaller W (1979a) Erstnachweis eines Skorpions in Dominikanischem Bernstein (Stuttgarter Bernsteinsammlung: Arachnida, Scorpionida). Stuttg Beitr Naturkd Ser B (Geol Paläontol) 45:1–15

Schawaller W (1979b) Erstnachweis der Ordnung Geiβelspinnen in Dominikanischem Bernstein (Stuttgarter Bernsteinsammlung: Arachnida, Amblypygi). Stuttg Beitr Naturkd Ser B (Geol Paläontol) 50:1–12

Schawaller W (1982) Zwei weitere Skorpione in Dominikanischem Bernstein (Stuttgarter Bernsteinsammlung: Arachnida, Scorpionida). Stuttg Beitr Naturkd Ser B (Geol Paläontol) 82:1–14

Selden PA, Jeram AJ (1989) Palaeophysiology of terrestrialisation in the Chelicerata. Trans R Soc Edinb Earth Sci 80:303–310

Selden PA, Penney D (2017) Imaging techniques in the study of fossil spiders. Earth Sci Rev 166:111–131

Selden PA, Ren D (2017) A review of Burmese amber arachnids. J Arachnol 45:324–343

Selden PA, Shear WA (1992) A myriapod identity for the Devonian "Scorpion" *Tiphoscorpio hueberi*. Ber. Nat.-Med. Verein Innsbruck, Suppl 10:35–36

Serrano-Sánchez ML, Hegna TA, Schaaf P, Pérez L, Centeno-García E, Vega FJ (2015) The aquatic and semiaquatic biota in Miocene amber from the Campo LA Granja mine (Chiapas, Mexico): Paleoenvironmental implications. J S Am Earth Sci 62:243–256

Sharma PP, Gavish-Regev E (2025) The evolutionary biology of Chelicerata. Annu Rev Entomol 70:143–163. https://doi.org/10.1146/annurev-ento-022024-011250
Sharma PP, Kaluziak ST, Pérez-Porro AR, González VL, Hormiga G, Wheeler WC, Giribet G (2014) Phylogenomic interrogation of Arachnida reveals systemic conflicts in phylogenetic signal. Mol Biol Evol 31:2963–2984. https://doi.org/10.1093/molbev/msu235
Sharma PP, Baker CM, Cosgrove JG, Johnson JE, Oberski JT, Raven RJ, Harvey MS, Boyer SL, Giribet G (2018) A revised dated phylogeny of scorpions: phylogenomic support for ancient divergence of the temperate Gondwanan family Bothriuridae. Mol Phylogenet Evol 122:37–45. https://doi.org/10.1016/j.ympev.2018.01.003
Shear WA, Gensel PG, Jeram AJ (1996) Fossils of large terrestrial arthropods from the Lower Devonian of Canada. Nature 384:555–557
Shultz JW (2007) A phylogenetic analysis of the arachnid orders based on morphological characters. Zool J Linnean Soc 150:221–265. https://doi.org/10.1111/j.1096-3642.2007.00284.x
Soleglad ME, Fet V, Kovařík F (2005) The systematic position of the scorpion genera *Heteroscorpion* Birula, 1903 and *Urodacus* Peters, 1861 (Scorpiones: Scorpionoidea). Euscorpius 20:1–37
Solórzano-Kraemer MM (2010) Mexican amber. In: Penney D (ed) Biodiversity of fossils in amber from the major world deposits. Siri Scientific Press, Manchester, pp 42–56
Solórzano-Kraemer MM, Delclòs X, Clapham ME, Arillo A, Peris D, Jäger P, Stebner F, Peñalver E (2018) Arthropods in modern resins reveal if amber accurately recorded forest arthropod communities. Proc Natl Acad Sci 115:6739–6744. https://doi.org/10.1073/pnas.1802138115
Solórzano-Kraemer MM, Delclòs X, Engel MS, Peñalver E (2020) A revised definition for copal and its significance for palaeontological and Anthropocene biodiversity-loss studies. Sci Rep 10:19904. https://doi.org/10.1038/s41598-020-76808-6
Stach P, Natkaniec-Nowak L, Dumańska-Słowik M, Kosakowski P, Naglik B, Drzewicz P, Misiak J, Pršek J, George C, Ramírez-Gómez RE (2021) The study of Dominican amber-bearing sediments from Siete Cañadas and La Cumbre with a discussion on their origin. Sci Rep 11:18556. https://doi.org/10.1038/s41598-021-96520-3
Stockwell SA (1989) Revision of the phylogeny and higher classification of scorpions (Chelicerata). Ph.D. thesis, University of California, Berkeley. Published by University Microfilms International, Ann Arbor, Michigan
Størmer L (1963) *Gigantoscorpio willsi*, a new scorpion from the Lower Carboniferous of Scotland and its associated preying microorganisms. Skrifter Utgitt av det Norske Videnskaps-Akademi I Oslo. Mathematisk-Naturvidenskabelig Klasse 8:1–171
Størmer L (1970) Arthropods from the Lower Devonian (Lower Emsian) of Alken an der Mosel, Germany. Part 1: Arachnida. Senckenb Lethaea 51:335–369
Teruel R (2017) Some taxonomic corrections to the genus *Tityus* C. L. Koch, 1836 (Scorpiones: Buthidae) in Hispaniola. Greater Antilles Euscorpius 242:1–9
Teruel R (2019) A new semidesert *Microtityus* Kjellesvig-Waering, 1966 (Scorpiones: Buthidae) from Southeast Cuba, Greater Antilles. Euscorpius 273:1–15
Teruel R, de los Santos G (2018) Two new *Tityus* CL Koch, 1836 (Scorpiones: Buthidae) from Hispaniola, Greater Antilles. Euscorpius 257:1–16
Teruel R, de Armas LF, Kovařík F (2015) A new species of *Centruroides* Marx, 1890 (Scorpiones: Buthidae) from southern Hispaniola, Greater Antilles. Euscorpius 198:1–18
Waddington J, Rudkin DM, Dunlop JA (2015) A new mid-Silurian aquatic scorpion—one step closer to land? Biol Lett 11:20140815. https://doi.org/10.1111/j.1096-3642.2007.00284.x
Wills LJ (1947) A monograph of the British Triassic scorpions. The Palaeontographical Society, London, 100 & 101, 137 pp
Wolfe JM, Daley AC, Legg DA, Edgecombe GD (2016) Fossil calibrations for the arthropod Tree of Life. Earth Sci Rev 160:43–110. https://doi.org/10.1016/j.earscirev.2016.06.00
Weygoldt P, Paulus HF (1979) Untersuchungen zur Morphologie, Taxonomie und Phylogeny der Chelicerata. Z Zool Syst Evolutionsforsch 17:177–200. https://doi.org/10.1111/j.1439-0469.1979.tb00694.x

Chapter 8
Biogeography of Neotropical Scorpions: An Overview

Adriano Medeiros DeSouza and Alessandra Rodrigues Santos de Andrade

Abstract Scorpions, like other arachnids, play a vital role in environmental equilibrium by regulating the populations of their prey. Their significance, however, extends beyond ecological balance. Certain species pose a notable risk to human health due to their venom, which in some cases can be lethal. This duality underscores the necessity of understanding scorpions' biogeographical distribution and the evolutionary processes that have shaped these patterns over time. Here, we delve into the current state of knowledge surrounding the diversity and endemism of scorpions in the Neotropics. This chapter not only outlines the existing understanding but also highlights the lacunae in our knowledge, presenting both the challenges and opportunities that beckon future research in this field. We aim to provide a comprehensive view, drawing on recent studies and historical data to offer insights into the dynamic interplay of ecological, geographical, and evolutionary factors that define the distribution of Neotropical scorpions.

Keywords Distribution patterns · Endemism · Atlantic Rainforest · Amazonia Forest · Caatinga · Andes · Tepuis · Sierra Madre del Sur · Caribbean archipelago

8.1 Introduction

Biogeography is the field of knowledge that seeks to understand the patterns of diversity and distribution of life on Earth, as well as the processes responsible for their formation (Nelson and Platnick 1981; Brown and Lomolino 1998; Lomolino

A. M. DeSouza (✉)
Departamento de Farmácia, Centro de Ciências Biológicas e da Saúde, Universidade Estadual da Paraíba, Campina Grande, Paraíba, Brazil
e-mail: adriendrix@gmail.com

A. R. S. de Andrade
Laboratório de Mirmecologia, Comissão Executiva do Plano da Lavoura Cacaueira (CEPLAC), Itabuna, Bahia, Brazil

A. F. de Araujo Lira, E. González-Santillán (eds.), *The Biology of Neotropical Scorpions*, https://doi.org/10.1007/978-3-032-23194-9_8

et al. 2010; Sanmartín 2012). Over the centuries, researchers such as naturalist Alfred Russel Wallace (1823–1913) have sought to understand the diversity of Earth's species. The intriguing patterns of distribution and endemism of organisms continue to generate debate among researchers worldwide.

The Neotropical Region is undoubtedly one of the places that has attracted the most interest from researchers to this day. This region, which stretches from southern Mexico to northern Argentina, has a flora and fauna with many endemic species and a unique, complex evolutionary history (Morrone 2014a, b). With wide altitudinal and latitudinal variation, it is possible to find everything from tropical rainforests to savannas, Seasonally Dry Tropical Forests (SDTFs; see Pennington et al. 2000 for details), as well as transitional zones.

To understand the full diversity observed in the Neotropical Region, a multidisciplinary approach is necessary, incorporating contributions from various fields. Areas such as paleoclimatology, for example, provide extremely relevant information about changes in flora. Others, such as population genetics, reveal the genetic dynamics of populations of different organisms over generations (e.g., bottlenecks, population expansions). Therefore, considering only a single perspective on the origin of diversity in this Region can be misleading. Multiple lines of evidence are needed to understand the evolutionary history of this area.

Certainly, much of the observed diversity in the Neotropical Region is directly linked to geological history. Some of the patterns observed, for example, are impossible to interpret without the use of Plate Tectonics, proposed by Wegener in the twentieth century. Contrary to long-held beliefs, the continents are in constant motion. Over millions of years, this has resulted in the isolation and merging of large continental masses. Thus, the Neotropical biota has undergone numerous changes over time. Thus, it is clear that several factors have shaped the Neotropical biodiversity observed today, including scorpions, the subject of this book. Although these arachnids are often primarily remembered for their medical importance, they also perform several functions in Neotropical ecosystems. It is common to associate scorpions with xeric environments such as the Caatinga and Chaco, but there is also significant diversity in humid forest environments like the Atlantic Rainforest and the Amazon Rainforest. In this chapter, we will explore the main findings and conclusions of the most comprehensive studies on the diversity patterns observed in the Neotropical Region.

8.2 Neotropical Scorpions

Despite being considered a relatively low-diversity order of arachnids worldwide, with around 2900 valid species (Rein 2025), scorpions represent a very successful group of arachnids in the Neotropical Region (Table 8.1). We can find species that live in savannas, such as *Physoctonus amazonicus* Lourenço, 2017; species that inhabit dry forests, such as *Jaguajir rochae* (Borelli, 1910); or even species that have managed to colonize cave and island environments, such as *Troglotayosicus*

Table 8.1 Synopsis of the taxonomic status of neotropical scorpion families

Family	Endemic for Neotropical Region	Neotropical genera	Systematic revision	Phylogenetic hypothesis
Ananteridae	Not	2	Not	Not
Bothriuridae	Not	10	Yes[a]	Yes[b]
Buthidae	Not	14	Yes[a]	Yes[b]
Caraboctonidae	Yes	2	Yes	Not[c]
Chactidae	Not	13	Yes[a]	Yes[b]
Diplocentridae	Not	8	Yes[a]	Yes[b]
Euscorpiidae	Not	3	Not	Not
Hadruridae	Not	1	Yes	Yes
Hormuridae	Not	1	Yes	Yes
Troglotayosicidae	Yes	1	Yes	Yes
Typhlochactidae	Yes	4	Yes	Yes
Vaejovidae	Not	9	Yes[a]	Yes[b]

[a]Partially revised taxon
[b]Phylogenetic hypothesis for some taxa
[c]The phylogenies published to date do not discuss the internal relationships of Caraboctonidae, only the taxonomic status of the family

vachoni Lourenço, 1981, and *Heteroctenus junceus* (Herbst, 1800). Therefore, we will provide only the main biogeographic information related to each family below.

Regarding the taxa discussed, the purpose of this text is not to debate the taxonomic validity of disputed clades; therefore, the definitions adopted here for the taxa always follow the most recent publications on the systematics of the group. This approach, for example, is our criterion for treating *Ananteris* Thorell, 1891 as belonging to Ananteridae Pocock, 1900, a distinct family from Buthidae C.L. Koch, 1837. The same criterion is applied to genera and species. We will not discuss groups considered *incertae sedis*, such as *Uroctonus* Thorell, 1876 (see Santibáñez-López et al. 2019 for details).

It is also worth mentioning that the definition of the Neotropical Region adopted in this chapter follows the boundaries proposed by Morrone (2014b). Thus, taxa whose distributions lie mostly outside these limits will not be discussed here, as they fall outside the scope of this work. Likewise, the distribution of invasive or synanthropic species, such as *Tityus serrulatus* Lutz and Melo, 1922, will not be covered here because their distributions relate to broader issues beyond our objective.

8.2.1 *Ananteridae Pocock, 1900*

Ananteridae was recently elevated to family status to include seven extant scorpion genera and one fossil genus that were previously considered part of the Buthidae scorpions (Ythier 2024). The family is found worldwide, with two Neotropical genera: *Ananteris* Thorell, 1891 (with 100 valid species; see Rein 2025 for details) and *Microananteris* Lourenço, 2003 (with 4 species). Neotropical ananterids are represented by small scorpions (ranging from 9 to 41 mm) that inhabit the leaf litter of rainforests, dry forests, and savannah areas such as the Brazilian Cerrado (Lourenço 1982, 2002; Lira et al. 2018; Ythier 2024).

Although they represent such a diverse group, we still know very little about the processes responsible for the diversification of Ananteridae in the Neotropical Region. Ythier (2024) discussed the family's global distribution pattern, which, although not tested by biogeographic methods, appears to align with separation events of the Pangea and Gondwana continents. However, more robust biogeographic studies are needed on Neotropical ananterids. Many species are known only from their type localities, which may indicate a high level of endemism but also highlight a gap in our knowledge.

8.2.2 Bothriuridae Simon, 1880

Bothriuridae is one of the most diverse scorpion families in the world, with over 160 valid species (Rein 2025). In the Neotropical Region, the following genera are recorded: *Bothriurus* Peters, 1861; *Brazilobothriurus* Lourenço and Monod, 2000; *Brachistosternus* Pocock, 1893; *Mauryius* Ojanguren-Affilastro and Mattoni, 2017; *Orobothriurus* Maury, 1975; *Pachakutej* Ochoa, 2004; *Thestylus* Simon, 1880; *Timogenes* Simon, 1880; *Urophonius* Pocock, 1893; and *Vachonia* Abalos, 1954. There are also endemic representatives of the Andean region, which will not be addressed here as they fall outside the scope of this text.

The systematics of Neotropical bothriurids is mostly resolved, with many revisions and phylogenetic hypotheses published in recent decades (Ochoa 2004; Ojanguren-Affilastro and Ramırez 2009; Ojanguren-Affilastro et al. 2010, 2011, 2012, 2016, 2020; Ochoa et al. 2011; Mattoni et al. 2012). Thanks to these studies, we have a good understanding of the distribution and history of most Neotropical genera. Among the most diverse genera, only *Bothriurus* still lacks studies investigating phylogenetic relationships between species.

Neotropical bothriurid species occur in Argentina, Bolivia, Brazil, Chile, Ecuador, Peru, Paraguay, and Uruguay. *Bothriurus* and *Brachistosternus* have wide distributions, with species inhabiting arid regions, Seasonally Dry Tropical Forests (Caatinga, Chaco, etc.), savanna formations (Brazilian Cerrado), and even rainforests (such as the Brazilian Atlantic Rainforest). The distribution pattern of Bothriuridae is connected to various geological events, both ancient and more recent. Sharma et al. (2018) showed that the lineage has a very old history, with diversification occurring between the Devonian and Carboniferous periods, consistent with the separation of the Gondwana continent.

Other geological events apparently had a determining effect on the evolutionary history of Neotropical Bothriuridae. The distribution pattern of *Orobothriurus* is directly associated with the formation of the Andes (Ochoa et al. 2011). The uplift of the Andes was also a major factor in the diversification of *Brachistosternus*, likely by promoting the creation of new habitats (Ceccarelli et al. 2016). The family also includes genera with restricted distributions, such as *Mauryius*, *Pachakutej* (endemic to the Prepuna Province—see Morrone 2014b for details about the Province), and *Thestylus* (endemic to the Brazilian Atlantic Rainforest).

8.2.3 Buthidae C.L. Koch, 1837

Undoubtedly, Buthidae is the most important family of scorpions due to several aspects that will be briefly discussed here. The group presents the greatest species diversity, with over 1300 valid species worldwide (Esposito et al. 2018; Rein 2025), representing almost half of all known scorpion species. In addition, buthids receive significant attention because of the medically important accidents caused by some species within the group (Stockmann and Ythier 2010). Buthid species occur throughout the world, except in Antarctica and New Zealand (Fet and Lowe 2000). This great diversity is also evident in the Neotropical Region.

About one-third of all known Buthidae are endemic to the Neotropics and are grouped into 14 genera: *Alayotityus* Armas, 1973; *Centruroides* Marx, 1890; *Chaneke* Francke, Teruel and Santibanez-López, 2014; *Heteroctenus* Pocock, 1893; *Ischnotelson* Esposito, Yamaguti, Souza, Pinto-da-Rocha and Prendini, 2017; *Jaguajir* Esposito, Yamaguti, Souza, Pinto-da-Rocha and Prendini, 2017; *Mesotityus* González-Sponga, 1981; *Microtityus* Kjellesvig-Waering, 1966; *Physoctonus* Mello-Leitão, 1934; *Rhopalurus* Thorell, 1876; *Tityopsis* Armas, 1974; *Tityus* C. L. Koch, 1836; *Troglorhopalurus* Lourenço, Baptista and Giupponi, 2004; and *Zabius* Thorell, 1893. *Darchenia* Vachon, 1977 was not included due to uncertainty about the true distribution area of this taxon (see Lourenço 1995 for more details).

The great diversity of Neotropical Buthidae species is also reflected in the ecology of the group: buthids are found in all Neotropical ecosystems. There are species that live in savannas, such as *Physoctonus amazonicus* Lourenço, 2017; in humid forests, such as *Tityus brazilae* Lourenço and Eickstedt, 1984; and also species adapted to caves, such as *Troglorhopalurus translucidus* Lourenço, Baptista and Giupponi, 2004. This diversity is also reflected in the size of the animals, ranging from small scorpions, such as *Microtityus* Kjellesvig-Waering, 1966 (around 12 mm), to large ones, such as *Centruroides* Marx, 1890 (reaching over 130 mm) (Stockmann and Ythier 2010).

Despite the great importance of the group and its species diversity, Neotropical buthids still have gaps in taxonomic and sampling knowledge. *Centruroides*, *Microtityus*, and *Tityus*, which represent almost 90% of Neotropical buthids, lack comprehensive systematic reviews or have only been partially reviewed (e.g., Goodman et al. 2021a). In addition, many species are recorded only from their type localities, which may indicate endemism but might also reflect sampling bias. These gaps make it difficult to draw further inferences about the distribution patterns of most Neotropical buthids. However, the taxonomic resolution of the family has been gradually improving in recent years (Esposito et al. 2017; Ojanguren-Affilastro et al. 2017; Esposito et al. 2018; Teruel and Rodríguez-Cabrera 2020; Goodman et al. 2021a, b).

The increased taxonomic resolution of Neotropical Buthidae has enabled advances in understanding the distribution patterns of some species, such as the scorpion fauna of the Caribbean archipelago. Esposito and Prendini (2019) investigated the biogeographic diversification of Centruroidinae, seeking to reconstruct the ancestral distributions of the species and to understand the colonization process of the Caribbean archipelago by these scorpions. The authors concluded that Centruroidinae scorpions colonized the Caribbean in two independent dispersal events. In addition, the study identified a "reverse colonization" event, in which an ancestor from an island dispersed to the mainland.

8.2.4 *Caraboctonidae Kraepelin, 1905*

The family was proposed by Soleglad and Fet (2003) to initially group the genera *Caraboctonus* Pocock, 1893, *Hadruroides* Pocock, 1893 (both in the subfamily Caraboctoninae Kraepelin, 1905), and also *Hadrurus* Thorell, 1876 (included in the subfamily Hadrurinae Stahnke, 1974). Recently, after a phylogenomic analysis of the family, Hadrurinae was elevated to family status (Santibáñez-López et al. 2020), giving Caraboctonidae its current configuration with two valid genera.

Successive studies in recent decades (Ochoa and Chaparro 2008; Ochoa and Prendini 2010; Rossi 2012; Ythier 2021; Ythier and Lourenço 2023) have increased our taxonomic knowledge of the group. Currently, Caraboctonidae includes 26 valid species (almost all belonging to the genus *Hadruroides*) (Rein 2025), represented by small to medium-sized scorpions, with sizes ranging from 29 to 80 mm (Ythier 2021).

Caraboctonidae has a distribution restricted to arid environments in Chile, Ecuador, and Peru (where the greatest species diversity is recorded), as well as in transitional areas with Patagonia. The group also includes species recorded in island environments, such as the Galapagos archipelago. Although many species have restricted distributions, others, such as *Caraboctonus keyserlingi* Pocock, 1893, have widespread distributions and can even adapt to urban environments.

8.2.5 *Chactidae Pocock 1893*

This is a highly diverse taxon, with over 200 valid species (Rein 2025), most of which are endemic to the Neotropical Region, with the only exception being *Nullibrotheas allenii* (Wood, 1863), which occurs in Baja California. They range from small to large scorpions and mainly inhabit the Amazon Rainforest. Chactidae is a taxon that still presents numerous gaps in taxonomic knowledge—only

Chactopsis Kraepelin, 1912, *Chactopsoides* Ochoa, Rojas-Runjaic, Pinto-da-Rocha and Prendini, 2013, and *Megachactops* Ochoa, Rojas-Runjaic, Pinto-da-Rocha and Prendini, 2013 have been recently revised (see Ochoa et al. 2013 for more details). Thus, the genera discussed below may undergo changes in the future.

Lourenço and Duhem (2009) commented on the distribution of *Vachoniochactas* González-Sponga, 1978. According to the study, this is an endemic genus to the Amazon Rainforest, restricted to the Tepuis, which are "table-top" formations dating back to the Proterozoic, exclusive to the Guiana Shield. The five valid species of *Vachoniochactas* known to date are associated with this type of formation, which suggests that the evolutionary history of the group is most likely directly linked to these formations.

Within Chactidae, we also observe genera with disjunct distributions, as is the case with *Hadrurochactas* Pocock, 1893. All nine known species of the genus occur in humid forests (Lourenço 2010; Ythier and Lourenço 2024), with most recorded in the northern Amazon Rainforest. However, *Hadrurochactas brejo* (Lourenço, 1988) and *Hadrurochactas araripe* Lourenço, 2010 occur in formations called "Brejos de Altitude." These formations are isolated enclaves of humid forest within the Caatinga dry forest in Northeastern Brazil and represent evidence of past connections between the Amazon Rainforest and the Brazilian Atlantic Rainforest (Andrade-Lima 1982; Prance 1982; De Oliveira et al. 1999; Werneck et al. 2011; Sobral-Souza et al. 2015).

Another genus that also presents a disjunct distribution is *Teuthraustes* Simon, 1878, with many species recorded in Ecuador and Venezuela, in addition to a few others in the Brazilian and Peruvian Amazon Rainforest (Lourenço and Duhem 2010; Ythier and Lourenço 2017; Lourenço and Ythier 2022). According to Ythier and Lourenço (2017), the observed distribution of the 28 species of the genus, with a large number of species in Ecuador and Venezuela, is linked to an explosive speciation pattern of the group, occurring in the Andes (for species in Ecuador) and in the Tepuis (for species in Venezuela). However, it is also worth highlighting that the group lacks a systematic review and a phylogenetic proposal clarifying the relationships among species, so the observed distribution pattern may represent an artifact caused by gaps in taxonomic knowledge.

8.2.6 Diplocentridae Karsch, 1880

This family is composed of small to large scorpions (20–90 mm), with more than 130 valid species of Diplocentridae (Rein 2025), distributed across the Palearctic, Nearctic, and Neotropical regions (Sissom and Fet 2000). The greatest diversity of the group occurs in the Neotropical Region, with many endemic species. The species recorded for this region are grouped into eight genera: *Cazierius* Francke, 1978; *Cryptoiclus* Teruel and Kovarik, 2012; *Didymocentrus* Kraepelin, 1905; *Diplocentrus* Peters, 1861; *Heteronebo* Pocock, 1899; *Kolotl* Santibáñez-López, Francke and Prendini, 2014; *Oiclus* Simon, 1880; and *Tarsoporosus* Francke, 1978.

The systematics of the group presented many inconsistencies until Francke (1978) revised it, establishing new genera and redefining others. Recently, the family has been undergoing revisions, and consequently, the taxonomic resolution of diplocentid scorpions has been gradually increasing (Santibáñez-López et al. 2013, 2014a, b, 2017). Despite these advances, Diplocentridae still lacks studies aimed at recognizing species distribution patterns and understanding the processes responsible for the observed patterns of diversity and endemism.

8.2.7 Euscorpiidae Laurie, 1896

Euscorpiidae represents a moderate diverse scorpion family in the Neotropical Region, being represented by 13 species distributed in three valid genera: *Megacormus* Karsch, 1880, *Plesiochactas* Pocock, 1900, and *Troglocormus* Francke, 1981 (Rein 2025). Although Euscorpiidae is primarily a Holarctic lineage, its Neotropical representatives constitute a small and geographically restricted component of the regional scorpion fauna (e.g., Trujillo and Armas 2012; Kovarik 2019; Teruel et al. 2023). The taxonomic placement and internal relationships of the family have been the subject of long-standing debate (see Sharma et al. 2018; Santibáñez-López et al. 2022 for details).

Members of Neotropical Euscorpiidae are generally small-medium sized scorpions, characterized by conservative external morphology, cryptic coloration, and a strong association with mesic or montane environments (e.g., Trujillo and Armas 2012; González-Santillán et al. 2017; Kovarik 2019, Teruel et al. 2023). In the Neotropics, both *Megacormus* and *Plesiochactas* are largely confined to montane regions of Mesoamerica, particularly southern Mexico and northern Central America, where they inhabit leaf litter, rocky substrates, and forested habitats (Trujillo and Armas 2012; González-Santillán et al. 2017; Teruel et al. 2023). *Troglocormus* species are restricted to caves in Mexico (Santibáñez-López et al. 2015). Compared to other neotropical scorpion families, euscorpiids exhibit low species richness and relatively narrow ecological tolerances.

From a systematic perspective, species-level taxonomy remains incompletely resolved, and several taxa have historically been described based on limited material (e.g., Zárate-Gálvez and Francke 2009; Trujillo and Armas 2012). Phylogenetic hypotheses suggest that the Neotropical representatives of Euscorpiidae form relictual lineages, likely reflecting ancient dispersal or vicariance events associated with the complex geological history of Mesoamerica.

Despite their restricted diversity, the biogeography of *Megacormus, Plesiochactas,* and *Troglocormus* remains poorly understood. Their patchy and often allopatric distributions suggest a strong influence of historical isolation and habitat specialization. Further integrative studies combining morphology, phylogenomics, and detailed distributional data are needed to clarify species boundaries, evolutionary relationships, and the biogeographic processes that shaped the current distribution of Neotropical Euscorpiidae.

8.2.8 *Hadruridae* Stahnke, 1974

Hadruridae represents one of the least diverse scorpion families recorded in the Neotropics, with nine valid species to date, grouped into two genera: *Hadrurus* Thorell, 1876 and *Hoffmannihadrurus* Fet and Soleglad, 2004. Soleglad and Fet (2003) proposed that the clade constituted a subfamily within Caraboctonidae. However, this view was recently revised based on phylogenomic analyses, and the taxon was subsequently elevated to family status by Santibáñez-López et al. (2020).

The family is composed of large scorpions (up to 130 mm) (Francke and Prendini 2008) typical of arid environments in the United States (which will not be discussed here as they are beyond the scope of this chapter) and Mexico. In general, the genera and species are well-resolved from a taxonomic point of view, with extensive systematic revisions and phylogenetic proposals for the group (Fet and Soleglad 2008; Francke and Prendini 2008; Soleglad and Fet 2010). In the Neotropical Region, only two species occur: *Hoffmannihadrurus aztecus* (Pocock, 1902) and *Hoffmannihadrurus gertschi* (Soleglad, 1976), both restricted to Mexico (Fet and Soleglad 2008) (Fig. 8.1).

Fet and Soleglad (2008) commented on the distribution of *Hoffmannihadrurus* species. Both species exhibit a disjunct distribution relative to the rest of the family. Both *H. aztecus* and *H. gertschi* occur in the southern mainland of Mexico (Fet and Soleglad 2008); however, both species have an allopatric distribution (Fig. 8.1). We still have much to understand about the biogeographic processes that gave rise to these two lineages, and further studies are needed to investigate this topic.

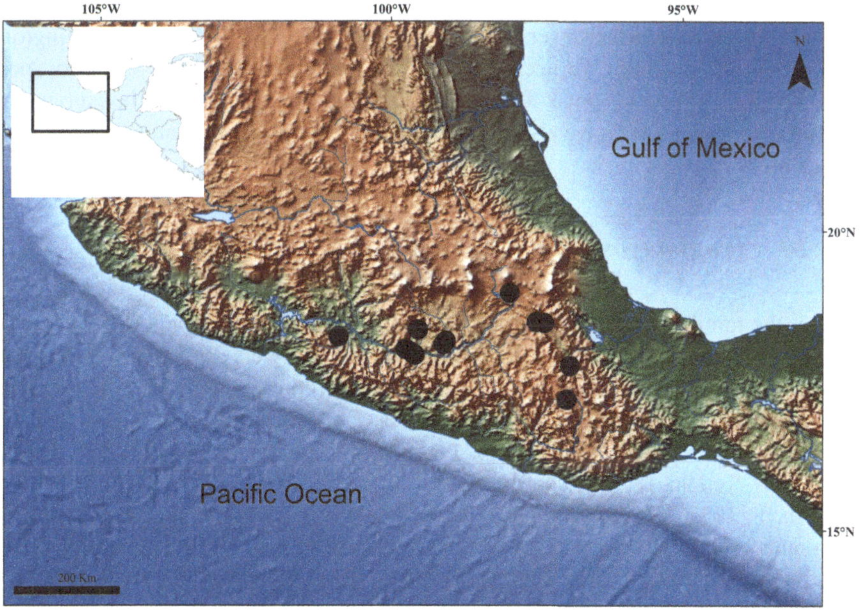

Fig. 8.1 Current distribution of Neotropical Hadruridae scorpion species

8.2.9 Hormuridae Laurie, 1896

This family is composed of medium to large scorpions (some species can reach more than 20 cm) distributed in the Neotropical Region, Africa, Madagascar, and tropical forests in the Indo-Pacific Region (Lourenço 2002; Monod and Prendini 2015; Lourenço et al. 2018). To date, the family comprises 11 genera and 117 valid species (Rein 2025), and in the Neotropical Region, it is represented by 10 species of the genus *Opisthacanthus* Peters, 1861. The Neotropical species belong to two groups within the subgenus *Opisthacanthus*: the cayaporum group (with four species recorded in savannah areas) and the lepturus group (with six species recorded in rainforests) (Lourenço et al. 2018).

The genus *Opisthacanthus* exhibits a Gondwanan distribution pattern, with many species on the African continent and on the island of Madagascar (Lourenço 2002; Lourenço et al. 2018). Regarding the Neotropical *Opisthacanthus*, much remains to be learned about the processes responsible for their diversification in this region. There are still knowledge gaps concerning the distribution of these species, with many known only from their type localities. In addition, hypotheses regarding the origin of the group need to be tested using analytical methods, making it difficult to infer possible biogeographic barriers.

According to Lourenço (2002), the current disjunct distribution of Neotropical *Opisthacanthus* species represents a relict of a past, broader distribution. Thus, the genus's range was probably much larger in the past, shifting according to climatic fluctuations throughout Earth's geological history, especially during the Quaternary period (Lourenço 2017). Such fluctuations—particularly glaciation events—led to the contraction of forest formations adapted to humid environments and the expansion of formations adapted to xeric environments (Prado and Gibbs 1993; Pennington et al. 2000; Mayle 2006; Werneck 2011; Werneck et al. 2011). In this scenario, species adapted to these open formations, such as scorpions of the subgenus *Opisthacanthus*, would have also expanded their distributions (Lourenço 2017).

8.2.10 Troglotayosicidae Lourenço, 1998

This is the family with the lowest species diversity in the Neotropical Region, comprising only one genus and seven valid species to date (Rein 2025). It constitutes a group of scorpions that for more than 20 years was represented by a single species, *Troglotayosicus vachoni* Lourenço, 1981, described from the Los Tayos cave in Ecuador (Moreno-González et al. 2024). Initially, a phylogenetic relationship with *Belisarius* Simon, 1879, a scorpion genus from the Pyrenees, was proposed by Lourenço (1981). Thus, *T. vachoni* was considered to belong to the Chactidae (Lourenço 1981). However, the taxonomic position of *T. vachoni* within Chactidae was questioned until Lourenço (1998) finally created Troglotayosicidae to accommodate two genera: *Troglotayosicus* and *Belisarius*. The latter was recently removed from the family, resulting in the current taxonomic arrangement.

Troglotayosicidae is composed of small (15–32 mm), troglomorphic scorpions that inhabit caves or are directly associated with the soil (Botero-Trujillo et al. 2021; Moreno-González et al. 2024). The family has a restricted distribution, being recorded only in Colombia and Ecuador (Moreno-González et al. 2024) (Fig. 8.2). Only one species has been recorded in more than one location, and there is no sympatry among the species. The phylogenetic relationships were recently elucidated by Moreno et al. (2024). Some species, such as *T. vachoni*, may have extremely restricted distributions; however, as these scorpions are difficult to sample, there are likely still gaps in our knowledge of this enigmatic taxon.

8.2.11 Typhlochactidae Mitchell, 1971

Typhlochactidae constitutes one of the least diverse and most enigmatic families of scorpions in the Neotropical Region, with animals ranging from small to large (10–70 mm) (Vignoli and Prendini 2009). The family is composed of 11 species (Rein 2025), grouped into two subfamilies: Alacraninae Vignoli and Prendini, 2009, which has one valid genus, *Alacran* Francke, 1982 (with three species), and Typhlochactinae Mitchell, 1971, which includes the genera *Typhlochactas* Mitchell, 1971 (with six species), *Sotanochactas* Francke, 1986, and *Stygochactas* Vignoli and Prendini, 2009, the latter two being monotypic genera (Vignoli and Prendini 2009; Prendini et al. 2010; Santibáñez-López et al. 2014c).

The taxon as a whole has very few records, with few specimens collected to date. All species of Typhlochactidae recorded so far are endemic to eastern Mexico and exhibit varying degrees of troglomorphism (Vignoli and Prendini 2009; Prendini et al. 2010) (Fig. 8.3). *Alacran*, *Sotanochactas*, and *Stygochactas* are exclusively hypogean, while *Typhlochactas* includes both hypogean and endogean species (Vignoli and Prendini 2009). Troglomorphism in Typhlochactidae seems to be directly related to the evolutionary history of the group, as discussed below.

The species of Typhlochactidae were reviewed by Vignoli and Prendini (2009), and subsequently, a proposal for the phylogenetic relationships among species was presented by Prendini et al. (2010) (until then, nine species were recognized). In addition to confirming the monophyly of the group, Prendini et al. (2010) demonstrated that the endogean species of Typhlochactidae evolved from lineages with hypogean habits, and that this event occurred more than once. Contrary to previous assumptions, the hypogean condition is ancestral within the family (Prendini et al. 2010).

When describing a third species of *Alacran*, Santibáñez-López et al. (2014c) proposed a hypothesis of phylogenetic relationships among the valid species of the genus and suggested a hypothesis to explain their distribution pattern. According to the authors, the three valid species of the genus evolved from a lineage widely distributed in the cave system of the Sierra Madre del Sur, in eastern Mexico. Initially, the ancestral population was panmictic, using as a corridor an extensive groundwater system present in the region. Then, following several vicariance events, this

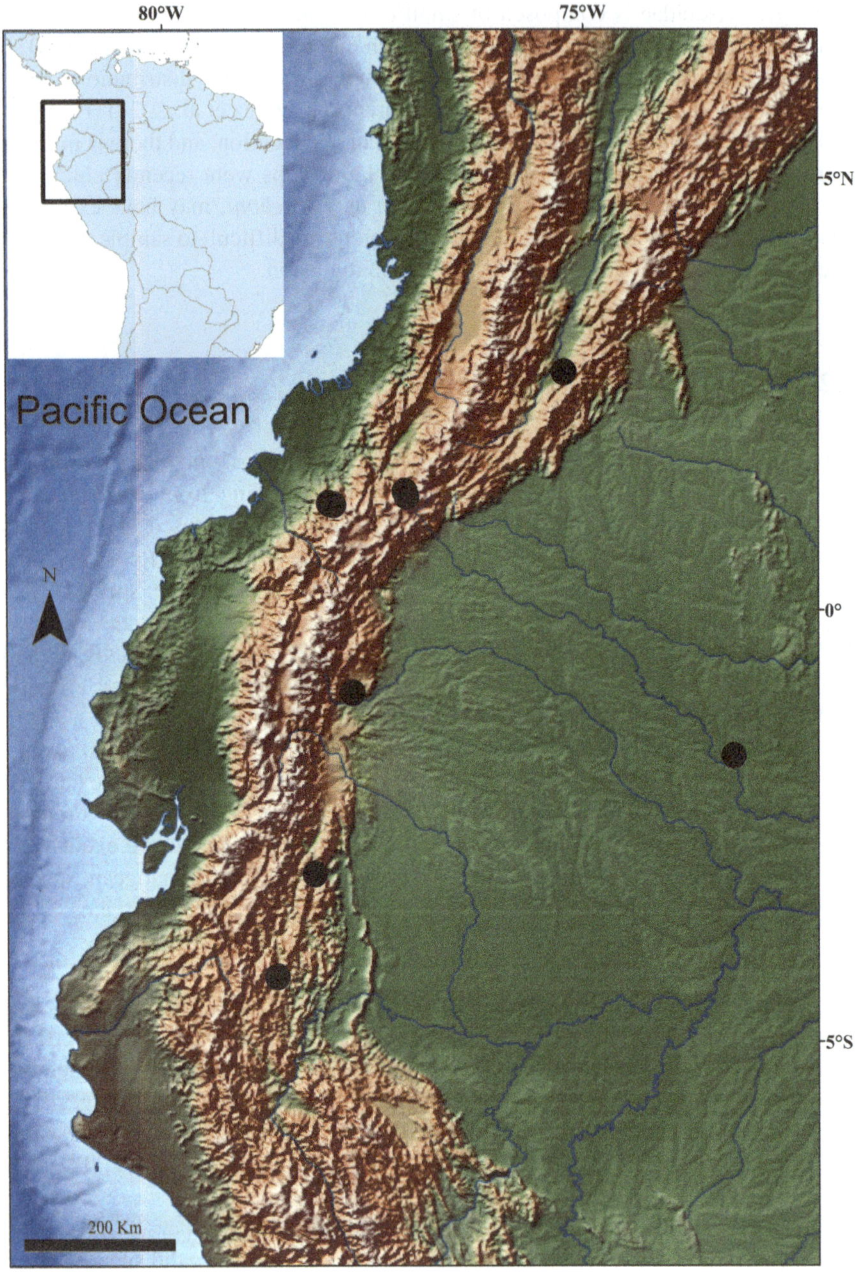

Fig. 8.2 Current distribution of Troglotayosicidae scorpion species

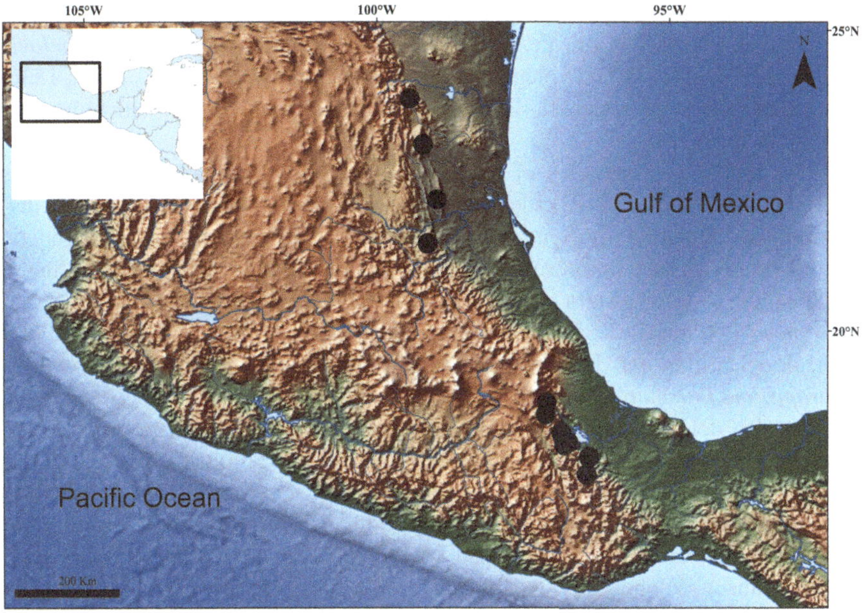

Fig. 8.3 Current distribution of Typhlochactidae scorpion species

corridor was interrupted, isolating the populations that subsequently diverged and gave rise to the currently known *Alacran* species (Santibáñez-López et al. 2014c).

8.2.12 *Vaejovidae Thorell, 1876*

Vaejovidae comprises 25 genera and more than 230 species distributed from southwestern Canada to Guatemala, occurring in arid, semi-arid environments, highland forests, caves, and deserts (Bryson et al. 2013). The species in this family vary greatly in size, with representatives ranging from just 14 mm to almost 94 mm in length, reflecting their broad ecological and morphological diversity (González-Santillán and Prendini 2013). The family is predominantly represented by species with a Nearctic distribution, but studies in recent decades have shown that its geographic distribution extends to the Neotropical Region (Santibáñez-López et al. 2019), according to the biogeographic divisions of Morrone (2014b).

The diversity of Vaejovidae is organized into three subfamilies: Syntropinae, Smeringurinae, and Vaejovinae, although their monophyly requires additional phylogenetic confirmation (Santibáñez-López et al. 2019). In the Neotropical Region, more than 50 species are known, distributed among nine genera: *Balsateres* González-Santillán and Prendini, 2013; *Chihuahuanus* González-Santillán and Prendini, 2013; *Franckeus* Soleglad and Fet, 2005; *Konetontli*

González-Santillán and Prendini, 2013; *Kuarapu* Francke and Ponce-Saavedra, 2010; *Mesomexovis* González-Santillán and Prendini, 2013; *Paravaejovis* Williams, 1980; *Thorellius* Soleglad and Fet, 2008; and *Vaejovis* C.L. Koch, 1836. Among the Neotropical genera, *Vaejovis* is the most diverse taxon, with more than 25 known species.

Regarding the systematics of the family, the study conducted by González-Santillán and Prendini (2013) advanced the understanding of the diversity and distribution of the subfamily Syntropinae, establishing its monophyly. Phylogenetic analyses demonstrated that several previous classifications were artificial, leading to the creation of new genera such as *Balsateres*, *Chihuahuanus*, *Konetontli*, and *Mesomexovis* (González-Santillán and Prendini 2013). Subsequently, the study by González-Santillán and Prendini (2018) consolidated the validity of *Thorellius*, *Kuarapu*, and *Balsateres*. When designating the neotype of *Balsateres cisnerosi*, the authors also delimited the species' distribution in the Balsas Depression, in the states of Michoacán and State of Mexico. The geographic limits of *Thorellius cristimanus* and *T. intrepidus* were also redefined. For *T. cristimanus*, the distribution was confirmed in the states of Colima, south-central Jalisco, and northeastern Michoacán. *Thorellius intrepidus* had its Neotropical distribution confirmed in Colima, Jalisco, Michoacán, State of Mexico, Guerrero, and Nayarit.

Taxonomic advances in Syntropinae have revealed an endemic fauna, with many species associated with specific habitats, such as tropical deciduous forests, arid zones, mountainous regions, and isolated geological formations. The current distribution pattern of Syntropinae species reflects a complex biogeographic history, marked by events of geographic isolation (González-Santillán and Prendini 2013). For González-Santillán and Prendini (2013), the distribution of the ancestral clade that includes *Balsateres*, *Kuarapu*, and *Thorellius* probably covered central and northern Mexico to the southwestern United States, being later fragmented by geological events such as the separation of the Baja California Peninsula. Furthermore, the extreme isolation of the Balsas Depression may have promoted the evolution of an endemic scorpion fauna, including the stenotopic genera *Kuarapu* (lithophile) and *Balsateres* (psamophile) (González-Santillán and Prendini 2013).

Despite its diversity, Vaejovidae still lacks biogeographic studies focused specifically on Neotropical species. Most of the available interpretations derive from specific analyses or concentrate on transition zones, leaving large gaps regarding the historical and ecological processes that shaped the diversity of the clade as a whole. Thus, comprehensive studies are needed, incorporating multiple sources of evidence (spatial, phylogenetic, ecological data, etc.), to reconstruct the evolutionary history of the group in the region. This will allow us to understand the patterns of diversity and endemism of Neotropical Vaejovidae, as well as which processes shaped this taxon of scorpions in the Neotropical Region.

8.3 Final Considerations

As presented so far, it is possible to observe the great diversity of Neotropical scorpions. With approximately one-third of all known species worldwide, we still have much to understand about the distribution patterns of scorpions living in the Neotropical Region. Taxa such as Chactidae (Table 8.1) still lack systematic reviews and robust phylogenetic proposals. Thus, taxonomic uncertainties are among the main obstacles impeding the advancement of knowledge about the group. In addition, many taxa exhibit varying degrees of sampling; many species are undersampled, which makes it impossible to accurately assess their degree of endemism. Therefore, increasing our knowledge of endemism patterns in Neotropical scorpions is directly linked to improvements in taxonomic understanding as well as enhanced sampling efforts for the group.

Acknowledgments We are grateful to the editors for the invitation. We also thank André Lira for revision, his valuable comments and suggestions on the manuscript.

References

Andrade-Lima D (1982) Present day forest refuges in Northeastern Brazil. In: Prance GT (ed) Biological diversification in the tropics. Columbia University Press, New York, pp 245–254

Botero-Trujillo R, Ochoa JA, Prendini L (2021) A new troglomorphic, leaf-litter scorpion from Ecuador (Troglotayosicidae: *Troglotayosicus*). Am Mus Novit 3981:1–24. https://doi.org/10.1206/3981.1

Brown JH, Lomolino MV (1998) Biogeography. Sinauer Associates, Sunderland

Bryson RW Jr, Savary WE, Prendini L (2013) Biogeography of scorpions in the *Pseudouroctonus minimus* complex (Vaejovidae) from South-Western North America: implications of ecological specialization for pre-Quaternary diversification. J Biogeogr 40(10):1850–1860. https://doi.org/10.1111/jbi.12134

Ceccarelli FS, Ojanguren-Affilastro AA, Ramírez MJ, Ochoa JA, Mattoni CI, Prendini L (2016) Andean uplift drives diversification of the bothriurid scorpion genus *Brachistosternus*. J Biogeogr 43(10):1942–1954. https://doi.org/10.1111/jbi.12760

De Oliveira PE, Barreto AMF, Suguio K (1999) Late Pleistocene/Holocene climatic and vegetational history of the Brazilian Caatinga: the fossil dunes of the middle São Francisco River. Palaeogeogr Palaeoclimatol Palaeoecol 152:319–337. https://doi.org/10.1016/S0031-0182(99)00061-9

Esposito LA, Prendini L (2019) Island ancestors and new world biogeography: a case study from the scorpions (Buthidae: Centruroidinae). Sci Rep 9(1):3500. https://doi.org/10.1038/s41598-018-33754-8

Esposito LA, Yamaguti HY, Souza CA, Pinto-da-Rocha R, Prendini L (2017) Systematic revision of the Neotropical club-tailed scorpions, *Physoctonus*, *Rhopalurus*, and *Troglorhopalurus*, revalidation of *Heteroctenus*, and descriptions of two new genera and three new species (Buthidae: Rhopalurusinae). Bull Am Mus Nat Hist 415:1–136

Esposito LA, Yamaguti HY, Pinto-da-Rocha R, Prendini L (2018) Plucking with the plectrum: phylogeny of the New World buthid scorpion subfamily Centruroidinae Kraus, 1955 (Scorpiones: Buthidae) reveals evolution of three pecten-sternite stridulation organs. Arthropod Syst Phylogeny 76:87–122. https://doi.org/10.3897/asp.76.e31942

Fet V, Lowe G (2000) Family Buthidae. In: Fet V, Sissom WD, Lowe G, Braunwalder ME (eds) Catalog of the scorpions of the world (1758–1998). New York Entomological Society, New York

Fet V, Soleglad ME (2008) Cladistic analysis of superfamily Iuroidea, with emphasis on subfamily Hadrurinae (Scorpiones: Iurida). Bol Soc Entomol Aragon 43:255–281

Francke OF (1978) Systematic revision of diplocentrid scorpions (Diplocentridae) from circum-Caribbean lands. Special Publications The Museum Texas Tech University 14:1–92

Francke OF, Prendini L (2008) Phylogeny and classification of the giant hairy scorpions, *Hadrurus* Thorell (Iuridae Thorell): a reappraisal. Syst Biodivers 6(2):205–223. https://doi.org/10.1017/S1477200008002697

González-Santillán E, Prendini L (2013) Redefinition and generic revision of the North American vaejovid scorpion subfamily Syntropinae Kraepelin, 1905, with descriptions of six new genera. Bull Am Mus Nat Hist 382:1–71

González-Santillán E, Prendini L (2018) Systematic revision of the North American Syntropine Vaejovid scorpion genera *Balsateres*, *Kuarapu*, and *Thorellius*, with descriptions of three new species. Bull Am Mus Nat Hist 420:1–81

González-Santillán E, González-Ruiz JM, Escobedo-Morales LA (2017) A new species of *Megacormus* (Scorpiones, Euscorpiidae) from an oak-pine forest in Guanajuato, Mexico with an identification key to the species in the genus. Zootaxa 4299:221–237

Goodman A, Prendini L, Francke OF, Esposito LA (2021a) Systematic revision of the arboreal Neotropical "*thorellii*" clade of *Centruroides* Marx, 1890, bark scorpions (Buthidae C.L. Koch, 1837) with descriptions of six new species. Bull Am Mus Nat Hist 452:1–92

Goodman A, Prendini L, Esposito LA (2021b) Systematics of the arboreal Neotropical '*thorellii*' clade of *Centruroides* bark scorpions (Buthidae) and the efficacy of mini-barcodes for museum specimens. Diversity 13:441. https://doi.org/10.3390/d13090441

Kovarik F (2019) Review of *Megacormus* Karsch, 1881, with description of a new species (Scorpiones: Euscorpiidae). Euscorpius 296:1–46

Lira AFA, DeSouza AM, Albuquerque CMR (2018) Environmental variation and seasonal changes as determinants of the spatial distribution of scorpions (Arachnida: Scorpiones) in Neotropical forests. Can J Zool 96(9):963-972. https://doi.org/10.1139/cjz-2017-0251

Lomolino MV, Riddle BR, Whittaker RJ, Brown JH (2010) Biogeography, 4th edn. Sinauer Associates, Sunderland

Lourenço WR (1981) Scorpions cavernicoles de l'Equateur: *Tityus demangei* n. sp. et *Ananteris ashmolei* n. sp. (Buthidae): *Troglotayosicus vachoni* n. gen., n. sp. (Chactidae), scorpion troglobie. Bull Mus Natl Hist Nat Sect A Zool Biol Ecol Anim 2:635–662

Lourenço WR (1982) Révision du genre *Ananteris* Thorell, 1891 (Scorpiones, Buthidae) et description de six espèces nouvelles. Bull Mus Natl Hist Nat Paris, 4e sér 4(A1/2):119–151

Lourenço WR (1995) *Darchenia*, a genus of scorpion which was not American. Biogeogr C R Séances Soc Biogéogr 71(4):197–199

Lourenço WR (1998) Panbiogéographie, les distributions disjointes et le concept de famille relictuelle chez les scorpions. Biogeographica 74:133–144

Lourenço WR (2002) Scorpions of Brazil. Les Édition de l'If, Paris. 308p

Lourenço WR (2010) The disrupted pattern of distribution of the genus *Hadrurochactas* Pocock; evidence of past connections between Amazon and the Brazilian Atlantic forest. C R Biol 333:41–47. https://doi.org/10.1016/j.crvi.2009.11.002

Lourenço WR (2017) Description of a new species of *Opisthacanthus* Peters (Scorpiones: Hormuridae) from Suriname/Brazil border with some biogeographic considerations. Acta Biol Paranaense 46(1–2):9–22

Lourenço WR, Duhem B (2009) The genus *Vachoniochactas* Gonzalez-Sponga (Scorpiones, Chactidae), a model of relictual distribution in past refugia of the Guayana region of South America. C R Biol 332:1085–1091. https://doi.org/10.1016/j.crvi.2009.09.006

Lourenço WR, Duhem B (2010) The geographical pattern of distribution of the genus *Teuthraustes* Simon (Scorpiones, Chactidae) in South America and description of a new species. C R Biol 333:858–863. https://doi.org/10.1016/j.crvi.2010.09.005

Lourenço WR, Ythier E (2022) The genus *Teuthraustes* Simon, 1878 (Scorpiones: Chactidae) in Brazil with description of a new species. Rev Ibér Aracnol 40:119–125

Lourenço WR, Wilmé L, Waeber PO (2018) The genus *Opisthacanthus* Peters, 1861 (Scorpiones: Hormuridae), a remarkable Gondwanian group of scorpions. C R Biol 341(2):131–143. https://doi.org/10.1016/j.crvi.2017.12.003

Mattoni CI, Ochoa JA, Ojanguren-Affilastro AA, Prendini L (2012) *Orobothriurus* (Scorpiones: Bothriuridae) phylogeny, Andean biogeography, and the relative importance of genitalic and somatic characters. Zool Scr 41:160–176. https://doi.org/10.1111/j.1463-6409.2011.00508.x

Mayle FE (2006) The late Quaternary biogeographical history of South American Seasonally Dry Tropical Forests: insights from palaeo-ecological data. In: Pennington RT, Lewis GP, Ratter JA (eds) Neotropical savannas and seasonally dry forests: plant diversity, biogeography and conservation. CRC Press Taylor & Francis Group, Boca Raton, pp 395–416

Monod L, Prendini L (2015) Evidence for Eurogondwana: the roles of dispersal, extinction and vicariance in the evolution and biogeography of Indo-Pacific Hormuridae (Scorpiones: Scorpionoidea). Cladistics 31(1):71–111. https://doi.org/10.1111/cla.12067

Moreno-González JA, Luna-Sarmiento DA, Prendini L (2024) Phylogeny of the troglomorphic scorpion genus *Troglotayosicus* (Scorpiones: Troglotayosicidae) with description of a new species from Colombia. Am Mus Novit 4011:1–39

Morrone JJ (2014a) Cladistic biogeography of the Neotropical region: identifying the main events in the diversification of the terrestrial biota. Cladistics 30:202–214. https://doi.org/10.1111/cla.12039

Morrone JJ (2014b) Biogeographical regionalisation of the Neotropical region. Zootaxa 3782(1):1–110. https://doi.org/10.11646/zootaxa.3782.1.1

Nelson G, Platnick N (1981) Systematics and biogeography —cladistics and Vicariance. Columbia University Press, New York

Ochoa JA (2004) Filogenia del género *Orobothriurus* y descripción de un nuevo género de Bothriuridae (Scorpiones). Rev Ibér Aracnol 9:43–73

Ochoa JA, Chaparro JC (2008) Nueva especie de escorpión del género *Hadruroides* (Scorpiones: Caraboctoninae) de los vales interandinos de Perú. Rev Peru Biol 15(1):5–10

Ochoa JA, Prendini L (2010) The genus *Hadruroides* Pocock, 1893 (Scorpiones, Iuridae), in Peru: new records and descriptions of six new species. American Museum of Natural History 3687:1–56

Ochoa JA, Ojanguren-Affilastro AA, Mattoni CI, Prendini L (2011) Systematic revision of the Andean scorpion genus *Orobothriurus* Maury, 1976 (Bothriuridae), with discussion of the altitude record for scorpions. Bull Am Mus Nat Hist 359:1–90. https://doi.org/10.1206/359.1

Ochoa JA, Rojas-Runjaic FJ, Pinto-da-Rocha R, Prendini L (2013) Systematic revision of the Neotropical scorpion genus *Chactopsis* Kraepelin, 1912 (Chactoidea: Chactidae), with descriptions of two new genera and four new species. Bull Am Mus Nat Hist 378:1–121

Ojanguren-Affilastro AA, Mattoni CI (2017) *Mauryius* n. gen. (Scorpiones: Bothriuridae), a new neotropical scorpion genus. Arthropod Syst Phylogeny 75(1):125–139. https://doi.org/10.3897/asp.75.e31879

Ojanguren-Affilastro AA, Ramırez MJ (2009) Phylogenetic analysis of the scorpion genus *Brachistosternus* (Arachnida, Scorpiones, Bothriuridae). Zool Scr 38:183–198. https://doi.org/10.1111/j.1463-6409.2008.00367.x

Ojanguren-Affilastro AA, Ochoa JA, Mattoni CI, Prendini L (2010) Systematic revision of the granulatus group of *Urophonius* Pocock, 1893 (Scorpiones, Bothriuridae), with description of a new species from central Chile. Am Mus Novit 3695:1–40. https://doi.org/10.1206/3695.2

Ojanguren-Affilastro AA, Pizarro-Araya J, Prendini L (2011) New data on Chilean Urophonius Pocock 1893, with description of a new species. Am Mus Novit 3725:1–44. https://doi.org/10.1206/3725.2

Ojanguren-Affilastro AA, Ochoa JA, Mattoni CI, Prendini L (2012) *Rumikiru*, n. gen. (Scorpiones, Bothriuridae). A new Scorpion genus from the Atacama Desert. Am Mus Novit 3731:1–43. https://doi.org/10.1206/3695.2

Ojanguren-Affilastro AA, Mattoni CI, Ochoa JA, Ramírez M, Ceccarelli FS, Prendini L (2016) Phylogeny, species delimitation and convergence in the South American bothriurid scorpion genus *Brachistosternus* Pocock 1893: integrating morphology, nuclear and mitochondrial DNA. Mol Phylogenet Evol 94:159–170. https://doi.org/10.1016/j.ympev.2015.08.007

Ojanguren-Affilastro AA, Adilardi RS, Mattoni CI, Ramírez MJ, Ceccarelli FS (2017) Dated phylogenetic studies of the southernmost American buthids (Scorpiones; Buthidae). Mol Phylogenet Evol 110:39–49. https://doi.org/10.1016/j.ympev.2017.02.018

Ojanguren-Affilastro AA, Ramírez MJ, Pizarro-Araya J (2020) Phylogenetic analysis of the winter and southernmost scorpion genus *Urophonius* (Bothriuridae), with the description of two new Patagonian species. Zool Anz 289:50–66. https://doi.org/10.1016/j.jcz.2020.09.003

Pennington RT, Prado DE, Pendry CA (2000) Neotropical seasonally dry forests and Quaternary vegetation changes. J Biogeogr 27:261–273. https://doi.org/10.1046/j.1365-2699.2000.00397.x

Prado DE, Gibbs PE (1993) Patterns of species distributions in the dry seasonal forests of South America. Ann Mo Bot Gard 80:902–927. https://doi.org/10.2307/2399937

Prance GT (1982) Forest refuges: evidence from woody angiosperms. In: Prance GT (ed) Biological diversification in the tropics. Columbia University Press, New York, pp 137–158

Prendini L, Francke OF, Vignoli V (2010) Troglomorphism, trichobothriotaxy and typhlochactid phylogeny (Scorpiones, Chactoidea): more evidence that troglobitism is not an evolutionary dead-end. Cladistics 26:117–142. https://doi.org/10.1111/j.1096-0031.2009.00277.x

Rein JO (2025) The scorpion files. Available online: https://www.ntnu.no/ub/scorpion-files/. Accessed 30 June 2025

Rossi A (2012) Three new species of the genus *Hadruroides* Pocock, 1893 from central Peru. Onychium 9:10–26

Sanmartín I (2012) Historical biogeography: evolution in time and space. Evol: Educ Outreach 5:555–568. https://doi.org/10.1007/s12052-012-0421-2

Santibáñez-López CE, Francke OF, Prendini L (2013) Systematics of the *keyserlingii* group of *Diplocentrus* Peters, 1861 (Scorpiones: Diplocentridae), with descriptions of three new species from Oaxaca, Mexico. Am Mus Novit 3777:1–48. https://doi.org/10.1206/3777.2

Santibáñez-López CE, Francke OF, Prendini L (2014a) Phylogeny of the North American scorpion genus *Diplocentrus* Peters, 1861 (Scorpiones: Diplocentridae) based on morphology, nuclear and mitochondrial DNA. Arthropod Syst Phylogeny 72:257–279. https://doi.org/10.3897/asp.72.e31789

Santibáñez-López CE, Francke OF, Prendini L (2014b) *Kolotl*, n. gen. (Scorpiones: Diplocentridae), a new scorpion genus from Mexico. Am Mus Novit 3815:1–14. https://doi.org/10.1206/3815.1

Santibáñez-López CE, Francke OF, Prendini L (2014c) Shining a light into the world's deepest caves: phylogenetic systematics of the troglobiotic scorpion genus *Alacran* Francke, 1982 (Typhlochactidae: Alacraninae). Invertebr Syst 28:643–664. https://doi.org/10.1071/IS14035

Santibáñez-López CE, Francke OF, Ureta C, Possani LD (2015) Scorpions from Mexico: from species diversity to venom complexity. Toxins 8:1–18. https://doi.org/10.3390/toxins8010002

Santibáñez-López CE, Kriebel R, Sharma PP (2017) *eadem figura manet*: measuring morphological convergence in diplocentrid scorpions (Arachnida: Scorpiones: Diplocentridae) under a multilocus phylogenetic framework. Invertebr Syst 31:233–248. https://doi.org/10.1071/IS16078

Santibáñez-López CE, González-Santillán E, Monod L, Sharma PP (2019) Phylogenomics facilitates stable scorpion systematics: reassessing the relationships of Vaejovidae and a new higher-level classification of Scorpiones (Arachnida). Mol Phylogenet Evol 135:22–30. https://doi.org/10.1016/j.ympev.2019.02.021

Santibáñez-López CE, Ojanguren-Affilastro AA, Sharma PP (2020) Another one bites the dust: taxonomic sampling of a key genus in phylogenomic datasets reveals more non-monophyletic groups in traditional scorpion classification. Invertebr Syst 34:133–143. https://doi.org/10.1071/IS19033

Santibáñez-López CE, Aharon S, Ballesteros JA, Gainett G, Baker CM, González-Santillán E, Harvey MS, Hassan MK, Abu Almaaty AH, Aldeyarbi SM, Monod L, Ojanguren-Affilastro A, Pinto-da-Rocha R, Zvik Y, Gavish-Regev E, Sharma PP (2022) Phylogenomics of scorpions reveal contemporaneous diversification of scorpion mammalian predators and mammal-active sodium channel toxins. Syst Biol 71:1281–1289. https://doi.org/10.1093/sysbio/syac021

Sharma PP, Baker CM, Cosgrove JG, Johnson JE, Oberski JT, Raven RJ, Harvey MS, Boyer SL, Giribet G (2018) A revised dated phylogeny of scorpions: phylogenomic support for ancient divergence of the temperate Gondwanan family Bothriuridae. Mol Phylogenet Evol 122:37–45. https://doi.org/10.1016/j.ympev.2023.107886

Sissom WD, Fet V (2000) Family Diplocentridae. In: Fet V, Sissom WD, Lowe G, Braunwalder ME (eds) Catalog of the scorpions of the world (1758–1998). New York Entomological Society, New York

Sobral-Souza T, Lima-Ribeiro MS, Solferini VN (2015) Biogeography of Neotropical Rainforests: past connections between Amazon and Atlantic Forest detected by ecological niche modeling. Evol Ecol 29:643–655. https://doi.org/10.1007/s10682-015-9780-9

Soleglad ME, Fet V (2003) High-level systematics and phylogeny of the extant scorpions (Scorpiones: Orthosterni). Euscorpius 2003:1–56. https://doi.org/10.18590/euscorpius.2003.vol2003.iss11.1

Soleglad ME, Fet V (2010) Further observations on scorpion genera *Hadrurus* and *Hoffmannihadrurus* (Scorpiones, Caraboctonidae). ZooKeys 59:1–13. https://doi.org/10.3897/zookeys.59.546

Stockmann R, Ythier E (2010) Scorpions of the world. N.A.P. Editions, Verrières-le-Buisson, 565 pp

Teruel R, Rodríguez-Cabrera TM (2020) Revision of the genus *Tityopsis* Armas, 1974 (Scorpiones: Buthidae). Part 1. General updates and description of four new species. Euscorpius 304:1–40

Teruel R, Kovarik F, Lowe G, Stahlavsky F (2023) Two new species of the remarkable scorpion genus *Megacormus* Karsch, 1881 (Scorpiones: Euscorpiidae). Euscorpius 375:1–24

Trujillo RE, Armas LF (2012) Nueva especie de *Plesiochactas* Pocock, 1900 (Scorpiones: Euscorpiidae) de Guatemala. Bol Soc Entomol Aragon 50:263–266

Vignoli V, Prendini L (2009) Systematic revision of the troglomorphic North American scorpion family Typhlochactidae (Scorpiones: Chactoidea). Bull Am Mus Nat Hist 326:1–94. https://doi.org/10.1206/570.1

Werneck FP (2011) The diversification of eastern South American open vegetation biomes: historical biogeography and perspectives. Quat Sci Rev 30:1630–1648. https://doi.org/10.1016/j.quascirev.2011.03.009

Werneck FP, Costa GC, Colli GR, Prado DE, Sites JW Jr (2011) Revisiting the historical distribution of Seasonally Dry Tropical Forests: new insights based on palaeodistribution modelling and palynological evidence. Glob Ecol Biogeogr 20:272–288. https://doi.org/10.1111/j.1466-8238.2010.00596.x

Ythier E (2021) Two new species of *Hadruroides* Pocock, 1893 from Peru and Ecuador (Scorpiones, Caraboctonidae). Faunitaxys 9(11):1–8. https://doi.org/10.57800/faunitaxys-9(11)

Ythier E (2024) A new high-altitude scorpion species of the genus *Ananteris* Thorell, 1891 (Scorpiones: Ananteridae) from the Pico da Neblina, Brazil. Faunitaxys 12(19):1–9. https://doi.org/10.57800/faunitaxys-12(19)

Ythier E, Lourenço WR (2017) The geographical patterns of distribution of the genus *Teuthraustes* Simon, 1878 in Ecuador and description of three new species (Scorpiones, Chactidae). ZooKeys 721:45–63. https://doi.org/10.3897/zookeys.721.21529

Ythier E, Lourenço WR (2023) A new species of *Hadruroides* Pocock, 1893 from Peru (Scorpiones: Caraboctonidae). Faunitaxys 11(76):1–7. https://doi.org/10.57800/faunitaxys-11(76)

Ythier E, Lourenço WR (2024) A new species of *Hadrurochactas* Pocock, 1893 (Scorpiones: Chactidae) from the Mitaraka Massif in French Guiana. Faunitaxys 12(59):1–9. https://doi.org/10.57800/faunitaxys-12(59)

Zárate-Gálvez K, Francke OF (2009) Redescription of *Plesiochactas mitchelli* (Scorpiones: Euscorpiidae): a rare scorpion from Central America. J Arachnol 37:338–345

Chapter 9
Subterranean and Troglomorphic Scorpions from the Neotropics

Edmundo González-Santillán, Javier Blasco-Aróstegui, and Jairo A. Moreno-González

Abstract The Neotropics are home to at least one-third of the world's biodiversity. Cavernicolous taxa constitute a significant portion, yet many new species are still undiscovered. After a century of debate on how to classify subterranean organisms into different ecological categories, no consensus has been reached. The Schiner--Racovitza system is still the most widely used, although several authors have proposed modifications, in some cases tailored to particular groups. Such is the case of scorpions. Among other inconsistencies, it has been noted that troglomorphism alone is not a necessary criterion for defining troglobiont scorpions, as some endogean species also exhibit troglomorphic traits despite not inhabiting hypogean environments. As has been suggested elsewhere, these classifications are likely to remain contentious unless ecology, evolutionary history, ecomorphological adaptations, and natural history are jointly considered. To date, 68% of the world's troglobites and an overwhelming 85% of endogean species occur in the Neotropics. These figures underscore the urgent need to evaluate the conservation status of these organisms and their habitats.

Keywords Hypogean ecosystems · Biodiversity shortfalls · Racovitzan impediment · Arachnida · Scorpiones · Conservation biology

E. González-Santillán (✉)
Colección Nacional de Arácnidos, Departamento de Zoología, Instituto de Biología, Universidad Nacional Autónoma de México, Coyoacán, Mexico City, Mexico
e-mail: edmundo.gonzalez@ib.unam.mx

J. Blasco-Aróstegui
Centre for Ecology, Evolution and Environmental Changes, Departamento de Biologia Animal, Faculdade de Ciências, Universidade de Lisboa, Lisbon, Portugal
e-mail: javierblasco77@gmail.com

J. A. Moreno-González
Division of Invertebrate Zoology, American Museum of Natural History, New York, NY, USA
e-mail: jmorenogonzalez@amnh.org

A. F. de Araujo Lira, E. González-Santillán (eds.), *The Biology of Neotropical Scorpions*, https://doi.org/10.1007/978-3-032-23194-9_9

9.1 Introduction

The Neotropical region is one of the most vast and biodiverse regions on Earth. It extends from North America (central Mexico) to South America (central Argentina), including Central America and the Antilles (Morrone 2018). Its complex topography and altitude range—from sea level to over 5000 m a.s.l.—support an exceptional diversity of biomes (Rull 2020): from deserts to tropical rainforests, and a diverse array of aquatic and hypogean ecosystems. For more than 150 years, all these Neotropical ecosystems had been classified using biogeographical regionalization, with contrasting proposals based on phytogeographical, zoogeographical, and ecoregional frameworks (Morrone 2014). Most recent analyses have delimited 57 provinces (Morrone et al. 2022) and have incorporated previously defined ecoregions (i.e., Olson et al. 2001).

The Neotropics possess at least a third of the World's biota (Morrone 2014; Antonelli et al. 2018), harboring more species of amphibians, ants, birds, butterflies, lizards, and mammals than tropical Africa and Southeast Asia combined (Raven et al. 2020). Such an extraordinary biodiversity is widely regarded as the product of complex interactions among biotic (e.g., adaptive radiation, extinction, colonization, and range expansion) and abiotic factors (e.g., geology, climate, catastrophic events) (Antonelli 2022). These dynamics, along with the huge variety of niches to exploit, promoted the adaptation and specialization of Neotropical fauna, thereby generating a close association between phenotype and ecology which favored the evolution of key innovations. Among the most conspicuous and remarkable phenotypic specializations that have captivated naturalists throughout history are those associated with subterranean environments.

The subterranean habitats are networks of heterogeneous intercommunicated spaces of the subsoil with poor to no light, relatively low and constant temperatures throughout the year, and high humidity often resulting in standing pools, dripping water, and water film on cave walls (Hesselberg 2023). These subterranean environments are essential for life on Earth as they provide ecological services such as regulation (e.g., hydrological, biotic, and external ecosystems), support (e.g., soils, water, surface/subsurface rocks), provision (e.g., nutrients, food, water, materials, shelter, etc.), and cultural services (e.g., well-being and health, recreation, historical and scientific value) (Menin and Bacci 2023). From the biological perspective, subterranean habitats serve both as models and frontiers for modern research in the fields of ecology, ethology, and evolution (Mammola et al. 2021).

Although subterranean ecosystems are among the most widespread environments on Earth, they remain largely unexplored, and our comprehension of their biodiversity is still limited (Nicolosi and Gerovasileiou 2024). Documentation efforts have historically been biased toward more conspicuous organisms such as vertebrates (e.g., bats, fishes) (Mammola et al. 2021). In contrast, subterranean arthropods—despite a growing body of works in recent decades— remain underrepresented in most surveys due to challenges in sampling and, in some cases, their low population densities (Harrison et al. 2024). Within them, arachnids stand out as both diverse and abundant in subterranean environments, where they play a central role as predators (Agnarsson 2023).

Among subterranean arachnids, spiders (Araneae) are the most abundant, ubiquitous, and best-documented group (Mammola and Isaia 2017). Other arachnid orders commonly present in subterranean environments include Amblypygi (whip--spiders), Palpigradi (micro whip-scorpions), Ricinulei (hooded-thicks), Schizomida (short-tailed whip-scorpions), Thelyphonida (whip-scorpions), and Scorpiones (scorpions). When arachnid populations are well-established and associated with subterranean ecosystems, they often display remarkable troglomorphic adaptations such as reduction of sclerotization and pigmentation, attenuation of appendages, and reduction or complete loss of eyes or ocelli (Volschenk and Prendini 2008; Prendini et al. 2021). However, interpreting troglomorphisms in arachnids is particularly challenging. These traits are not exclusive to obligate hypogean species but can also occur in endogean species inhabiting soil layers or the leaf-litter. As a result, distinguishing truly cave-adapted arachnids from similar epigean or endogean species remains a significant taxonomic and ecological challenge.

9.2 Classification of Subterranean and Troglomorphic Scorpions

The classification of organisms inhabiting subterranean environments has been a subject of debate within the subterranean biology community for over a century (Trajano and Carvalho 2017). The first and most widely used system was proposed by Schiner (1854) and later revised by Racovitza (1907), which delineated three primary categories: troglobites, species strictly adapted to subterranean life; troglophiles, species that inhabit subterranean ecosystems but also interact with epigean environments; and trogloxenes, species that use subterranean habitats only occasionally or for specific purposes. Known as the Schiner-Racovitza system, this pioneering framework was subsequently adopted by many authors, often with modifications (Sket 2008; Volschenk and Prendini 2008; Trajano 2012; Trajano and Carvalho 2017; Prendini et al. 2021). A multitude of subcategories have been postulated since, giving rise to various schools of thought. While some classification systems were considered suitable for certain taxa, they appeared inadequate for others, with critics arguing that they introduced "unnecessary and confusing" categories vaguely defined along a continuum of epigean to hypogean microhabitats (Sket 2008; Trajano and Carvalho 2017). Moreover, none of these early schemes incorporated key aspects of life in hypogean habitats, including ecology, ecomorphological adaptations, microhabitat use, natural history, and phylogenetic relationships, which would have facilitated a more consolidated and robust definition of the categories within them.

The Schiner-Racovitza system was subsequently refined by Trajano (2012) and Trajano and Carvalho (2017) through the incorporation of the population source-sink model (Pulliam 1988). In this model, the ecological perspective was incorporated, but other lines of evidence such as those provided by ecomorphotypes and phylogeny are often inextricably connected and are necessary to make decisions. In addition, the methodological challenges associated with acquiring ecological data for each of the categories in the Schiner-Racovitza framework hinder accurate classification, and the distinction between troglophiles and trogloxenes remains contingent on the availability of natural history data. As noted by Trajano and Carvalho (2017), certain limitations in the Schiner-Racovitza system persist to this day even in the newer systems that have been implemented.

Arachnids are among the organisms that have been routinely classified using the traditional Schiner-Racovitza system, including subsequent modifications by Trajano (2012) and Trajano and Carvalho (2017). Nevertheless, scorpions can be regarded as the sole Order for which a contemporary reevaluation of the classification of cavernicolous species has been conducted. A seminal classification of cave scorpions was published by Volschenk and Prendini (2008), who were the first to point out that scorpions could exhibit distinctive troglomorphic characteristics (e.g., thin cuticle, elongated appendages and telson, absence of ocelli) that are not inherently associated with living in a subterranean environment. Consequently, troglomorphies alone proved insufficient for accurately assigning a species to any of the three main categories, and further information is needed regarding its ecological habits. Thus, correlation between the presence of troglomorphisms as an approach to infer the troglobitic status of a species as suggested by previous authors (i.e., Christiansen 1962; Howarth 1983) was no longer reliable (i.e., Volschenk and Prendini 2008; Trajano and Carvalho 2017). Troglomorphism can be the result of either modifications within a lineage (i.e., autapomorphies) or modifications leading to separate lineages (i.e., synapomorphies) (Trajano and Carvalho 2017).

In a recent study, Prendini et al. (2021) undertook a comprehensive reclassification of epigean, endogean, and hypogean scorpions. Departing from previous conceptual advances by Trajano (2012) and Trajano and Carvalho (2017), the authors presented a key to ecological classification of the cavernicolous and troglomorphic scorpions. The system encompassed seven entries, the initial three categories delineated in relation to the scorpion's habitat:

1. Epigean: species in this category inhabit any above-ground habitat, including microhabitats such as crevices in rocks or logs. They do not exhibit troglomorphisms and display a variety of ecomorphotypes (e.g., psammophilous, lithophilous, corticolous, pelophilous) (Prendini 2001). Illustrative examples from the Neotropics include the genera *Bothriurus* Peters, 1861 and *Franckeus* Soleglad & Fet, 2005.
2. Endogean: species in this category inhabit within soil or leaf litter. They may or may not exhibit troglomorphisms (e.g., reduction of sclerotization and pigmentation, reduction of median and/or lateral ocelli) and generally display a humicolous ecomorphotype (Prendini 2001). Notable examples from the Neotropics include species of the genus *Troglotayosicus* Lourenço, 1981 and *Typhlochactas* Mitchell, 1971.

3. Hypogean: species in this category inhabit subterranean environments, ranging from large cave chambers to the *Milieu Souterrain Superficiel* (MSS) (i.e., not only the large chambers of caves but the network of narrow crevices and fissures interconnecting them) (Juberthie et al. 1980; Mammola et al. 2016; Blasco--Aróstegui and Prendini 2024). They typically exhibit pronounced troglomorphies (e.g., reduction of sclerotization and pigmentation, elongation of appendages and telson, and absence or reduction of ocelli) and display either a lithophilous or cavernicolous ecomorphotype (Prendini 2001). Notable examples from the Neotropics include *Alacran* Francke, 1982 and *Troglorhopalurus* Lourenço, Baptista and Giupponi, 2004.

   A second triad of entries in the key incorporates the Trajano and Carvalho (2017) reinterpretation of Schiner-Racovitza's classical three-category system, along with the addition of a fourth category not included in the Schiner-Racovitza ecological classification:

4. Trogloxene: species in this category can establish viable populations in both epigean and hypogean habitats, having epigean source populations but potentially forming hypogean sink populations (Mammola 2019). This concept may also encompass species confined inside caves during a specific season of the year and, in contrast to troglophile scorpions, other closely related species exist outside caves. This, combined with the fact that trogloxenes often exhibit weak or no troglomorphies, can create a certain degree of overlap with accidentals. It naturally prompted debate on whether the two should be treated as distinct categories. Sket (2008) suggested merging them, whereas Trajano (2012) argued for their separation, excluding accidentals from the Schiner-Racovitza system but still considering both accidentals and trogloxenes as necessary terms. Later, Prendini et al. (2021) continued to treat them as separate concepts, providing several arguments in favor while also acknowledging the difficulties and the importance of distinguishing accidental records from trogloxenes. Neotropical trogloxenes include *Tityus grottoedensis* Botero-Trujillo & Flórez, 2014 and *Pseudouroctonus reddelli* Gertsch & Soleglad, 1972 (Table 9.1).
5. Troglophile: species in this category occur exclusively within subterranean ecosystems, either near the cave entrance or deep inside, throughout the entire annual cycle, and have no closely related phylogenetic counterparts outside caves. Some authors suggest that, unlike troglobites, troglophiles maintain source populations in both hypogean and epigean environments, regularly migrating between these habitats and thereby promoting gene flow (Mammola 2019). Troglophiles may exhibit weak or absent troglomorphic traits. With the exception of phylogenetic relationships, most of the diagnostic features for this category can be determined during field expeditions. However, obtaining robust ecological and natural history data of subterranean environments is often challenging (Mammola et al. 2021), given the rarity of hypogean scorpions. Examples from the Neotropical scorpion fauna include *Tityopsis sheylae* Teruel and Rodríguez, 2020 and *Vaejovis davidi* Soleglad & Fet, 2005 (Table 9.1).
6. Troglobite: species in this category occur exclusively within subterranean ecosystems, at varying depths within caves and the MSS. Some authors suggest that,

Table 9.1 Updated list (2025) of cavernicolous and troglomorphic scorpions in the Neotropical Realm (Morrone et al. 2022) with their distribution, degree of troglomorphism, and their classification according to Schiner-Racovitza system as presented in Prendini et al. (2021). The order of the seven troglomorphic features observed in each species is ordered as follows: median eyes, lateral eyes, pigmentation, sclerotization, legs and pedipalps, metasoma, and telson

Family	Species	Distribution	Habitat	Troglomorphism	Classification
Bothriuridae Simon, 1880	*Bothriurus araguayae* Vellard, 1934	Brazil	IC; S	NR, NR, NR, NR, NA, NA, NE	Epigean: accidental
	Thestylus aurantiurus Yamaguti and Pinto-da-Rocha, 2003	Brazil	IC; S	NR, NR, NR, NR, NA, NA, NE	Epigean: accidental
Buthidae C. L. Koch, 1837	*Alayotityus delacruzi* Armas, 1973	Cuba	IC	R, R, SR, SR, NA, NA, NE	Hypogean: troglophile
	Alayotityus juraguaensis Armas, 1973	Cuba	S	NR, NR, NR, NR, NA, NA, NE	Epigean: accidental
	Centruroides gracilis (Latreille, 1804)	Belice, Guatemala, Mexico, U.S.A. etc.	IC; S	NR, NR, NR, NR, NA, NA, NE	Epigean: accidental
	Heteroctenus junceus (Herbst, 1800)	Cuba	IC; S	NR, NR, NR, NR, NA, NA, NE	Epigean: accidental
	Ischnotelson peruassu Esposito et al., 2017	Brazil	S	NR, NR, NR, NR, NA, NA, NE	Epigean: accidental
	Tityopsis inexpectata (Moreno, 1940)	Cuba	IC; S	NR, NR, NR, NR, NA, NA, NE	Epigean: accidental
	Tityopsis sheylae Teruel and Rodríguez, 2020	Cuba	IC	NR, NR, SR, SR, At, At, NE	Hypogean: troglophile
	Tityus blaseri Mello-Leitão, 1931	Brazil	IC; S	NR, NR, NR, NR, NA, NA, NE	Epigean: accidental
	Tityus confluens Borelli, 1899	Argentina, Bolivia, Paraguay	IC; S	NR, NR, NR, NR, NA, NA, NE	Epigean: accidental

(continued)

Table 9.1 (continued)

Family	Species	Distribution	Habitat	Troglomorphism	Classification
	Tityus confluens bodoquena Lourenço et al., 2004	Brazil	IC	NR, NR, NR, NR, NA, NA, NE	Hypogean: trogloxene
	Tityus demangei Lourenço, 1981	Ecuador	IC; S	NR, NR, NR, NR, NA, NA, NE	Epigean: accidental
	Tityus jussarae Lourenço, 1988	Ecuador	IC	NR, NR, NR, NR, NA, NA, NE	Hypogean: trogloxene
	Tityus grottoedensis Botero-Trujillo and Flórez, 2014	Colombia	IC; S	NR, NR, NR, NR, NA, NA, NE	Epigean, Hypogean: trogloxene
	Tityus magnimanus Pocock, 1897	Venezuela	IC; S	NR, NR, NR, NR, NA, NA, NE	Epigean, Hypogean: accidental
	Tityus monaguensis González-Sponga, 1974	Venezuela	IC; S	NR, NR, NR, NR, NA, NA, NE	Epigean, Hypogean: accidental
	Tityus obscurus Gervais, 1843	Brazil, French, Guiana, Suriname	IC; S	NR, NR, NR, NR, NA, NA, NE	Epigean: accidental
	Tityus spelaeus Moreno-González et al., 2021	Brazil	IC	NR, NR, SR, SR, NA, NA, NE	Hypogean: troglophile
	Tityus stigmurus (Thorell, 1876)	Brazil	IC; S	NR, NR, NR, NR, NA, NA, NE	Epigean: accidental
	Troglorhopalurus lacrau (Lourenço and Pinto-da-Rocha, 1997)	Brazil	IC	NR, NR, SR, SR, NA, NA, NE	Hypogean: troglophile
	Troglorhopalurus translucidus Lourenço et al., 2004	Brazil	IC	R, R, SR, SR, At, At, NE	Hypogean: troglobite
	Troglorhopalurus iuiu Sousa-Carvalho, Lopes-Silva et al. 2025	Brazil	IC	NR, NR, NR, NR, NA, NA, NE	Hypogean: troglobite

(continued)

Table 9.1 (continued)

Family	Species	Distribution	Habitat	Troglomorphism	Classification
Chactidae Pocock, 1893	*Broteochactas trezzii* (Vignoli and Kovařík, 2003)	Venezuela	IC	R, NR, SR, SR, At, At, NE	Endogean
	Chactas viloriai Rojas-Runjaic, 2004	Venezuela	IC	NR, NR, NR, NR, NA, NA, NE	Hypogean: troglophile
	Chactas vanbenedenii (Gervais, 1843)	Colombia Venezuela	IC; S	NR, NR, NR, NR, NA, NA, NE	Epigean: accidental
	Chactas yupai González-- Sponga, 1994	Venezuela	S	NR, NR, NR, NR, NA, NA, NE	Epigean: accidental
Diplocentridae Karsch, 1880	*Cazierius cayacoa* Teruel et al., 2021	Dominican Republic	IC	Ab, Ab, R, R, At, At, E	Hypogean: troglobite
	Cazierius ciguayo Teruel et al., 2021	Dominican Republic	IC	Ab, Ab, R, R, At, At, E	Hypogean: troglobite
	Cryptoiclus rodriguezi Teruel and Kovařík, 2012	Cuba	IC (entrance)	R, R, R, R, NA, NA, NE	Endogean
	Diplocentrus actun Armas and Palacios-Vargas, 2002	Mexico	IC	Ab, R, Ab, Ab, At, NA, E	Hypogean: troglobite
	Diplocentrus anophthalmus Francke, 1977	Mexico	IC	Ab, R, Ab, Ab, At, NA, E	Hypogean: troglobite
	Diplocentrus cueva Francke, 1978	Mexico	IC	R, R, SR, SR, At, NA, NE	Hypogean: troglobite
	Diplocentrus mitchelii Francke, 1977	Mexico	IC	NR, NR, NR, NR, NA, NA, NE	Epigean: accidental
	Diplocentrus reddelii Francke, 1977	Mexico	inside sinkhole; S	NR, NR, NR, NR, NA, NA, NE	Epigean: accidental
	Kolotl magnus (Beutelspacher and López-- Forment, 1991)	Mexico	IC; S	NR, NR, NR, NR, NA, NA, NE	Epigean: accidental
	Heteronebo clareae Armas, 2001	U.S.A.	IC	R, R, Ab, Ab, At, NA, NE	Hypogean: troglophile or trogloxene

(continued)

Table 9.1 (continued)

Family	Species	Distribution	Habitat	Troglomorphism	Classification
Euscorpiidae Laurie, 1896	*Megacormus gertschi* Díaz-Najera, 1966	Mexico	IC; S	NR, NR, NR, NR, NA, NA, NE	Epigean: accidental
	Troglocormus ciego Francke, 1981	Mexico	IC	Ab, R, NR, NR, At, NA, E	Hypogean: troglobite
	Troglocormus willis Francke, 1981	Mexico	IC	Ab, R, R, R, At, NA, E	Hypogean: troglobite
Troglotayosicidae Lourenço, 1998	*Troglotayosicus akaido* Moreno--González et al., 2024	Colombia	LL	Ab, R (second pair), R, NR, NA, NA, NE	Endogean
	Troglotayosicus ballvei Botero--Trujillo et al., 2021	Ecuador	LL	Ab, R (second pair), R, NR, NA, NA, NE	Endogean
	Troglotayosicus hirsutus Botero-Trujillo et al., 2012	Colombia	LL	Ab, R (second pair), R, NR, NA, NA, NE	Endogean
	Troglotayosicus humiculum Botero-Trujillo and Francke, 2009	Colombia	LL	Ab, R (second pair), R, NR, NA, NA, NE	Endogean
	Troglotayosicus meijdeni Botero-Trujillo et al., 2017	Colombia	LL	Ab, R (second pair), R, NR, NA, NA, NE	Endogean
	Troglotayosicus muranunkae Lourenço et al., 2020	Ecuador	S	Ab, R (second pair), R, NR, NA, NA, NE	Endogean
	Troglotayosicus vachoni Lourenço, 1981	Ecuador	IC	Ab, R (second pair), R, NR, NA, NA, NE	Hypogean: troglobite
Typhlochactidae Mitchell, 1971	*Alacran chamuco* Francke, 2009	Mexico	IC	Ab, Ab, Ab, Ab, At, At, GE	Hypogean: troglobite
	Alacran tartatus Francke, 1982	Mexico	IC	Ab, Ab, Ab, Ab, At, At, GE	Hypogean: troglobite
	Alacran triquimera Santibáñez--López et al., 2014	Mexico	IC	Ab, Ab, Ab, Ab, At, At, GE	Hypogean: troglobite

(continued)

Table 9.1 (continued)

Family	Species	Distribution	Habitat	Troglomorphism	Classification
	Sotanochactas elliotti (Mitchell, 1971)	Mexico	IC	Ab, Ab, Ab, Ab, At, At, GE	Hypogean: troglobite
	Stygochactas granulosus (Sissom and Cokendolpher, 1998)	Mexico	IC	Ab, Ab, Ab, Ab, At, At, E	Hypogean: troglobite
	Typhlochactas cavicola Francke, 1986	Mexico	IC	Ab, Ab, Ab, Ab, At, NA, E	Hypogean: troglobite
	Typhlochactas mitchelli Sissom, 1988	Mexico	LL	Ab, Ab, Ab, Ab, NA, NA, E	Endogean
	Typhlochactas reddelli Mitchell, 1968	Mexico	IC	Ab, Ab, Ab, Ab, At, NA, E	Hypogean: troglobite
	Typhlochactas rhodesi Mitchell, 1968	Mexico	IC	Ab, Ab, Ab, Ab, At, NA, E	Hypogean: troglobite
	Typhlochactas sissomi Francke et al., 2009	Mexico	S	Ab, Ab, Ab, Ab, NA, NA, E	Endogean
	Typhlochactas sylvestris Mitchell and Peck, 1977	Mexico	LL	Ab, Ab, Ab, Ab, NA, NA, E	Endogean
Vaejovidae Thorell, 1876	*Franckeus minckleyi* (Williams, 1968)	Mexico	IC	NR, NR, NR, NR, NA, NA, NE	Epigean: accidental
	Pseudouroctonus apacheanus (Gertsch and Soleglad, 1972)	Mexico, U.S.A.	IC; S	NR, NR, NR, NR, NA, NA, NE	Epigean: accidental
	Pseudouroctonus reddelli Gertsch and Soleglad, 1972	Mexico, U.S.A.	IC; S	NR, NR, NR, NR, NA, NA, NE	Epigean: trogloxene
	Pseudouroctonus savvasi Francke, 2009	Mexico	IC	SR, SR, R, R, At, At, NE	Hypogean: troglophile
	Pseudouroctonus sprousei Francke and Savary, 2006	Mexico	IC	R, R, NR, NR, At, At, E	Hypogean: troglophile
	Vaejovis davidi Soleglad and Fet, 2005	Mexico	IC	NR, NR, NR, NR, At, At, NE	Hypogean: troglophile

(continued)

Table 9.1 (continued)

Family	Species	Distribution	Habitat	Troglomorphism	Classification
	Vaejovis gracilis Gertsch and Soleglad, 1972	Mexico	IC	R, R, R, R, At, At, NE	Hypogean: troglobite
	Vaejovis mitchelli Sissom, 1991	Mexico	IC; S	NR, NR, NR, NR, NA, NA, NE	Epigean: accidental
	Vaejovis norteno Sissom and González-- Santillán, 2004	Mexico	IC	NR, NR, NR, NR, NA, NA, SE	Hypogean: troglophile
	Vaejovis rossmani Sissom, 1989	Mexico	IC; S	NR, NR, NR, NR, NA, NA, NE	Epigean: accidental
	Vaejovis sprousei Sissom, 1990	Mexico	IC; S	NR, NR, NR, NR, NA, NA, NE	Epigean: accidental

Ab absent, *At* attenuate, *E* elongated, *GE* greatly enlarged, *IC* inside cave, *LL* leaf litter, *NA* not attenuated, *NE* not elongated, *NR* not reduced, *R* reduced, *S* surface, *SE* slightly elongated, *SR* slightly reduced

unlike troglophiles, troglobites or troglobionts have subterranean source populations and may have sink populations in epigean habitats (Trajano and Carvalho 2017; Mammola 2019). These scorpions exhibit pronounced troglomorphic characteristics (e.g., reduction or loss of ocelli and pigmentation, appendage attenuation, and reduction of sclerotization). To a certain degree, this term could be used as a synonym for hypogean species (Volschenk and Prendini 2008). The Neotropical genera *Troglorhopalurus* and *Alacran* comprise perfect examples to underscore the interchangeability of these terms.

7. Accidental: Prendini et al. (2021) use this term, excluded from the Schiner--Racovitza system by Trajano (2012), to include scorpion species that inhabit epigean habitats but occasionally enter caves, either by mishap or in search of milder climatic conditions. They do not exhibit any troglomorphic adaptation and are usually collected far from caves. As noted above, *Bothriurus* and *Franckeus* can serve as Neotropical examples of this category.

Achieving an appropriate ecological classification of subterranean organisms is comparable to the long-standing challenge of defining boundaries between live entities. These debates have persisted for decades, underscoring the difficulty of categorizing or defining a continuum in nature. Within this context, the definition of categories in the Schiner-Racovitza system may be confusing or overlap when scarce information is available on natural history of the organisms (Sket 2008; Trajano 2012; Trajano and Carvalho 2017; Mammola 2019). The collection of additional evidence and field observations is critical to accurately classify some ambiguous cases. In scorpions, *Heteronebo clareae* Armas, 2001, which was described from three juvenile specimens collected in a cave in Navassa Island, USA comprise a perfect example. In the absence of further information extracted from exhaustive surface surveys, biological observations, or adult specimens, Prendini et al. (2021) categorized it as an endogean/hypogean trogloxene species, although it was recognized as troglophile or trogloxene in the present investigation (Table 9.1). Thus, the practical application of these ecological classification systems of subterranean organisms in the Neotropics is only likely to succeed after a comprehensive census and phylogeny of cavernicolous species is conducted, and when additional data on ecology, population dynamics, and evolution of subterranean scorpions become available.

9.3 Diversity and Distribution

The present revision uses the most up to date delineation of America's biogeographical areas to assess the diversity and distribution of its subterranean and troglomorphic scorpions (Morrone et al. 2022). According to this proposal, the Neotropical region is delineated to the north by the Mexican provinces of the Sierra Madre Oriental, Sierra Madre Occidental, Pacific Lowlands, Transmexican Volcanic Belt, and Veracruzan (Fig. 9.1). The southernmost region is delineated by the American transition zone, characterized by boreal environments, including the Monte, Cuyan, and High Andean provinces, as well as the Atacama province (Fig. 9.2).

As illustrated in Table 9.1, the diversity and categories are in accordance with the classification proposed by Prendini et al. (2021). A noteworthy pattern that merits attention is that higher taxa exhibit greater diversity globally. This finding lends further credence to the hypothesis that the Old World is home to a greater number of ancient lineages, with at least twice the number of families and genera observed in the Neotropics. However, a comparative analysis of troglobite, endogean, and troglomorphic species shows that they make up 68%, 85%, and 67%, respectively, of the world's scorpion fauna (Table 9.2). These numbers led to the hypothesis that this may be the result of recent radiation in the Neotropics, while other authors considered it the product of a more splitter taxonomy (Fet and Lowe 2000). In the following paragraphs, the number of genera and species within each scorpion family in the Neotropics and the respective regions they inhabit is summarized, referring to a main Table 9.1 for additional information on ecological classification and taxonomic classifications.

Interestingly, the family Bothriuridae Simon, 1880 is quite diverse and ubiquitous in South America, inhabiting both arid and humid environments, as well as low and high altitudes. To date, however, no hypogean species have been recorded, and only two accidental occurrences have been documented in the literature (Table 9.1). The reasons behind this pattern remain unclear. The bothriurids are globally distributed, with its members inhabiting the Old and New World. This suggests an ancient origin and may indicate a recent faunal turnover in the Neotropics, which could explain why few species have colonized hypogean environments.

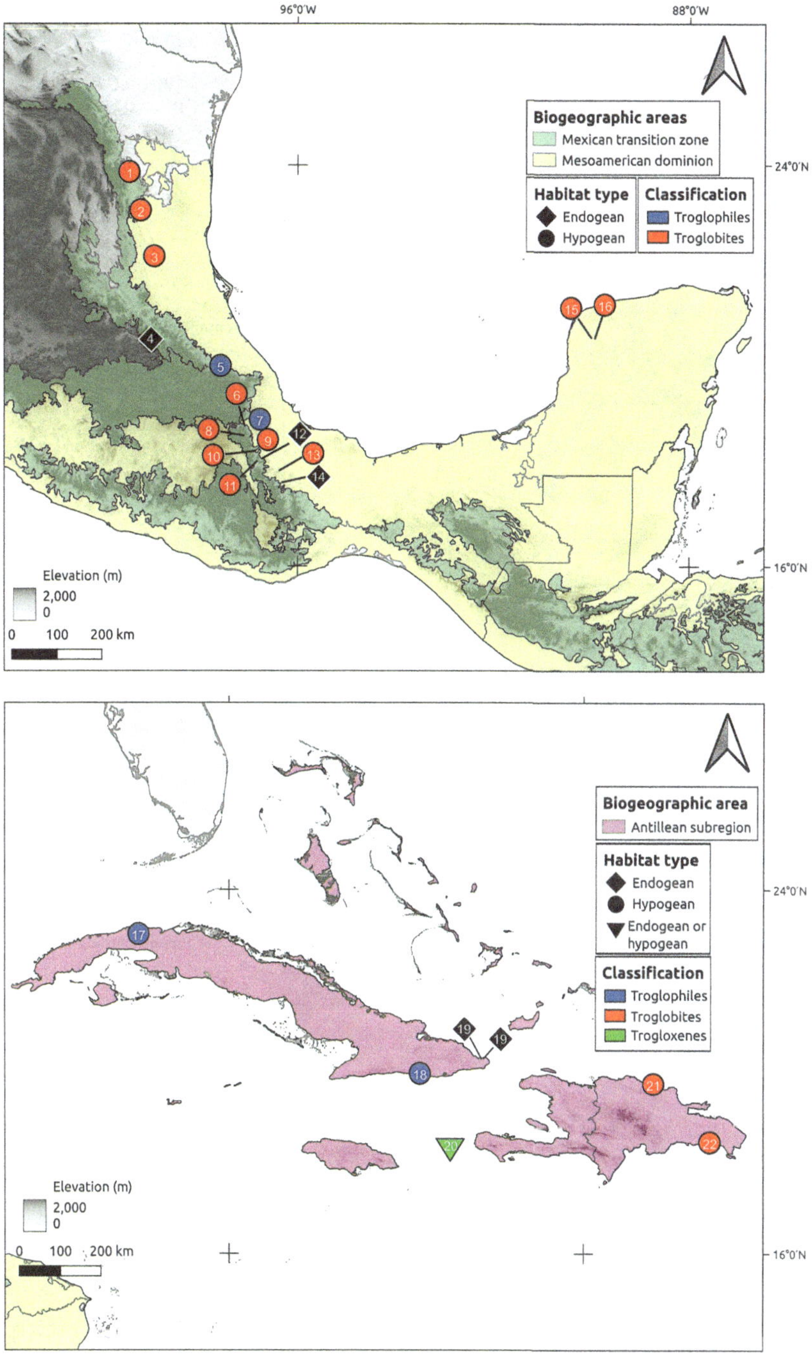

Fig. 9.1 Distribution of Neotropical cave-dwelling scorpions from North America (top) to the Antilles (bottom), showing biogeographic areas according to Morrone et al. (2022). Numbers: (1) *Typhlochactas cavicola*; (2) *Typhlochactas rhodesi*; (3) *Sotanochactas elliotti*; (4) *Typhlochactas sissomi*; (5) *Vaejovis davidi*; (6) *Typhlochactas reddelli*; (7) *Vaejovis gracilis*; (8) *Stygochactas granulosus*; (9) *Diplocentrus cueva*; (10) *Alacran triquimera*; (11) *Alacran tartarus*; (12) *Typhlochactas mitchelli*; (13) *Alacran chamuco*; (14) *Typhlochactas sylvestris*; (15) *Diplocentrus anophthalmus*; (16) *Diplocentrus actun*; (17) *Tityopsis sheylae*; (18) *Alayotityus delacruzi*; (19) *Cryptoiclus rodriguezi*; (20) *Heteronebo clarae*; (21) *Cazierius ciguayo*; (22) *Cazierius cayacoa*

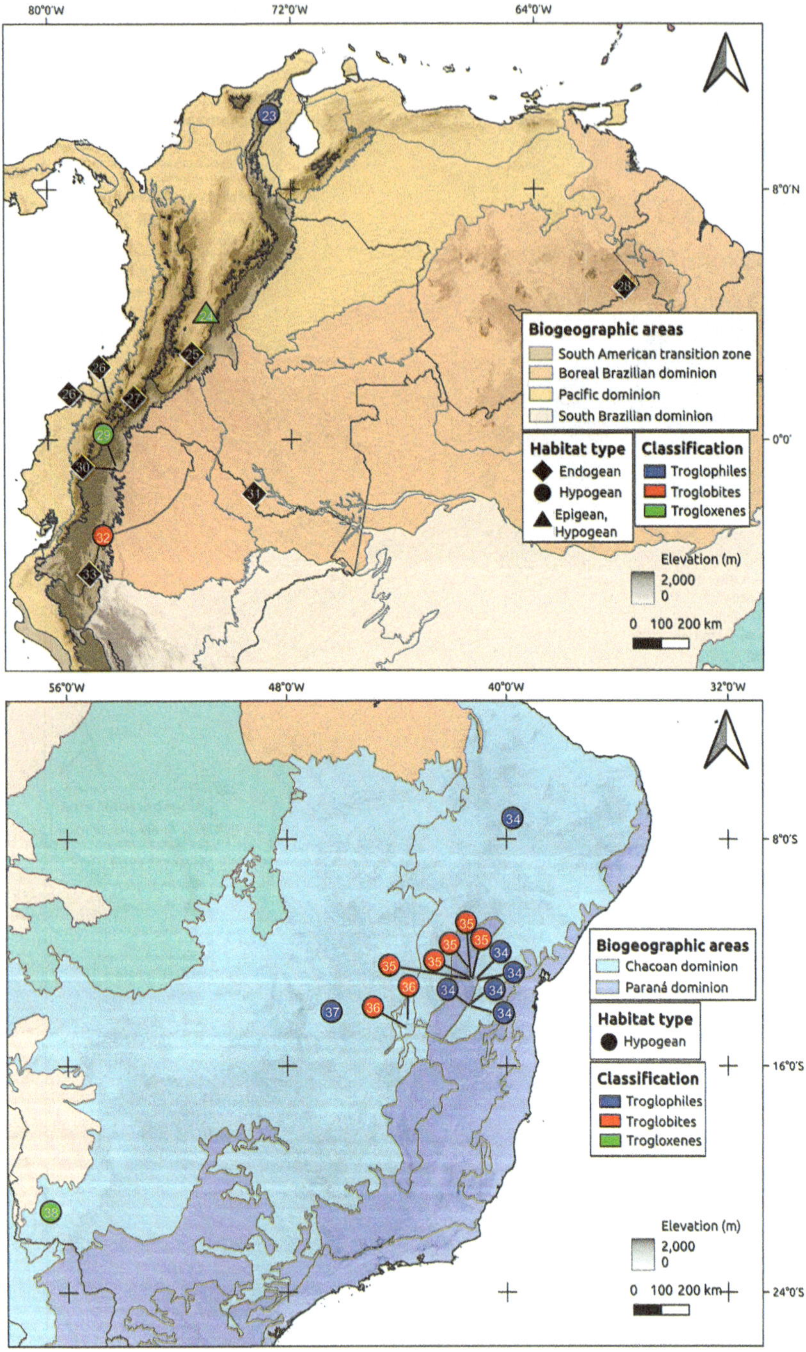

Fig. 9.2 Distribution of Neotropical cave-dwelling scorpions from Northern South America (top) to Eastern South America (bottom), showing biogeographic areas according to Morrone et al. (2022). Numbers: (23) *Chactas vilorai*; (24) *Tityus grottoedensis*; (25) *Troglotayosicus meijdeni*; (26) *Troglotayosicus humiculum*; (27) *Troglotayosicus hirsutus*; (28) *Broteochactas trezzii*; (29) *Tityus jussarae*; (30) *Troglotayosicus ballvei*; (31) *Troglotayosicus akaido*; (32) *Troglotayosicus vachoni*; (33) *Troglotayosicus muranunkae*; (34) *Troglorhopalurus lacrau*; (35) *Troglorhopalurus translucidus*; (36) *Troglorhopalurus iuiu*; (37) *Tityus spelaeus*; (38) *Tityus confluens bodoquena*

Table 9.2 Diversity of troglomorphic and cavernicolous scorpions from the Neotropical region updated (2025) after Prendini et al. (2021). The bracketed numbers in the troglobite, endogean, and troglomorphic taxa section indicate the global richness and the percentage of the Neotropical region in relation to the worldwide totals

	Hypogean			Endogean	Epigean	Troglomorphic
	Troglobite	Troglophile	Trogloxene	Humicolous	Accidentals	
Family	3 [13, 23%]	2	3	2 [14, 14%]	2	7 [15, 46%]
Genera	7 [17, 41%]	3	5	3 [6, 50%]	8	16 [29, 55%]
Species	19 [28, 68%]	5	9	12 [14, 85%]	20	39 [58, 67%]

In the family Buthidae C.L. Koch, 1837, only two troglobite and four troglophile species are currently known (Table 9.1). Most reports correspond to trogloxenes and accidentals, which is not surprising given the errant habits typical of most species in this family. However, two Cuban genera and two Brazilian species display varying degrees of troglomorphism (Table 9.1).

The family Chactidae Pocock, 1893 only includes one endogean species and one troglophile species; other cases appear to be accidental.

The family Diplocentridae Karsch, 1880 is notable and diverse in the Neotropics, with two genera and five species exhibiting pronounced troglomorphism (Table 9.1). Classified as troglobites, these species are found in the Caribbean islands of Cuba and the Dominican Republic, as well as in Mexico.

The family Euscorpiidae Laurie, 1896 has not been systematically evaluated over the past 25 years in the Neotropics, in contrast to the sustained research on its Palearctic counterparts (Soleglad and Sissom 2001; Podnar et al. 2021; Blasco-Aróstegui and Prendini 2023, 2024). In the present work, we adopt the classification of Fet and Sissom (2000) in Euscorpiidae. Accordingly, two troglobite euscorpiid species are recognized here (Table 9.1), with several accidental records observed in the field by one of the authors (EGS, pers. obs.) and a single case reported in the literature (Sissom and Reddell 2009).

The Neotropical scorpion families Troglotayosicidae Lourenço, 1981 and Typhlochactidae Mitchell, 1971 are characterized by the pronounced troglomorphism exhibited by all their species. Taxa within both families exhibit a clear eco-morphological association with endogean (e.g., subsoil, leaf-litter) and hypogean habitats (e.g., complex subterranean environments). Phylogenetic analyses of Troglotayosicidae (Moreno-González et al. 2024) and Typhlochactidae (Prendini et al. 2010) revealed an opposite pattern of ancestral habitat origin. One hypothesis posits that the ancestors of Troglotayosicidae were leaf-litter dwellers and subsequently colonized hypogean habitats, such as the Cueva de Los Tayos in Ecuador. In contrast, the descendants of the ancestral Typhlochactidae appear to have recently colonized endogean habitats in Mexico's Sierra de Juárez region. These cases exemplify the remarkable adaptability of scorpions in the Neotropical region and highlight their potential as living models for understanding the evolutionary processes underlying cavernicolous adaptation. Moreover, these two scorpion families alone account for five of the seven genera classified within the troglobite and endogean categories in the Neotropics (Table 9.1), together encompassing nine troglobitic and nine endogean species.

The family Vaejovidae Thorell, 1876, a predominantly Nearctic scorpion family, has colonized the northernmost regions of the Neotropics and exhibits all known ecomorphotypes in scorpions (González-Santillán and Prendini 2013). It includes one troglobitic species and seven troglophilic species (Table 9.1). We reevaluated *Vaejovis gracilis* (Gertsch & Soleglad, 1972), a species that was previously classified as troglophilic. Based on Prendini et al. (2021) criteria, we determined that *V. gracilis* is troglobitic, however. This species has only been collected inside caves and exhibits marked troglomorphic traits, including reduced median and lateral ocelli, diminished pigmentation, and attenuated appendages, just like another undescribed species picture in Fig. 9.3a. The presence of several troglophile species in Vaejovidae may indicate a recent radiation, potentially leading toward isolation and eventual evolution into cavernicolous specialization (Fig. 9.3b).

The diversity of subterranean and troglomorphic scorpions in the Neotropical region contrasts with that of other regions also possessing extensive and complex cave systems, such as the Western Palearctic. For example, the Balkans are considered a global hotspot for troglobitic fauna, yet only eleven scorpion species have been reported from caves, all of them belonging to the family Euscorpiidae and two genera: *Alpiscorpius* Gantenbein et al., 1999 and *Euscorpius* Thorell, 1876 (Blasco-Aróstegui and Prendini 2024). Among these, all exhibit varying degrees of cave adaptation and merely five species are exclusively cavernicolous: three are accidentals, two are trogloxenes, five are troglophiles, and only one is a true troglobiont (Karaman 2020; Blasco-Aróstegui and Prendini 2024). In the Iberian Peninsula and Sardinia, three additional troglomorphic scorpion species occur, all belonging to the endemic family Belisaridae Lourenço, 1998. Of these, two species have completely lost their ocelli, constituting the genus *Belisarius* Simon, 1879, and are primarily associated with highly humid leaf-litter and subsoil habitats (Lourenço 2015). The third species, *Sardoscorpius troglophilus* Tropea and Onnis, 2020 still possesses

Fig. 9.3 Neotropical scorpions associated with hypogean ecosystems. (**a**) *Vaejovis* sp. adult ♂ from a karstic cave system near Córdoba, Veracruz. Mexico (troglobite). (**b**) *Vaejovis* sp. adult ♂ from a volcanic cave system near Jalapa, Veracruz, Mexico (troglophile). (Credit photos: (**a**) Jorge I. Mendoza, (**b**) Alberto Hernández Lozano)

ocelli but has lost pigmentation, exhibits reduced sclerotization, and has not been reported outside the cave systems of Sardinia (Karaman 2020; Tropea and Onnis 2020).

The relatively low number of cavernicolous scorpions known from the Balkans diverges from the larger and more complex radiations observed in other cave-adapted invertebrates of the region, such as amphipods, pseudoscorpions, and beetles (Njunjić et al. 2016; Bedek et al. 2019; Borko et al. 2021; Hlebec et al. 2023). This discrepancy may be explained by the broad ecological plasticity of scorpions and the availability of alternative refugia, which together limited the extent of cave colonization in Europe, with only a few species becoming fully cavernicolous (Médail and Diadema 2009; Podnar et al. 2021; Blasco-Aróstegui and Prendini 2023, 2024). The historical tropical climate of the ancient Balkans, with its abundance of humid habitats, likely promoted the diversification of early humicolous and hygrophilous scorpions during the Tertiary (Ćurčić and Jovanović 2004). Only a small part of this "proto-European scorpion fauna" survived successive and prolonged climatic transitions—including Miocene aridification and Quaternary glaciations—by retreating into hypogean refugia, where they persisted as climate relicts and eventually became fully adapted to cave life (Karaman 2020; Tropea and Onnis 2020; Blasco-Aróstegui and Prendini 2024). The remaining lineages endured by colonizing alternative refugial habitats, such as relictual forests, mountains, and other mesic environments (Podnar et al. 2021; Blasco-Aróstegui and Prendini 2023, 2024). Besides, temporary occupation of caves during glaciation events may not have restricted some scorpion lineages from returning to the surface, which would likely explain the lithophilous habits of most *Euscorpius* species, as lithophily could be an evolutionary precursor to troglophily (Volschenk and Prendini 2008).

In addition to the similar glacial and aridification stages that took place during the Neogene, other distinct factors have contributed to the greater diversification of cavernicolous scorpions in the Neotropical region. The infamous Chicxulub crater was formed when a meteorite impacted and devastated nearly 80% of the planet's biodiversity, triggering the Cretaceous-Paleogene (K-Pg) mass extinction (Alvarez et al. 1980). Prendini et al. (2010) proposed that during this extinction event, cave systems may have acted as refugia for epigean fauna, including the ancestors of the family Typhlochactidae. They hypothesized that ancestral epigean relatives may have invaded the Sierra Madre Oriental prior to the meteorite impact at the K-Pg boundary (ca. 66 Mya) with deep cave habitats providing refuge for some lineages. Over millions of years, these isolated subterranean populations gradually acquired troglomorphic features, ultimately giving rise to highly specialized troglobiont endemic fauna within the region.

Evidence for the use of caves as refugia has, in fact, been demonstrated in populations of the trogloxene species *Pseudouroctonus reddelli* Gertsch and Soleglad, 1972 from Texas (Bryson et al. 2014). Prendini et al. further argue that the K-Pg extinction eliminated potential epigean competitors and opened numerous ecological niches, thereby promoting the recolonization of endogean habitats first. Today, species such as *Typhlochactas mitchelli* Sissom, 1988, *Typhlochactas sissomi* Francke et al., 2009, and *Typhlochactas sylvestris* Mitchell and Peck, 1977, more modern than their hypogean counterparts according to Prendini et al. (2010) phylogeny, can be found inhabiting leaf-litter (Volschenk and Prendini 2008; Prendini et al. 2010).

This catastrophic event must have also affected the ancestors of the Caribbean and Mexican diplocentrids, euscorpiids, and, to a lesser extent, the South American troglotayosicids and some buthid species, likely forcing part of these epigean taxa into subterranean habitats. One major constraint of this hypothesis, however, is that the hypogean habitats in question may not have existed at the time of their supposed invasion by typhlochactids. The karstic material that constitutes most of the Sierra Madre Oriental, which is home to a high diversity of Mexican troglobiont scorpions, was submerged during the Cretaceous period. The uplift of this terrain, resulting from the Laramide orogeny, occurred much later, during the late Eocene, approximately 35–33 Mya (Eguiluz-de-Antuñano 2007). Thus, the timing of the region's geological uplift and formation of karstic ecosystems would contradict the early invasion hypothesis.

Although the troglobitic Typhlochatidae lineages may be considered ancestral, this scorpion family of the Sierra Madre Oriental may have undergone more recent radiation than assumed. Nonetheless, this idea still requires testing and validation through more precise geological dating and molecular phylogenies. For now, another point in favor of the relictual hypothesis is that the hemispermatophore of typhlochactids has been classified as a "one-fold" bauplan in Monod et al. (2017), a morphological character that positions the family Typhlochactidae as an ancient group at the base of the Parvorder Iurida phylogeny.

Despite the difficulty of establishing a direct correlation between the meteorite impact and the diversification of typhlochactids after the Cretaceous-Paleogene mass extinction, it is nevertheless noteworthy that among the few dated Neotropical scorpions associated with hypogean habitats in a phylogeny is *Troglorhopalurus lacrau* (Lourenço and Pinto-da-Rocha 1997), a troglophile species from Brazil. Its divergence indeed coincides with a burst of diversification in most New World Buthidae and Vaejovidae in the Neotropics shortly after the K-Pg extinction (Santibáñez-López et al. 2022). This pattern suggests that the event may have exerted a stronger and clearer impact on epigean scorpions than on hypogean forms. A striking example is provided by vaejovids, which are entirely absent from the Yucatán Peninsula, the region closest to the impact center (see Chap. 5).

9.4 Ecology and Conservation Biology

Biodiversity knowledge is hampered by the so-called "shortfalls". Three of the most widely recognized, and applicable on a global scale, are the Linnaean shortfall (i.e., the gap between the number of species formally described and the actual number of species that exist), the Wallacean shortfall (i.e., the gap between the known and actual geographic distributions of species), and the Darwinian shortfall (i.e., the gap in our understanding of species' evolutionary histories and phylogenetic relationships) (Hortal et al. 2015). In subterranean ecosystems, these shortfalls are even more evident, with an additional challenge: the difficulty of sampling and access to caves or other hypogean habitats, which limits our ability to reduce these knowledge gaps (Mammola et al. 2021). This phenomenon is known as the "Racovitzan impediment" (Ficetola et al. 2019).

In the Neotropical region, most subterranean habitats remain poorly explored or are yet awaiting discovery. Locating a suitable entrance to a cave system is often challenging, particularly in dense forests or at high-elevation mountain ranges. Even once accessed, the fraction of the habitat that biospeleologists or speleologists can survey represents only a small portion of the environment available to organisms, which may create a misleading impression of rarity in some species that only occasionally appear in near-surface chambers (Mammola et al. 2021). Beyond the three main biodiversity knowledge gaps mentioned above, the Eltonian shortfall (i.e., the gap between the known and actual interactions among species or among groups of species) is perhaps the most pronounced in subterranean ecosystems (Vignoli and Prendini 2009; Hortal et al. 2015; Trajano and Carvalho 2017; Ficetola et al. 2019; Blasco-Aróstegui and Prendini 2024). The inherent inaccessibility and remoteness of most caves, combined with the logistical constraints of extended fieldwork, makes it extremely difficult to study the biota thoroughly or to conduct population monitoring as effectively as on the surface.

Consequently, the paucity of ecological data is noticeable across numerous cave-dwelling organisms, including scorpions (Vignoli and Prendini 2009; Souza-Silva et al. 2021; Blasco-Aróstegui and Prendini 2024). Only a few remarkable ecological habits have been documented in cavernicolous Neotropical scorpions, aside from their frequent association with moist microhabitats within caves. For instance, *Alacran tartarus*—the scorpion recorded at the greatest depths of any cave worldwide (ca. 720–916 m)—appears to be amphibious: some individuals were collected underwater in small cave streams (Vignoli and Prendini 2009).. Indirect observations of an undescribed species of troglophile *Vaejovis* (Fig. 9.3b) have indicated that subadults use epiphytic plants as a refuge, climbing several meters above the soil to mature and eventually returning to caves in the Cuetzalan karstic area of Puebla, Mexico. Adults and juveniles have only been collected inside the caves (E.G.-S. per. obs.).

Despite extensive taxonomic efforts, no standardized sampling studies of subterranean scorpions have yet been conducted in the Neotropics, and encounters with these organisms remain largely opportunistic (Lourenço 1981; Carvalho et al. 2025). Many of the cave-specialized scorpion species, particularly the obligate cave dwellers or troglobites, are assumed to follow a K-selected strategy, characterized by traits such as low reproductive output, slow growth, delayed maturity, and energy conservation in resource-poor environments (Karaman 2020). However, no direct observations of litter size or juvenile development exist, as no individuals have been reported carrying nymphs, nor has any study of ontogenetic development been conducted for any troglobite species. Similarly, little is known about their diets, interactions with other cave organisms, metabolic rates, or population sizes and dynamics.

The vast and complex networks of crevices, secondary galleries, and screes, combined with the apparent rarity of these species, currently allow only speculative inferences. Further research is urgently needed for reducing the Eltonian shortfall in subterranean environments by improving our understanding of scorpion species' ecology, population dynamics, and interactions is therefore a central task for contemporary arachnologists.

The conservation of hypogean ecosystems in the Neotropics is also still at an early stage, with most efforts concentrated in Brazil as recent changes in Brazilian legislation weakened cave protection, allowing "less relevant" (i.e., physically and/or biologically) caves to be destroyed (Cardoso et al. 2021; Junta et al. 2025). From bats to numerous invertebrates, endemic subterranean organisms have increasingly drawn conservation attention over the last decade (Pellegrini and Lopes Ferreira 2012; Simões et al. 2014; Jaffé et al. 2018; de Sousa Barros et al. 2020; Torres and Bichuette 2024; Duarte et al. 2025). However, two major issues remain unsolved: (1) most subterranean environments remain inadequately surveyed and comprehensive faunistic inventories are largely lacking, making it impossible to accurately estimate the true diversity of these habitats or assess the conservation status of their inhabitants; and (2) national red lists, where most of these species receive protection, are not formally recognized by global conservation bodies, limiting assessments of the actual number of threatened species worldwide.

Consequently, organizations such as the International Union for Conservation of Nature (IUCN) and the Convention on International Trade in Endangered Species of Wild Fauna and Flora (CITES) do not fully recognize the fragility and threatened status of most subterranean fauna mainly due to the lack of precise knowledge regarding population sizes and trends. No subterranean scorpion species have yet been formally included in any threatened category or CITES-listed, despite formal petitions (Karaman 2020; Prendini et al. 2021; Blasco-Aróstegui and Prendini 2024; Blasco-Aróstegui et al. 2025; Carvalho et al. 2025). Incorporating endemic troglobitic scorpions into global Red Lists or other protection frameworks could be a crucial first step toward ensuring their conservation and preventing their tentative extinction. At present, only eight cave-dwelling scorpion species in the Neotropics, each representing a distinct genus, are known to occur within natural protected areas (Table 9.3). Together, they span four different families but they only account for a 24% of the total described diversity of cavernicolous scorpions in the region (i.e., trogloxenes, troglophiles, and troglobites). Brazil, Cuba, and Mexico each harbor two species, whereas the Dominican Republic and Venezuela host a single species each.

Besides, cave systems are particularly sensitive to anthropogenic pressures and climate change, as they are closely and delicately linked to surface (epigean) habitats. The expansion of recreational tourism, pollution, and the fragmentation of previously continuous habitats for public access are severely affecting the structure and stability of these ecosystems (Mammola et al. 2019a, b; Nanni et al. 2023). Thus, the inclusion of further subterranean systems within natural protected areas, alongside their overlying ecosystems, is essential for an adequate conservation of the fauna they shelter. In alignment with the National Biodiversity Strategies and Action

Table 9.3 Cave-dwelling scorpion species from the Neotropics occurring within natural protected areas, with their taxonomic classification, countries of distribution, and specific protected areas of occurrence

Family	Species	Country	Natural protected area
Buthidae C. L. Koch, 1837	*Alayotityus delacruzi* Armas, 1973	Cuba	Siboney-Juticí Ecological Reserve
	Tityopsis sheylae Teruel and Rodríguez, 2020	Cuba	Parque Escales de Jaruco
	Tityus confluens bodoquena Lourenço et al., 2004	Brazil	Parque Nacional Sierra da Bodoquena
	Troglorhopalurus translucidus Lourenço et al., 2004	Brazil	Parque Nacional da Chapada Diamantina
Chactidae Pocock, 1893	*Broteochactas trezzi* (Vignoli and Kovařík, 2003)	Venezuela	Parque Nacional Natural Canaima
Diplocentridae Karsch, 1880	*Cazierius ciguayo* Teruel et al. 2021	Dominican Republic	Parque Nacional El Choco y Las Cuevas de Cabarete
Typhlochactidae Mitchell, 1971	*Sotanochactas elliotti* (Mitchell, 1971)	Mexico	Reserva de la Biosfera Sierra del Abra Tanchipa
	Typhlochactas rhodesi Mitchell, 1968	Mexico	Reserva de la Biosfera El Cielo

Plans (NBSAPs) and the 2030 targets of the Convention on Biological Diversity (CBD), caves and other hypogean habitats should be granted the same levels of protection as surface ecosystems. Such actions are essential not only for maintaining biodiversity, but also for safeguarding ecosystem functions and services, ecological integrity, and landscape connectivity.

A systematic assessment of the conservation status of cavernicolous scorpions—many of which represent ancient evolutionary relicts—should also be conducted. Without such efforts, these species risk being lost to environmental crises they are poorly equipped to withstand. Preserving their unique genetic heritage is critical, as they play an irreplaceable role in sustaining ecological balance within subterranean ecosystems. Strengthening their conservation would therefore contribute directly to reducing biodiversity shortfalls and advancing the global objectives set under the NBSAPs and CBD 2030 framework (Blasco-Aróstegui et al. 2025; Convention on Biological Diversity 2025a, b).

9.5 Conclusions

The scorpion fauna associated with subterranean habitats in the Neotropics remains poorly studied. A primary factor contributing to the dearth of long-term studies on the natural history, ecology, and evolution of troglobionts is the inaccessibility to the vast and complex networks of crevices, and secondary galleries, further complicated by floodings during the rainy season.

This chapter presented a comprehensive overview, synthesizing all available information on the remarkable Neotropical troglomorphic and cave-dwelling scorpion fauna documented in the existing literature. Notwithstanding, the present study proffers informative insights into the diversity of cavernicolous and troglomorphic scorpions. Our analyses identified patterns in regional diversity, emphasizing the interplay of geological, ecological, and evolutionary factors that shaped the biodiversity of these animals in the Neotropics, with comparative references to the Western Palearctic. For instance, the pervasive diversity of troglobionts in the Mesoamerican dominion (Fig. 9.1) and the recurrent use of endogean habitats in the South American transition zone (Fig. 9.2) are highlighted. While the Antillean subregion is home to a relatively few numbers of hypogean scorpions (Fig. 9.1), the Chocoan and Paraná dominions exhibit biodiversity levels comparable to those of the Mesoamerican dominion (Fig. 9.2).

Evidence shows that this pattern is in constant flux, as ongoing discoveries and documentation continue to refine our understanding of species distributions and add new ones to the record. The chapter also exemplifies how to correctly classify troglomorphic and cave-dwelling Neotropical scorpion species into specific ecological categories of the subterranean realm, based on information on their natural history, phylogeny, ecology, and morphology, following classification systems proposed by previous authors and derived from the well-known Schiner-Racovitza framework.

Lastly, the chapter highlights biodiversity gaps, specifically the Linnaean (i.e., species yet to be described), Darwinian (i.e., gaps in understanding evolutionary relationships), and Eltonian shortfalls (i.e., limited ecological information), while identifying the factors contributing to these gaps in the current knowledge of subterranean scorpions and outlining steps for future research. Similarly, it also addresses the conservation challenges both scorpions and their subterranean habitats face in an unforeseeable future that is increasingly affected by climate change and human activity, which directly threaten these fragile ecosystems. It summarizes the few species currently under some form of protection and proposes measures to mitigate these threats, including expanding protected areas to encompass subterranean environments and increasing the number of species listed under IUCN and CITES.

References

Agnarsson I (2023) Grand challenges in research on arachnid diversity, conservation, and biogeography. Fronti Arachn Sci 2:1101141

Alvarez LW, Alvarez W, Asaro F, Michel HV (1980) Extraterrestrial cause for the Cretaceous--Tertiary extinction. Science 208(4448):1095–1108

Antonelli A (2022) The rise and fall of Neotropical biodiversity. Bot J Linn Soc 199(1):8–24

Antonelli A, Zizka A, Carvalho FA, Scharn R, Bacon CD, Silvestro D, Condamine FL (2018) Amazonia is the primary source of Neotropical biodiversity. Proc Natl Acad Sci 115(23):6034–6039

Bedek J, Taiti S, Bilandžija H, Ristori E, Baratti M (2019) Molecular and taxonomic analyses in troglobiotic *Alpioniscus* (Illyrionethes) species from the Dinaric Karst (Isopoda: Trichoniscidae). Zool J Linnean Soc 187(3):539–584
Blasco-Aróstegui J, Prendini L (2023) Glacial relicts? A new scorpion from Mount Olympus, Greece (Euscorpiidae: *Euscorpius*). Am Mus Novit 2023(4003):1–36
Blasco-Aróstegui J, Prendini L (2024) Redescription of *Euscorpius studentium* based on adult specimens; updated classification of cavernicolous Euscorpiidae; and review of cavernicolous scorpions in the Balkans. Diversity 16(12):737
Blasco-Aróstegui J, Moreno-González JA, Loria SF, Carvalho LS, Bird TL, Nguyen AD, Barahoei H, Haddad CR, Prendini L (2025) Include scorpions in global conservation plans. Science 389(6765): 1099–1100. https://doi.org/10.1126/science.adz4176
Borko Š, Trontelj P, Seehausen O, Moškrič A, Fišer C (2021) A subterranean adaptive radiation of amphipods in Europe. Nat Commun 12(1):3688
Bryson RW Jr, Prendini L, Savary WE, Pearman PB (2014) Caves as microrefugia: pleistocene phylogeography of the troglophilic North American scorpion *Pseudouroctonus reddelli*. BMC Evol Biol 14(1):9
Cardoso RC, Ferreira RL, Souza-Silva M (2021) Priorities for cave fauna conservation in the Iuiú karst landscape, northeastern Brazil: a threatened spot of troglobitic species diversity. Biodivers Conserv 30(5):1433–1455
Carvalho LS, Silva MIL, Souza PE, Ferreira RL (2025) Taxonomy and distribution of the cave-dwelling scorpions *Troglorhopalurus* (Scorpiones, Buthidae), with the description of a new cave-restricted species. Arthropoda 3(2):6
Christiansen K (1962) Proposition pour la classification des animaux cavernicoles. Spelunca 2:76–78
Convention on Biological Diversity (2025a) 2030 Targets (with Guidance Notes). https://www.cbd.int/gbf/targets
Convention on Biological Diversity (2025b) National Biodiversity Strategies and Action Plans (NBSAPs). https://www.cbd.int/nbsap
Ćurčić BPM, Jovanović VM (2004) The cave-dwelling fauna of the Balkan Peninsula: its origin and diversification. Acta Entomol Slovenica 12:35–56
de Sousa Barros J, Bernard E, Ferreira RL (2020) Ecological preferences of neotropical cave bats in roost site selection and their implications for conservation. Basic Appl Ecol 45:31–41
Duarte MVSA, Gallão JE, Bichuette ME (2025) Brazilian Troglophilic Biodiversity: a neglected fauna in conservation actions. Biota Neotropica 25(3):e20241711
Eguiluz-de-Antuñano S (2007) Laramide deformation in the Burgos Basin, Northeastern Mexico. In: 27th annual GCSSEPM foundation Bob F. Perkins research conference. Gulf Coast Section SEPM (Society for Sedimentary Geology), pp 688–702
Fet V, Lowe G (2000) Family Buthidae C. L. Koch, 1837. In: Fet V, Sissom WD, Lowe G, Braunwalder ME (eds) Catalog of the scorpions of the world. The New York Entomological Society, New York, pp 54–286
Fet V, Sissom WD (2000) Family Euscorpiidae Laurie, 1896. In: Fet V, Sissom WD, Lowe G, Braunwalder ME (eds) Catalog of the scorpions of the world (1758–1998). The New York Entomological Society, New York, pp 355–380
Ficetola GF, Canedoli C, Stoch F (2019) The Racovitzan impediment and the hidden biodiversity of unexplored environments. Conserv Biol 33(1):214–216
González-Santillán E, Prendini L (2013) Redefinition and generic revision of the North American vaejovid scorpion subfamily Syntropinae Kraepelin, 1905, with descriptions of six new genera. Bull Am Mus Nat Hist 2013(382):1–71
Harrison GT, Dunleavy HP, Vasquez-Valverde LF, Del Pozo-Valdivia AI, Ivanov K, Marek PE (2024) Arthropod diversity in shallow subterranean habitats of the Appalachian Mountains. Subterranean Biology 49:75–95
Hesselberg T (2023) The biomimetic potential of novel adaptations in subterranean animals. Front Ecol Evol 11:1127728

Hlebec D, Podnar M, Kučinić M, Harms D (2023) Molecular analyses of pseudoscorpions in a subterranean biodiversity hotspot reveal cryptic diversity and microendemism. Sci Rep 13(1):430

Hortal J, De Bello F, Diniz-Filho JAF, Lewinsohn TM, Lobo JM, Ladle RJ (2015) Seven shortfalls that beset large-scale knowledge of biodiversity. Annu Rev Ecol Evol Syst 46(1):523–549

Howarth FG (1983) Ecology of cave arthropods. Ann Rev Entomol 28(1):365–389

Jaffé R, Prous X, Calux A, Gastauer M, Nicacio G, Zampaulo R, Souza-Filho PW, Oliveira G, Brandi IV, Siqueira JO (2018) Conserving relics from ancient underground worlds: assessing the influence of cave and landscape features on obligate iron cave dwellers from the Eastern Amazon. PeerJ 6:e4531

Juberthie C, Delay B, Bouillon M (1980) Sur l'existence d'un milieu souterrain superficiel en zone non calcaire. C R Acad Sci 290:49–52

Junta VGP, Souza-Silva M, Vaz GAS, Ferreira RL (2025) One cave, multiple worlds: cave zonation, habitat selection and conservation of cave-dwelling fauna in a new hotspot of subterranean biodiversity in South America. Community Ecol 26(2):287–302

Karaman I (2020) A new *Euscorpius* species (Scorpiones: Euscorpiidae) from a Dinaric cave-the first record of troglobite scorpion in European fauna. Biol Serb 42(1):14–31

Lourenço WR (1981) Scorpions cavernicoles de l'Equateur: *Tityus demangei* n. sp. et *Ananteris ashmolei* n. sp. (Buthidae): *Troglotayosicus vachoni* n. gen., n. sp. (Chactidae), Scorpion troglobie. Bull Mus Natl Hist Nat Sect A Zool Biol Ecol Anim 2:635–662

Lourenço WR (2015) The genus *Belisarius* Simon, 1879 (Scorpiones: Troglotayosicidae), with the description of a new vicariant species from the south of Spain. C R Biol 338(5):362–367

Lourenço WR, Pinto-da-Rocha R (1997) A reappraisal of the geographic distribution of the genus *Rhopalurus* Thorell (Scorpiones, Buthidae) and description of two new species. Biogeographica 73:181–191

Mammola S (2019) Finding answers in the dark: caves as models in ecology fifty years after Poulson and White. Ecography 42(7):1331–1351

Mammola S, Isaia M (2017) Spiders in caves. Proc R Soc B Biol Sci 284(1853):20170193. https://doi.org/10.1098/rspb.2017.0193

Mammola S, Giachino PM, Piano E, Jones A, Barberis M, Badino G, Isaia M (2016) Ecology and sampling techniques of an understudied subterranean habitat: the Milieu Souterrain Superficiel (MSS). Sci Nat 103(11–12):88

Mammola S, Cardoso P, Culver DC, Deharveng L, Ferreira RL, Fišer C, Galassi DM, Griebler C, Halse S, Humphreys WF (2019a) Scientists' warning on the conservation of subterranean ecosystems. Bioscience 69(8):641–650

Mammola S, Piano E, Cardoso P, Vernon P, Domínguez-Villar D, Culver DC, Pipan T, Isaia M (2019b) Climate change going deep: the effects of global climatic alterations on cave ecosystems. Anthropocene Rev 6(1–2):98–116

Mammola S, Lunghi E, Bilandžija H, Cardoso P, Grimm V, Schmidt SI, Hesselberg T, Martínez A (2021) Collecting eco-evolutionary data in the dark: impediments to subterranean research and how to overcome them. Ecol Evol 11(11):5911–5926. https://doi.org/10.1002/ece3.7556

Médail F, Diadema K (2009) Glacial refugia influence plant diversity patterns in the Mediterranean Basin. J Biogeogr 36(7):1333–1345

Menin DDS, Bacci DLC (2023) Speleosystemic Services: how to characterize caves from the perspective of Ecological Economics and Geosystemic Services? Geol USP Sér Cient 23(3):121–139

Monod L, Cauwet L, González-Santillán E, Huber S (2017) The male sexual apparatus in the order Scorpiones (Arachnida): a comparative study of functional morphology as a tool to define hypotheses of homology. Front Zool 14(1):1–48

Moreno-González JA, Luna-Sarmiento DA, Prendini L (2024) Phylogeny of the troglomorphic scorpion genus *Troglotayosicus* (Scorpiones: Troglotayosicidae) with description of a new species from Colombia. Am Mus Novit 2024(4011):1–39

Morrone JJ (2014) Biogeographical regionalization of the Neotropical region. Zootaxa 3782(1):1–110

Morrone JJ (2018) Evolutionary biogeography of the Andean region. CRC Press, Boca Raton

Morrone JJ, Escalante T, Rodríguez-Tapia G, Carmona A, Arana M, Mercado-Gómez JD (2022) Biogeographic regionalization of the Neotropical region: new map and shapefile. An Acad Bras Cienc 94

Nanni V, Piano E, Cardoso P, Isaia M, Mammola S (2023) An expert-based global assessment of threats and conservation measures for subterranean ecosystems. Biol Conserv 283:110136

Nicolosi G, Gerovasileiou V (2024) Towards invasion ecology for subterranean ecosystems. Biodivers Conserv 33(4):1561–1569. https://doi.org/10.1007/s10531-024-02820-1

Njunjić I, Perreau M, Hendriks K, Schilthuizen M, Deharveng L (2016) The cave beetle genus *Anthroherpon* is polyphyletic; molecular phylogenetics and description of *Graciliella* n. Gen. (Leiodidae, Leptodirini). Contrib Zool 85(3):337–359

Olson DM, Dinerstein E, Wikramanayake ED, Burgess ND, Powell GV, Underwood EC, D'amico JA, Itoua I, Strand HE, Morrison JC (2001) Terrestrial ecoregions of the World: a new map of life on earth: a new global map of terrestrial ecoregions provides an innovative tool for conserving biodiversity. Bioscience 51(11):933–938

Pellegrini TG, Lopes Ferreira R (2012) Management in a neotropical show cave: planning for invertebrates conservation. Int J Speleol 41(2):359–366

Podnar M, Grbac I, Tvrtković N, Hörweg C, Haring E (2021) Hidden diversity, ancient divergences, and tentative Pleistocene microrefugia of European scorpions (Euscorpiidae: Euscorpiinae) in the eastern Adriatic region. J Zool Syst Evol Res 59(8):1824–1849

Prendini L (2001) Substratum specialization and speciation in southern African scorpions: the Effect Hypothesis revisited. In: Fet V, Selden PA (eds) Scorpions 2001. In Memoriam Gary A. Polis. British Arachnological Society, Burnham Beeches, Bucks, pp 113–118

Prendini L, Francke OF, Vignoli V (2010) Troglomorphism, trichobothriotaxy and typhlochactid phylogeny (Scorpiones, Chactoidea): more evidence that troglobitism is not an evolutionary dead-end. Cladistics 26(2):117–142

Prendini L, Ehrenthal V, Loria SF (2021) Systematics of the relictual Asian scorpion family Pseudochactidae Gromov, 1998, with a review of cavernicolous, troglobitic, and troglomorphic scorpions. Bull Am Mus Nat Hist 453:1–149

Pulliam HR (1988) Sources, sinks, and population regulation. Am Nat 132(5):652–661

Racovitza EG (1907) Essai sur les problemes biospeologiques. Archives des Maladies du Coeur et des Vaisseaux 4e Serie 6: 371–48

Raven PH, Gereau RE, Phillipson PB, Chatelain C, Jenkins CN, Ulloa Ulloa C (2020) The distribution of biodiversity richness in the tropics. Science. Advances 6(37):eabc6228

Rull V (2020) Neotropical diversification: historical overview and conceptual insights. In: Rull V, Carnaval AC (eds) Neotropical diversification: patterns and processes. Springer, Cham, pp 13–49

Santibáñez-López, C.E., Aharon, S., Ballesteros, J.A., Gainett, G., Baker, C.M., González-Santillán, E., Harvey, M.S., Hassan, M.K., Abu Almaaty, A.H., Aldeyarbi, S.M., Monod, L., Ojanguren-Affilastro, A., Pinto-da-Rocha, R., Zvik, Y., Gavish-Regev, E., Sharma, P.P. (2022). Phylogenomics of scorpions reveal contemporaneous diversification of scorpion mammalian predators and mammal-active sodium channel toxins. Systematic Biology 71(6): 1281–1289. https://doi.org/10.1093/sysbio/syac021

Schiner JR (1854) Fauna der Adelsberger-, Lueger- und Magdalenen Grotte. In: Schmidl A (ed) Zur Höhlenkunde des Karstes. Die Grotten und Höhlen von Adelsberg, Lueg, Planina und Laas, Braumüller, pp 231–272

Simões MH, Souza-Silva M, Ferreira RL (2014) Cave invertebrates in northwestern Minas Gerais state, Brazil: endemism, threats and conservation priorities. Acta Carsologica 43(1):59–174

Sissom WD, Reddell JR (2009) Cave scorpions of Mexico and the United States. In: Cokendolpher JC, Reddell JR (eds) Studies on the cave and endogean fauna of North America, V, Texas Memorial Museum, Speleological Monographs, vol 7. Texas Memorial Museum, University of Texas at Austin, Austin, pp 19–32

Sket B (2008) Can we agree on an ecological classification of subterranean animals? J Nat Hist 42(21–22):1549–1563
Soleglad ME, Sissom WD (2001) Phylogeny of the family Euscorpiidae Laurie, 1896: a major revision. In: Fet V, Selden PA (eds) Scorpions 2001: in memoriam Gary A. Polis. British Arachnological Society, Burnham Beeches, pp 25–111
Souza-Silva M, Cerqueira RFV, Pellegrini TG, Ferreira RL (2021) Habitat selection of cave-- restricted fauna in a new hotspot of subterranean biodiversity in Neotropics. Biodivers Conserv 30(14):4223–4250
Torres DF, Bichuette ME (2024) Iron Islands: the importance of iron caves in the eastern Amazon for bat conservation. Austral Ecol 49(6):e13550. https://doi.org/10.1111/aec.13550
Trajano E (2012) Ecological classification of subterranean organisms. Encycl Caves 1:275–277
Trajano E, Carvalho MR (2017) Towards a biologically meaningful classification of subterranean organisms: a critical analysis of the Schiner-Racovitza system from a historical perspective, difficulties of its application and implications for conservation. Subterr Biol 22:1–26
Tropea G, Onnis C (2020) A remarkable discovery of a new scorpion genus and species from Sardinia (Scorpiones: Chactoidea: Belisariidae). Arachn Riv Aracnol Italiana 26:3–25
Vignoli V, Prendini L (2009) Systematic revision of the troglomorphic North American scorpion family Typhlochactidae (Scorpiones: Chactoidea). Bull Am Mus Nat Hist 326:1–94
Volschenk ES, Prendini L (2008) *Aops oncodactylus*, gen. et sp. nov., the first troglobitic urodacid (Urodacidae: Scorpiones), with a re-assessment of cavernicolous, troglobitic and troglomorphic scorpions. Invertebr Syst 22(2):235–257

Chapter 10
Functional Morphology of Predatory and Defensive Behaviour in Neotropical Scorpions

Yuri Simone

Abstract Morphological diversity in Neotropical scorpions is remarkable. The traits accounting for the largest morphological variance across scorpion families are the pincers (chelae), the venom delivery system, and the venom itself. These traits are associated with different performance in crucial ecological tasks like predatory or defensive behaviours. Bearing different sets of weapon systems led Neotropical scorpions to develop predatory strategies and behaviours aiming to simultaneously maximise prey capture success and self-defence from predators or attackers. This chapter will introduce the importance of functional morphology for comparative studies and how morphology influences individual fitness through trait performance (ecomorphological paradigm). Successively, the main differences in chelae, metasoma, and venom will be characterised for the main families. Lastly, two extensive paragraphs will be dedicated to literature examples of how differences in functional traits affect predatory and defensive behaviours in Neotropical scorpions.

Keywords Chelae • Performance • Prey-predator interaction • Venom

10.1 Introduction

This chapter explores the predatory and defensive behaviours of Neotropical scorpions, with a focus on the relationship between form and function and an emphasis on the functional morphology of their weapon systems. It begins with an overview of functional morphology, highlighting its role in understanding why scorpions exhibit certain shapes and adaptations. The concept of trait variability as a reflection of differences in organismal performance is introduced, briefly touching on the ecomorphological paradigm proposed by Arnold (1983).

Y. Simone (✉)
Laboratory of Functional Morphology, Department of Biology, University of Antwerp, Wilrijk, Belgium

A. F. de Araujo Lira, E. González-Santillán (eds.), *The Biology of Neotropical Scorpions*, https://doi.org/10.1007/978-3-032-23194-9_10

The primary focus is a comparative analysis of the predatory and defensive strategies employed by Neotropical scorpions, examined through the lens of their morphology. To build this foundation, the chapter delves into the structural features of scorpion weapon systems, including the chelae and stinger. While venom variation and its ecological roles in predation and defence are acknowledged, these topics are only briefly addressed, as they are covered extensively in other chapters.

The core of the chapter provides a structured exploration of predatory and defensive behaviours, consistently linking observed patterns to morphological features. While extensive literature exists on these topics, this chapter deliberately emphasises studies and examples specific to Neotropical scorpions, the focus of this book. References to non-Neotropical species are included only when absolutely necessary to provide context or contrast. Consequently, some well-known and frequently cited studies on scorpions outside the Neotropics have been intentionally omitted to maintain a clear and targeted discussion on Neotropical scorpions.

10.2 What Is Functional Morphology

For centuries, naturalists have been fascinated by the shapes and forms of animals, seeking to understand the underlying causes of the remarkable diversity in animal morphology. This curiosity about the mechanisms of nature, particularly in explaining how specific body structures function within ecological contexts, forms the foundation of functional morphology. Functional morphology studies the relationship between an organism's structure and its function, addressing fundamental questions about the diversity of life and how particular forms are adapted to perform specific ecological roles (Bock 1994; Stevens 2005; Ruff 2018). By analysing these adaptations, scientists can infer patterns in the evolutionary processes that shape organisms, thereby deepening our understanding of the natural world and the intricate links between form and function.

The forms that a body structure can take are virtually infinite. However, the boundary between the possible and the feasible is constrained by universal physical laws (Alexander 1985). Even when structures are physically feasible, they may still disappear if they negatively impact an organism's fitness. Only structures that are both feasible and positively affect an individual's fitness are likely to be selected and passed to future generations. But how can we assess whether a structure positively or negatively affects fitness? According to evolutionary theory, for a trait to be considered functional, it must exhibit variability across individuals, and this variability must correlate with Darwinian fitness, which measures an organism's ability to produce offspring that survive and reproduce. In birds, for instance, fitness can be quantified using metrics such as clutch size or the number of fledglings produced per reproductive event.

However, gathering information about fitness descriptors across all species can be challenging and time-consuming. To address this, Arnold (1983) proposed an innovative framework that links trait performance to morphological diversity and

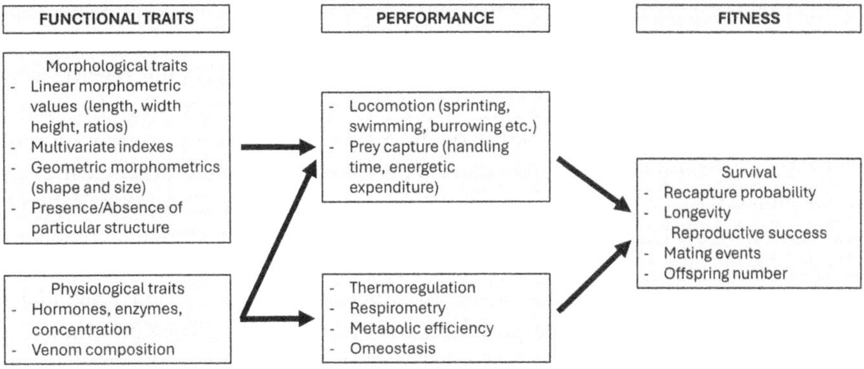

Fig. 10.1 Schematic representation of the morphology-performance-fitness paradigm proposed by Arnold (1983). Variation in a functional trait should be reflected in a differential performance outcome. In this regard, performance must have an ecological meaning in order to be associated with a positive or negative effect on fitness

fitness (see Fig. 10.1). This framework posits that differently shaped structures perform ecological tasks, such as capturing prey or defending against predators, at varying levels of efficiency, and individuals with higher-performing traits are likely to achieve greater fitness. This "performance paradigm" revolutionised comparative morphology by quantitatively linking trait performance to ecological function, allowing for objective comparisons across populations and species.

10.3 Scorpion Ecomorphology

Despite their conserved body plan, scorpions exhibit a remarkable morphological variability, particularly in their anterior appendages (legs and pedipalps) and in the metasoma including the telson. However, the reasons behind this phenotypic diversity remain unclear. Almost 50 years ago, the works led by Lamoral and Newland linked the variation in the morphology of scorpion pedipalps, metasoma, and ambulatory legs to microhabitat use, classifying scorpions into five ecomorphs corticolous (vegetation dwellers), lithophilous (rock climbers), pelophilous (hard soil fossorials), psammophilous (sand fossorials), and lapidicolous (ground surface vagrants) (Newlands 1972; Lamoral 1979). This classification, rather qualitative at that time, was recently revisited with a more quantitative approach leading to the reassessing of those ecomorphs into lithophilous, psammophilous, and pelophilous, phytophilous as equivalent for corticolous and the poor supported lapidocoluous was suppressed (Coelho et al. 2022). For example, scorpions with slender pedipalps and longer legs are typically associated with sand-dwelling habitats (psammophilous), while those with more robust pedipalps and shorter legs are found in rock-dwelling environments (pelophilous and litophilous).

Microhabitat usage is likely not the only evolutionary driver of scorpion morphological diversity. Intra- and interspecific interactions surely play an important, but overlooked, role in shaping scorpion's morphological diversity. In this regard, scorpions are particularly well-suited for studying the relationships between functional traits, their performance, and their ecological roles due to the clear separation of these functions into different body parts. This functional and morphological compartmentalisation, known as modularity (Wagner 1996; Klingenberg 2008; Zelditch and Goswami 2021), allows researchers to assign specific, unambiguous functions to different morphological modules. For instance, while many predators such as spiders or snakes rely on an integrated body structure (e.g., jaws) to grasp and inject venom into their prey, scorpions perform these tasks with three separate structures: they grasp prey with their pincers (chelae), inject venom through the stinger, and chew the prey with their chelicerae.

10.4 Metasoma and Telson Functional Morphology

The morphological diversity of the metasoma and telson, as well as their correlation with different stinging behaviours, has been extensively studied (Coelho et al. 2017; van der Meijden and Kleinteich 2017). In all scorpions, the metasoma consists of five cylindrical segments that tend to lengthen distally from the body. Each segment possesses an almost circular opening posteriorly, where the next segment fits in. Anteriorly, these segments possess two saddle-like protrusions laterally, which are able to rotate in two directions on the rim of the posterior circular opening of the previous segment allowing for a bending movement in the dorsoventral direction (up and down) in combination with a twisting movement about the metasoma longitudinal axis (Günther et al. 2021a). These flexible, dual-motion joints give the scorpion a wide range of movement, making it easier to sting prey, predators, or other scorpions from different angles (Bowerman 1972). This versatility may also help reduce the risk of damaging the aculeus (see below). The musculature of these segments has been well described (Lankester 1885; Günther et al. 2021b; Simone and van der Meijden 2021) and is responsible for the strikes observed during defensive instances.

Coelho et al. (2017) explored the striking shape and kinematics in different scorpion species, identifying two primary strike trajectories: "open" and "closed." In open trajectories, where the starting and ending points are further apart, the telson travels a greater distance at higher velocity. In closed trajectories, where the starting and ending points are closer, the telson moves through a smaller space but is quickly reset for repeated strikes. These differences in metasomal morphology translate into distinct striking performances, likely influencing telson shape and stinging strategy. However, it is not clear which is the advantage of one or the other stinging trajectory, since the stinging impact force has not been measured yet.

The telson, composed of the venom vesicle and the aculeus, is also extremely variable across scorpions. A comprehensive comparative analysis of telson shape across scorpion species has shown that more venomous species tend to have longer aculei with a larger radius of curvature (van der Meijden and Kleinteich 2017). Finite element analysis (FEA) of these longer aculei revealed basal stress concentrations, suggesting a higher risk of breakage at the base due to their shape. In contrast, shorter and thicker aculei displayed stress concentrations at the tip, as already shown for viper fangs (Broeckhoven and Du Plessis 2017). This finding is somewhat counterintuitive, as more venomous species use their telson more frequently, generating higher impact forces that could increase the risk of breakage. Basal ruptures are more severe than tip breakages, potentially restricting venomous species to a limited range of optimal stinging angles where strain energy is minimised (Zhao et al. 2016). A more conical aculeus shape would reduce basal stress concentrations, decreasing the likelihood of breakage (Bar-On et al. 2014). Interestingly, while more venomous species exhibit elongated aculei, which increases the risk of breakage, they also have thicker aculei, which reduces this risk. This balance between shape and structural integrity likely plays a key role in the evolution of stinging strategies in venomous scorpions.

Another morphological structure that may be present on the telson is the sub-aculear tubercle, a rounded or sometimes more conical protuberance located just before the start of the aculeus. This structure evolved independently multiple times in three scorpion families, Buthidae, Diplocentridae, and Vaejovidae (González-Santillán and Prendini 2015). The role of this structure is unknown. The hypothesis that it serves as a secondary pivot point during stinging was rejected by van der Meijden et al. (2017). However, Lourenço (2020) proposed that in species with particularly long aculei, the sub-aculear tubercle may act as a brake, potentially reducing the risk of aculeus breakage by stabilising the stinging motion.

10.5 Chela Functional Morphology

As the metasoma and the telson, also chelae in scorpions have a remarkable morphological variability (van der Meijden et al. 2010, 2012). The chelae form the last two segments of the pedipalps, the second pair of appendages extending from the prosoma. Structurally, the chelae consist of an immovable portion called the *manus* (or tibia), and a movable part known as the movable finger (or tarsus). These are connected by a single hinge joint that allows the movable finger to rotate in one plane opening and closing the chela. Chela closing is controlled by the contraction of the closing muscles located within the *manus* and the preceding segment, the *patella* (Fig. 10.2) (Snodgrass 1952; Dubale and Vyas 1968; Gilai and Parnas 1970). The opening, on the other hand, occurs through the relaxation of these muscles, combined with the release of elastic energy stored in the resilin component of the joint and the reduction of the internal haemolymph pressure (Mathew 1965; Alexander 1967; Govindarajan and Rajulu 1974; Sensenig and Shultz 2004).

To date, the internal anatomy of chela-closing muscles has not been thoroughly examined in a comparative framework. Existing detailed data are limited to a few

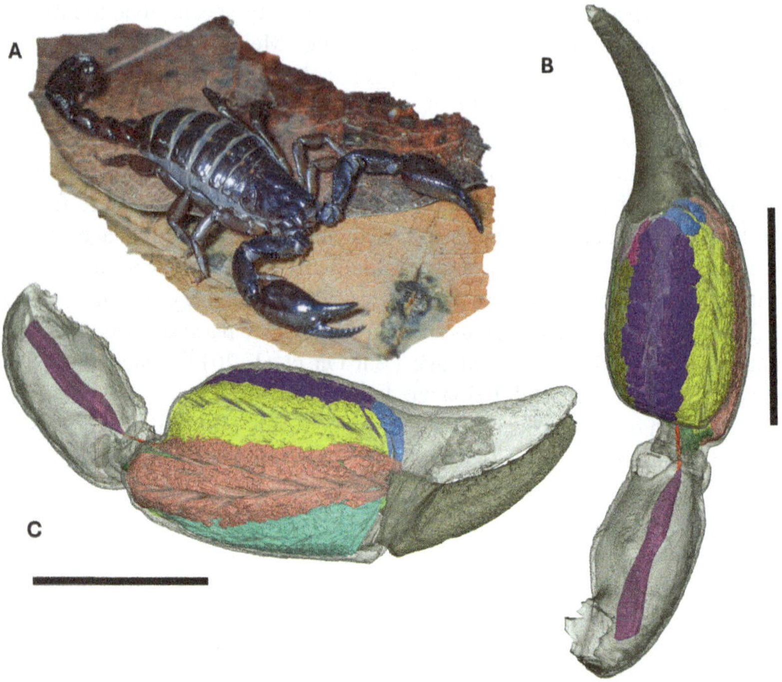

Fig. 10.2 A female individual of *Chactas* sp. (**a**) is shown resting on a leaf. Detailed visualisations of its patella and chela are provided using CT scanning, with (**b**) showing the dorsal view and (**c**) the lateral view. Muscle bundles within the chela and patella are depicted in various colours, while the cuticular surface of the manus has been rendered transparent to emphasise the internal arrangement of the closing muscles. Scale bar = 5 mm. (Photo of the female *Chactas* by Dr. Arie van der Meijden)

studies. Gilai and Parnas (1970) analysed the pedipalp of *Leiurus quinquestriatus* (Ehrenberg, 1828) and reported three muscle bundles in the manus and two in the patella. Dubale and Vyas (1968), studying the pedipalp of *Heterometrus fulvipes* (Koch, 1837), described eight closing muscle bundles in the manus and a single parallel muscle in the patella. Snodgrass (1952) provided additional but less detailed observations, illustrating the pedipalps of *Pandinus* and *Centruroides* without specifying the number of muscle bundles involved. Another recent comprehensive anatomical study is by Bicknell et al. (2024), which examined the pedipalps of *Urodacus manicatus* (Thorell, 1876). This work identified a single, parallel-fibred muscle bundle in the patella, but the number of bundles in the manus-closing muscle was not specified. Finally, in 2025, Simone et al. (2025) analysed the anatomy of the pedipalps in two species of scorpions at the extreme of the morphological gradient *Scorpio palmatus* (Ehrenberg, 1828) and *Hottentotta gentili* (Pallary, 1924) retrieving respectively fourteen and twelve bundles of the closing muscle within the *manus*. The different number of bundles and architecture revealed a morphological adaptation to enhance exclusively either closing force or velocity. *Scorpio palmatus* has short-fingered chela and maximises closing force through increased mechanical advantage and longer sarcomeres, enhancing muscle contraction efficiency. In contrast, *H. gentili*, the slender-chela species, optimises closing velocity through muscle orientations that favour rapid acceleration. While additional functional demands likely influence these designs, one morphology appears specialised for quickly capturing prey, while the other seems to be adapted for prey crushing.

Preliminary analyses of the muscle architecture suggest distinct functional roles for the patellar and manus muscles. The patellar muscle, with its long, parallel-oriented fibres aligned with the force's line of action and the short sarcomeres, likely facilitates rapid closing movements. In contrast, the shorter, pennate fibres with longer sarcomeres in the manus muscle appear specialised for generating force (Simone et al. 2025). It is interesting to notice that in a fast-closing scorpion species, the patella muscle represents almost 50% of the volume of all closing muscles while in a strong-closing species the same muscle is only the 1.4% of all closing muscles. While the contribution of the patellar muscle is nearly negligible in the species with strong chela closing, this muscle in the species with fast closing performance seems to play a multiple role depending on the performance needs. The greater length and short sarcomeres of this muscle suggest it contracts fast, however, its relatively large cross-sectional area, combined with a relatively longer moment arm, would still allow a fair (static) force to be generated when muscles contract nearly isometrically, e.g., while holding a prey.

Unfortunately, these functional considerations are yet to be proved. To resolve these uncertainties, further studies involving 3D reconstructions of internal musculature across the full range of morphological diversity are needed. Coupling these investigations with detailed analyses of muscle physiology and architecture will clarify the respective roles of the patellar and manus muscles in chela-closing performance.

10.6 Predatory Behaviour

Scorpions are nocturnal terrestrial predators that play a significant role in shaping arthropod communities, particularly in desert environments (Polis and Yamashita 1991). Neotropical scorpions have a wide range of arthropod prey (Miranda et al. 2015; Davison et al. 2020; Rodríguez-Cabrera et al. 2020) and intraguild predation events are also quite common including Opiliones (Albín and Toscano-Gadea 2015; Dionisio-da-Silva et al. 2024), amblypygids (Armas 1995; Armas et al. 2013; Teruel and Toledo 2014; Cubas-Rodríguez et al. 2024), solifuges (Armas 1995; Teruel 2015), spiders, other scorpions (Peretti et al. 1999; Rojas-Runjaic et al. 2008; Moreno-González and Hazzi 2012; Miranda et al. 2015; Rodríguez-Cabrera et al. 2020; Cubas-Rodríguez et al. 2024; Dionisio-da-Silva et al. 2024), and seldom also vertebrates (Armas 2001; Rodríguez-Cabrera et al. 2020; Silva-Júnior et al. 2023a). Unlike other taxa, Neotropical scorpions do not exhibit clear cases of prey specialisation. While it has been suggested that *Tityus uruguayensis* Borelli, 1901, might be an araneophagic species (Toscano-Gadea et al. 2006), the only study on this topic indicates a mere preference for spiders rather than true specialisation.

Scorpion predatory strategies generally fall into two categories: sit-and-wait predation and wandering predation. In the sit-and-wait strategy, scorpions remain stationary, waiting for prey to come within reach. When a prey approaches, they rapidly project their pedipalps to seize it and may inject venom if necessary. In the wandering strategy, scorpions actively search for prey by following chemical or mechanical soil-borne cues. Upon locating prey, they grasp it with their chelae and may inject venom as needed. A study on the feeding behaviour of *Tityus serrulatus* Lutz and Mello, 1922, demonstrated the critical role of the chelae in prey handling. When the chelae were immobilised with wax, the scorpions experienced greater difficulty managing their prey, relying instead on their chelicerae to attack and secure it (Colombo 2015). Interestingly, the dichotomy between sit-and-wait and wandering strategies is not rigid and one strategy can shift into another in response to external factors such as predator density or the presence of competing species. Studies on the foraging behaviour of *Tityus pusillus* Pocock, 1893 and *Ananteris mauryi* Lourenco, 1982 reveal intriguing behavioural flexibility. For example, the typical sit-and-wait predator *Tityus pusillus* adopts a wandering strategy when *A. mauryi* is present in the same area. Conversely, the normally active-wandering *A. mauryi* becomes more reliant on sit-and-wait predation when sharing its habitat with *T. pusillus*. (Dionisio-da-Silva et al. 2019 but also see Pessoa et al. 2022).

10.6.1 Functional Morphology in Feeding Behaviour of Neotropical Scorpions

According to optimal foraging theory, the ideal prey should maximise energy intake while minimising the effort required for capture (Pyke 1984). A key factor in this efficiency is the precise adaptation of body structures used to capture prey. Among these, the design of biting structures plays a crucial role and has been linked to prey characteristics across various taxa.

In general, biting structures designed to generate large forces are associated with a high mechanical advantage. Mechanical advantage is defined as the ratio between the in-lever (the shortest perpendicular distance from the applied force to the joint) and the out-lever (the shortest perpendicular distance from the joint to the point where force is exerted). Forceful biting structures typically have shorter out-levers and muscles with specific architectural adaptations. These muscles are often large, with shorter fibres arranged at an angle (pennate muscles), longer sarcomeres, and attachment points further from the axis of rotation (long moment arms). While these adaptations enhance biting force, they reduce speed (Simone et al. 2025). As a result, animals with high mechanical advantage biting structures are better suited to capturing slow-moving prey.

Conversely, biting structures with a low mechanical advantage are optimised for speed, making them ideal for capturing fast and elusive prey. These structures feature long out-levers, long muscle fibres that are either parallel to the line of force or only slightly pinnate, shorter sarcomeres, and shorter moment arms. Although these adaptations favour speed, they come at the expense of force generation, making these structures less effective against hard-bodied prey. Instead, they are better suited for soft-bodied prey, which requires less force to subdue.

Scorpions' biting structures, the chelae, exhibit remarkable morphological diversity (Fig. 10.3). However, it remains unclear how their evolution correlates with variations in prey-type consumption, unlike the well-established relationship observed in crabs (Seed and Hughes 1995, Yamada and Boulding 1998). However, it is fair to consider that the different chela closing performance coupled with different chela morphology influences both prey capture efficiency and handling time when dealing with prey of differing characteristics. Baerg (1954), for example, notes that the Jamaican scorpion *Centruroides insulanus* (Thorell, 1876), a species with slender and weak pedipalps, can only feed on carabid beetles if their hard elytra are removed. Conversely, the more robustly pincered scorpion *Bothriurus bonariensis* (Koch, 1842) exhibits low acceptance rates for fast-reacting prey such as spiders (Simone et al. 2018). The absence of a generalised pattern in prey preferences among scorpions can be attributed to several factors. Scorpions feed infrequently, making field observations of predation rare, and information on their diets is extremely limited. Additionally, molecular tools for analysing scorpion gut contents have only recently been developed (Simone et al. 2022). This lack of systematic dietary data represents one of the most significant gaps in our understanding,

Fig. 10.3 Morphological variation of Neotropical scorpion chelae. From left to right: *Troglotayosicus meijdeni*, *Hadruroides juanchaparroi*, *Caraboctonus keyserlingi*, *Centruroides gracilis*, *Tityus pachyurus* (female), *Ananteris* sp. The pictures of *T. meijdeni*, *T. pachyurus*, and *Ananteris* sp. were provided by Dr. Julio Cesar Gonzalez Gomez

preventing the identification of clear trophic patterns. Another potential explanation for the lack of a direct relationship between chela morphology and diet is the complementary role of venom in prey capture. Venom provides scorpions, especially those with weaker pedipalps, with an additional tool to subdue a wider variety of prey.

10.6.2 Pinching or Stinging?

All scorpions grasp their prey using the chelae; however, the use of stinger depends on chela morphology and prey resistance (Rein 1993; Rein 2003; Edmunds and Sibly 2010; García et al. 2022). Species with stouter chelae tend to subdue their prey with the sole use of their strong chelae while species with slender chelae tend to almost always sting their prey relying on the paralytic and quick effects of their potent venoms. García et al. (2022) investigated the predatory behaviour of four Neotropical scorpion species with varying chela morphologies from the genera *Chactas*, *Centruroides*, *Tityus*, and *Opistacanthus*. The study examined their responses to two types of prey: crickets (soft-bodied) and cockroaches (hard-bodied). Species with more elongated chelae almost always stung both prey types to incapacitate them, whereas species with more robust chelae never stung soft-bodied prey and only occasionally stung hard-bodied prey. Interestingly, the use of the stinger did not affect immobilisation time across species and prey types. This suggests that for species that do not inject venom, their pedipalps are as effective in subduing prey as the venom of species with slender chelae. Regarding *Chactas*, the species studied exhibits notable sexual dimorphism in both pedipalp morphology and venom toxicity against invertebrates. Females had stronger pedipalps but less lethal venom against invertebrates compared to males (González-Gómez et al. 2020). Unfortunately, no details are provided regarding whether males and females used the stinger differently, nor were any significant differences observed. This suggests a possible stereotyped behaviour that is independent of chela morphology and venom toxicity. This result would contrast with what has been observed in *Bothriurus bonarensis* where females adapted their behavioural units according to the defensive responses of the prey faced (Simone et al. 2018).

Another factor influencing stinger use in Neotropical scorpions is venom lethality toward different prey types. Scorpion venom exhibits varying toxicity depending on the target species (van der Meijden et al. 2017), and there appears to be a positive correlation between the frequency of stinger use and venom LD_{50}. Specifically, the higher the LD_{50} (indicating less lethal venom), the more frequently the stinger is employed during prey capture. This pattern was observed in *Tityus fuhrmanni* Kraepelin, 1914, where the stinger was used more often against cockroach prey and less frequently against spiders. Analysis of the venom's LD_{50} revealed that *T. fuhrmanni* venom is more lethal to spiders (i.e., lower LD_{50}) than to cockroaches (Arroyave-Munoz et al. 2022).

The lethality of venom toward a specific prey target is not the only factor influencing scorpion predatory behaviour; the amount of venom available for use also plays a crucial role in determining which prey to attack or avoid. Scorpions never completely deplete their venom glands under natural conditions, even during repeated attacks when one might expect them to use all their venom to survive. However, scorpions may have recently used their venom and are in the process of regenerating its depleted components, a process that significantly increases basal metabolism and is energetically costly. Research has long explored behavioural changes resulting from the unnatural complete depletion of venom. For instance, Dionisio-da-Silva et al. (2019) found that in *Jaguajir rochae* (Borelli, 1910), scorpions with empty or partially replenished venom glands tended to avoid attacking larger prey, instead favouring smaller and less challenging targets.

In conclusion, the interplay between chela morphology, venom potency, and behavioural flexibility reflects the complex evolutionary strategies Neotropical scorpions employ to optimise prey capture. While robust chelae can compensate for reduced venom use, species with slender chelae rely more heavily on stinging, particularly against resistant prey. However, this balance is influenced not only by the prey type but also by venom availability and energetic constraints. The current lack of detailed ecological and behavioural data, particularly regarding feeding strategies and prey specificity, underscores the need for integrative research combining morphology, venom biology, and fine-scale behavioural observations to unravel the adaptive significance of functional traits in scorpion predation.

10.7 Defensive Behaviour in Neotropical Scorpions

Neotropical scorpions are ferocious predators but occasionally they can be often preyed by other arthropods like assassins bugs (Reduvidaee) (Lira et al. 2016a; Silva-Júnior et al. 2023b), ants (Höfer et al. 1996; Dionisio-da-Silva and Lira 2020), katydid (Tettigonidae) (Almeida et al. 2022), spiders (Lourenço et al. 2006; Lira and Costa 2014; Lira et al. 2016b; Faúndez and Albornoz 2017; Taucare-Ríos and Iuri 2021; Cubas-Rodríguez et al. 2024), amblypygids (Forcelledo and Armas 2014), scorpions (Moreno-González and Hazzi 2012; Cubas-Rodríguez et al. 2024) but also vertebrates like amphibians (Botero-Trujillo 2007; Jared et al. 2020), lizards (Ray et al. 2021; Silva-Júnior et al. 2023a), and snakes (Solórzano and Greene 2012). A more comprehensive list of scorpion predators is available in Dupre (2020). Coupled with their role in trophic chains, Neotropical scorpions have remarkable morphological diversity. This variation includes differences in overall size, pedipalps, and venom delivery systems (metasoma and telson), both across families and often within the same species. In many cases, these structures exhibit strong sexual dimorphism (McLean et al. 2018). This diversity in forms is mirrored by distinct uses of their iconic weapons in both predatory and defensive contexts.

Defensive behaviour can be elicited in different ways, body part restriction (Nisani and Hayes 2011; van der Meijden et al. 2013, 2015; Carlson et al. 2014;

Rasko et al. 2018; Lira et al. 2020) and electro stimulation through implanted electrodes (Miller et al. 2016). In particularly aggressive species, even a light puff of air is sufficient to provoke a defensive response (Nisani and Hayes 2015). The perceived threat level plays a critical role in determining whether a scorpion's defensive response will involve fleeing or escalating to a more aggressive behaviour (van der Meijden et al. 2013; Miller et al. 2016; Lira et al. 2017). The more intense the perceived threat, the higher the frequency of stinging or pinching, with this response varying according to the scorpion's morphology (Heatwole 1967; van der Meijden et al. 2013; Lira et al. 2017).

10.7.1 Fleeing

One of the primary defence mechanisms for many Neotropical scorpions is fleeing in order to avoid contact with attackers. In *T. pusillus*, juveniles and females primarily opt to flee when provoked into a defensive response. However, the variation in defensive behaviour is influenced by which part of the body is restrained, and consequently, the level of perceived threat (Lira et al. 2020). In *Centruroides*, the trend is the opposite, with males more likely to resort to fleeing as their main defence (Carlson et al. 2014). In this group, males often have longer legs, which enable faster locomotion. This trait is linked to their more ground-based, errant lifestyle, as males actively search for females during the reproductive season, while females tend to be more arboreal (Miller et al. 2016). Interestingly at least 14 species within the genus *Ananteris* autotomise the metasoma before fleeing (Lira et al. 2014; Mattoni et al. 2015; Bhukal 2016; García-Hernández and Machado 2020). Autotomy typically occurs between the third and fourth metasomal segments, though in rare cases it may occur between the first and second or second and third segments (Mattoni et al. 2015). As a result of autotomy, individuals lose nearly 25% of their body mass, along with the last portion of the digestive tract, including the anus, which leads to constipation and weight gain over time. Despite this substantial loss, the reduction in body load due to metasomal autotomy does not result in a significant difference in sprinting speed between autotomised males or females and non-autotomised individuals (García-Hernández and Machado 2022). In the long term, locomotion performance declines as severe constipation symptoms develop, affecting males more than females (García-Hernández and Machado 2022). Autotomised individuals, though more vulnerable to predation and less effective at capturing prey, can survive up to a year and even reproduce (García-Hernández and Machado 2020, 2021). While male reproductive success remains unaffected by autotomy and constipation, the fitness of autotomised females can decrease by as much as 35% (García-Hernández and Machado 2021).

To date, experiments on fleeing performance have only been performed on members of the family Buthidae and we lack information about all the other families. It can be inferred that fleeing behaviour is likely linked to leg morphology and the more mobile lifestyle of these litter- and ground-dwelling species. Species with fossorial, pelophilous, or lithophilous habits tend to have shorter, wider legs compared to errant species (Newlands 1972; Coelho et al. 2022) and typically adopt a sit-and-wait predatory strategy, spending most of their time near or at the entrance of their burrows. The shorter, more muscular legs of these species may facilitate quick retreats into their burrows rather than promoting fast escape across open ground. Comparative data from other Neotropical families, such as Hormuridae, Chactidae, and Bothriuridae, would help clarify whether leg morphology, rather than factors like sex or ontogenetic stage, plays a primary role in determining fleeing behaviour.

10.7.2 Stinging and Pinching Behaviour

When fleeing and retreating is no longer an option, scorpions violently face their attackers by using their best weapons. Scorpions with robust chelae tend to primarily use their chelae for self-defence while scorpions with elongated and slender chelae will opt for stinging as main defensive behaviour (Heatwole 1967; van der Meijden et al. 2013). Chelae that have low aspect ratio (wider than longer) have larger pinchforce than chelae with higher aspect ratio (longer than wider) (van der Meijden et al. 2010, 2012; Simone and van der Meijden 2018). This difference in pinching performance is counterbalanced by different venom potency where scorpions with long-fingered chelae have more lethal venoms (Forde et al. 2022).

Stinging frequency in defensive situations has been well-documented in Neotropical buthids specifically in the *Tityus* and *Centruroides* genera. In *T. stigmurus* (Thorell, 1876) and *T. pusillus*, a higher threat level is associated with an increase in stinging frequency (Lira et al. 2017, 2020; Albuquerque and Lira 2021). In the genus *Centruroides* stinging behaviour shows intersexual differences with females being more aggressive than males showing higher stinging frequency (Carlson et al. 2014; Miller et al. 2016).

However, there is a remarkable knowledge gap regarding the defensive behaviours of other Neotropical scorpion families. To date, only two non-buthid species, *Caraboctonus keyserlingi* (Pocock, 1893) and *Opisthacanthus lepturus* (Beauvois, 1805), have documented stinging behaviours in defensive contexts. The first species has a chela shape that does not suggest either force or speed specialisation (van der Meijden et al. 2012; Simone and van der Meijden 2018). As a consequence out of 150 defensive events distributed across ten specimens, the most frequent defensive reaction was using simultaneously the chelae and the metasoma (76.6%) followed by using the metasoma only (23.3%) (van der Meijden et al. 2013). *Opisthacanthus lepturus*, which has robust pincers and a short metasoma, used only its chelae in 92.8% of 42 defensive responses, employing both pincers and metasoma in just 7.2% of instances (Heatwole 1967).

Due to the high energy cost of venom production, scorpions may use "dry stings" (stings without venom injection) in defence. Dry stings have been reported in *T. stigmurus*, although without clear differences in frequency between high and low-threat situations (Lira et al. 2017). In *Hadrurus arizonensis* Ewing, 1928, dry stings were more frequent under high-threat conditions (Rasko et al. 2018) whereas the opposite pattern was observed in *Parabuthus transvaalicus* Purcell 1899 (Nisani and Hayes 2011). These findings suggest that scorpions can modulate venom release during defensive stings, allowing them to conserve this costly resource while still delivering a painful deterrent and inflicting tissue damage (Evans et al. 2019). Indeed, an immediate painful, rather than a slowly lethal venom is probably the best strategy for a more effective defence (Nelsen et al. 2014; Niermann et al. 2020).

In this regard, in an intriguing recent discovery, Laborieux described a new *Tityus* species in the Cordillera Oriental region of Colombia capable of venom-spitting behaviour (Laborieux 2024). This finding is particularly notable, as venom-spitting was previously known only in a few species of the South African genus *Parabuthus* (Nisani and Hayes 2011, 2015), making *Tityus* the first genus of Neotropical scorpions showing such a defensive behaviour. Venom-spitting behaviour is a common defensive mechanism in several arthropods (Eisner and Meinwald 1966; Eisner et al. 2005; Nisani and Hayes 2011, 2015). In all known cases, this behaviour targets attackers or predators rather than natural prey. When the venom reaches mucous membranes, it causes an intense burning pain, and if it contacts the eyes, it can even lead to blindness (Nisani and Hayes 2015). Although the venom of this new *Tityus* species has yet to be analysed, it warrants particular attention to better understand the nature of this rare behaviour in scorpions.

10.7.3 Stridulation

Stridulation is a distinctive non-aggressive defensive behaviour used by various scorpion species as a predation deterrent (Alexander 1958). While the sound produced is generally similar across species, the structures involved in sound production vary. Dumortier (1963) identified four main types of stridulatory structures:

1. **Chelicera-Cephalothorax Type**: In African scorpionids like *Opisthopthalmus*, stridulation occurs as bristles on the dorso-mesial surface of the first cheliceral segment rub against a ridged area on the inner carapace. Additionally, chitinous platelets on each chelicera rub together, producing a characteristic sound (Alexander 1958)
2. **Pedipalp-Walking Leg Type**: In *Pandinus* and *Heterometrus* scorpions, and the Diplocentridae *Oiclus* distributed in the Lesser Antilles islands, stridulation results from friction between bristles and a striated cuticle. In *Pandinus*, the bristles are located on the coxa of the first leg, which rubs against a striated cuticle on the pedipalp coxa. In *Heterometrus*, the arrangement is reversed, with bristles on the pedipalp coxa rubbing against the striated cuticle on the first leg coxa (Alexander 1960; Constantinou and Cloudsley-Thompson 1984).
3. **Tail-Sting Type**: In the buthid *Parabuthus*, stridulation occurs through the stinger's tip rubbing against metasomal segments or the last mesosomal tergite (Alexander 1958).

4. **Pectine-Sternite Type**: Unlike the above, some Neotropical scorpions produce stridulation by rubbing the pectines against plates on the third sternite. This behaviour has been observed in *Rhopalurus* and other buthid genera such as *Jaguajir*, *Heteroctenus*, and *Troglorhopalurus* (Pocock 1904; Lourenço and Cloudsley-Thompson 1995; Lourenço et al. 2000; Esposito et al. 2018).

In addition, **Pretergite-Tergite Type** stridulation occurs in Bothriuridae species such as *Timogenes* and *Brachistosternus*, where granules on the pretergites create sound by rubbing against the posterior tergites when the mesosoma arches (Acosta and Maury 1990; Ochoa and Ojanguren-Affilastro 2007).

10.8 Concluding Remarks

Despite significant progress in understanding the macroevolutionary patterns of morphological diversification in Neotropical scorpions, substantial gaps remain in uncovering the evolutionary processes behind these traits, particularly those associated with predatory and defensive behaviours. The variety of morphological adaptations, including chela shapes, venom composition, and metasoma structure, is central to their ecological success. However, the selective pressures and evolutionary mechanisms that drive these adaptations remain poorly understood.

A major limitation lies in the lack of comprehensive behavioural data linking morphological traits to specific ecological functions. While predatory strategies and defensive tactics are crucial for the survival and diversification of scorpions, there is a scarcity of empirical studies connecting these behaviours with morphology across phylogenetic scales. For instance, the influence of venom complexity on dietary breadth or prey specialisation remains largely unexplored, partly because dietary data are often limited to field observations, overlooking unobserved interactions. Similarly, the relationship between chela morphology and prey-handling efficiency has been inadequately studied, particularly in non-buthid scorpions.

Methodological limitations also contribute to these knowledge gaps. The integration of high-resolution phylogenies with biomechanical models and functional ecology has the potential to reveal the evolutionary trajectories of morphological traits. However, such integrative frameworks are still underdeveloped in the study of Neotropical scorpions. Advances in molecular tools, coupled with expanded sampling efforts across understudied regions, could provide the necessary data to address these gaps.

Future research should prioritise a multidisciplinary framework that combines field observations, experimental studies, and advanced computational methods. By bridging the gaps in our understanding of the ecological and evolutionary drivers of predatory and defensive behaviours, we can gain deeper insights into the macroevolutionary processes shaping the remarkable morphological diversity of Neotropical scorpions.

References

Acosta LE, Maury EA (1990) Estridulacion en *Timogenes elegans* (Mello-Leitão) (Scorpiones, Bothriuridae). Bol Soc Biol Concepc 61:29–37

Albín A, Toscano-Gadea CA (2015) Predation among armored arachnids: *Bothriurus bonariensis* (Scorpions, Bothriuridae) versus four species of harvestmen (Harvestmen, Gonyleptidae). Behav Process 121:1–7. https://doi.org/10.1016/j.beproc.2015.10.003

Albuquerque KBC, Lira AFA (2021) Sex-based defensive behavior influenced by threat level in the scorpion *Tityus pusillus* (Scorpiones: Buthidae). J Arachnol 49(3):402–406. https://doi.org/10.1636/JoA-S-20-074

Alexander AJ (1958) On the stridulation of scorpions. Behaviour 12:339–352

Alexander AJ (1960) A note on the evolution of stridulation within the family scorpionidae. Proc Zool Soc London 133:391–399

Alexander AJ (1967) Problems of limb extension in the scorpion, *Opisthophthalmus latimanus* Koch. Trans R Soc S Afr 37:165–181

Alexander RMN (1985) The ideal and the feasible: physical constraints on evolution. Biol J Linn Soc 26:345–358

Almeida MRN, Nascimento JAF, Machado EO, Lira AFA (2022) Once a prey, now a predator: an unusual record of a scorpion (Scorpiones: Chactidae) predated by a katydid (Orthoptera: Tettigoniidae) in the western Brazilian Amazon. Acta Amazon 52:229–231. https://doi.org/10.1590/1809-4392202200151

Armas LF (1995) Informe de viaje Breve crónica de una expedición aracnológica a Nicaragua expedición Luis F. de Armas. Cocuyo 4:2–3

Armas LF (2001) Frogs and Lizards as prey of some Greater Antillean arachnids. Rev Ibér Aracnol 3:87–88

Armas LF, Rodríguez TM, Teruel R (2013) Depredación de *Phrynus damonidaensis* (Amblypygi: Phrynidae) por *Alayotityus sierramaestrae* (Scorpiones: Buthidae) y lista de los enemigos naturales de los amblipigios. Rev Ibér Aracnol 22:107–108

Arnold SJ (1983) Morphology, performance and fitness. Integr Comp Biol 23:347–361

Arroyave-Munoz A, van der Meijden A, Estrada-Gomez S, Garcia LF (2022) Linking toxicity and predation in a venomous arthropod: the case of *Tityus fuhrmanni* (Scorpiones: Buthidae), a generalist predator scorpion. J Venom Anim Toxins Incl Trop Dis 28:1–10. https://doi.org/10.1590/1678-9199-JVATITD-2021-0036

Baerg WJ (1954) Regarding the biology of the common Jamaican scorpion. Ann Entomol Soc Am 47:272–276

Bar-On B, Barth FG, Fratzl P, Politi Y (2014) Multiscale structural gradients enhance the biomechanical functionality of the spider fang. Nat Commun 5:3894. https://doi.org/10.1038/ncomms4894

Bhukal R (2016) First record of autotomy in the neotropical scorpion *Ananteris cussinii* Borelli, 1910 in Trinidad, W.I. Living World, Journal of the Trinidad Tobago Field Natutalists' Club 39–40

Bicknell RDC, Edgecombe GD, Goatley CHR, Charlton G, Paterson JR (2024) Pedipalp anatomy of the Australian black rock scorpion, *Urodacus manicatus*, with implications for functional morphology. Aust J Zool 72:ZO2304. https://doi.org/10.1071/ZO23044

Bock WJ (1994) Concepts and methods in ecomoprhology. J Biosci 19:403–413

Botero-Trujillo R (2007) Anuran predators of scorpions: *Bufo marinus* (Linnaeus, 1758) (Anura: Bufonidae), first known natural enemy of *Tityus nematochirus* Mello-Leitão, 1940 (Scorpiones: Buthidae). Rev Ibér Aracnol 13:199–202

Bowerman RF (1972) A muscle receptor organ in the scorpion postabdomen I. The sensory system. J Comp Physiol A 81:133–146

Broeckhoven C, du Plessis A (2017) Has snake fang evolution lost its bite? New insights from a structural mechanics viewpoint. Biol Lett 13(8). https://doi.org/10.1098/rsbl.2017.0293

Carlson BE, McGinley S, Rowe MP (2014) Meek males and fighting females: sexually-dimorphic antipredator behavior and locomotor performance is explained by morphology in bark scorpions (*Centruroides vittatus*). PLoS One 9:1–8. https://doi.org/10.1371/journal.pone.0097648

Coelho P, Kaliontzopoulou A, Rasko M, van der Meijden A (2017) A 'striking' relationship: scorpion defensive behaviour and its relation to morphology and performance. Funct Ecol 31(7):1390–1404. https://doi.org/10.1111/1365-2435.12855

Coelho P, Kaliontzopoulou A, Sousa P, Stockmann M, van der Meijden A (2022) Reevaluating scorpion ecomorphs using a naïve approach. BMC Ecol Evol 22:1–10. https://doi.org/10.1186/s12862-022-01968-0

Colombo W (2015) Behavior of *Tityus serrulatus* (Scorpiones, Buthidae) in captivity. Rev Ibér Aracnol 27:105–108

Constantinou C, Cloudsley-Thompson JL (1984) Stridulatory structures in scorpions of the families Scorpionidae and Diplocentridae. J Arid Environ 7(4):359–364. https://doi.org/10.1016/S0140-1963(18)31352-1

Cubas-Rodríguez AM, Cotoras DD, Taucare-Ríos A, Armas LF, Brescovit AD, Gómez G, Stice TL (2024) When the predator becomes the prey: new records of intraguild predation among Central American and Caribbean arachnids (Arachnida: Amblypygi, Araneae, Scorpiones). Stud Neotropical Fauna Environ 59(3):1–11. https://doi.org/10.1080/01650521.2024.2353007

Davison A, Brown T, Arrivillaga C (2020) Notes on the diet and reproduction of the bark scorpion *Centruroides gracilis* (Scorpiones: Buthidae) on Utila Island, Honduras. Euscorpius 314:1–7

Dionisio-da-Silva W, Lira AFA (2020) Record of *Ananteris mauryi* (Scorpiones: Buthidae) preyed upon by *Ectatomma planidens* (Hymenoptera: Formicidae) in the Brazilian Atlantic rainforest. Entomol News 129(5):497–503. https://doi.org/10.3157/021.128.0508

Dionisio-da-Silva W, Lira AFA, Albuquerque CMR (2019) Prey-predator interactions between two intraguild predators modulate their behavioral decisions. Acta Ethologica 22:195–201. https://doi.org/10.1007/s10211-019-00326-6

Dionisio-da-Silva W, Foerster SÍA, Gallão JE, Lira AFA (2024) What's for dinner? Prey consumption by Neotropical scorpions across contrasting environments. J Arachnol 52(1):26–30. https://doi.org/10.1636/JoA-S-22-060

Dubale MS, Vyas AB (1968) The structure of the chela of *Heterometrus* sp. and its mode of operation. Bull South Calif Acad Sci 67:240–244

Dumortier B (1963) Morphology of sound emission apparatus in Arthropoda. In: Busnel R (ed) Acoustic behaviour of animals. Elsevier, New York, pp 277–345

Dupre G (2020) Les predateurs des scorpions (Arachnida: Scorpiones). Arachnides 94:1–40

Edmunds MC, Sibly RM (2010) Optimal sting use in the feeding behavior of the scorpion *Hadrurus spadix*. J Arachnol 38:123–125

Eisner T, Meinwald J (1966) Defensive secretions of arthropods. Science 153:1341–1350

Eisner T, Eisner M, Siegler M (2005) Secret weapons: defenses of insects, spiders, scorpions, and other many-legged creatures. (ed. Michael F) Harvard University Press, Harward

Esposito LA, Yamaguti HY, Pinto-da-Rocha R, Prendini L (2018) Plucking with the plectrum: phylogeny of the New World buthid scorpion subfamily Centruroidinae Kraus, 1955 (Scorpiones: Buthidae) reveals evolution of three pecten-sternite stridulation organs. Arthropod Syst Phylogeny 76:87–122. https://doi.org/10.3897/asp.76.e31942

Evans ERJ, Northfield TD, Daly NL, Wilson DT (2019) Venom costs and optimization in scorpions. Front Ecol Evol 7:1–7. https://doi.org/10.3389/fevo.2019.00196

Faúndez EI, Albornoz M (2017) Sobre un registro de predación de *Steatoda triangulosa* (Walckenaer, 1802) (Araneae: Theridiidae) sobre *Tityus trivittatus* Kraepelin, 1898 (Scorpiones: Buthidae) en Argentina. Rev Ibér Aracnol 30:165–166

Forcelledo LJ, Armas LF (2014) Depredación de *Centruroides gracilis* (Scorpiones: Buthidae) por *Paraphrynus cubensis* (Amblypygi: Phrynidae). Rev Ibér Aracnol 25:111–112

Forde A, Jacobsen A, Dugon MM, Healy K (2022) Scorpion species with smaller body sizes and narrower chelae have the highest venom potency. Toxins 14:1–11. https://doi.org/10.3390/toxins14030219

García LF, Valenzuela-Rojas JC, González-Gómez JC, Lacava M, van der Meijden A (2022) Pinching or stinging? Comparing prey capture among scorpions with contrasting morphologies. J Venom Anim Toxins Incl Trop Dis 28:1–9. https://doi.org/10.1590/1678-9199-JVATITD-2021-0037

García-Hernández S, Machado G (2020) 'Tail' autotomy and consequent stinger loss decrease predation success in scorpions. Anim Behav 169:157–167. https://doi.org/10.1016/j.anbehav.2020.08.019

García-Hernández S, Machado G (2021) Fitness implications of nonlethal injuries in scorpions: females, but not males, pay reproductive costs. Am Nat 197:379–389. https://doi.org/10.1086/712759

García-Hernández S, Machado G (2022) Short- and long-term effects of an extreme case of autotomy: does "tail" loss and subsequent constipation decrease the locomotor performance of male and female scorpions? Integr Zool 17(5):672–688. https://doi.org/10.1111/1749-4877.12604

Gilai BYA, Parnas I (1970) Neuromuscular physiology of the closer muscles in the pedipalp of the scorpion *Leiurus quinquestriatus*. J Exp Biol 52:325–344

González-Gómez JC, Valenzuela-Rojas JC, García LF, Franco-Pérez LM, Guevara G, Buitrago S, Cubillos A, van der Meijden A (2020) Sexual dimorphism in the biomechanical and toxicological performance in prey incapacitation of two morphologically distinct scorpion species (*Chactas* sp. and *Centruroides* sp.). Biol J Linn Soc 129:190–198. https://doi.org/10.1093/biolinnean/blz146

González-Santillán E, Prendini L (2015) Systematic revision of the North American Syntropine vaejovid scorpions with a subaculear tubercle, *Konetontli* González-Santillán and Prendini, 2013. Bull Am Mus Nat Hist 2015:1–78

Govindarajan S, Rajulu GS (1974) Presence of resilin in a scorpion *Palamnaeus swammerdami* and its role in the food-capturing and sound-producing mechanisms. Experientia 30:908–909

Günther A, Drack M, Monod L, Wirkner CS (2021a) A unique yet technically simple type of joint allows for the high mobility of scorpion tails. J R Soc Interface 18:1–8. https://doi.org/10.1098/rsif.2021.0388

Günther A, Monod L, Wirkner CS (2021b) Comparative morphology of scorpion metasomata: muscles and cuticle. Arthropod Struct Dev 60:101003. https://doi.org/10.1016/j.asd.2020.101003

Heatwole H (1967) Defensive behaviour of some Panamanian scorpions. Caribb J Sci 7:15–17

Höfer H, Wollscheid E, Gasnier T (1996) The relative abundance of *Brotheas amazonicus* (Chactidae, Scorpiones) in different habitat types of a central Amazon rainforest. J Arachnol 24:34–38

Jared C, Alexandre C, Mailho-Fontana PL, Pimenta DC, Brodie ED, Antoniazzi MM (2020) Toads prey upon scorpions and are resistant to their venom: a biological and ecological approach to scorpionism. Toxicon 178:4–7. https://doi.org/10.1016/j.toxicon.2020.02.013

Klingenberg CP (2008) Morphological integration and developmental modularity. Annu Rev Ecol Evol Syst 39:115–132. https://doi.org/10.1146/annurev.ecolsys.37.091305.110054

Laborieux L (2024) Biomechanics of venom delivery in South America's first toxungen-spraying scorpion. Zool J Linnean Soc 202:zlae161. https://doi.org/10.1093/zoolinnean/zlae161

Lamoral B (1979) The scorpions of Namibia (Arachnida: Scorpionida). Ann Natal Mus 23:497–784

Lankester ER (1885) On the muscular and endoskeletal systems of Limulus and Scorpio; with some notes on the anatomy and generic characters of scorpions. Zool Soc Lond:311–384

Lira AFA, Costa AA (2014) First record of a brown widow spider *Latrodectus geometricus* Koch, 1841 (Araneae, Theridiidae) feeding scorpion (Scorpiones, Bothriuridae) in a Brazilian Atlantic Forest. Braz J Biol 74:1011. https://doi.org/10.1590/1519-6984.18013

Lira AFA, Sarinho NMS, DeSouza AM, Albuquerque CMR (2014) Metasomal autotomy in *Ananteris mauryi* Lourenço, 1982 (Scorpiones: Buthidae). J Insect Behav 27:279–282. https://doi.org/10.1007/s10905-013-9431-3

Lira AFA, Araújo VLN, Albuquerque CMR (2016a) Predation of a scorpion (Scorpiones: Buthidae) by an assassin bug (Heteroptera: Reduviidae) in the Brazilian Atlantic forest. Turk J Zool 40:294–296. https://doi.org/10.3906/zoo-1504-27

Lira AFA, Foerster SÍA, Costa Silva AA (2016b) Reports of scorpion predation by spiders in the Brazilian Atlantic forest and Caatinga (Arachnida: Scorpiones, Araneae). Rev Ibér Aracnol 28:87–89

Lira AFA, Santos AB, Silva NA, Martins RD (2017) Threat level influences the use of venom in a scorpion species, *Tityus stigmurus* (Scorpiones, Buthidae). Acta Ethol 20:291–295. https://doi.org/10.1007/s10211-017-0274-3

Lira AFA, Almeida FMF, Albuquerque CMR (2020) Reaction under the risk of predation: effects of age and sexual plasticity on defensive behavior in scorpion *Tityus pusillus* (Scorpiones: Buthidae). J Ethol 38:13–19. https://doi.org/10.1007/s10164-019-00615-4

Lourenço WR (2020) The coevolution between telson morphology and venom glands in scorpions (Arachnida). J Venom Anim Toxins Incl Trop Dis 26. https://doi.org/10.1590/1678-9199-JVATITD-2020-0128

Lourenço WR, Cloudsley-Thompson JL (1995) Stridulatory apparatus and the evolutionary significance of sound production in Rhopalurus species (Scorpiones: Buthidae). J Arid Environ 31:423–429. https://doi.org/10.1016/S0140-1963(05)80125-9

Lourenço WR, Huber D, Cloudsley-Thompson JL (2000) Description of the stridulatory apparatus in some species of the genus *Rhopalurus* Thorell (Scorpiones: Buthidae). Ekol Bratislava 19:141–144

Lourenço WR, Brescovit AD, Rheims CA, Cloudsley-Thompson JL (2006) First record of a crab spider (Thomisidae) preying on a scorpion. Bol Soc Entomol Aragonesa 39:404–405

Mathew AP (1965) On the movable claw of the pedipalps in Heterometrus scaber. J Anim Morphol Physiol 12:271–275

Mattoni CI, García-Hernández S, Botero-Trujillo R, Ochoa JA, Ojanguren-Affilastro AA, Pinto-da-Rocha R, Prendini L (2015) Scorpion sheds "tail" to escape: consequences and implications of autotomy in scorpions (Buthidae: *Ananteris*). PLoS One 10:1–15. https://doi.org/10.1371/journal.pone.0116639

McLean CJ, Garwood RJ, Brassey CA (2018) Sexual dimorphism in the Arachnid orders. PeerJ 2018:1–54. https://doi.org/10.7717/peerj.5751

Miller DW, Jones AD, Goldston JS, Rowe MP, Rowe AH (2016) Sex differences in defensive behavior and venom of the striped bark scorpion *Centruroides vittatus* (Scorpiones: Buthidae). Integr Comp Biol 56:1022–1031. https://doi.org/10.1093/icb/icw098

Miranda R, Bermúdez S, Cleghorn J, Cambra RA (2015) Presas de escorpiones (Arachnida: Scorpiones) de Panamá, con observaciones sobre el comportamiento de depredación. Rev Ibér Aracnol 27:115–123

Moreno-González JA, Hazzi NA (2012) Intraguild predation case: *Tityus forcipula* Gervais, 1843 (Scorpiones, Buthidae) feeding on *Chactas vanbenedeni* Gervais, 1843 (Scorpiones, Chactidae) in Colombia. Rev Ibér Aracnol 20:117–120

Nelsen DR, Nisani Z, Cooper AM, Fox GA, Gren ECK, Corbit AG, Hayes WK (2014) Poisons, toxungens, and venoms: redefining and classifying toxic biological secretions and the organisms that employ them. Biol Rev 89:450–465. https://doi.org/10.1111/brv.12062

Newlands G (1972) Ecological adaptations of Kruger National Park scorpionids (Arachnida: Scorpionides). Koedoe 15:37–48

Niermann CN, Tate TG, Suto AL, Barajas R, White HA, Guswiler OD, Secor SM, Rowe AH, Rowe MP (2020) Defensive venoms: is pain suffcient for predator deterrence? Toxins 12:1–17. https://doi.org/10.3390/toxins12040260

Nisani Z, Hayes WK (2011) Defensive stinging by *Parabuthus transvaalicus* scorpions: risk assessment and venom metering. Anim Behav 81:627–633. https://doi.org/10.1016/j.anbehav.2010.12.010

Nisani Z, Hayes WK (2015) Venom-spraying behavior of the scorpion *Parabuthus transvaalicus* (Arachnida: Buthidae). Behav Process 115:46–52. https://doi.org/10.1016/j.beproc.2015.03.002

Ochoa JA, Ojanguren-Affilastro AA (2007) Systematics and distribution of *Brachistosternus* (*Brachistosternus*) *ehrenbergii* (Gervais, 1841), with the first record of stridulation in this genus *Brachistosternus* (Scorpiones: Bothriuridae). Stud Neotropical Fauna Environ 42:61–69. https://doi.org/10.1080/01650520600800310

Peretti AV, Acosta LE, Benton TG (1999) Sexual cannibalism in scorpions: fact or fiction? Biol J Linn Soc 68:485–496. https://doi.org/10.1006/bijl.1999.0347

Pessoa VB, Moura GJB, Lira AFA (2022) Influence of predator density in foraging behavior and microhabitat choice on litter-dwelling scorpions. Neotrop Biodivers 8:297–300. https://doi.org/10.1080/23766808.2022.2126020

Pocock RI (1904) On a new stridulating-organ in scorpions discovered by W.J. Burchell in Brazil in 1828. Ann Mag Nat Hist 7:56–62

Polis GA, Yamashita T (1991) The ecology and importance of predaceous arthropods in desert communities. In: Polis GA (ed) The ecology of desert communities. University of Arizona Press, Tucson, pp 180–222

Pyke GH (1984) Optimal foraging theory: a critical review. Annu Rev Ecol Evol Syst 15:523–575

Rasko M, Coelho P, Simone Y, van der Meijden A (2018) How to attack a scorpion: venom metering during a repeated attack. Anim Behav 145:125–129. https://doi.org/10.1016/j.anbehav.2018.09.019

Ray G, Burum B, Sherwood D (2021) Predation of *Tityus ocelote* (Scorpiones: Buthidae) by the lizard *Norops polylepis* (Sauria: Dactyloidae) in Costa Rica. Rev Ibér Aracnol 38:187–188

Rein JO (1993) Sting use in two species of *Parabuthus* scorpions (Buthidae). J Arachnol 21:60–63

Rein JO (2003) Prey capture behavior in the East African scorpions *Parabuthus leiosoma* (Ehrenberg, 1828) and *P. pallidus* Pocock, 1895 (Scorpiones: Buthidae). Euscorpius 1895:1–8

Rodríguez-Cabrera T, Teruel R, Morell-Savall E (2020) Scorpion predation in Cuba: new cases and a review. Euscorpius 2020:1–7

Rojas-Runjaic FJM, Borges A, Armas LF (2008) Nueva especie de *Opisthacanthus* Peters, 1861 (Scorpiones, Hemiscorpiidae) de la Sierra de Perijá, Venezuela, basada en criterios morfológicos y moleculares. Bol la Soc Entomol Aragonesa 43:49–59

Ruff CB (2018) Functional morphology in the pages of the AJPA. Am J Phys Antropol 165:688–704

Seed R, Hughes RN (1995) Criteria for prey size-selection in molluscivorous crabs with contrasting claw morphologies. J Exp Mar Biol Ecol 193:177–195

Sensenig AT, Shultz JW (2004) Elastic energy storage in the pedipalpal joints of scorpions and sun-spiders (Arachnida, Scorpiones, Solifugae). J Arachnol 32:1–10

Silva-Júnior AO, Brito-Almeida TR, Feitosa MLB, Jansen RG, Peres M, Azevedo R, Moura GJB, Barbosa AR, Lira AFA (2023a) Can size make a difference? Cross-predation occurrences between lizards and scorpions in the Brazilian seasonal dry tropical forest. Invertebr Zool 20:343–347. https://doi.org/10.15298/invertzool.20.3.06

Silva-Júnior AO, Celante GL, Silva AM, Gil-Santana HR, Moura GJB, Lira AFA (2023b) Report of intraguild predation of the scorpion *Physoctonus debilis* (C. L. Koch, 1840) (Scorpiones: Buthidae) by the assassin bug *Microtomus tibialis* Stichel, 1926 (Hemiptera: Reduviidae). Rev Chil Entomol 49:267–270

Simone Y, van der Meijden A (2018) Fast and fine versus strong and stout: a trade-off between chela closing force and speed across nine scorpion species. Biol J Linn Soc 123:208–217. https://doi.org/10.1093/biolinnean/blx139

Simone Y, van der Meijden A (2021) Armed stem to stinger: a review of the ecological roles of scorpion weapons. J Venom Anim Toxins Incl Trop Dis 27:1–21. https://doi.org/10.1590/1678-9199-JVATITD-2021-0002

Simone Y, van der Meijden A, Garcia LF, Lacava M, Viera C (2018) Analysis of predatory versatility in females of the scorpion *Bothriurus bonariensis* (Koch, 1842) (Scorpiones: Bothriuridae): overcoming prey with different defensive mechanisms. J Insect Behav 31:402–415. https://doi.org/10.1007/s10905-018-9677-x

Simone Y, Chaves C, van der Meijden A, Egeter B (2022) Metabarcoding analysis of different portions of the digestive tract of scorpions (Scorpiones, Arachnida) following a controlled diet regime shows long prey DNA half-life. Environ DNA 4:1176–1186. https://doi.org/10.1002/edn3.311

Simone Y, Herrel A, Boistel R, van der Meijden A (2025) Functional divergence of scorpion pedipalps: musculoskeletal specialization toward opposing performance optima. J Morphol 286:e70055. https://doi.org/10.1002/jmor.70055
Snodgrass RE (1952) A textbook of arthropod anatomy. Comstock Publishing Associates, Ithaca
Solórzano A, Greene HW (2012) Predation in nature by a scorpion-hunter, *Stenorrhina freminvillei* (Serpentes, Colubridae). UNED Res J 4:31–32
Stevens RD (2005) Functional morphology meets macroecology: size and shape distributions of New World bats. Evol Ecol Res 7:837–851
Taucare-Ríos A, Iuri HA (2021) Primeros registros de depredación intragremio de *Sicarius thomisoides* Walckenaer, 1847 (Araneae: Sicariidae) sobre el escorpión *Brachistosternus mattoni* (Ojanguren-Affilastro, 2005) (Scorpiones: Bothriuridae) y el solífugo Pseudocleobis sp. (Solifugae: Ammotrechidae) en Chile. Rev Chil Entomol 47:211–217
Teruel R (2015) A scorpion predating upon a solifuge in Mexico (Arachnida: Scorpiones, Solifugae). Rev Ibér Aracnol 26:93–94
Teruel R, Toledo A (2014) Yet another case of scorpions preying upon amblypygids in nature (Scorpiones, Amblypygi). Rev Ibér Aracnol 24:111–112
Toscano-Gadea CA, Costa FG, State S, Province ER, Keyserling L, Costa L (2006) Is *Tityus uruguayensis* (Buthidae) an araneophagic scorpion? An experimental analysis of its predatory behaviour on spiders and insects. Bull Br Arachnol Soc 13:256–264
van der Meijden A, Kleinteich T (2017) A biomechanical view on stinger diversity in scorpions. J Anat 230:497–509. https://doi.org/10.1111/joa.12582
van der Meijden A, Herrel A, Summers A (2010) Comparison of chela size and pincer force in scorpions; getting a first grip. J Zool 280:319–325
van der Meijden A, Kleinteich T, Coelho P (2012) Packing a pinch: functional implications of chela shapes in scorpions using finite element analysis. J Anat 220:423–434. https://doi.org/10.1111/j.1469-7580.2012.01485.x
van der Meijden A, Coelho PL, Sousa P, Herrel A (2013) Choose your weapon: defensive behavior is associated with morphology and performance in scorpions. PLoS One 8:e78955. https://doi.org/10.1371/journal.pone.0078955
van der Meijden A, Coelho P, Rasko M (2015) Variability in venom volume, flow rate and duration in defensive stings of five scorpion species. Toxicon 100:60–66. https://doi.org/10.1016/j.toxicon.2015.04.011
van der Meijden A, Koch B, van der Valk T, Vargas-Muñoz LJ, Estrada-Gómez S (2017) Target-specificity in scorpions; comparing lethality of scorpion venoms across arthropods and vertebrates. Toxins 9:312. https://doi.org/10.3390/toxins9100312
Wagner GP (1996) Homologues, natural kinds and the evolution of modularity. Am Zool 36:36–43
Yamada SB, Boulding EG (1998) Claw morphology, prey size selection and foraging efficiency in generalist and specialist shell-breaking crabs. J Exp Mar Biol Ecol 220:191–211
Zelditch ML, Goswami A (2021) What does modularity mean? Evol Dev 23:377–403. https://doi.org/10.1111/ede.12390
Zhao ZL, Shu T, Feng XQ (2016) Study of biomechanical, anatomical, and physiological properties of scorpion stingers for developing biomimetic materials. Mater Sci Eng C 58:1112–1121. https://doi.org/10.1016/j.msec.2015.09.082

Chapter 11
Sampling Techniques for Scorpions

Nancy Lo-Man-Hung and Andria de Paula Santos-da-Silva

Abstract This chapter provides a detailed review of methods for collecting, capturing, and preserving scorpions, emphasizing their ecological and epidemiological importance. Sampling techniques include active daytime and nighttime searches, with ultraviolet light at night being particularly effective due to scorpion fluorescence. Passive methods like pitfall traps and reports of scorpion stings are also utilized. The choice of collection method depends on habitat, species behavior, and research goals. Preservation techniques vary from chemical methods, such as using alcohol and formaldehyde, to dry methods like freeze-drying, with specific protocols for molecular studies employing high-concentration ethanol and RNA stabilizers to maintain DNA and RNA integrity. Legal regulations govern scorpion collection to ensure conservation and scientific use, with permits required from environmental agencies. Combining multiple collection and preservation strategies enhances the efficiency and quality of scorpion research, supporting advances in the understanding of their diversity, ecology, and evolutionary biology.

Keywords Arachnida · Collection methods · Biodiversity studies

11.1 Introduction

Scorpions are present in various ecosystems, making them significant for ecological, conservation, and biogeographic studies (Polis 1990). As arthropod predators, scorpions regulate the populations of insects and other arthropods, contributing to the maintenance of their habitat dynamics (Fattorini et al. 2011). Additionally, scorpions are valuable indicators of environmental changes as their diversity and distribution

N. Lo-Man-Hung (✉)
Universidade de São Paulo, Instituto de Biociências, Departamento de Ecologia, São Paulo, Brazil

A. d. P. Santos-da-Silva
Instituto Butantan, São Paulo, Brazil

A. F. de Araujo Lira, E. González-Santillán (eds.), *The Biology of Neotropical Scorpions*, https://doi.org/10.1007/978-3-032-23194-9_11

patterns reflect the health of ecosystems (Sadine et al. 2023, Lira et al. 2018, 2023; Feitosa et al. 2024; Viana-Junior et al. 2024; Brito-Almeida et al. 2025). Another example is that climate change and habitat degradation can cause changes in the populations, species composition, and distribution of scorpions, especially in opportunistic species, which can impact the number of accidents involving these animals (Guerrra-Duarte et al. 2023; Feitosa et al. 2024). It is important to mention that such changes often affect the species composition of a given area through the phenomenon of substitution, i.e., micro-habitats/habitat specialist species are replaced by opportunistic/generalist species that are better able to survive the anthropogenic effects of urban areas (Lourenço and Eickstedt 2009; Feitosa et al. 2024).

Approximately 2840 species of scorpions are described and distributed in 24 families worldwide (Rein 2024). However, there is no consensus and there are different hypotheses about the systematic classification of scorpions (see Prendini and Wheeler 2005; Santibáñez-López et al. 2019). Of these numbers, around 50% of the world's scorpion fauna were documented in the Neotropical region (Lourenço 2002), with Mexico and Brazil being the countries with the greatest diversity of scorpions in the world, with 280 and 190 species respectively (Santibáñez-López et al. 2015; Sharma et al. 2015; Silva et al. 2023). This biodiversity is the result of several years of evolutionary processes shaped by the region's complex geological history, diverse habitats, climate changes, and ecological interactions (Antonelli and Sanmartín 2011; Rull 2011; Hughes et al. 2013; Antonelli 2022). Understanding scorpion ecological roles can provide valuable insights for conservation strategies and help manage biodiversity in their respective regions (Sadine et al. 2023; Lira et al. 2023) and can contribute to anti-scorpion serum development (Lucas et al. 2010).

Traditionally, scorpions have been studied by turning over surfaces like rocks, trash, and bricks. However, the data recovered from these methods can be limited and difficult to interpret from an ecological perspective (Tourinho and Lo-Man-Hung 2021). Furthermore, some species have specialized lifestyles, such as burrowing, which makes them harder to locate and study using surface-level methods (Williams 1968; Polis 1990). Besides, some species have specialized lifestyles, such as burrowing, which makes them harder to locate and study using surface-level methods (Williams 1968; Polis 1990). Therefore, collecting and preserving samples for research requires specific techniques tailored to animal behavior and habitat (Tourinho and Lo-Man-Hung 2021). This chapter reviews various methods for capturing and sampling scorpions, comparing them based on study type and objectives, while also addressing preservation techniques crucial for both morphological and molecular studies.

11.2 Overview of Most Used Sampling Techniques

Before learning about the different collection techniques for scorpions, we ask the question: Where can we collect these animals? However, knowledge about the habits and types of micro-habitats most likely to shelter these animals is essential. Most scorpions are nocturnal and crepuscular; during the day, they can be found in burrows or cracks, under rocks, under trunks, and fallen leaves (such as leaf litter), which they use as refuge and shelter (Polis 1990; Candido 1999; Stockman and Ythier 2010). However, there are reports of some diurnal species in caves, tropical forests, and coastal areas (Polis 1990; Candido 1999). They can be found during all seasons of the year (even under snow); however, they are more active and abundant in the warmer months of the year. But in tropical regions, they remain active all year round (Candido 1999; Stockman and Ythier 2010). For sampling scorpions in a natural environment, there are different collection methods that can be carried out during the daylight and/or night, as well as conducted passively (pitfall traps) or through active search (Candido and Porto 2010).

11.2.1 Daylight Active Search

The selective collection method is characterized by the search for scorpions in their different locations of occurrence. During daylight hours, several techniques can be employed to locate animals:

Digging Nests or a Canal Physically digging nests allows for direct scorpion collection, especially in arid environments where scorpions may be burrowed deeply (Dehghani et al. 2019). Another digging method involves creating a canal about 15–20 cm deep with flat walls, filling it with water, and maintaining a high-water level throughout the night (Dehghani et al. 2016a). Scorpions are attracted to the wet edges and can fall into the canal, drowning. This method is also used to prevent scorpions from entering residential areas (Keegan 1980).

Inspecting Trees Scorpions may also inhabit tree bark or foliage, making thorough inspections important. It can involve peeling back trees in the daytime to search for scorpions (Druce et al. 2004). As well as inspecting the roots of palm trees where some species of scorpions can be found, such as, for example, the chactid species *Neochactas parvulus* Pocock, 1897 (Santos-da-Silva obs. pers.). On the other hand, during the night, scorpions come out to forage or reproduce and, consequently, can be seen on the trunks of some trees (Fig. 11.1) (e.g., *Tityus obscurus* (Gervais, 1843) and *T. strandi* Werner, 1939 have been seen and collected at approximately three meters high in the western Amazon of Pará, Santos-da-Silva obs. pers.).

Looking Under Rocks and Trash Piles Like rock-rolling, this method involves searching under various debris where scorpions may be hiding (Ennik and Madon 2002). In this case, it is also recommended to use an entomological pick, geological hammer, and/or sometimes a hoe is necessary to explore burrows and/or termites (Stockman and Ythier 2010).

Fig. 11.1 Buthidae, juvenile specimen of *Tityus metuendus* Pocock, 1897. Left: natural coloration; right: under UV light. Specimen from Porto Velho, Rondônia, Brazil. (Photo credit: Rafael P. Indicatti)

Old Mud Walls Destruction The destruction of old mud walls or stone walls, especially in rural or agricultural areas, near water channels, can reveal hidden scorpion populations (Dehghani et al. 2019). These walls offer ideal shelters for scorpions, providing moisture and protection in the cracks and holes at ground level (Dehghani et al. 2016a). Destroying these structures allows researchers to uncover and collect scorpions that reside within them (Dehghani and Bigdelli 2007; Dehghani et al. 2019).

Pouring Water into Nests Some species of scorpions can build/excavate nests in the ground or leaf litter. Pouring water into these nests can encourage scorpions to emerge (Rojas et al. 2019). Pouring water slowly into these nests through a thin tube, continuously, until scorpions emerge, but with care to avoid drowning (Dehghani et al. 2016a; Hamid and Ramadan 2022).

Rock-Rolling Scorpions can easily find suitable shelter spots in rocky areas, where they often use cracks and holes that match the size of their bodies, preventing larger animals from reaching them. The method of rolling over rocks carefully to uncover scorpions beneath is commonly used in such habitats (Tourtlotte 1974; Druce et al. 2004; Shehab et al. 2011).

Rubber Band A rubber band is gently inserted into the scorpion's nest and twisted slightly. Once the band reaches the end of the nest and contacts the scorpion, the scorpion instinctively grips it with its pincers. Slowly pulling the band out will also bring the scorpion out of the nest (Gholizadeh et al. 2016).

Using a Wet Cloth or Gunny Sack Scorpions are attracted to humid places during their nocturnal activities, especially in the hot season, and may attach themselves to wet surfaces, such as clothes or gunny sacks, to absorb moisture through their mouthparts (Dehghani et al. 2016b). Moisture needs vary by species, and this behavior can be used to collect scorpions by placing wet materials at night and gathering them in the morning (Dehghani et al. 2016a, b).

Entomological Umbrella Consists of a 1 × 1 m square of cloth, connected at the vertices by two wooden rods arranged in a cross (or forming an "X") and fixed in the center by a screw. This artifact is placed under the branches of the bushes (up to 2 m high), which are shaken with a wooden stick. Animals that fall onto the fabric are captured and stored in a collector containing 70% or 80% alcohol (Saturnino, and Tourinho 2011; Carvalho et al. 2020). This method is not very common for collecting scorpions, being more used for sampling arboreal arachnids, especially spiders, present in the shrub layer. However, we recommend its use as a complementary collection method, since Lira et al. (2018) found that *Tityus pusilus* Pocock, 1893 also occurs in vegetation. Furthermore, the collection of scorpions of the genus *Ananteris* Thorell, 1891 and young *Tityus* C. L. Koch, 1836 using this technique has been witnessed (Santos-da-Silva, obs. pers.).

11.2.2 Night Active Search with Ultraviolet Light

Scorpions are solitary and nocturnal (Polis 1990) whose exoskeleton light shines fluoresces cyan-blue-green at a wavelength of 420–700 nm under ultraviolet, UV-A radiation (315–400 nm) (Fig. 11.1) (Lawrence 1954; Pavan 1954; Pavan and Vachon 1954; Koehler Jr 1979; Sissom et al. 1990; Lourenco 2012; López-Cabrera et al. 2020). Although the purpose of this fluorescence remains unclear, Gaffin et al. (2012) tested scorpion responses to UV and green light. They proposed that the scorpion's cuticle may function as a light collector, converting UV light to cyan-green and transmitting this signal to the central nervous system, possibly aiding in locating shelter.

In this way, one effective method for sampling scorpions is conducting a night active search using UV light (Russell 1969; Tourtlotte 1974; Candido and Porto 2010; Dehghani et al. 2016a). As scorpions fluoresce under UV light, holding a UV light at a correct distance – about 15–30 cm from the distance – on ground, brick, or muddy walls, and keeping it moving around, makes it easy to spot them against the dark background of their environment. They can then be captured using forceps (Fig. 11.2). The duration of each sampling varies depending on the study's aim and

Fig. 11.2 Chactid scorpion *Brotheas amazonicus* Lourenço, 1988, collected using forceps. Specimen from Juruti, Pará, Brazil. (Photo credit: Flávio Pimenta)

work field logistics. A variety of light-emitting diode (LED) lamps have been introduced for detecting and collecting scorpions (e.g., Lowe et al. 2003; Ramires et al. 2013).

For all active search methods (night and daylight) mentioned above, the use of Personal Protective Equipment (EPIs) is recommended, such as closed shoes, gloves, and long forceps (curved or straight), to ensure safety when handling scorpions (Stockman and Ythier 2010; Candido and Porto 2010).

11.3 Passive Method of Collecting by Scorpion-Stung People

This method relies on individuals who have been stung by scorpions to report their encounters. It is used in epidemiological studies and provides valuable data on scorpion distribution and behavior (Dehghani et al. 2016a). With this method, the occurrence of scorpion stings can be estimated at a determined area. Along with identifying associated factors such as sociodemographic data, characteristics of the dwelling, efforts to avoid scorpion stings, and individual information regarding scorpion stings that were experienced through the years (Trinidad-Porfirio et al. 2023).

11.4 Pitfall Traps

Pitfall traps are passive sampling devices that capture animals as they move through the environment. It needs to be used with care, because traps can be influenced by abundance, foraging activity, and body size of target species (Hancock and Legg 2012), color of trap (Buchholz and Möller 2018), collection period time, design, liquid preservative, etc. (Uetz and Unzicker 1976; Tourinho and Lo-Man-Hung

Fig. 11.3 Pitfall traps (20-liter buckets) buried at ground level to collect scorpions in Floresta National of Tapajós, Belterra, Pará, Brazil (left); Installation of a guide fence (plastic tarps) (right). (Photo credit: Hipócrates M. Chalkidis)

2021). These methods of capture are known by three types: pitfall traps, pitfall traps with drift fence, and trapping ditches (Fet 1980; Yamaguti and Pinto-da-Rocha 2006; Candido and Porto 2010).

The traps consist of containers (cans, or plastic cups (usually 500 ml) and buckets whose size can vary up to 90 l) placed at specific intervals based on study objectives, buried flush with the ground, and filled with or without a preservative solution (Druce et al. 2004; Candido and Porto 2010; Carvalho et al. 2020; Tourinho and Lo-Man-Hung 2021). When no preservative solution is used, such as when traps use buckets (Fig. 11.3), the animals cannot climb to escape once they fall into the trap (e.g., Yamaguti and Pinto-da-Rocha 2006; Lo-Man-Hung et al. 2008). Pitfall traps with drift fence are characterized by the installation of plastic tarps (Fig. 11.3) which sometimes influence the animal's displacement toward the trap (Candido and Porto 2010).

11.5 Trunk Refugia

Trunk refugia is a method for sampling tree-dwelling arthropods, providing a non-destructive and cost-effective alternative to traditional sampling methods like visual searching, branch shaking, or peeling tree bark (Salman et al. 2020). The trap is made from corrugated cardboard, designed as a cylindrical shelter attached to tree trunks, and can remain in place for varying periods. As a nonlethal method, it allows animals to be collected alive (Salman et al. 2020).

11.6 Vibration

This is a method to detect scorpions by modeling substrate vibrations associated with their movement. The approach involves eight stages: capturing the scorpion's behavior through a sinusoidal signal, amplifying it, and transmitting it to a nearby receiver (Aibinu et al. 2014). The signal is then filtered to remove noise, followed by modeling to determine key coefficients, and the coefficients are analyzed through clustering to identify behavioral patterns, i.e., early detection of scorpions, potentially preventing stings (Aibinu et al. 2014).

11.7 Preservation Methods

11.7.1 Chemical Preservation

Specimens collected in the field require appropriate preservation to maintain their integrity. Common chemicals include:

70–80% Alcohol Effective for general preservation, but can lead to tissue degradation over time (Marquina et al. 2021).

Isopropanol A suitable alternative for preserving specimens, especially for specimens for mounting (King and Porter 2004).

Formaldehyde Often used for initial fixation (Thavarajah et al. 2012) but requires careful handling due to toxicity (Dan et al. 2020).

11.7.2 Dry Preservation

For some research purposes, dry preservation methods are employed. This can include:

Desiccation Specimens are air-dried or stored with desiccants to remove moisture (Ungureanu 1972).

Freeze-drying (Lyophilization) Specimens are rapidly frozen and then subjected to a vacuum to remove moisture, preserving tissue structure and preventing degradation (Molnar et al. 2021).

Silica Gel Specimens are stored with silica gel, which acts as a desiccant to absorb moisture, maintaining the sample in a stable dry state (Allison et al. 2021).

Oven Drying Specimens are exposed to controlled heat in an oven to gradually remove moisture while preserving certain morphological features, though freeze-drying or desiccation is often preferred to better retain color (Ungureanu 1972; Molnar et al. 2021).

11.7.3 Preservation for Molecular Studies

When preparing specimens for molecular analysis, specific preservation methods are crucial:

95% or Absolute 100% Ethanol Recommended for DNA and RNA preservation due to its ability to penetrate cellular membranes and deactivate DNase activity more rapidly than other primary alcohols (King and Porter 2004; Marquina et al. 2021).

RNA Later A specialized reagent that stabilizes RNA for future analyses (Pedersen et al. 2019).

11.8 Legislation

In general, collecting scorpions, like any other wild animal, from their natural shelters is prohibited by law. Violation of these rules constitutes an environmental crime that can result in various consequences, such as fines, criminal and administrative sanctions. However, this activity is subject to specific regulations in the case of obtaining material for scientific purposes, production of antivenoms (anti-scorpion serums), and population control of scorpions of medical importance in urban areas (Brazil 2009; Candido and Porto 2010).

For example, in Brazil, it is important to mention that the collection, capture, and transportation of scorpions are regulated by the State, and must be authorized by the Brazilian Institute of the Environment and Renewable Natural Resources (Ibama), according to Ordinance No. 332/90 and Normative Instruction 109/97 (IBAMA 1997), which states that only authorized professionals can collect scorpions (Brazil 2009). The collection of scorpions for the development of scientific research generally requires a request for a collection and transportation license that must be issued/authorized by the Chico Mendes Institute for Biodiversity Conservation (ICMBio), through the Biodiversity Authorization and Information System (SISBIO) for the researcher (Candido and Porto 2010). Meanwhile, the collection of harmful synanthropic species (including scorpions) is authorized by normative instruction No. 141/2006 of IBAMA. Whose control, environmental management and elimination of this scorpions that represent a risk to human health are the responsibility of health municipal bodies (e.g., the Centro de Controle de Zoonoses (CCZ)) (Brazil 2009; Candido and Porto 2010), in accordance with Ordinances No. 1172/OM of June 15, 2014 and No. 1,138, of May 23, 2014 (Brasil 2014a, b).

11.9 Conclusion

Sampling and preserving scorpions, like other arachnids, involves various techniques tailored to specific research objectives, ensuring specimen integrity for later analysis (Tourinho and Lo-Man-Hung 2021). For over 70 years, ultraviolet (UV) lamps have been used to detect scorpions (Lawrence 1954; Pavan 1954; Pavan and Vachon 1954). Along with pitfall traps and rock-rolling, these are the major capture methods widely applied (Williams 1968; Dias et al. 2006; Araújo et al. 2010; Dehghani et al. 2016a; Nazari and Hassan 2016; Lira et al. 2020).

Rock-rolling and black light methods are particularly useful, and combining these techniques across different regions enhances sampling efficiency (Dehghani et al. 2019). Although pitfall traps are considered advantageous over methods like rock-rolling or burrow excavation (Williams 1968), they lack some benefits provided by UV detection. All these techniques must be adapted according to habitat type, topography, and study goals (William 1968). In some cases, multiple methods are employed in multi-taxonomic fieldwork to capture scorpions, myriapods, lizards, and other species (Druce et al. 2004; Lo-Man-Hung et al. 2008). Researchers may also develop new techniques for capturing scorpions in specific areas (e.g., Salman et al. 2020). The combination of techniques offers valuable insights into fauna diversity, ecology, and evolution (Tourinho and Lo-Man-Hung 2021).

In the same way, selecting the appropriate preservation method for scorpion specimens depends on the research objectives. Chemical preservation, such as using 70–80% alcohol, is widely used but may lead to tissue degradation over time, whereas isopropanol and formaldehyde offer alternative solutions for specific purposes, though formaldehyde requires careful handling due to its toxicity (King and Porter 2004; Thavarajah et al. 2012; Dan et al. 2020; Marquina et al. 2021). Dry preservation methods, including desiccation, freeze-drying, and silica gel storage, are effective in maintaining specimen integrity for morphological studies, with freeze-drying often preferred to retain tissue structure and color (Ungureanu 1972; Allison et al. 2021; Molnar et al. 2021). For molecular studies, using 95–100% ethanol or specialized reagents like RNA Later ensures the preservation of DNA and RNA, critical for genetic analysis (King and Porter 2004; Pedersen et al. 2019; Marquina et al. 2021). The choice of preservation method is thus determined by the balance between maintaining specimen morphology and preserving genetic material for future research.

References

Aibinu AM, Sadiq BA, Joseph E, Bello Salau H, Salami MJE (2014) Development of an intelligent scorpion detection technique using vibration analysis. In: 2014 International conference on computer and information sciences (ICCOINS), Kuala Lumpur, Malaysia, 1–4. https://doi.org/10.1109/ICCOINS.2014.6868374

Allison MJ, Round JM, Bergman L, Mirabzadeh A, Allen H, Weir A, Helbing CC (2021) The effect of silica desiccation under different storage conditions on filter-immobilized environmental DNA. BMC Res Notes 14. https://doi.org/10.1186/s13104-021-05530-x

Antonelli A (2022) The rise and fall of Neotropical biodiversity. Bot J Linn Soc 199(1):8–24. https://doi.org/10.1093/botlinnean/boab061

Antonelli A, Sanmartín I (2011) Why are there so many plant species in the Neotropics? Taxon 60:403–414. https://doi.org/10.1002/tax.602010

Araújo CS, Candido DM, Araújo HFP, Dias SC, Vasconcellos A (2010) Seasonal variations in scorpion activities (Arachnida: Scorpiones) in an area of Caatinga vegetation in northeastern Brazil. Zoologia 27(3):372–376. https://doi.org/10.1590/S1984-46702010000300008

Brasil (2009) Ministry of Health. Manual de controle de escorpiões. Brasília/DF: Secretaria of Vigilância em Saúde. Departamento de Vigilância Epidemiológica. (Série B. Textos Básicos de Saúde)

Brasil (2014a) Portaria n° 1.138, de 23 de maio de 2014—define as ações e os serviços de saúde voltados para vigilância, prevenção e controle de zoonoses e de acidentes causados por animais peçonhentos e venenosos, de relevância para a saúde pública

Brasil (2014b) Portaria n° 1172/OM de 15 de junho de 2014—referente a organização do SUS—compete ao município o registro, a captura, apreensão e eliminação de animais que representem risco à saúde do homem, cabendo ao estado a supervisão, acompanhamento e orientação dessas ações

Brito-Almeida TR, Foerster SÍA, Lima JR, Silva MA, Moura GJB, Lira AFA (2025) Free pasture of exotic goats reduces diversity and negatively affects body condition in scorpions (Arachnida: Scorpiones) assemblage from Brazilian seasonal dry tropical forest. Ecol Evol 15:e70804. https://doi.org/10.1002/ece3.70804

Buchholz S, Möller M (2018) Assessing spider diversity in grasslands—does pitfall trap color matter? J Arachnol 46(2):376–379. https://doi.org/10.1636/JoA-S-16-062.1

Candido DM (1999) Escorpiões. In: Brandão CRF, Cancello EM (eds) Biodiversidade do Estado de São Paulo. Volume 5. Invertebrados Terrestres. FAPESP, São Paulo, pp 23–34

Candido DM, Porto TJ (2010) Técnicas de Coleta e Manejo de Escorpiões em Cativeiro. In: Brazil TK, Porto TJ (eds) Os escorpiões. EDUFBA, Salvador, pp 33–45

Carvalho LS, Oliveira BB, Dias DF, Silva ISS, Soares JF, Silva MIL, Santos TC, Silva PRR, Paranhos JDN (2020) Biodiversidade e conservação dos aracnídeos do Parque Nacional da Serra das Confusões, sul do Piauí; p. 119-149 In: MMM Ivanov (Org.). Unidades de Conservação do Estado do Piauí.Teresina: EDUFPI

Dan S, Pant M, Kaur T, Pant S (2020) Toxic effect of formaldehyde: a systematic review. Int Res J Mod Eng Technol Sci 2(9):179–189

Dehghani R, Bigdelli S (2007) Surveying the habitats on *Hemiscorpius lepturus* scorpion in Khuzestan province. Pajouhesh va Sazandegi Anim Fish Sci 20:81–87

Dehghani R, Tehrani AM, Ghadami F, Sanaei-Zadeh H (2016a) Methods for collecting and capturing scorpions: a review. Res Rev J Zool Sci 4(3):65–72

Dehghani R, Sanaei-Zadeh H, Mostafaii G, Iranshahi L (2016b) Feeding behavior of the Iranian dangerous scorpion species in the laboratory. J Entomol Zool Stud 4:1156–1159

Dehghani R, Kassiri H, Mohammadzadeh N (2019) Comparison of various methods of collecting scorpions (Arachnida, Scorpiones) in Khuzestan Province, Southwestern Iran. Archives of Clinical Infectious Diseases 14(4):e84452. https://doi.org/10.5812/archcid.84452

Dias SC, Candido DM, Brescovit AD (2006) Scorpions from Mata do Buraquinho, João Pessoa, Paraíba, Brazil, with ecological notes on a population of *Ananteris mauryi* Lourenço (Scorpiones, Buthidae). Rev Bras Zool 23(3):707–710. https://doi.org/10.1590/S0101-81752006000300014

Druce D, Hamer M, Slotow R (2004) Sampling strategies for millipedes (Diplopoda), centipedes (Chilopoda) and scorpions (Scorpionida) in savanna habitats. Afr Zool 39(2):293–304. https://doi.org/10.1080/15627020.2004.11657225

Ennik F, Madon MB (2002) Scorpions; p. 135-138 In RP Meyer, MB Madon (Eds.). Arthropods of Public Health Significance in California. A Training Manual for Vector Control Technician's Certification Examination Administered by the California Department of Health Services. Sacramento: MVCAC

Fattorini S, Dennis RLH, Cook LM (2011) Conserving organisms over large regions requires multi-taxa indicators: one taxon's diversity-vacant area is another taxon's diversity zone. Biol Conserv 144:1690–1701. https://doi.org/10.1016/j.biocon.2011.03.002

Feitosa MLB, Barbosa-da-Siva HR, Salomão RP, DeSouza AM, Moura GJB, Lira AFA (2024) Effects of landscape metrics on scorpion (Arachnida: Scorpiones) assemblage in a tropical urban ecosystem. Ecol Evol 14:e11026. https://doi.org/10.1002/ece3.11026

Fet V (1980) Ecology of the scorpion (Arachnida, Scorpiones) of the Southeastern Kara-Kum. Entomol Rev 59(1):223–228

Gaffin DD, Bumm LA, Taylor MS, Popokina NV, Mann S (2012) Scorpion fluorescence and reaction to light. Anim Behav 83(2):429–436. https://doi.org/10.1016/j.anbehav.2011.11.014

Gholizadeh S, Lalehzari E, Bavani MM, Hosseini A, Khalkhali HR, Rafinejad J (2016) Bioecology and scorpion envenomation in Ramshir district, Khuzestan Province, southwestern Iran. Appl Entomol Zool 51:37–42. https://doi.org/10.1007/s13355-015-0367-2

Guerrra-Duarte C, Saavedra-Langer R, Matavel A, Oliveira-Mendes BBR, Chavez-Olortegui C, Paiva ALB (2023) Scorpion envenomation in Brazil: current scenario and perspectives for containing an increasing health problem. PLoS Negl Trop Dis 17(2):e0011069. https://doi.org/10.1371/journal.pntd.0011069

Hamid K, Ramadan ES (2022) The first data on the BioEcology of scorpions (Arachnida: Scorpiones) in Bavi County, Southwestern Iran (2016–2017). Int Arch Health Sci 9(1):25–29. https://doi.org/10.4103/iahs.iahs_179_21

Hancock MH, Legg CJ (2012) Pitfall trapping bias and arthropod body mass. Insect Conserv. Divers. 5: 312–318

Hughes CE, Pennington RT, Antonelli A (2013) Neotropical plant evolution: assembling the big picture. Bot J Linn Soc 171:1–18

IBAMA (1997) Instrução Normativa IBAMA n. 109, de 12 de setembro de 1997. Destina-se a estabelecer e uniformizar os procedimentos de expedição de licença de pesquisa para realização de atividades científicas em Unidades de Conservação Federais de Uso indireto, definidas como Parques Nacionais, Reservas Biológicas, Estações Ecológicas e Reservas Ecológicas. Diário Oficial da União, Poder Executivo, Brasília, DF, 25 set. 1997

Keegan HL (1980) Scorpions of medical importance. University Press of Mississippi, Jakson, p 140

Koehler FH Jr (1979) Physical characteristics of the fluorescence spectra of scorpions. Master's thesis. California State University, Jackson, MI

King JR, Porter SD (2004) Recommendations on the use of alcohols for preservation of ant specimens (Hymenoptera, Formicidae). Insect. Soc. 51:197–202. https://doi.org/10.1007/s00040-003-0709-x

Lawrence RF (1954) Fluorescence in Arthropoda. J Entomol Soc South Afr 17:167–170

Lira AFA, DeSouza AM, Albuquerque CMR (2018) Environmental variation and seasonal changes as determinants of the spatial distribution of scorpions (Arachnida: Scorpiones) in Neotropical forests. Can J Zool 96:963–972. https://doi.org/10.1139/cjz-2017-0251

Lira AFA, Vieira AGT, Oliveira RF (2020) Seasonal influence on foraging activity of scorpion species (Arachnida: Scorpiones) in a seasonal tropical dry forest remnant in Brazil. Stud Neotropical Fauna Environ 55(3):226–232. https://doi.org/10.1080/01650521.2020.1724497

Lira AFA, Moura GJB, Foerster SIA (2023) Scorpion assemblages in threatened Brazilian forests: the role of environmental factors in explaining beta-diversity patterns. Insect Conserv Divers 17:128–138. https://doi.org/10.1111/icad.12699

Lo-Man-Hung NF, Gardner TA, Ribeiro-Junior MA, Barlow J, Bonaldo AB (2008) The value of primary, secondary, and plantation forests for Neotropical epigeic arachnids. J Arachnol 36:394–401. https://doi.org/10.1636/CT07-136.1

López-Cabrera D, Ramos-Ortiz G, González-Santillán E, Espinosa-Luna R (2020) Characterization of the fluorescence intensity and color tonality in the exoskeleton of scorpions. J Photochem Photobiol B Biol 209:111945. https://doi.org/10.1016/j.jphotobiol.2020.111945

Lourenço WR (2002) Scorpiones. In: Adis J (ed) Amazonian Arachnida and Myriapoda: identification keys to all classes, orders, families, some genera and lists of known terrestrial species. Pensoft Publishers Publishes, Moscow, pp 399–438

Lourenço WR, Eickstedt VR (2009) Escorpiões de importância médica; p. 198–213 In JLC Cardoso, FDS França, FHWen, CMS Malaque, V Haddad (Eds.). Animais peçonhentos no Brasil: biologia, clínica e terapêutica dos acidentes. São Paulo: Sarvier

Lourenco WR (2012) Fluorescence in scorpions under UV light; can chaerilids be a possible exception? C R Biol 335(12):731–734. https://doi.org/10.1016/j.crvi.2012.11.001

Lowe G, Kutcher SR, Edwards D (2003) A powerful new light source for ultraviolet detection of scorpions in the field. Euscorpius 8:1–7

Lucas S, Goldoni PA, Candido DM, Knysak I (2010) Butantan Institute: strategies to obtain scorpions for the production of anti-scorpion serum. J Venom Anim Toxins Incl Trop Dis 16(4):530–533. https://doi.org/10.1590/S1678-91992010000400002

Marquina D, Buczek M, Ronquist F, Łukasik P (2021) The effect of ethanol concentration on the morphological and molecular preservation of insects for biodiversity studies. PeerJ 9:e10799. https://doi.org/10.7717/peerj.10799

Molnar A, Lakat T, Hosszu A, Szebeni B, Balogh A, Orfi L, Szabo AJ, Fekete A, Hodrea J (2021) Lyophilization and homogenization of biological samples improves reproducibility and reduces standard deviation in molecular biology techniques. Amino Acids 53(6):917–928. https://doi.org/10.1007/s00726-021-02994-w

Nazari M, Hassan R (2016) Study on distribution of scorpions to provide prevention and interventions in combating scorpionism in Poldokhtar county, Lorestan province. Iran J Clin Diagn Res 10(12):LC05. https://doi.org/10.7860/JCDR/2016/22759.8960

Pavan P (1954) Presenza e distribuzione di una sostanza fluorescente nel tegumento degli scorpioni. Boll Soc Ital Biol Sper 30:801–803

Pavan M, Vachon M (1954) Sur l'existence d' une substance fluorescente dans les teguments des scorpions. C R Acad Sci 239:1700–1702

Pedersen KB, Williams A, Watt J, Ronis MJ (2019) Improved method for isolating high-quality RNA from mouse bone with RNAlater at room temperature. Bone Rep 11:100211. https://doi.org/10.1016/j.bonr.2019.100211

Polis GA (1990) The biology of scorpions. Stanford University Press, Redwood City

Prendini L, Wheeler WC (2005) Scorpion higher phylogeny and classification, taxonomic anarchy, and standards for peer review in online publishing. Cladistics (5):446–494. https://doi.org/10.1111/j.1096-0031.2005.00073.x

Ramires EN, Peracetta LF, Nogas C, Navarro-Silva MA, Paladini EP (2013) Equipment based on high power UV and white light LEDs to collect and observe scorpions (Arachnida: Scorpiones) and other fluorescent organisms. Zoologia 30(4):463–466. https://doi.org/10.1590/S1984-46702013000400015

Rein JO (2024) The scorpion files. Norwegian University of Science and Technology, Trondheim; Available online: https://www.ntnu.no/ub/scorpion-files. Accessed October 2024

Rojas A, Solano-Brenes D, Muniz DG, Machado G (2019) Gone with the rain: negative effects of rainfall on male reproductive success in a nest-building arachnid, Behavioral Ecology, 30(4):1145–1156, https://doi.org/10.1093/beheco/arz063

Rull V (2011) Neotropical biodiversity: timing and potential drivers. Trends Ecol Evol 26:508–513

Russell FE (1969) Scorpion collecting. Toxicon 6:307–308

Sadine SE, Souilem Z, Belgaid Y, Chedad A, Chebihi B, Zouaimia A, Bensakhri Z, Houhamdi M, Zebsa R (2023) Effects of climate on scorpion diversity in arid ecosystems of the Sahara desert of Algeria. Diversity 15(4):541. https://doi.org/10.3390/d15040541

Salman IN, Ferrante M, Möller DM, Lubin Y (2020) Trunk refugia: a simple, inexpensive method for sampling tree trunk arthropods. J Insect Sci 20(2):1–6. https://doi.org/10.1093/jisesa/ieaa012

Santibáñez-López CE, Francke OF, Ureta C, Possani LD (2015) Scorpions from Mexico: From Species Diversity to Venom Complexity. Toxins (Basel). 8(1):2. https://doi.org/10.3390/toxins8010002

Santibáñez-López CE, González-Santillán E, Monod L, Sharma PP (2019) Phylogenomics facilitates stable scorpion systematics: Reassessing the relationships of Vaejovidae and a new higher-level classification of Scorpiones (Arachnida). Mol Phylogenet Evol (35):22–30. https://doi.org/10.1016/j.ympev.2019.02.021

Sharma PP, Fernández R, Esposito LA, González-Santillán E, Monod L (2015) Phylogenomic resolution of scorpions reveals multilevel discordance with morphological phylogenetic signals. Proc R Soc Publi B 282:20142953. https://doi.org/10.1098/rspb.2014.2953

Shehab AH, Amr SZ, Lindsell AJ (2011) Ecology and biology of scorpions in Palmyra. Syria Turk J Zool 35(3):333–341

Silva GV, Trandafilov, GC, Polidoro BR, Antonangelo LR, de Oliveira CF (2023) Escorpionismo no estado de São Paulo: um problema de saúde pública em ascensão. Brazilian Journal of Health Review 6(6):6973–26982. https://doi.org/10.34119/bjhrv6n6-034

Sissom WD, Polis GA, Watt DD (1990) Field and laboratory methods. In: Polis GA (ed) The biology of scorpions. Stanford University Press, Stanford, pp 215–221

Stockman R, Ythier E (2010) Scorpions of the world. N. A. P. Editions, Verrières-le-Buisson, p 572

Saturnino R, Tourinho AL (2011) Apostila curso de treinamento em "Aracnologia: Sistemática, Coleta, Fixação e Gerenciamento de Dados". Programa de Pesquisa em Biodiversidade – PPBio, Sinop, 73

Thavarajah R, Mudimbaimannar VK, Elizabeth J, Rao UK, Ranganathan K (2012) Chemical and physical basics of routine formaldehyde fixation. J Oral Maxillofac Pathol 16(3):400–405. https://doi.org/10.4103/0973-029X.102496

Tourinho AL, Lo-Man-Hung N (2021) Standardized sampling methods and protocols for harvestman and spider assemblages. In: Santos JC, Fernandes GW (eds) Measuring arthropod biodiversity, 1st edn. Springer International Publishing, Cham, pp 365–400

Tourtlotte GI (1974) Studies on the biology and ecology of the northern scorpion, *Paruroctonus boreus* (Girard). Great Basin Nat 30:167–179

Trinidad-Porfirio BE, Morales-Pérez A, Nava-Aguilera E, Flores-Moreno M, Morales-Nava L, García-Leyva J, Silva-Domínguez R, Cortés-Guzmán AJ, Fernández-Salas I, Andersson N (2023) Occurrence of scorpion sting and associated factors in a highly marginalized municipality in Guerrero, Mexico: a cross-sectional study. PLoS Negl Trop Dis 17(5):e0011271. https://doi.org/10.1371/journal.pntd.0011271

Uetz GW, Unzicker JD (1976) Pitfall trapping in ecological studies of wandering spiders. J Arachnol 3:101–111

Ungureanu EM (1972) Methods for dissecting dry insects and insects preserved in fixative solutions or by refrigeration. Bull World Health Organ 47(2):239–244

Viana-Junior AB, Silva LFS, Pinheiro-Júnior EC, Santos EKN, Araújo MC, Alves IEC, Martins BS, Pereira JS, Santana-Santos R, Santos BA, Bezerra-Gusmão MA (2024) Temperature and tree number drive tree crown-dwelling arthropod diversity in Brazilian semi-arid cities. Insect Conserv Divers 18:206–218. https://doi.org/10.1111/icad.12795

Williams SC (1968) Methods of sampling scorpion populations. Proceedings of the California Academy of Sciences, 4th series 36(4):221–230

Yamaguti HY, Pinto-da-Rocha R (2006) Ecology of *Thestylus aurantiurus* of the Parque Estadual da Serra da Cantareira, São Paulo, Brazil (Scorpiones, Bothriuridae). J Arachnol 34:214–220

Chapter 12
Cytogenetics of Neotropical Scorpions

Bruno Rafael Ribeiro de Almeida, Jonas Gama Martins, and Renata Coelho Rodrigues Noronha

Abstract This chapter explores the cytogenetic aspects of Neotropical scorpions. Initially, the main chromosomal rearrangements involved in the evolution of the karyotype in the different families of the region are discussed. Then, the genomic structures of holocentric and monocentric chromosomes are examined, as well as atypical meiotic processes, such as achiasmatic meiosis. In addition, phenomena such as polyploidy, parthenogenesis, and the diversity of sex chromosome systems in these arthropods are addressed. Finally, the chapter presents examples of chromosomal variations and their applications in the taxonomy of these arachnids, highlighting the relevance of cytogenetic data for evolutionary and systematic studies in this group.

Keywords Scorpiones · Chromosome · Neotropical region

12.1 Introduction

The Neotropical region has the greatest diversity of scorpions in the world, corresponding to approximately 8 families, 48 genera, and 800 species (Lourenço 2002). However, only 67 species (8.3% of this total) have been studied cytogenetically (Schneider et al. 2024). This gap can be attributed to factors such as the scarcity of

B. R. R. de Almeida (✉)
Universidade do Estado do Pará, Campus XVIII, Cametá, Pará, Brazil
e-mail: bruno.rafael@uepa.br

J. G. Martins
Pós-Graduação em Genética, Conservação e Biologia Evolutiva (PPGGCBEv), Instituto Nacional de Pesquisas da Amazônia (INPA), Amazonas, Brazil
e-mail: jonas.martins@posgrad.inpa.gov.br

R. C. R. Noronha
Laboratório de Genética e Biologia Celular, Centro de Estudos Avançados da Biodiversidade, Universidade Federal do Pará, Belém, Pará, Brazil
e-mail: renatarcrn@gmail.com

A. F. de Araujo Lira, E. González-Santillán (eds.), *The Biology of Neotropical Scorpions*, https://doi.org/10.1007/978-3-032-23194-9_12

experts, technical and experimental challenges, and limitations in research investment. Despite these obstacles, recent studies highlight the value of cytogenetics as an essential tool for the systematics and delimitation of species in Scorpiones, as well as for the elucidation of the evolutionary mechanisms and genomic organization of these arachnids. These analyses have allowed significant advances in the understanding of the chromosomal biology of this group.

The first cytogenetic studies on Neotropical scorpions date back to the 1930s, with the pioneering description of the diploid number of *Opisthacanthus elatus* (Gervais, 1844) (Wilson 1931). In the following decades (1940–1950), Salvador Toledo Piza Jr. published a series of fundamental works on the karyotypes of Buthidae and Bothriuridae scorpions in Brazil, highlighting polymorphisms and meiotic behavior, especially in cytotypes of *Tityus bahiensis* (Perty, 1833) (Piza 1939a, b, 1940, 1943a, 1944, 1947a, b, 1948b, 1950a, 1957). Later, other researchers identified new chromosome races in this same species (Brieger and Graner 1943; Cunha and Pavan 1954; Takahashi 1976), while the first studies in Chactidae were initiated by Ferreira (1968). Most of the cytogenetic studies mentioned used chromosome preparations obtained from histological sections of paraffin-embedded gonads. This approach predominated until 2009, when new techniques revolutionized the field: hypotonization with KCl 0.075 M and tissue digestion with 45% acetic acid allowed significantly improved chromosome dispersion (Schneider et al. 2009a, b). Between 2010 and 2024, the introduction of fluorescent in situ hybridization (FISH) allowed the physical mapping of repetitive sequences, offering new insights into chromosomal and genomic evolution in scorpions (Schneider and Cella 2010; Adilardi et al. 2014). More recently, from 2019 onwards, immunocytogenetics was incorporated into the studies, with the application of specific antibodies to investigate proteins and epigenetic modifications involved in the control of the meiotic cycle in Neotropical scorpions (Almeida et al. 2019).

The main chromosomal characteristics (diploid number, distribution of heterochromatin, and nucleolar organizing regions) of the Neotropical families of the order Scorpiones studied to date are described below. Detailed information on each genus and species can be seen in Table 12.1.

12.2 Buthidae

Among the Neotropical scorpions, Buthidae stands out as the best characterized cytogenetically, with chromosomal data available for 51 species. The variation in diploid number in this family is remarkable, ranging from $2n = 5$ in *T. bahiensis* to $2n = 32$ in *Tityus curupi* Ojanguren-Affilastro et al. 2017 and *Tityus neibae* Armas, 1999 (Table 12.1).

The visualization of constitutive heterochromatin (HC) by conventional C-banding or by staining with the fluorochrome 4,6-diamidine-2-phenylindole (DAPI) was performed in some members of Buthidae from the Neotropical region. In most of the species analyzed, HC blocks with variable sizes are located mainly in

Table 12.1 Cytogenetic data of Neotropical scorpions

Family	Species	Country	2n	Morphology	Meiotic configuration	Heterochromatin	NOR	Reference
Ananteridae	*Ananteris balzanii*	Brazil	12,14	Holocentric	2n = 14 (5II + CIV; 7II)	C bands absent	1 pair, terminal	Piza (1947a) and Mattos et al. (2013)
Buthidae	*Centruroides baracoae*	Cuba	26	Holocentric	13II	–	1 pair, terminal	Šťáhlavský et al. (2020)
	Centruroides gracilis	Cuba	24	Holocentric	10II + CIV	–	1 pair, terminal	Šťáhlavský et al. (2020)
	Heteroctenus abudi	Dominican Republic	16	Holocentric	8II	–	1 pair, terminal	Šťáhlavský et al. (2020)
	Heteroctenus garridoi	Cuba	23	Holocentric	10II + CIII	–	1 pair, terminal	Šťáhlavský et al. (2020)
	Heteroctenus princeps	Dominican Republic	16	Holocentric	8II	–	1 pair, terminal	Šťáhlavský et al. (2020)
	Ischnotelson guanambiensis	Brazil	25	Holocentric	11II + CIII; II + VN	–	1 pair, terminal	Ubinski et al. (2018)
	Ischnotelson peruassu	Brazil	26	Holocentric	13II	Terminal	1 pair, terminal	Ubinski et al. (2018)
	Ischnotelson sp.	Brazil	26	Holocentric	13 II	–	1 pair, terminal	Ubinski et al. (2018)
	Isometrus maculatus	Brazil	12	Holocentric	1CVIII + 2II	–	1 pair, terminal	Piza (1947b) and Piza (1955)
	Jaguajir agamemnon	Brazil	28	Holocentric	14II; CXXVIII	Terminal and subterminal	1 pair, terminal	Mattos et al. (2013) and Ubinski et al. (2018)
	Jaguajir pintoi	Brazil	28	Holocentric	14II; 13II + CIV	Terminal and subterminal	1 pair, terminal	Ubinski et al. (2018)

(continued)

Table 12.1 (continued)

Family	Species	Country	2n	Morphology	Meiotic configuration	Heterochromatin	NOR	Reference
	Jaguajir rochae	Brazil	28	Holocentric	8II + CIV + CVII 8II + CIV + CVIII 8II + 1CXII 9II + CX 10II + CVIII	Terminal	1 pair (terminal) or 3 chromosomes (2 terminal and 1 interstitial)	Piza (1957), Mattos et al. (2013), Ubinski et al. (2018) and Schneider et al. (2015)
	Physoctonus debilis	Brazil	26	Holocentric	–	–	1 pair, terminal	Ubinski et al. (2018)
	Rhopalurus laticauda	Brazil	22	Holocentric	9II + CIV	–	1 pair, terminal	Ubinski et al. (2018)
	Tityopsis rolandoi	Cuba	20	Holocentric	10II	–	–	Kovařík et al. (2024)
	Tityopsis sheylae	Cuba	22	Holocentric	11II	–	–	Kovařík et al. (2024)
	Tityus clathratus	Brazil	19,20	Holocentric	**2n = 19** (1II + CXIII; 2II + CXI;3II + CXI; 3II + CXIII; 4II + CX; 4II + CXI; 5II + CIX; 5II + CX;5II + CXI; VN) **2n = 20** (10II; 3II + CX; 3II + CXI;4II + CIX; 4II + CX; 5II + CX; VN)	–	1 pair, terminal	Mattos et al. (2018)
	Tityus maranhensis	Brazil	20	Holocentric	10II; 9II; 8II + CIV	Terminal	1 pair, terminal	Mattos et al. (2013) and Almeida et al. (2022)

	Tityus mattogrossensis	Brazil	20,19	Holocentric	**2n = 19** (6II + CIII+CIV; 8II + CIII; VN) **2n = 20** (10II; 9II; 8II + CIV; VN)	–	1 pair, terminal	Piza (1947a) and Mattos et al. (2018)
	Tityus paraguayensis	Brazil	16,17,18	Holocentric	**2n = 16** (8II; 5II + CIV; 5II + CVII+1CVIII; VN) **2n = 17** (5II + CVIII; 5II + VIII; VN) **2n = 18** (9II; 14II; VN)	Terminal	1 pair, terminal	Mattos et al. (2013) and Mattos et al. (2018)
	Tityus pusillus	Brazil	19,20	Holocentric	**2n = 19** (7II + CIV; 7II + CV; 7II + CVI; 8II + CIII; PO; VN). **2n = 20** (10II; 7II + CVI; VN).	–	1 pair, terminal	Mattos et al. (2018)
	Tityus silvestris	Brazil	16,24	Holocentric	**2n = 16** (2II + CXII; VN). **2n = 24** (12II; VN).	Terminal	1 pair, terminal	Mattos et al. (2018), Almeida et al. (2019) and Almeida et al. (2022)
	Tityus mana	French Guiana	20	Holocentric	7II + CVI	–	1 pair, terminal	Šťáhlavský et al. (2020)
	Tityus aba	Brazil	18	Holocentric	9II	Terminal and interstitial	–	Lima et al. (2023)
	Tityus argentinus	Argentina	9,10	–	5II; 1II + 1CVII	–	–	Mola et al. (2011)

(continued)

Table 12.1 (continued)

Family	Species	Country	2n	Morphology	Meiotic configuration	Heterochromatin	NOR	Reference
	Tityus bahiensis	Brazil, Argentina	5, 6, 7, 8, 9, 10,12, 13, 14, 15 17, 18, 19	Holocentric	**2n = 5** (III + II; 2II + I) **2n = 6** (3II; II + CIV) **2n = 7** (CVII) **2n = 8** (2II + CIV) **2n = 9** (II + CVII; CIX; 3II + CIII; II + CIII+CIV) **2n = 10** (3II + CIV; II + 2CIV; 2II + CVI; 5II; 2II + 2CIII; CIV + 3II; CVI + 2II; CVIII+II; CX) **2n = 12** (2II + CIII+CV) **2n = 13** (3II + CVII; 2II + IX; 4II + CV) **2n = 14** (4II + CV) **2n = 15** (6II + 1CIII) **2n = 17** (6II + CV; 5II + CVII) **2n = 18** (5II + CIII+CV) **2n = 19** (7 II + CV) **2n = 20** (7II + 2CIII)	Terminal, subterminal, and interstitial	1 pair, terminal or 1 pair (terminal and subterminal) or 1 pair (terminal and interstitial)	Piza (1939a, b, 1940, 1943a, b, 1944, 1947a, c, 1948b, c, 1949, 1950a, b), Brieger and Graner (1943), Cunha and Pavan (1954), Takahashi (1976), Benavente (1982), Schneider et al. (2009b), Mattos et al. (2013), Mattos et al. (2014), Lima et al. (2023) and Adilardi et al. (2020)
	Tityus carrilloi	6	Argentina	Holocentric	–	Terminal, subterminal, and interstitial	1 pair, terminal	Adilardi et al. (2013)
	Tityus confluens	13	Brazil, Argentina	Holocentric	2n = 13 (5II + 1CIII) 2n = 6 (1II + 1CIV; 1CVI) 2n = 5 (1CV)	Terminal and intestitial	1 pair, terminal or 1 pair, terminal and interstitial	Mattos et al. (2013) and Adilardi et al. (2016)
	Tityus costatus	16	Brazil	Holocentric	–	Terminal and intestitial	1 pair, terminal	Mattos et al. (2013)

	Tityus crassimanus	29	Dominican Republic	Holocentric	12II + CV	–	1 pair, terminal	Šťáhlavský et al. (2020)
	Tityus curupi	32,31	Argentina	Holocentric	16II; 14II + 1CIII	Terminal and intestitial	1 pair, terminal	Ojanguren-Affilastro et al. (2017)
	Tityus neglectus	26,27	Brazil	Holocentric	12 II + 1CIII	–	–	Piza (1950b)
	Tityus ecuadoriensis	24	Ecuador	Holocentric	12 II	–	1 pair, terminal	Šťáhlavský et al. (2020)
	Tityus eiii	22	Dominican Republic		7II + CVIII	–	1 pair, terminal	Šťáhlavský et al. (2020)
	Tityus trivittatus	14	Brazil	Holocentric	5II+ 1CIV	–	–	Piza (1948a)
	Tityus uruguayensis	31	Argentina	Holocentric	12II + CVII; 10II + CIV + CVII	Terminal	2 pairs, terminal	Ojanguren-Affilastro et al. (2017)
	Tityus fasciolatus	14	Brazil	Holocentric	14II	Terminal and intestitial	1 pair, terminal	Mattos et al. (2013) and Mattos et al. (2014)
	Tityus gaffini	22	Colombia	Holocentric	11II	–	2 pairs, terminal	Šťáhlavský et al. (2020)
	Tityus smithii	6	Grenada		3II	–	3 pair, terminal	Šťáhlavský et al. (2020)
	Tityus martinpaechi	6	Brazil	Holocentric	CVI	Terminal and intestitial	1 pair, terminal	Mattos et al. (2013) and Mattos et al. (2014)
	Tityus stigmurus	16	Brazil	Holocentric	–	Terminal	1 pair, terminal	Mattos et al. (2013)

(continued)

Table 12.1 (continued)

Family	Species	Country	2n	Morphology	Meiotic configuration	Heterochromatin	NOR	Reference
	Tityus serrulatus	12	Brazil	Holocentric	12II	Terminal	1 pair (terminal) in male or 1 chromosome (terminal) in female	Schneider and Cella (2010) and Lima et al. (2020)
	Tityus neibae	32	Dominican Republic	Holocentric	10II + III + III + VI	–	1 pair, terminal	Šťáhlavský et al. (2020)
	Tityus magnimanus	20	–	Holocentric	10II	–	–	Kovařík et al. (2009)
	Tityus metuendus	14,15,16	Brazil	Holocentric	2n = 14 (7II; 5II + 1CIV) 2n = 15 (2II + 1CIII + 1CVIII) 2n = 16 (4II + +1CVIII)	Terminal	–	Piza (1950c) and Lima et al. (2023)
	Tityus obscurus	11,12,13, 14,16	Brazil	Holocentric	2n = 16 (8II; 6II + 1CIV) 2n = 14 (7II; 4II + 1CIII) 2n = 13 (5II + 1CIII) 2n = 12 (6II; 3II + 2CIII) 2n = 11 (4II + 1CIII)	Terminal and intestitial	2 pairs, Terminal	Almeida et al. (2017)
	Tityus apiacas	14	Brazil	Holocentric	7II; VN	Terminal and Interstitial	1 pair, Terminal	Lima et al. (2023)
	Tityus fuhrmanni	22	Colombia	Holocentric	11II	–	1 pair, Terminal	Šťáhlavský et al. (2020)
	Tityus ythieri	20	–	Holocentric	10II	–	–	Kovařík et al. (2009)
	Zabius fuscus	18	Argentina	Holocentric	9II; 7II + 1CIV	Terminal	1 pair, Terminal	Adilardi et al. (2015)

Chactidae	*Brotheas amazonicus*	50, 52	Brazil	**2n = 50**: 30M/SM+ 20A **2n = 50:** 24M + 4SM + 22A **2n = 52:** 20M + 10SM + 22A	2n = 50 (25II; 23II + IV) 2n = 52 (26II; 23II + 1CVI; 24II + 1CIV)	Terminal and interstitial	1 pair (Terminal) or 1 chromosome with NORs in both ends	Ferreira (1968), Almeida et al. (2023) and Lima et al. (2023)
	Brotheas paraenses	48	Brazil	14M + 4SM + 30A	24II	–	1 pair, Interstitial	Almeida et al. (2023)
	Brotheas silvestris	40	Brazil	28M + 6SM + 6A	20II	–	1 pair, Terminal	Almeida et al. (2023)
	Chactopsis amazônica	36	Brazil	10M/SM + 26A	18II; 16II + 1CIV	–	1 pair, Terminal	Lima et al. (2023)
	Neochactas parvulus	54	Brazil	4M + 26SM + 6A + 18Micro	27II	Pericentromeric and intestitial	1 pair, Interstitial	Almeida et al. (2023)
	Neochactas sp.	30	Brazil	20M/SM + 10A	15II	–	–	Lima et al. (2023)
Bothriuridae	*Bothriurus araguayae*	42,44	Brazil	(32ST + 8SM + 2M)	21II; 19II + 1CIV	–	3 pairs, Terminal	Ferreira (1968) and Schneider et al. (2009a)
	Bothriurus rochensis	46	Brazil	16ST + 16SM + 14MT	23II	Pericentromeric and terminal	2 pairs, Terminal	Schneider et al. (2009a)
	Bothriurus flavidus	48	Argentina	–	–	–	–	Giacomozzi (1977)
	Bothriurus prospicuus	50	Argentina	–	–	–	–	Giacomozzi (1977)
	Bothriurus sp.	36	Brazil	–	–	–	–	Piza (1947a)
	Brachistosternus alienus	28,46	Argentina	–	2n = 46 (23II)	Terminal and interstitial	–	Adilardi et al. (2005)
	Brachistosternus ferrugineus	46	Argentina	M + SM + A	–	–	–	Rodríguez-Gil et al. (2009)

(continued)

Table 12.1 (continued)

Family	Species	Country	2n	Morphology	Meiotic configuration	Heterochromatin	NOR	Reference
	Brachistosternus montanus	46	Argentina	M + SM + A	–	–	–	Rodríguez-Gil et al. (2009)
	Brachistosternus pentheri	42,46	Argentina	–	–	–	–	Rodríguez-Gil et al. (2009)
	Timogenes elegans	48	Argentina	–	–	–	–	Giacomozzi (1977)
Hormuridae	*Opisthacanthus elatus*	60,62	–	–	–	–	–	Wilson (1931)

the terminal, subterminal, and/or interstitial region of two or more chromosomes (Table 12.1). *Ananteris balzanii* Thorell, 1891 is an exception to this pattern, since in specimens from Brazil no heterochromatic regions were observed by C-banding or DAPI (Mattos et al. 2013). In *Tityus obscurus* (Gervais, 1843) and *Tityus confluens* Borelli, 1899, there is the occurrence of extra-large chromosomes originated by fusion, with interstitial C-bands that may correspond to the point of union of the chromosomes involved in this rearrangement (Almeida et al. 2017; Adilardi et al. 2016). According to Mattos et al. (2013), large heterochromatic blocks are commonly observed in species that present a low frequency of structural alterations, such as fusions/fissions, suggesting a role of HC in the genomic stability of this family.

Nucleolar organizer regions (NORs) were mapped in the chromosomes of neotropical Buthidae using both silver ion impregnation (Ag-NOR banding) and FISH with probes specific for 18S rDNA, 28S rDNA, and 45S rDNA sequences (Šťáhlavský et al. 2020). In most species analyzed, the ribosomal sites were located in the terminal region of a chromosome pair, although exceptions to this pattern were identified in *Jaguajir rochae* (Borelli, 1910), *Tityus uruguayensis* Borelli, 1901, *Tityus gaffini* Lourenço, 2000, *Tityus smithii* Pocock, 1893 and *T. obscurus*, which presented rDNA clusters in three or more chromosomes of the karyotype (Table 12.1). The interstitial and/or subterminal location of NORs in some Buthidae species results from fusion/fission-type chromosome rearrangements, as observed in *T. bahiensis* and *T. confluens* (Schneider et al. 2009a; Adilardi et al. 2016), or from reciprocal translocation, as exemplified in *J. rochae* (Mattos et al. 2014).

12.3 Chactidae

To date, the Chactidae family has six species with described karyotypes, whose diploid numbers vary from 2n = 30 (*Neochactas* sp.) to 2n = 54 (*Neochactas parvulus* (Parvulus, 1897)) (Table 12.1).

Regarding constitutive heterochromatin (HC), its presence was observed in only two species of this family (Table 12.1). In *Brotheas amazonicus* Lourenço, 1988, C bands were detected exclusively in the terminal region of 10 chromosome pairs (Lima et al. 2023). In *N. parvulus*, large blocks of HC were identified mainly in the pericentromeric region of macrochromosomes (Almeida et al. 2023). Furthermore, an interstitial C band was observed in pair 2 of this species, colocalizing with the nucleolus organizer region. It is worth mentioning that some microchromosomes of *N. parvulus* also presented positive C-bands (Almeida et al. 2023).

Regarding NOR, in this family, it is generally present in a single pair of chromosomes, located in a terminal or interstitial position (Table 12.1). However, an exception was observed in cytotype B (2n = 52) of *B. amazonicus*, in which variations in the number and size of ribosomal sites were detected (see topic Sect. 12.9.3).

12.4 Bothriuridae

To date, only ten species of Bothriuridae have been cytogenetically characterized. However, in most of them, only the diploid number has been determined (Table 12.1). In this family, chromosomal variation is wide, ranging from 2n = 28 (*Brachistosternus alienus* Lonnberg, 1898) to 2n = 50 (*Bothriurus prospicuus* Mello-Leitão, 1934) (Table 12.1). In some cases, such as in *Bothriurus araguayae* Vellard, 1934, chromosome pairs composed of heteromorphic chromosomes, originating from reciprocal translocation, have been recorded (Schneider et al. 2009a).

The distribution of HC was studied only in *Bothriurus rochensis* San Martin, 1965 and *B. alienus* (Table 12.1). In *B. rochensis,* C bands were observed in the pericentromeric and terminal regions of some pairs of the karyotype (Schneider et al. 2009a); when *B. rochensis* metaphases were stained with DAPI, two bivalents exhibited DAPI+ terminal regions (Schneider et al. 2009a). In *B. alienus*, HC blocks were observed in the terminal and interstitial regions of several chromosome pairs (Adilardi et al. 2013).

NOR was mapped only in the karyotype of two *Bothriurus* species by Ag-NOR banding: in pairs 18 and 22 in *B. rochensis*, and pairs 1, 4, and 9 of *B. araguayae* (Schneider et al. 2009a). In both cases, this rDNA is located in the terminal region (Table 12.1).

12.5 Hormuridae

In this family, to date, only the diploid number of *O. elatus* is known, which corresponds to 2n = 60.62 (Table 12.1).

12.6 Chromosome Structure and Karyotype

Buthidae karyotypes are characterized by the presence of holocentric chromosomes, which do not have a single localized centromere, but rather centromeric proteins distributed along the entire length of the chromatids (Marques and Pedrosa-Harand 2016). This organization can be identified by the absence of primary constrictions in metaphase chromosomes and by the holokinetic behavior during anaphase, in which sister chromatids segregate in parallel (Mandrioli and Manicardi 2020). An important consequence of holocentricity is the retention of chromosome fragments after fissions, since these fragments maintain the ability to bind to microtubules and behave as independent chromosomes, leading to an increase in the diploid number (Zedek and Bureš 2017). Furthermore, fission/fusion rearrangements are frequent in Buthidae, especially in heterozygous conditions (affecting only one of the homologs), which explains the occurrence of odd diploid numbers and the variation in chromosome size in several species (Table 12.1).

Despite this karyotypic plasticity, the diploid number is conserved in some Neotropical lineages of Buthidae. For example, all three species of the genus *Jaguajir* have 2n = 28 (Table 12.1). Likewise, conservation is observed even at the population level: Brazilian populations of *Isometrus maculatus* (DeGeer, 1778), separated by more than 5000 km (Bahia and Acre), maintain the diploid number 2n = 12 (Piza 1955).

The structural differences between holocentric and monocentric chromosomes in Scorpiones are reflected in the organization of the kinetochore plate. In species of Bothriuridae (*B. araguayae* and *B. rochae* Mello-Leitão, 1932), which have monocentric chromosomes, analyses of synaptonemal complexes demonstrated that kinetochores occupy 7.4% and 5.4% of the total chromosome length, respectively, during prophase I (Schneider et al. 2009a).

In contrast, in representatives of Buthidae (such as *Tityus fasciolatus* Pessoa, 1935, *T. bahiensis, J. rochae,* and *J. agamemnon* (C. L. Koch, 1839)), kinetochores were not detected in the synaptonemal complexes in prophase I (Schneider et al. 2015), which corroborates the holocentric nature of these chromosomes. Benavente (1982), through transmission electron microscopy, observed that the kinetochores in *T. bahiensis* bivalents only partially cover the extension of the chromosomes, being detectable exclusively in metaphase I and metaphase II phases, and always oriented toward the poles.

From a morphological point of view, the karyotypes of Neotropical scorpions generally present a gradual distribution in chromosome size, with chromosomes arranged in decreasing order, except in cases of pronounced heteromorphism resulting from fusion or fission events. However, in *N. parvulus* (Chactidae), a bimodal karyotype has been documented (Almeida et al. 2023), composed of two morphologically distinct chromosome categories: macrochromosomes (regular-sized chromosomes) and extremely reduced, punctiform microchromosomes, whose morphology cannot be accurately determined by conventional optical microscopy. Almeida et al. (2023) proposed that the microchromosomes of *N. parvulus* may have originated from successive macrochromosome fission events.

12.7 What Do We Know About Achiasmatic Meiosis in Neotropical Scorpions?

Meiosis is the process responsible for the formation of gametes in eukaryotes, divided into two stages: meiosis I (reductional) and meiosis II (equational), resulting in four haploid cells. In neotropical scorpions, meiosis presents peculiar characteristics, such as achiasmia, formation of multivalent chains, and telokinetic activity, becoming the target of intense investigation (Mattos et al. 2018; Almeida et al. 2022).

In several species of Neotropical scorpions, the rate of chromosome rearrangement is considered high, mainly due to events such as reciprocal translocations, Robertsonian translocations, fusions/fissions, and deletions/duplications. When only one of the homologs presents the structural alteration, the chromosome pair is considered heterozygous for the rearrangement. During meiosis I, these heterozygous pairs, when pairing and synapsing with their respective non-rearranged homologs, often form multichromosomal associations, also called multivalent meiotic associations or, morphologically, chromosome chains and rings, visible in metaphase I. For example, the occurrence of a chromosome fusion or fission can result in the formation of trivalents (CIII), as observed in *Heteroctenus garridoi* (Armas, 1974) (Šťáhlavský et al. 2020). In *Chactopis amazonica* Lourenço & Francke, 1986, a reciprocal translocation between two chromosome pairs gave rise to a quadrivalent (CIV) (Lima et al. 2023). Even more complex configurations have been described, such as chains of eight chromosomes (CVIII) in *T. metuendus* Pocock, 1897 (Piza 1950), or multivalent associations with 10 (CX), 11 (CXI) and even 13 (CXIII) chromosomes in *T. clathratus* C. L. Koch, 1844 (Mattos et al. 2018); in addition to multivalent associations with a variable number of components (VN), whose total number of chromosomes involved could not be determined.

These configurations can vary at different levels interpopulational, intrapopulational, and even intra-individual (Schneider et al. 2009b; Mattos et al. 2014). The chromosome composition of multivalent associations can vary among individuals. In *T. confluens*, for example, differences were observed in the constitution of meiotic trivalents, in which the chromosome carrying the 45S rDNA ribosomal locus was involved in the association in a given individual, while in another it did not integrate the same multivalent structure (Mattos et al. 2014). Interestingly, despite the complexity in metaphase I, metaphase II presents the expected haploid number, indicating efficient alternate segregation (Schneider et al. 2009b; Adilardi et al. 2015; Almeida et al. 2022; Lima et al. 2023).

As mentioned previously, the accurate identification of the number of chromosomes involved in multivalent associations during meiosis can be difficult in some species. To overcome this limitation, Schneider et al. (2015) employed ultrastructural analysis of synaptonemal complexes (SC), a technique that allows direct visualization of interactions between homologous chromosomes during meiotic prophase I. The SC is a tripartite protein structure composed of two lateral elements, associated with the chromosome axes of each homolog, and a central element that connects both lateral elements, promoting chromosome synapsis (Lisachov et al. 2024). Its assembly occurs in zygotene, persisting throughout pachytene, a stage in which it facilitates gene alignment and recombination between homologs. From diplotene onwards, the CS is progressively unstructured, and the bivalents remain cohesive only in the regions where crossing-over events occurred, morphologically evidenced as chiasmata.

Using this high-resolution cytogenetic approach, Schneider et al. (2015) demonstrated that the meiotic multivalents observed in *J. rochae* were composed of 12 homologous chromosomes. Additionally, the authors described the presence of gaps (discontinuities in the structure of the SC) in *T. fasciolatus*, which indicate the absence of synapsis between certain segments, possibly related to the occurrence of heterozygous rearrangements in that specific chromosomal region. The analysis also revealed the presence of interlocks—structural intertwining between pairs of distinct homologs during synapsis—in *T. bahiensis* and *T. fasciolatus*, which reflect complexities in the three-dimensional organization of chromatin during the early stages of prophase I.

Although the meiotic recombination process plays an essential role in generating genetic variability in populations, male meiosis in scorpions has been widely described as achiasmatic, that is, with the absence of chiasmata. In Scorpiones, achiasmatic meiosis is interpreted as a cytogenetic adaptation that favors the regular segregation of chromosomes involved in multivalent associations, such as quadrivalents and other complex configurations, resulting from chromosomal rearrangements. However, this characteristic raises questions about the occurrence and functionality of crossing-over in male Neotropical scorpions. Studies conducted by Almeida et al. (2022, 2023), using immunolocalization techniques of proteins associated with meiotic recombination in *T. silvestris* Pocock, 1897 and *T. maranhensis* Lourenço et al., 2006, demonstrated that the initial steps of the crossing-over process occur normally (such as the formation of DNA double-strand breaks (DSBs) and the action of Rad51), allowing the physical approximation of homologs (pairing).

The final step of recombination, responsible for the repair of the broken strand in a reciprocal manner, is normally mediated by the mismatch repair system (MMR) enzymes, such as MLH1, MLH3, MSH4, and MSH5, or by the endonuclease MUS81. In *T. silvestris*, gene expression analyses by RT-PCR revealed the constitutive expression of these genes in ovaries, embryos, and testes. Despite the expression of MMR system genes (MLH1/3, MSH4/5), recombination nodes (MLH1/MLH3 foci in the axes of the synaptonemal complex during pachytene) and chiasmata are absent in *Tityus* and *Jaguajir* (Schneider et al. 2015), indicating that recombination may not occur, maintaining only its structural role in the alignment and synapsis between homologous chromosomes, even in the absence of effective recombination (Almeida et al. 2022).

Chiasms, in addition to representing the points of genetic recombination between homologous chromatids, play a crucial role in maintaining the cohesion of bivalents until the transition to anaphase I during meiosis. In neotropical scorpions, this stabilizing function appears to be compensated by the persistence of the synaptonemal complex and the cohesin protein SMC3 throughout advanced stages of prophase I (Schneider et al. 2009b, 2015; Almeida et al. 2019). Almeida et al. (2019) proposed that, in *T. silvestris*, specific epigenetic chromatin modifications, such as trimethylation of lysine 27 of histone H3 (H3K27me3), may be involved in the maintenance of these protein structures. According to Almeida et al. (2019), in *T. silvestris*, H3K27me3 signals are colocalized with chromosomal regions in the initial phases of synapsis, with the intensity of this epigenetic mark increasing progressively throughout the bivalents, following the dynamics of the synaptic process and remaining detectable until metaphase I.

Female meiotic cells have been observed in only a limited number of Neotropical scorpion species. In *Zabius fuscus* (Thorell, 1876), Adilardi et al. (2015) documented female meiotic chromosomes in the diffuse stage, some of which presented morphology compatible with bivalents connected by two chiasmata; however, the data obtained were not conclusive regarding the chiasmatic nature of these associations. In *Troglorhopalurus translucidus* Lourenço et al., 2004, analyses of a female specimen revealed post-pachytene cells with bivalents aligned in parallel, but without morphological evidence of chiasmata (Ubinski et al. 2018). These findings suggest the possibility of achiasmatic meiosis in both sexes in Scorpiones, a hypothesis supported by similar observations made in Old World species (Shanahan and Hayman 1990).

In organisms with holocentric systems, the attachment of mitotic spindle microtubules occurs along the entire length of the chromosomes, enabling their proper migration to the cell poles during anaphase. In *T. maranhensis*, a species that presents the 8II + 1CIV chromosome configuration in meiosis I, it was observed that this holocentric pattern of microtubule/kinetochore interaction is maintained during metaphase I and anaphase I, with the insertion of microtubules along the entire length of the bivalents and quadrivalents (Almeida et al. 2022). However, during metaphase II, microtubule anchoring occurs exclusively in the terminal regions of the chromosomes, characterizing a telokinetic behavior (Almeida et al. 2022). This terminal interaction pattern has already been described in other arthropods with holocentric chromosomes and is suggested as an adaptive mechanism that favors accurate chromosome segregation during meiosis.

12.8 Genomic Organization of Repetitive DNA

The main repetitive sequence mapped in Neotropical scorpion chromosomes corresponds to the 45S rDNA ribosomal locus, which includes the 18S and 28S rDNA subunits (Schneider et al. 2009a; Adilardi et al. 2015; Almeida et al. 2017, 2023; Ubinski et al. 2018; Lima et al. 2023). As discussed in Sects. 12.2, 12.3, and 12.4, most species present one or two pairs of chromosomes with rDNA sites, predominantly in a terminal position (Mattos et al. 2014; Adilardi et al. 2014; Almeida et al. 2017; Šťáhlavský et al. 2020). Furthermore, these sites are often associated with adjacent regions of heterochromatin (Schneider et al. 2009b; Adilardi et al. 2015; Almeida et al. 2019, 2023). According to Schneider et al. (2009b), these heterochromatic blocks may contribute to the structural stability of NORs, preventing chromosomal breaks and rearrangements. In *T. silvestris*, epigenetic modifications, such as the acetylation of lysine 9 of histone H3 (H3K9ac), appear to regulate the transcriptional activity of NORs during leptotene and zygotene, maintaining their expression in these early phases of prophase I (Almeida et al. 2019).

Telomeres in Neotropical scorpions are composed of the sequence TTAGG, typical of arthropods (Adilardi et al. 2015; Almeida et al. 2022; Lima et al. 2023). Interestingly, although chromosome fusions have been documented in Buthidae, there is no evidence of interstitial telomeric sequences (ITSs) at the junction points. Adilardi et al. (2016) and Almeida et al. (2017) suggest that telomeric regions may be completely eliminated during the breakage and fusion processes, preventing the retention of ITSs.

C_0t-1 DNA, rich in highly and moderately repetitive sequences (transposable elements, satellite DNAs, microsatellites, etc.), was mapped by FISH in only four species: three Buthidae (*T. obscurus, T. metuendus,* and *Ischnotelson peruassu* Esposito et al., 2017) and one Chactidae (*B. amazonicus*). In *T. obscurus*, C0t-1 DNA is located at the ends of chromosomes, suggesting the absence of typical centromeric sequences, a common pattern in organisms with holocentric chromosomes (Almeida et al. 2017).

Lima et al. (2023) performed two independent experiments with the C_0t-1 DNA probe: first, they tested the probe on chromosomes of the same species from which the C_0t-1 DNA was obtained, which resulted in strong species-specific signals in *B. amazonicus, T. metuendus,* and *I. peruassu*, located mainly in the terminal region; in a second step, these authors attempted to hybridize the C_0t-1 DNA probe of *T. metuendus* on chromosomes of the other species mentioned. In this case, the hybridization signals were scarce and of low intensity, located mainly in the interstitial region of chromosomes of *B. amazonicus, T. bahiensis,* and *I. peruassu*. These results suggest low conservation of repetitive sequences among different scorpion species, evidencing a pattern of rapid and species-specific evolution of these genomic elements.

The chromosomal organization of histones and U snDNA has been studied in few species. In *T. obscurus*, histone H3 clusters are located at the ends of all chromosomes, while U2 snDNA occupies an interstitial region in pair 2 (Almeida et al. 2017). In Chactidae (*N. parvullus, B. paraensis* Simon 1880, *B. amazonicus*), U2 snDNA is also interstitially distributed in a single pair (Almeida et al. 2022), highlighting its potential as a cytogenetic marker for evolutionary studies.

Regarding transposable elements, only the *Mariner* transposon was analyzed in *T. obscurus*, where it was abundant along the chromosomes (Almeida et al. 2017). Its presence in euchromatin was associated with low genetic recombination. Only in the *T. obscurus* cytotype A individual, carrying a reciprocal translocation in heterozygosity, *Mariner* accumulation was observed in distal heterochromatin (Almeida et al. 2017).

In *T. obscurus*, the concentration of repetitive sequences (heterochromatin, C_0t-1 DNA, histone H3, and 45S rDNA) in the terminal region allowed Almeida et al. (2017) to suggest that the chromosome ends in this scorpion would be analogous to pericentromeres of monocentric chromosomes, which flank the kinetochore, but are not covered by it.

12.9 Chromosomal Variations and Meiosis

Some species of Neotropical scorpions present extensive variation in diploid number (2n) and chromosome morphology due to chromosomal rearrangements. Below, we highlight the main cases observed, with emphasis on *T. bahiensis*.

12.9.1 Tityus bahiensis

This scorpion is the most cytogenetically studied Neotropical species, presenting chromosomal polymorphism and a wide geographic distribution, covering northeastern Argentina, southeastern Brazil, and Paraguay (Lourenço 2015). This species presents remarkable intraspecific variation in diploid number (2n = 5 to 19), mainly due to fusion/fission rearrangements and reciprocal translocation, which occur both at the interpopulation and intrapopulation levels (Table 12.1).

A striking example of this diversity was described by Schneider et al. (2009b) in a population from Piracicaba (SP, Brazil): while 12 males presented 2n = 6, one male and one female presented 2n = 5. The latter karyotype included an extra-large chromosome, resulting from the fusion between an autosome and a chromosome carrying the NOR (nucleolar organizer region). These findings reinforce the karyotypic instability in *T. bahiensis*, highlighting its potential as a model for studies of chromosome evolution in scorpions.

The high structural heterozygosity observed in *T. bahiensis* is correlated with significant changes in chromosome synapsis during the meiotic cycle of this species. In the early phases of prophase I and metaphase I, for example, the presence of bivalents with poorly condensed chromatin, gaps, formations compatible with loops and synaptonemal complexes with discontinuous regions were observed, indicating that the rearrangements that originated these chromosomes occur in a heterozygous state (Schneider et al. 2009b, 2015). Furthermore, in metaphase I, the formation of extensive multi-chromosomal associations is observed. These multivalents can exhibit different configurations, which can be trivalent, quadrivalent (sometimes two quadrivalents joined by one of the elements), open or closed chromosome chains involving 7, 8, 9, and 10 chromosomes (Piza 1939a, 1943a, Schneider et al. 2009b, Mattos et al. 2013, Adilardi et al. 2020). In all cases, cells in metaphase II always present expected haploid numbers, indicating that the segregation of chromosomes involved in multivalents occurs by alternate means, resulting in the formation of balanced gametes. According to Schneider et al. (2009b), the holocentric morphology and achiasmatic meiosis favor the alternate segregation of multivalents, while common ecological factors in Buthidae may facilitate hybridization between divergent cytotypes, expanding karyotypic diversity.

Adilardi et al. (2020) proposed a North-South polymorphism model to explain the variation in *T. bahiensis*. In this case, the species would have dispersed via floating vegetation along rivers, with the ancestral karyotype being 2n = 20. The diploid numbers were classified into three groups:

- Northern Form (Minas Gerais, Brazil): 2n = 17–20 (high values).
- Intermediate Region (São Paulo/Paraná, Brazil): 2n = 5–10 (low values).
- Southern Form (Argentina): 2n = 12–15 (intermediate values).

According to Adilardi et al. (2020), the intermediate region (group 2) would be a hybridization zone between the Northern and Southern forms, where the reduced 2n would have been fixed by genetic drift (founder effect). However, additional studies are needed to validate this hypothesis and elucidate the evolutionary dynamics of this chromosomal system.

12.9.2 Tityus obscurus

In *T. obscurus*—a scorpion widely distributed in some states of the Brazilian Amazon—the diploid number ranges from 2n = 11 to 2n = 16. Cytogenetic studies conducted by Almeida et al. (2017) in populations from the state of Pará, Brazil, revealed distinct patterns of karyotypic variation.

The populations of Belém (2n = 16), Afuá (predominantly 2n = 14), and Rurópolis (2n = 12) present greater chromosome stability, with relatively conserved diploid numbers. In contrast, the population of Santarém presented high intraspecific diversity, with cytotypes ranging from 2n = 11 to 2n = 14. Some of these cytotypes presented extra-large chromosomes, interstitial C bands, and heteromorphic pairs, indicating heterozygous fusions (Almeida et al. 2017). These rearrangements were confirmed by analysis of meiotic cells: In cytotype F (2n = 12) from Santarém, two trivalent associations (one including the extra-large chromosome) and three regular bivalents were observed in metaphase I. In cytotype G (2n = 11), one trivalent and four bivalents were identified.

Although the Belém population presented predominantly 8II, one individual exhibited a chain of four chromosomes (6II + 1CIV), resulting from a reciprocal translocation in heterozygosity. Based on these data, Almeida et al. (2017) proposed a central-marginal model for the geographic structuring of the *T. obscurus* karyotype. In this model, Santarém represents the central area, due to the high chromosomal variability and multiple structural rearrangements. Afuá, Belém, and Rurópolis correspond to the marginal areas, as they present more conserved karyotypes.

12.9.3 Brotheas amazonicus

Brotheas amazonicus is a scorpion of the Chactidae family, distributed in most states of the Brazilian Amazon. Almeida et al. (2023) demonstrated geographic variation in the karyotype of this species, comparing populations from Manaus and Parintins (Amazonas, Brazil). The population from Manaus (cytotype A) presented greater karyotypic stability, with a constant diploid number (2n = 50) and formation of 25 regular bivalents in meiosis I. In contrast, individuals from Parintins (cytotype B) presented 2n = 52, with the majority exhibiting 26 bivalents, although Almeida et al. (2023) reported the occasional occurrence of quadrivalents and hexavalents in some specimens—attributed to reciprocal translocations in heterozygosity.

In addition to the numerical differences, the cytotypes diverged in the organization of the 45S rDNA sites. Cytotype A (2n = 50): presented rDNA in pair 16. Cytotype B (2n = 52): presented rDNA in pair 4, with polymorphism in the location of this sequence on the chromosome (one homolog presented a single terminal cluster, while the other presented two sites, one at each end) (Almeida et al. 2023). Notably, pair 4 of cytotype B was involved in the formation of multivalents, including the hexavalent—in which only the homolog with two rDNA sites was observed (Almeida et al. 2023). These variations suggest the action of structural rearrangements, such as translocations and deletion/duplication events, which would explain the polymorphism in the nucleolar organizer region (NOR) in cytotype B.

12.9.4 Bothriurus araguayae

Most individuals studied by Schneider et al. (2009a) of *B. araguayae* showed a karyotype that gradually decreases in size, with 2n = 42. However, a male from Rio Claro (São Paulo state, Brazil) exhibited a karyotype with size heteromorphisms in pairs 3, 5, and 21. In pair 3, a striking morphological difference was observed: one homolog was an extra-large metacentric, while the other corresponded to a standard-sized subtelocentric. During meiosis, 19 regular bivalents and one quadrivalent were identified, the latter resulting from a reciprocal translocation in heterozygosity. This rearrangement explains the heteromorphisms detected in the karyotype.

12.10 Polyploidy in *Tityus*

Endopolyploidy consists of a process in which DNA replication occurs without subsequent cell division, resulting in cells with an increased chromosome number in relation to the diploid standard. In Neotropical scorpions, endopolyploidy has been identified in some members of Buthidae, such as *I. guanambiensis* Leornaducci et al., 2005 (Ubinski et al. 2018), *T. paraguayensis* Kraepelin, 1895 (Mattos et al. 2014), *T. pusillus* Pocock, 1893 (Mattos et al. 2018), and *T. bahiensis* (Piza 1940).

In *T. paraguayensis*, whose diploid number is 2n = 16, most of the cells analyzed exhibited the expected formation of eight bivalents during meiosis. However, some cells exhibited aberrant meiotic configurations that did not correspond to the diploid karyotype of this scorpion; for example, nine meiotic cells presented 5 bivalents, three chains of 4 chromosomes, and one chain of 8 chromosomes (5II + 3CIV + CVIII), whose sum results in 2n = 30 (Mattos et al. 2014). Statistical analyses comparing the measurements of the total diploid set confirmed significant differences between diploid and polyploid cells, corroborating the hypothesis of endopolyploidy. This phenomenon may have been favored by cytogenetic characteristics of Buthidae, such as the presence of holocentric morphology, achiasmatic meiosis, and the ability to segregate the chromosomes involved in meiotic chains in an alternate manner (Mattos et al. 2014). Furthermore, the occurrence of only regular haploid numbers in metaphase II suggests that polyploid cells may have undergone degeneration throughout meiosis (Mattos et al. 2014).

12.11 Sex Determination Systems in Neotropical Scorpions

In scorpions, heteromorphic sex chromosomes have been observed in non-neotropical species (Srivastava and Agrawal 1961; Qumsiyeh et al. 2013, 2014), suggesting an XX/XY sex determination chromosome system (SCDS).

Among Neotropical species, one of the first reports of SCDS was proposed by Piza (1947) in *T. bahiensis*. Males of this species, collected in São Joaquim (SP, Brazil), presented 2n = 9 (with one multivalent CVII and one bivalent), while females exhibited 2n = 10. During meiosis, males produced II spermatocytes with n = 4 or n = 5, indicating male heterogamy. According to Piza (1948b), the fertilization of an n = 5 egg by an n = 5 sperm would produce females, while an n = 4 sperm would produce males.

Another case of SCDS, like described by Piza (1947) for *T. bahiensis*, was suggested for *T. confluens* (Adilardi et al. 2016). In this species, the diploid number presents intraspecific variation, with females consistently exhibiting 2n = 6, while males present 2n = 5 or 6. This divergence is associated with the multivalents found during meiosis I in males, which allowed the distinction of four different cytotypes (A, B, C, and D, see Table 12.1). Adilardi et al. (2016) proposed that the ancestral karyotype of *T. confluens* had a pair of morphologically indistinguishable (homomorphic) XY chromosomes. Subsequently, a fusion involving the Y chromosome and an autosome would have occurred, followed by additional fusions with other chromosome pairs, which involved the original Y in the multivalents observed in males, giving rise to structurally heteromorphic chromosomes. In contrast, females of *T. confluens* demonstrate a symmetrical and homomorphic karyotype.

In cytotypes A, B, and D (2n = 6) of *T. confluens*, males heterozygous for the fusions presented multivalents during meiosis, resulting in two types of spermatocytes with three chromosomes each: one containing the X chromosome (homomorphic) and the other containing the Y chromosome fused to an autosome. Thus, when spermatocytes carrying the X fertilized an egg with n = 3 homomorphic chromosomes, they generated females, while spermatocytes carrying the Y reconstituted the heteromorphic karyotypes, giving rise to males. In the case of cytotype C (2n = 5), males would generate two types of spermatocytes, one with two chromosomes (which would carry the X) and the other with three (which would carry the Y). In this case, fertilization by gametes containing the X would result in females with 2n = 5 and a heteromorphic karyotype; however, this karyotype was not identified in the sample analyzed, which led the authors to suggest that this genotype is extremely rare.

In *T. serrulatus* Lutz and Mello 1922 (parthenogenetic), females have a single chromosome carrying the NOR (Schneider and Cella 2010), probably due to a deletion. In sexual populations, females maintain this pattern, while males have two NOR chromosomes, forming a normal bivalent (Lima et al. 2020). This suggests a ZZ/ZW system, with females as the heterogametic sex.

12.12 Parthenogenesis

Most Neotropical parthenogenetic scorpions belong to the Buthidae family, with a predominance of thelytoky (exclusive production of females without fertilization). Cytogenetic studies in species such as *T. carrilloi* Ojanguren-Affilastro, 2021 and *T. serrulatus* revealed macrokaryotypic conservation between geographically distant populations (Schneider and Cella 2010; Adilardi et al. 2014). In *T. serrulatus*, the presence of a single NOR chromosome in females suggests development by apomixis (mitosis), since meiosis would produce eggs without NOR—which was not observed (Schneider and Cella 2010). Regarding *T. carrilloi*, the conservation of karyotype characteristics among populations more than 900 km apart was attributed to the expansion of an ancestral lineage (parthenogenetic or not) with the capacity to colonize urban centers, which carried the 2n = 6 karyotype, and which was rapidly dispersed in northeastern Argentina due to human actions (Adilardi et al. 2014).

Arrhenotoky (asexual production of males) was only recorded in *T. metuendus* (Iquitos, Peru). Populations from Amazonas and Acre (Brazil) present karyotypes typical of sexual reproduction, with no evidence of haplody in males (Piza 1950; Lima et al. 2023).

12.13 Applications of Karyotype in the Taxonomy of Neotropical Scorpions

Scorpion taxonomy has traditionally been based on morphological characters. However, in recent years, an integrative approach—combining morphological, molecular, and cytogenetic data—has become essential for the precise delimitation of taxa and the identification of cryptic species, often indistinguishable by external characteristics alone. The main cytogenetic markers used include: diploid number, chromosome morphology, heterochromatin distribution, location of NORs, and meiotic configurations. Alterations in these parameters can affect chromosome pairing and segregation during meiosis, reducing hybrid fertility and acting as a reproductive barrier between closely related species.

The case of *B. pentheri* Mello-Leitão, 1931 illustrates the usefulness of cytogenetics in the taxonomy of Bothriuridae. This species, distributed in central-western Argentina, presented two geographically structured morphological forms (North and South). Rodríguez-Gil et al. (2009) demonstrated that the northern form has 2n = 46, while the southern form has 2n = 42, suggesting that they are distinct species. This hypothesis was later confirmed by Ojanguren-Affilastro et al. (2016) through molecular analyses.

In Buthidae, karyotypic differences also helped in distinguishing species. In the genus *Tityopsis*, *T. rolandoi* Kovarik et al., 2024 (2n = 20) and *T. sheylae* Teruel and Rodríguez-Cabrera, 2020 (2n = 22) were easily differentiated based on chromosome number (Kovařík et al. 2024). Similarly, Mattos et al. (2018) suggested that populations of *T. silvestris* from Amazonas (2n = 24, configuring 24II) and Mato

Grosso (2n = 16, with 2II + CXII), Brazil, present cytogenetic divergences consistent with separation into distinct species. Another example is *T. curupi*, originally confused with *T. uruguayensis* in Argentina. Cytogenetic analyses revealed that *T. curupi* has 2n = 32 (16II or 14II + 1CIII), while *T. uruguayensis* has 2n = 31 (12II + VII or 10II + CIV + VII), in addition to differences in the distribution of rDNA sites (Ojanguren-Affilastro et al. 2017). The congruence between molecular, morphological, and cytogenetic data validated *T. curupi* as a distinct species.

In groups in which species are differentiated only by sexual characteristics or geographic distribution, cytogenetics offers robust markers that are independent of these factors. An example is found in *B. silvestris* and *B. paraensis* (Chactidae), which occur in sympatry in Pará (Brazil) and are morphologically similar, distinguished mainly by coloration and size—characters subject to intraspecific variation (Lourenço 2002). However, Almeida et al. (2023) demonstrated striking karyotypic differences: *B. silvestris* has 2n = 40 (28m + 6sm + 6a), while *B. paraensis* has 2n = 48 (14m + 4sm + 30a), in addition to divergences in the location of rDNA sites (terminal in pair 5 in *B. paraensis* vs. interstitial in pair 16 in *B. silvestris*). These differences reflect extensive chromosomal rearrangements, reinforcing the value of cytogenetics as a taxonomic tool in Chactidae.

12.14 Final Considerations and Perspectives

Scorpion cytogenetics is an area of study that is still expanding, but has seen significant progress in recent years. However, to date, only four families from the Neotropical region have had karyotypic data described, which highlights the need for greater efforts to advance this field of knowledge. Chromosome analyses have revealed remarkable diversity among the different species, not only in relation to the number and morphology of chromosomes, but also in relation to the mechanisms of sexual determination, reproductive systems, and genomic organization of repetitive DNA in these arachnids.

As demonstrated in this chapter, cytogenetic analyses in Neotropical scorpions have demonstrated the plasticity of meiotic mechanisms in this order, including the absence of chiasmata, telokinetic activity, and the capacity to support extensive multivalents originated by rearrangements in the heterozygous state. In the future, new insights into the study of meiosis may contribute to understanding the origin and maintenance of chromosomal systems in Scorpiones.

Cytogenetic studies have played a fundamental role in elucidating taxonomic and evolutionary aspects of Neotropical scorpions. The examples cited in this chapter have shown that cytogenetics remains a powerful and complementary tool to molecular and morphological approaches, providing useful cytological markers for the classification and phylogeny of the group. It is hoped that the data reviewed in this chapter will encourage future research and species descriptions that integrate cytogenetic data with genomic, morphological, and ecological information, in order to further deepen our knowledge of the diversity and evolution of Neotropical scorpions.

References

Adilardi RS, Ojanguren-affilastro AA, Rodríguez Gil SG, Scioscia CL, Mola LM (2013) Meiotic studies in *Brachistosternus alienus* (Scorpiones; Bothriuridae). J Arachnol 41:222–226

Adilardi RS, Affilastro AO, Martí DA, Mola LM (2014) Cytogenetic analysis on geographically distant parthenogenetic populations of *Tityus trivitattus* Kraepelin, 1898 (Scorpiones, Buthidae): karyotype, constitutive heterochromatin and rDNA localization. Comp Cytogenet 8:81–92. https://doi.org/10.3897/compcytogen.v8i2.6461

Adilardi RS, Ojanguren-Affilastro AA, Mattoni CI, Mola LM (2015) Male and female meiosis in the mountain scorpion *Zabius fuscus* (Scorpiones, Buthidae): heterochromatin, rDNA and TTAGG telomeric repeats. Genetica 143:393–401. https://doi.org/10.1007/s10709-015-9838-1

Adilardi RS, Ojanguren-Affilastro AA, Mola LM (2016) Sex-linked chromosome heterozigosity in males of *Tityus confluens* (Buthidae): a clue about the presence of sex chromosomes in scorpions. PLoS One 11:1–18. https://doi.org/10.1371/journal.pone.0164427

Adilardi RS, Ojanguren-Affilastro AA, Martí DA, Mola LM (2020) Chromosome puzzle in the southernmost populations of the medically important scorpion *Tityus bahiensis* (Perty 1833) (Buthidae), a polymorphic species with striking structural rearrangements. Zool Anz 288:139–150. https://doi.org/10.1016/j.jcz.2020.08.001

Almeida BRR, Milhomem-Paixão SSR, Noronha RCR, Nagamachi CY, Costa MJR, Pardal PPO, Coelho JS, Pieczarka JC (2017) Karyotype diversity and chromosomal organization of repetitive DNA in *Tityus obscurus* (Scorpiones, Buthidae). BMC Genet 18:35. https://doi.org/10.1186/s12863-017-0494-6

Almeida BRR, Noronha RCR, Costa MRJ, Nagamachi CY, Pieczarka JC (2019) Meiosis in scorpion *Tityus silvestris*: new insigths into achiasmatic chromosome. Biol Open 8:bio040352. https://doi.org/10.1242/bio.040352

Almeida BRR, Noronha RCR, Cardoso AL, Martins C, Martins JG, Procópio REL, Nagamachi CY, Pieczarka JC (2022) Kinetic activity of chromosomes and expression of recombination genes in achiasmatic meiosis of *Tityus* (*Archaeotityus*) scorpions. Int J Mol Sci 23:9179. https://doi.org/10.3390/ijms23169179

Almeida BRR, Malcher S, Costa M, Martins J, Procópio R, Noronha R, Nagamachi C, Pieczarka J (2023) High chromosomal reorganization and presence of microchromosomes in Chactidae scorpions from the Brazilian Amazon. Biology 12:1–14. https://doi.org/10.3390/biology12040563

Benavente R (1982) Holocentric chromosomes of arachnids: presence of kinetochore plates during meiotic divisions. Genetica 59:23–27

Brieger FG, Graner EA (1943) On the cytology of *Tityus bahiensis* with special reference to meiotic prophase. Genetics 28:269–274

Cunha AB, Pavan C (1954) Duas novas configurações cromosômicas em *Tityus bahiensis* (Scorpiones—Buthidae). Ciên Cult 6:18–20

Ferreira A (1968) Contribution to the knowledge of cytology of two species of Brazilian scorpions: *Opisthacantus manauarensis*, Ferreira, 1967 (Scorpiones, Scorpionidae) and *Bothriurus asper araguaie* (Scorpiones, Bothriuridae). An Acad Bras Cienc 40:97–99

Giacomozzi RO (1977) Citotaxonomía de Bothriuridae (Scorpionida) de la Argentina. Physis (Buenos Aires) 37:236

Kovařík F, Stahlavský F, Korínková T, Král J, Ende T (2009) *Tityus ythieri* Lourenço, 2007 is a synonym of *Tityus magnimanus* Pocock, 1897 (Scorpiones: Buthidae): a combined approach using morphology, hybridization experiments, chromosomes, and mitochondrial DNA. Euscorpius 77:1–12

Kovařík F, Stockmann M, Šťáhlavský F, Yong S (2024) *Tityopsis rolandoi* sp. n. (Scorpiones: Buthidae) from Cuba. Euscorpius 400:1–16

Lima JF, Carvalho LS, Schneider MC (2020) The first chromosomal analysis of bisexual populations of the Brazilian scorpion *Tityus serrulatus* (Scorpiones: Buthidae). J Arachnol 48:77–83

Lima JF, Carvalho LS, Carvalho MA, Schneider MC (2023) Chromosome diversity in Buthidae and Chactidae scorpions from Brazilian fauna: diploid number and distribution of repetitive DNA sequences. Genet Mol Biol 46:e20220083. https://doi.org/10.1590/1678-4685-GMB-2022-0083

Lisachov A, Dedukh D, Simanovsky S, Panthum T, Singchat W, Srikulnath K (2024) Spaghetti connections: synaptonemal complexes as a tool to explore chromosome structure, evolution, and meiotic behavior in fish. Cytogenet Genome Res 164:1–15. https://doi.org/10.1159/000538238

Lourenço WR (2002) Scorpiones. In: Adis J (ed) Amazonian Arachnida and Miryapoda: identification keys to all classes, orders, families, some genera an list of know terrestrial species. Pensoft Publishes, Moscow, pp 399–438

Lourenço WR (2015) What do we know about some of the most conspicuous scorpion species of the genus *Tityus*? A historical approach. J Venom Anim Toxins Incl Trop Dis 21:1–20. https://doi.org/10.1186/s40409-015-0016-9

Mandrioli M, Manicardi GC (2020) Holocentric chromosomes. PLoS Genet 16:e1008918. https://doi.org/10.1371/journal.pgen.1008918

Marques A, Pedrosa-Harand (2016) Holocentromere identity: from the typical mitotic linear structure to the great plasticity of meiotic holocentromeres. Chromosoma 125:669–681. https://doi.org/10.1007/s00412-016-0612-7

Mattos VF, Cella DM, Carvalho LS, Candido DM, Schneider MC (2013) High chromosome variability and the presence of multivalent association in buthid scorpions. Chromosom Res 21:121–136. https://doi.org/10.1007/s10577-013-9342-3

Mattos VF, Carvalho LS, Cella DM, Schneider MC (2014) Location of 45S ribosomal genes in mitotic and meiotic chromosomes of Buthid scorpions. Zool Sci 31:603–607. https://doi.org/10.2108/zs140005

Mattos VF, Carvalho LS, Carvalho MA, Schneider MC (2018) Insights into the origin of the high variability of multivalent-meiotic associations in holocentric chromosomes of *Tityus* (*Archaeotityus*) scorpions. PLoS One 13:e0192070. https://doi.org/10.1371/journal.pone.0192070

Mola LM, Regagliati PJ, Rodríguez-Gil SG, Adilardi R (2011) Variaciones meióticas y evolución cromosómica en insectos y arácnidos con cromosomas holocinéticos. J Basic Appl Genet 22:1–6

Ojanguren-Affilastro AA, Mattoni CI, Ochoa JA, Ramírez MJ, Ceccarelli FS, Prendini L (2016) Phylogeny, species delimitation and convergence in the South American bothriurid scorpion genus *Brachistosternus* Pocock 1893: integrating morphology, nuclear and mitochondrial DNA. Mol Phylogenet Evol 94:159–170. https://doi.org/10.1016/j.ympev.2015.08.007

Ojanguren-Affilastro AA, Adilardi RS, Cajade R, Ramírez MJ, Ceccarelli FS, Mola LM (2017) Multiple approaches to understanding the taxonomic status of an enigmatic new scorpion species of the genus *Tityus* (Buthidae) from the biogeographic Island of Paraje Tres Cerros (Argentina). PLoS One 12:e0181337. https://doi.org/10.1371/journal.pone.0181337

Piza ST (1939a) Comportamento dos cromossomos na primeira divisão do espermatócito do *Tityus bahiensis*. Sci Genet 1:255–261

Piza ST (1939b) Considerações em torno da meiose do *Tityus bahiensis* (Scorpiones-Buthidae) e uma nova teoria sobre a movimentação dos cromossômios. J Agron 2:343–370

Piza ST (1940) Poliploidia natural em *Tityus bahiensis* (Scorpiones) associada a aberrações cromossômicas espontâneas. Rev Biol Hyg 10:143–155

Piza ST (1943a) Meiosis in the male of the Brazilian scorpion *Tityus bahiensis*. Rev Agric 18:249–276

Piza ST (1943b) A propósito da meiose de *Tityus bahiensis*. Rev Agric 18:351–369

Piza ST (1944) A case of spontaneous end-to-end permanent union of two non-homologous chromosomes in the Brazilian scorpion *Tityus bahiensis* accompanied by irregularities in pairing. Rev Agric 19:133–147

Piza ST (1947a) Notas sobre cromossômios de alguns escorpiões brasileiros. An Esc Super Agric "Luiz de Queiroz" 62:169–176

Piza ST (1947b) Interessante comportamento dos cromossômios na espermatogênese do escorpião *Isometrus maculatus* De Geer. An Esc Super Agric "Luiz de Queiroz" 62:177–182
Piza ST (1947c) Uma raça cromossômica natural de *Tityus bahiensis* (Scorpiones-Buthidae). An Esc Super Agric "Luiz de Queiroz" 62:183–192
Piza ST (1948a) Primeiras observações sobre os cromossômios de *Tityus trivittatus* Krpln (Scorpiones—Buthidae). Rev Agric São Paulo 24:177–180
Piza ST (1948b) Uma nova raça cromossômica natural de *Tityus bahiensis* (Scorpiones—Buthidae). Rev Agric São Paulo 24:181–186
Piza ST (1948c) Variações cromossômicas do *Tityus bahiensis* de São Joaquim. Rev Agric São Paulo 24:187–194
Piza ST (1949) "Ouro Preto", nova e interessante raça cromossômica de *Tityus bahiensis* (Scorpiones—Buthidae). Sci Genet 3:147–159
Piza ST (1950a) Variações cromossômicas do *Tityus bahiensis* de Ribeirão Preto. Ciên Cult 2:57–59
Piza ST (1950b) Observações cromossômicas em escorpiões brasileiros. Ciên Cult 2:202–206
Piza ST (1950c) Primeiras observações sôbre os cromossômios do *Tityus metuendus* Pocock. Sci Genet 4:162–167
Piza ST (1955) Notable geographical expansion of a peculiar type of chromosomal constitution of the cosmopolitan scorpion *Isometrus maculatus*. Arq Mus Nac 42:611–612
Piza ST (1957) The chromosomes of *Rhopalurus* (Scorpiones—Buthidae). Can Entomol 89:565–568
Qumsiyeh MB, Ina S, Salsaa M, Amr ZS (2013) Records of scorpions from of Palestinian Territories, with the first chromosomal data (Arachnida: Scorpiones). Zool Middle East 59:70–76
Qumsiyeh MB, Amr ZS, Srour KTA, Al-Fawaghra N (2014) Karyotype for *Nebo hierichonticus* (Simon 1872) from the Palestinian Territories (Scorpiones: Scorpionidae). Cytologia 79:277–280. https://doi.org/10.1508/cytologia.79.277
Rodríguez-Gil SG, Ojanguren-Affilastro AA, Barral LM, Scioscia CL, Mola LM (2009) Cytogenetic of three species of the genus *Brachistosternus* from Argentina (Scorpiones: Bothriuridae). J Arachnol 37:331–337
Schneider MC, Cella DM (2010) Karyotype conservation in 2 populations of the parthenogenetic scorpion *Tityus serrulatus* (Buthidae): rDNA and its associated heterochromatin are concentrated on only one chromosome. J Hered 101:491–496. https://doi.org/10.1093/jhered/esq004
Schneider MC, Zacaro AA, Pinto-da-Rocha R, Candido DM, Cella DM (2009a) A comparative cytogenetic analysis of two Bothriuridae species and overview of the chromosome data of scorpion. J Hered 100:545–555. https://doi.org/10.1093/jhered/esp023
Schneider MC, Zacaro AA, Pinto-da-Rocha R, Candido DM, Cella DM (2009b) Complex meiotic configuration of the holocentric chromosomes: the intriguing case of the scorpion *Tityus bahiensis*. Chromosom Res 17:883–898. https://doi.org/10.1007/s10577-009-9076-4
Schneider MC, Mattos VF, Carvalho LS, Cella DM (2015) Organization and behavior of the synaptonemal complex during achiasmatic meiosis of four buthid scorpions. Cytogenet Genome Res 144:341–347. https://doi.org/10.1159/000375388
Schneider MC, Mattos VF, Cella DM (2024) The scorpion cytogenetic database. www.arthropodacytogenetics.bio.br/scorpiondatabase. Accessed 19 Dec 2024
Shanahan CM, Hayman DL (1990) Synaptonemal complex formation in male scorpions exhibiting achiasmate meiosis and structural heterozygosity. Genome 33:914–926
Srivastava MDL, Agrawal U (1961) Absence of chiasmata and formation of a complex chromosomal body in the spermatogenesis of the scorpion *Palamnaeus longimanus*. Caryologia 14:63–77
Šťáhlavský F, Nguyen P, Sadílek D, Štundlová J, Just P, Haddad CR, Koc H, Ranawana KB, Stockmann M, Yagmur EA, Kovařík F (2020) Evolutionary dynamics of rDNA clusters on chromosomes of buthid scorpions (Chelicerata: Arachnida). Biol J Linn Soc 131:547–565. https://doi.org/10.1093/biolinnean/blaa118

Takahashi CS (1976) Cytogenetical studies on the effects of high natural radiation levels in *Tityus bahiensis* (Scorpiones, Buthidae) from Morro do Ferro, Brazil. Radiat Res 67:371–381. https://doi.org/10.2307/3574425

Ubinski CV, Carvalho LS, Schneider MC (2018) Mechanisms of karyotype evolution in the Brazilian scorpions of the subfamily Centruroidinae (Buthidae). Genetica 146:475–486. https://doi.org/10.1007/s10709-018-0038-7

Wilson EB (1931) The distribution of sperm-forming materials in scorpions. J Morphol 52:429–483. https://doi.org/10.1002/jmor.1050520205

Zedek F, Bureš P (2017) Holocentric chromosomes: from tolerance to fragmentation to colonization of the land. Ann Bot 121:9–16. https://doi.org/10.1093/aob/mcx118

Chapter 13
Neotropical Scorpions as Useful Models for Ecological Studies

André Felipe de Araujo Lira

Abstract Scorpions are an ancient group of invertebrates that have successfully colonized practically all terrestrial ecosystems. Their assemblage is typically composed of a few dominant species and habitat specialists. In addition, as generalist predators, they play an important role in the energy flow of their ecosystems by regulating their prey populations, including other scorpion species. Moreover, they are considered an important food resource for other predators, highlighting their integral role in food webs. However, due to their considered low species richness (generally 5–6 species per assemblage), these arachnids have often been overlooked in ecological studies in favor of more species-rich groups such as spiders and mites. Despite this, scorpions are responsive to changes in their habitat at both local and landscape scales, showing sensitivity to environmental variables that influence their population dynamics and assemblage structure. The objective of this chapter is to provide information that supports the use of scorpions as models for ecological studies. Facing a global biodiversity crisis, it is essential to incorporate species-poor organisms to improve our understanding of ecological patterns and processes. Thus, in this chapter, I will address population and assemblage aspects of scorpions with examples of studies where these animals were used as models, focusing on the Neotropical region, where unique ecological dynamics offer valuable insights for conservation and management.

Keywords Conservation • Anthropogenic impacts • Arachnids • Macroecology

---

A. F. de Araujo Lira (✉)
Colección Nacional de Arácnidos, Departamento de Zoología, Instituto de Biología, Universidad Nacional Autónoma de México, Coyoacán, Mexico City, Mexico
e-mail: andre.lira@st.ib.unam.mx

A. F. de Araujo Lira, E. González-Santillán (eds.), *The Biology of Neotropical Scorpions*, https://doi.org/10.1007/978-3-032-23194-9_13

13.1 Introduction

Ecology has been central to environmental conservation by deepening our understanding of the processes that underpin ecosystems and biodiversity. While conservation efforts primarily targeted isolated species or specific protected areas (Saout et al. 2013; Venter et al. 2014; Gillingham et al. 2015), advances in ecology have shifted this focus toward ecological processes and the interconnectedness between organisms and their surroundings, fostering more comprehensive and effective conservation strategies (Wallington et al. 2005; Bennet et al. 2009; Harvey et al. 2016). For example, ecological insights have been instrumental in understanding species' responses to environmental changes at both local and landscape scales (e.g., Lira et al. 2022; Oliveira et al. 2025; Villanueva-Bonilla et al. 2025), which is crucial for designing protected areas that maintain ecological connectivity and enhance ecosystem resilience to human disturbances (Hansen and DeFries 2007; Cumming 2016; Belote et al. 2017). Studying population, community, and functional patterns enables conservation approaches that account for diversity, dynamics, and threats across appropriate biological scales.

Many ecologists tend to focus their research on communities with high species richness (e.g., Tourinho et al. 2020; Almeida et al. 2023; Cooper et al. 2024), as these diverse groups offer greater opportunities for exploration of complex interspecific interactions. Conversely, species-poor groups are often overlooked, frequently perceived as less representative or insufficiently variable for certain studies (Landeiro et al. 2018; Guénard et al. 2025). This bias, however, risks underestimating the ecological and functional importance of low-diversity organisms and underscores the need to broaden ecological research to include these ecologically relevant but less species-rich groups.

Moreover, taxa traditionally regarded as having low species richness may serve as valuable bioindicators (e.g., Landeiro et al. 2018; Porto et al. 2018; Lira et al. 2021a, b). Effective bioindicator taxa must meet several criteria: ecological sensitivity and fidelity to environmental changes; stable and well-established taxonomy to ensure accurate identification; comprehensive knowledge of their biology and natural history for proper interpretation of their responses; broad geographic distributions and sufficient abundance for representative sampling; and ease of observation and collection in the field and laboratory (Gerlach et al. 2013). Additionally, ideal bioindicators typically play important ecological roles by correlating with other ecosystem processes and components (Gerlach et al. 2013; Puig-Gironès and Real 2022). Therefore, groups with few species can still provide crucial information for monitoring and understanding environmental health.

In the context of the ongoing global biodiversity crisis (Gerber and Iacona 2024; Saunders et al. 2025), incorporating a diverse range of taxa in conservation efforts is essential. Using multiple taxa as bioindicators or ecological models broadens and deepens our understanding of ecosystems, as different groups exhibit unique ecological functions, sensitivities, and responses to environmental changes (Toranza and Arim 2010; Nagy et al. 2018). This multifaceted approach improves detection of environmental impacts that might be missed if research is restricted to more species-rich or well-studied taxa. Furthermore, precise taxonomy and identification are vital for setting conservation priorities, ensuring effective protection of species and habitats, and developing evidence-based environmental policies (Mace 2004; Sandall et al. 2023). Amid accelerating species loss, expanding the conservation focus to include both taxonomically diverse and species-poor taxa is critical to preserving ecosystem integrity and fostering more resilient management strategies. Thus, the integrated application of varied taxa enhances conservation outcomes and supports efforts to address the complex challenges posed by the global biodiversity crisis.

13.2 Population Dynamics

Ecological studies involving scorpions have been facilitated through the discovery of fluorescence under ultraviolet light (see the Chap. 4), facilitating field observations to be conducted (see the Chap. 11). Despite this, advances in the ecological knowledge and natural history of these animals are still incipient, especially concerning the Neotropical species. Therefore, the text below will address the characteristics of scorpions that make them useful models for ecological studies, without providing an exhaustive review but a comprehensive presentation of the main topics.

13.2.1 Trophic Interactions

Scorpions are considered generalist predators (Fig. 13.1), consuming a wide variety of invertebrates, and even small vertebrates (e.g., Cubas-Rodríguez and Teruel 2022; Silva-Júnior et al. 2023; Cubas-Rodríguez et al. 2024; Dionisio-da-Silva et al. 2024). In a study conducted across three Brazilian biomes (Caatinga, Cerrado, and Atlantic Forest), which compiled occasional field records over 14 years, 19 groups of arthropods were reported as scorpion prey (Dionisio-da-Silva et al. 2024). According to these authors, the observed diet composition of the scorpions was mainly based on cockroaches, crickets, grasshoppers, spiders, and other scorpions, highlighting the generalist diet and cannibalism of scorpions, with many antagonistic events among the records of intraguild predation. However, it is worth noting that some scorpion species seem to show preferences for certain types of prey. For example, *Tityus uruguayensis* Borelli, 1901 demonstrated a preference when spiders were offered as a food resource over insects (Toscano-Gadea and Costa 2006). Similar results were found for other scorpion species, such as *Tityus brazilae* Lourenço and Eickstedt, 1984, which shows a preference for crickets (Lira, unpub. data), and *Diplocentrus* spp., which prefers adult beetles (Jimena Cid-Uribe, pers. comm.).

Fig. 13.1 Predation events involving Neotropical scorpion species. (**a**) *Jaguajir rochae* cannibalizing a juvenile, (**b**) *Tityus pusillus* cannibalizing a juvenile, (**c**) *Jaguajir rochae* predating a *Sicarius* sp. Spider, and (**d**) *Tityus pusillus* predating a cockroach. (Photos (**a**) and (**c**) by Gabriel Celante, (**b**) by Leticia Lins, and (**d**) Welton Dionisio)

Regarding cannibalism and intraguild predation, these appear to be important regulatory interactions for the population of these animals. Several records have been published reporting one of these interactions (cannibalism or intraguild predation) for Neotropical species (e.g., Moreno-Gonzalez and Hazzi 2012; Lira et al. 2017a, b; Silva-Júnior et al. 2021; Martins et al. 2024). In general, body size seems to play an important role in defining predator-prey relationships. Larger individuals act as predators of smaller ones (e.g., Moreno-Gonzalez and Hazzi 2012; Silva-Júnior et al. 2021). Although uncommon, predation between individuals of similar size can occur (e.g., Martins et al. 2024; Melo-dos-Santos et al. 2025). Another important factor for the occurrence of these interactions is environmental complexity. In a laboratory study, higher rates of both cannibalism and intraguild predation were reported in arenas with lower environmental complexity compared to arenas with higher complexity (Moreira et al. 2022). According to the authors, more complex areas limit encounters between animals and thus reduce the risk of antagonistic interactions among them. In addition, scorpions are capable of detecting chemical trails left on the substrate by other scorpions and can alter their behavior to either avoid or facilitate encounters (Dionisio-da-Silva et al. 2019; Feitosa et al. 2022; Dionisio-da-Silva et al. 2025). Therefore, body size and habitat complexity acts as a key factor in intra- and interspecific interactions among scorpions.

13.2.2 Spatial Distribution

Given their predatory nature and intolerance toward other individuals, scorpions do not exhibit a random distribution within their habitat (Fig. 13.2). Many species show a predictable pattern and are found in specific microhabitats (Álvarez et al. 2013; Lira and DeSouza 2014; Lira et al. 2018a; Lapinski 2025). For example, in a study investigating microhabitat preferences in a Brazilian mountain forest, three scorpion species were recorded, each occupying a distinct microhabitat without overlap (Lira and DeSouza 2014). Similar results were reported for scorpion species found in the Brazilian Atlantic Forest (Lira et al. 2018a). However, this pattern appears to be flexible and dependent on the ecosystem. In dry forest areas, there is

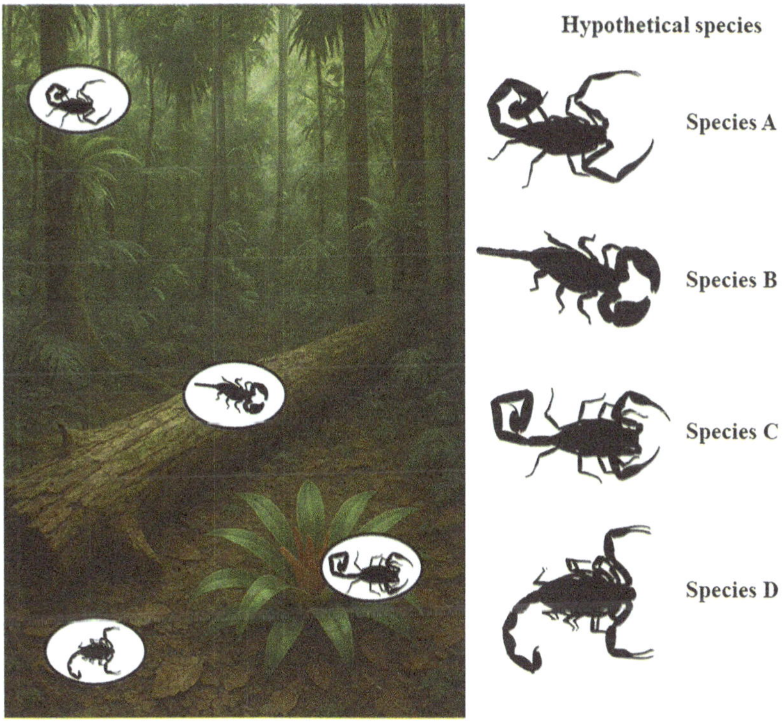

Fig. 13.2 Schematic illustration of the distribution of different hypothetical scorpion species in the habitat

greater niche overlap among species (Álvarez et al. 2013; Lira et al. 2018a). In dry forests of Colombia, four species were reported, three of which occupied microhabitats associated with the forest floor (Álvarez et al. 2013). Nevertheless, despite this apparent niche overlap, species employ mechanisms to avoid encounters with others and thus reduce antagonistic interactions. For example, both *Tityus pusillus* Pocock, 1893 and *Ananteris mauryi* Lourenço, 1982 are considered leaf litter-dwelling species; however, they occupy this substrate differently, with *T. pusillus* found in the upper layers and *A. mauryi* in the lower layers (Lira et al. 2013). It has also been reported that the arboreal species *Centruroides flavopictus* (Pocock, 1898) and *C. catemacoensis* Goodman, Prendini, Francke and Esposito, 2021 use vegetation differently, with the former being associated with *Astrocaryum mexicanum* palms and the latter found at elevated heights (higher than 4 m) (Goodman and Esposito 2020). Similar results were described in Costa Rica, where *Centruroides limbatus* (Pocock, 1898), *Tityus ocelote* Francke and Stockwell, 1987, and *Tityus pachyurus* Pocock, 1897 used nearly the entire available height range, with *T. ocelote* occurring significantly higher than the other two species (Lapinski 2025).

Aspects such as developmental stage and seasonality can also influence the distribution of scorpions in their microhabitat. Juvenile individuals avoid overlapping the same foraging sites used by adults in order to prevent being cannibalized (Lira et al. 2013). For example, juveniles of *Centruroides balsasensis* Ponce and Francke, 2004, *T. pusillus,* and *Brachistosternus ferrugineus* (Thorell, 1876) forage in the vegetation, while adults are commonly found in the leaf litter (Ponce-Saavedra et al. 2006; Lira et al. 2013, 2018a; Nime et al. 2016). Seasonality is also an important factor affecting the distribution patterns of scorpions in their microhabitat (Lira et al. 2013, 2018a; Santos et al. 2018). Lira et al. (2018a) investigated the effect of seasonality on 11 species belonging to the families Buthidae and Bothriuridae across two distinct ecosystems: Atlantic Forest (rainforest) and Caatinga (dry forest). According to the authors, changes in microhabitat use were more pronounced in the Caatinga assemblages than in those of the Atlantic Forest, suggesting that in markedly seasonal environments the differences in microhabitat use are more evident. Seasonality may influence not only microhabitat use but also affect spatial distribution patterns (Santos et al. 2018). However, this does not seem to be a general pattern for all species; for example, *T. pusillus* showed aggregated and random patterns during the dry and rainy seasons, respectively (Santos et al. 2018), whereas the species *Hoffmanihadrurus gertschi* (Soleglad, 1976) and *A. mauryi* exhibited aggregated patterns regardless of the season (Quijano-Ravell et al. 2012; Santos et al. 2018).

13.2.3 Activity Patterns

Overall, scorpions exhibit foraging activity patterns related to climatic variables such as temperature and precipitation. In bothriurids *Thestylus aurantiurus* Yamaguti and Pinto-da-Rocha, 2003 and *Bothriurus signatus* Pocock, 1893, foraging activity peaks during the months with more rainfall and temperature (spring and summer) (Yamaguti and Pinto-da-Rocha 2006; Ott and Ott 2014). Similar results were reported for buthids *T. pusillus* and *Microtityus jaumei* Armas, 1974, with increased foraging activity during summer in Brazil and Cuba, respectively (Cala-Riquelme and Colombo 2011; Dionisio-da-Silva et al. 2018). However, in strongly seasonal environments such as dry forests, foraging activity seems to be correlated with increased precipitation. Araújo et al. (2010) reported a peak in foraging activity of *Bothriurus asper* Pocock, 1893 and *Bothriurus rochai* Mello-Leitão, 1932 during the rainy months in a Caatinga area. This peak is generally associated with the reproductive period when males increase foraging activity to find females (Dionisio-da-Silva et al. 2018).

However, this pattern can vary due to local climatic conditions. For example, Dionisio-da-Silva et al. (2018) reported that the reproductive period of *T. pusillus* occurs between October and January in a Brazilian Atlantic forest. However, Moreira et al. (2024) studied a population 30 km away and found the reproductive period to occur between February and March. Variations in foraging activity periods have also been reported for the scorpion *A. mauryi*. According to Dias et al. (2006), *A. mauryi* showed greater activity during the dry season. However, later studies involving this species found higher foraging activity during the rainy season (Dionisio-da-Silva et al. 2018). Another important point to consider is that some species exhibit an increased foraging activity during the opposite period. For example, in a mountainous system on Pampean grasslands in central Argentina, the species *Bothriurus prospicuus* Mello-Leitão, 1934, *Bothriurus bonariensis* (C. L. Koch, 1842), and *Bothriurus flavidus* Kraepelin, 1911 have an activity peak between December and January, whereas *Urophonius mahuidensis* Mauryi, 1973 and *Urophonius iheringi* Pocock, 1893 are more active in June (Schwerdt et al. 2016). In the Atlantic Forest and Caatinga, the peak foraging activity occurs during the dry and rainy seasons, respectively (Araújo et al. 2010; Lira et al. 2018a). However, for the species *B. asper* and *Tityus stigmurus* (Thorell, 1876) in the Atlantic Forest, and for *Physoctonus debilis* (C. L. Koch, 1840) in the Caatinga, the peak foraging activity occurred in the rainy and dry seasons, respectively (Lira et al. 2018a). This differentiation in foraging activity suggests that species are avoiding being active during the same period, thereby reducing the risk of intraguild predation (Dionisio-da-Silva et al. 2018; Lira et al. 2018a). This indicates that the scorpion assemblage, in addition to spatial niche stratification, also exhibits temporal niche differentiation.

13.3 Scorpions as Model for Conservation Studies

The aspects addressed—ranging from trophic interactions and spatial patterns to activity behaviors linked to seasonality—demonstrate that scorpions exhibit a complex population dynamic influenced by multiple environmental and biological factors. This complexity, combined with their distribution across diverse Neotropical ecosystems and their intra- and interspecific interactions, reinforces the potential of scorpions as useful models for ecological studies. In the following section, concrete examples of research using scorpions as experimental and field systems will be presented to illustrate the applicability of these characteristics in ecological investigations focused on conservation.

13.3.1 Habitat Fragmentation

Habitat fragmentation, defined as the breaking up of large, continuous forests into smaller, isolated patches, is recognized as a principal driver of biodiversity loss globally (Fahrig 2003; Yuan et al. 2024). The process reduces overall habitat area, increases isolation, and exposes species to edge effects, consistently leading to reductions in species richness, abundance, and long-term ecosystem stability (Pinto et al. 2024; Sgarlata et al. 2025). These impacts are especially pronounced in the smallest and most isolated fragments and can worsen over time as ecological processes lag, causing phenomena such as "extinction debt" and changes in ecosystem functioning (Vellend et al. 2006; Pinto et al. 2024).

In Neotropical forests, studies directly addressing scorpion assemblages remain recent but have provided clear insights (Lira et al. 2015, 2016, 2021a, b). Specifically for *T. pusillus* in the Atlantic Forest, fragmentation has demonstrated sex-specific effects: while male body size remains unchanged, females tend to be smaller in more fragmented or lower-quality habitats (Lira et al. 2021a, b). This is likely linked to female *T. pusillus* exhibiting more cryptic behaviors and being more sensitive to reductions in habitat quality, such as decreased litter mass, which is critical for foraging and shelter. Research comparing *A. mauryi* and *T. pusillus*, both leaf litter dwellers, shows distinct responses to fragmentation (Lira et al. 2015). *Tityus pusillus* abundance decreases sharply in fragments with less dry litter mass, whereas *A. mauryi* appears more resilient and unaffected by this variable. Furthermore, at the assemblage level, habitat fragmentation in the Atlantic Forest results in reduced scorpion species richness and abundance (Lira et al. 2016). Smaller fragments support fewer species, confirming that fragment area is directly related to species richness.

Overall, these findings reinforce that forest fragmentation not only threatens scorpion diversity directly but also alters population structures and species interactions in ways that may reduce ecosystem resilience. Maintaining or restoring larger, connected forest patches is critical for conserving these specialized arthropods and the ecological roles they fulfill.

13.3.2 Habitat Restoration

In response to disturbances in the original forest vegetation, the structure of secondary forests is markedly different from that of pristine primary forests (Guariguata and Ostertag 2001). Since secondary forests regenerate mainly through natural processes after disturbances, they are generally unstable and represent successional stages (Lamb 2011; Rosenfield et al. 2023). Thus, various levels of secondary forest regeneration develop, each with its own characteristic biodiversity (Rosenfield et al. 2023).

Studies addressing the response of Neotropical scorpions to these conditions are still scarce (Nime et al. 2014; Lira et al. 2018b). Nime et al. (2014) compared scorpion diversity between a mature forest and a 15-year-old secondary forest undergoing restoration in the Argentine Arid Chaco. These authors found that species composition in the secondary forest resembles that of the primary forest after approximately 15 years, but the relative abundances of these species differ. In a study conducted in the Brazilian Atlantic Forest comparing scorpion diversity between primary vegetation and vegetation undergoing 65 years of recovery, no significant differences in species richness or abundance were detected (Lira et al. 2018b).

These studies show that secondary forests, despite structural differences from primary forests, can support scorpion assemblages with species compositions becoming similar over time. Some differences in relative abundance persist, reflecting ongoing succession. In long-term recovering sites, richness and abundance may match primary forests, indicating scorpions' resilience and adaptability. These findings highlight the importance of secondary forests for scorpion conservation and the need for further research on their population dynamics in regenerating habitats.

13.3.3 Urban Ecology

The rapid expansion of urban areas and the associated increase in population density have had a significant impact on local biodiversity (Sushinsky et al. 2012; Simkin et al. 2022). The consequences of rapid urbanization, including deforestation, fragmentation, and loss of connectivity, have resulted in many native species inhabiting areas in close proximity to or even within urban environments (McKinney 2002; McDonald et al. 2008). It is therefore widely acknowledged that the conversion of native forests into city cores represents a major factor in ecosystem disturbance (McKinney 2006; Tee et al. 2018).

Studies focused on scorpions in urban areas have traditionally been linked to public health concerns (e.g., Zanetta et al. 2020; Benzaken et al. 2024; Pucca et al. 2025). However, more recently, research has emerged analyzing the effects of urbanization on scorpion assemblages (Feitosa et al. 2024, 2025a, b). Feitosa et al. (2024) examined the role of urban forests in maintaining scorpion assemblages in an Atlantic Forest region. They found that landscape metrics related to vegetation were important for species composition, with forest-dwelling species occurring in areas of greater forest cover. Similarly, areas with higher forest cover harbored individuals in better body condition. In a subsequent study, forest cover was a key determinant of scorpion functional body trait sizes (Feitosa et al. 2025a). Areas with lower forest cover harbored species with larger functional body traits compared to areas with higher forest cover, which showed species with smaller functional body traits. In an urban area on Neotropical submontane forest, forest habitats exhibited greater abundance and species richness and were sensitive to surrounding land use (Feitosa et al. 2025b). Furthermore, forest habitats are important for maintaining better body condition, while city core areas promote the emergence of traits adapted to urban conditions (Feitosa et al. 2025b).

These studies highlight the critical role of urban forest remnants in sustaining diverse and healthy scorpion assemblages amid expanding urbanization. While native forest cover supports species richness, favorable body conditions, and smaller functional traits, urban core areas exert selective pressures that favor species adapted to highly modified environments. Understanding these dynamics is essential for informing urban biodiversity conservation strategies and mitigating the ecological impacts of urban expansion on scorpion assemblages.

13.3.4 Land Use

The transformation of forests into unnatural matrices such as croplands or pastures is a major driver of habitat disturbance in native ecosystems (Theobald et al. 2025). Depending on the type of natural ecosystem, the forest-matrix transition can affect the ability of native species to move across disturbed landscapes (Kruger et al. 2022). Consequently, in human-altered landscapes, species sensitive to habitat changes are often replaced by less-sensitive species that can successfully occupy modified sites (Kromer et al. 2025).

For example, in Neotropical forests and their adjacent non-natural matrices (planted forests or croplands) in Brazil and Mexico, differences in species assemblages between environments have been documented (Lira et al. 2019). These authors reported higher diversity and abundance in undisturbed habitats. Similarly, in three Caatinga vegetation areas with varying degradation levels, more complex and preserved areas supported higher species richness and abundance (Foerster et al. 2020). Lira et al. (2021a, b), evaluating scorpions as bioindicators of human disturbance in the Atlantic Forest and Caatinga, found that species composition and abundance were more sensitive to landscape configuration changes than species richness. Specifically, they highlighted that: (i) the Atlantic Forest assemblage exhibits an abundance-dependent response to landscape; (ii) species composition variation in the Atlantic Forest and Caatinga was influenced by landscape diversity and the presence of croplands, respectively; and (iii) assemblages in the Atlantic Forest are more sensitive to landscape changes than those in the Caatinga.

More recently, Brito-Almeida et al. (2025) assessed the effects of free goat pasture on scorpion assemblages in Caatinga areas, finding that: (i) scorpion diversity was negatively affected, with reductions in species richness and abundance in goat-impacted areas; (ii) species composition shifted, favoring habitat-generalist species over specialists; and (iii) sites with free goat pasture contained scorpions with reduced body condition (body mass and size) compared to sites without goats. These studies demonstrate that human-induced habitat transformations, including agricultural matrices and grazing, strongly influence Neotropical scorpion assemblages by altering species diversity, composition, and individual body condition. Such impacts vary between ecosystems but consistently reduce habitat specialists and favor more generalist, disturbance-tolerant species. Understanding these responses is critical for using scorpions as bioindicators and for developing conservation strategies that mitigate the effects of landscape modification on native arthropod biodiversity.

13.3.5 Climate Change

Together with habitat loss, climate change represents a major threat to the integrity of ecosystems and organisms, particularly those inhabiting vulnerable areas such as biodiversity hotspots (Hare et al. 2011; Bellard et al. 2014; Trew and Maclean 2021). Studies addressing the potential effects of climate change are relatively recent (Lira et al. 2020; Ureta et al. 2020) and mostly focus on medically important species (e.g., Martinez et al. 2018; Barroso et al. 2025; Moura et al. 2025).

In northeastern Brazil, all evaluated species ($n = 10$) showed a drastic reduction in their future potential distribution, ranging from 44% to 72% compared to their current range (Lira et al. 2020). Furthermore, these authors also reported a decline in scorpion species richness under future climate change scenarios. Ureta et al. (2020), projecting future scenarios for 24 Mexican scorpion species, found that these animals are insufficiently protected by existing natural protected areas under both current and future climatic conditions. They also concluded that species migration to climatically suitable areas in the future is unlikely to be feasible.

These findings underscore that climate change poses a severe risk to Neotropical scorpion biodiversity by drastically shrinking potential habitats and reducing species richness. The limited protection offered by current conservation areas and low likelihood of successful species migration emphasize the urgent need for adaptive management strategies to mitigate the impacts of climate change on these ecologically and medically important organisms.

13.4 Conclusion

This chapter has demonstrated that Neotropical scorpions possess complex ecological dynamics driven by multifaceted biotic and abiotic factors. Their trophic interactions, spatial distribution patterns, and activity behaviors reveal nuanced strategies to coexist and adapt within diverse ecosystems, making them valuable models for ecological research. Furthermore, their sensitivity to environmental disturbances—ranging from habitat fragmentation and restoration processes to urbanization, land-use change, and climate change—highlights their potential as effective bioindicators for monitoring ecosystem health.

Looking forward, expanding research on Neotropical scorpions can provide deeper insights into their ecological roles and responses to accelerating anthropogenic pressures. Future studies integrating molecular tools, long-term monitoring, and experimental manipulations would enhance understanding of their adaptability and resilience. Moreover, incorporating scorpions in multi-taxa biodiversity assessments and conservation planning can improve ecosystem management strategies in fragmented and changing landscapes. Ultimately, recognizing scorpions as integral components of ecological networks enriches both theoretical and applied ecology, fostering conservation approaches that encompass a broader spectrum of biodiversity in the Neotropics and beyond.

Acknowledgments I would like to thank the Dirección General de Asuntos del Personal Académico (DGAPA) for the postdoctoral fellowship from the Universidad Nacional Autónoma de México. I would also like to thank Drs. Gabriel Celante, Leticia Lins, and Welton Dioniosio for kindly providing the photos used in the chapter. Finally, I also thank Dr. Stênio Foerster for their valuable comments in a previous version of this chapter.

References

Almeida T, Arnan X, Capowiez Y, Hedde M, Mesléard F, Dutoit T, Blight O (2023) Ants in restoration ecology: why, what's and the way forward. Land Degrad Dev 35(4):1284–1295. https://doi.org/10.1002/ldr.5006

Álvarez D, Gómez D, Martínez J, Mendoza H, Ruiz S (2013) Microhabitat use by scorpions in a tropical dry forest relict of the Colombian Caribbean. Rev Colomb Entomol 39(2):301–304. https://doi.org/10.25100/socolen.v39i2.8256

Araújo CS, Candido DM, Araújo HFP, Dias SC, Vasconcellos A (2010) Seasonal variations in scorpion activities (Arachnida: Scorpiones) in an area of Caatinga vegetation in northeastern Brazil. Zoologia 27(3):372–376. https://doi.org/10.1590/S1984-46702010000300008

Barroso RF, Cardoso VL, Alves AG, Lira AFA, Ferreira RG, Costa LF, Tizo-Pedroso E (2025) Climate change increases public health risks from *Tityus* scorpion stings in Brazil. Toxicon 258:108326. https://doi.org/10.1016/j.toxicon.2025.108326

Bellard C, Leclerc C, Leroy B, Bakkenes M, Veloz S, Thuiller W, Courchamp F (2014) Vulnerability of biodiversity hotspots to global change. Glob Ecol Biogeogr 23(12):1376–1386. https://doi.org/10.1111/geb.12228

Belote RT, Dietz MS, Jenkins CN, McKinley PS, Irwin GH, Fullman TJ, Leppi JC, Aplet GH (2017) Wild, connected, and diverse: building a more resilient system of protected areas. Ecol Appl 27(4):1050–1056. https://doi.org/10.1002/eap.1527

Bennet AF, Haslem A, Cheal DC, Clarke MF, Jones RN, Koehn JD, Lake PS, Lumsden LF, Lunt IA, Mackey BG, Nally RM, Menkhorst PW, New TR, Newell GR, O'Hara T, Quinn GP, Radford JQ, Robinson D, Watson JEM, Yen AL (2009) Ecological processes: a key element in strategies for nature conservation. Ecol Manage Restor 10(3):192–199. https://doi.org/10.1111/j.1442-8903.2009.00489.x

Benzaken Z, Silva LL, Silva-Neto A, Sachett J, Fé N, Guerra MGVB, Monteiro W (2024) Urbanization and scorpion stings in Manaus, the largest metropolis in the Brazilian Amazon. Toxicon 247:107823. https://doi.org/10.1016/j.toxicon.2024.107823

Brito-Almeida TR, Foerster SÍA, Lima JR, Silva MA, Moura GJB, Lira AFA (2025) Free pasture of exotic goats reduces diversity and negatively affects body condition in scorpions (Arachnida: Scorpiones) assemblage from Brazilian seasonal dry tropical forest. Ecol Evol 15(1):e70804. https://doi.org/10.1002/ece3.70804

Cala-Riquelme F, Colombo M (2011) Ecology of the scorpion, *Microtityus jaumei* in Sierra Canasta. Cuba J Insect Sci 11(1):1–10. https://doi.org/10.1673/031.011.8601

Cooper JEJ, Plummer KE, Middlebrook I, Siriwardena GM (2024) Using butterfly survey data to model habitat associations in urban developments. J Appl Ecol 61(4):773–783. https://doi.org/10.1111/1365-2664.14583

Cubas-Rodríguez AM, Teruel R (2022) Predation by arachnids (Araneae, Scorpiones) on reptiles and amphibians (Anura, Squamata) in Costa Rica and Mexico. Rev Ibér Aracnol 41(1):153–157

Cubas-Rodríguez AM, Cotoras DD, Taucare-Rios A, de Armas LF, Brescovit AD, Gómez G, Stice TL (2024) When the predator becomes the prey: new records of intraguild predation among Central American and Caribbean arachnids (Arachnida: Amblypygi, Araneae, Scorpiones). Stud Neotropical Fauna Environ 59(3):1217–1227. https://doi.org/10.1080/01650521.2024.2353007

Cumming GS (2016) The relevance and resilience of protected areas in the Anthropocene. Anthropocene 13:46–56. https://doi.org/10.1016/j.ancene.2016.03.003

Dias SC, Candido DM, Brescovit AD (2006) Scorpions from Mata do Buraquinho, João Pessoa, Paraíba, Brazil, with ecological notes on a population of *Ananteris mauryi* Lourenço (Scorpiones, Buthidae). Rev Bras Zool 23(3):707–710. https://doi.org/10.1590/S0101-81752006000300014

Dionisio-da-Silva W, Lira AFA, Albuquerque CMR (2018) Distinct edge effects and reproductive periods of sympatric litter-dwelling scorpions (Arachnida: Scorpiones) in a Brazilian Atlantic forest. Zoology 129:17–24. https://doi.org/10.1016/j.zool.2018.06.001

Dionisio-da-Silva W, Lira AFA, Albuquerque CMR (2019) Prey-predator interactions between two intraguild predators modulate their behavioral decisions. Acta Ethol 22:195–201. https://doi.org/10.1007/s10211-019-00326-6

Dionisio-da-Silva W, Foerster SÍA, Gallão JE, Lira AFA (2024) What's for dinner? Prey consumption by Neotropical scorpions across contrasting environments. J Arachnol 52(1):26–30. https://doi.org/10.1636/JoA-S-22-060

Dionisio-da-Silva W, Rocha-da-Silva KLA, Veloso HMG, daSilva MB (2025) Hide and seek: chemical cues drive site preference among potential mates and intraguild competitors. Biol J Linn Soc 144(4):blae058. https://doi.org/10.1093/biolinnean/blae058

Fahrig L (2003) Effects of habitat fragmentation on biodiversity. Annu Rev Ecol Evol Syst 34:487–515. https://doi.org/10.1146/annurev.ecolsys.34.011802.132419

Feitosa MLB, Dionisio-da-Silva W, Lira AFA, Teles-Pontes WJ (2022) Fear as an enemy? Behavioral changes of *Ananteris mauryi* (Scorpiones: Buthidae) triggered by chemical cues from an intraguild predator. Can J Zool 100:488–493. https://doi.org/10.1139/cjz-2022-0027

Feitosa MLB, Barbosa-da-Silva HR, Salomão RP, DeSouza AM, Moura GJB, Lira AFA (2024) Effects of landscape metrics on scorpion (Arachnida: Scorpiones) assemblage in a tropical urban ecosystem. Ecol Evol 14(2):e11026. https://doi.org/10.1002/ece3.11026

Feitosa MLB, Correa CMA, Souza FS, Lira AFA (2025a) Functional traits as an indicator of urbanization impact on the scorpion assemblage in Neotropical forest. Can J Zool 103:1–7. https://doi.org/10.1139/cjz-2024-0037
Feitosa MLB, Alvarado F, Lima HVG, Moura GJB, Lira AFA (2025b) Scorpions in urbanized landscapes: influence of habitat type and landscape on body condition, diversity and functional traits in a Neotropical submontane forest. Acta Oecol 128:104087. https://doi.org/10.1016/j.actao.2025.104087
Foerster SÍA, Lira AFA, Almeida CG (2020) Vegetation structure as the main source of variability in scorpion assemblages at small spatial scales and further considerations for the conservation of Caatinga landscapes. Neotrop Biol Conserv 15(4):533–550. https://doi.org/10.3897/neotropical.15.e59000
Gerber LR, Iacona GD (2024) Aligning data with decision to address the biodiversity crisis. PLoS Biol 22(6):e3002683. https://doi.org/10.1371/journal.pbio.3002683
Gerlach J, Samways M, Pryke J (2013) Terrestrial invertebrates as bioindicators: an overview of available taxonomic groups. J Insect Conserv 17:831–850. https://doi.org/10.1007/s10841-013-9565-9
Gillingham PK, Bradbury RB, Roy DB, Alnderson BJ, Baxter JM, Bourn NAD, Crick HQP, Findon RA, Fox R, Franco A, Hill JK, Hodgson JA, Holt AR, Morecroft MD, O'Hanlon NJ, Oliver TH, Pearce-Higgins JW, Procter DA, Thomas JA, Walker KJ, Walmsley CA, Wilson RJ, Thomas CD (2015) The effectiveness of protected áreas in the conservation of species with changing geographical ranges. Biol J Linn Soc 115(3):707–717. https://doi.org/10.1111/bij.12506
Goodman A, Esposito L (2020) Niche partitioning in congeneric scorpions. Invertebr Biol 139(1):e12280. https://doi.org/10.1111/ivb.12280
Guariguata MR, Ostertag R (2001) Neotropical secondary forest succession: changes in structural and functional characteristics. For Ecol Manag 148(1–3):185–206. https://doi.org/10.1016/S0378-1127(00)00535-1
Guénard B, Hughes AC, Lainé C, Cannicci S, Russell BD, Williams GA (2025) Limited and biased global conservation funding means most threatened species remain unsupported. Proc Natl Acad Sci 122(9):e2412479122. https://doi.org/10.1073/pnas.2412479122
Hansen AJ, DeFries (2007) Ecological mechanisms linking protected areas to surrounding lands. Ecol Appl 17(4):974–988. https://doi.org/10.1890/05-1098
Hare WL, Cramer W, Schaeffer M, Battaglini A, Jaeger CC (2011) Climate hotspots: key vulnerable regions, climate change and limits to warming. Reg Environ Chang 11:1–13. https://doi.org/10.1007/s10113-010-0195-4
Harvey E, Gounand I, Ward CL, Altermatt F (2016) Bridging ecology and conservation: from ecological networks to ecosystem function. J Appl Ecol 54(2):371–379. https://doi.org/10.1111/1365-2664.12769
Kromer T, Einzmann HJR, Mendieta-Leiva G, Zotz G (2025) Impact of land-use change on vascular epiphytes: a review. Plants (Basel) 14(8). https://doi.org/10.3390/plants14081188
Kruger H, Jaatinen K, Holopainen S, Niemi M, Vehkaoja M, Virtanen J, Vaananen V-M, Nummi P (2022) Margins matter: the importance of field margins as avian brood-rearing habitat in an intensive agricultural landscape. J Ornithol 164:101–114. https://doi.org/10.1007/s10336-022-02014-y
Lamb D (2011) Natural regeneration and secondary forests. In: Regreening the bare hills. World Forests, vol 8. Springer, Dordrecht. https://doi.org/10.1007/978-90-481-9870-2_5
Landeiro VL, Franz B, Heino J, Siqueira T, Bini LM (2018) Species-poor and low-lying sites are more ecologically unique in a hyperdiverse Amazon region: evidence from multiple taxonomic groups. Divers Distrib 24(7):966–977. https://doi.org/10.1111/ddi.12734
Lapinski W (2025) Vertical distribution and habitat use in arboreal scorpions from a tropical rainforest in Costa Rica. Arachnology 20(2):303–312. https://doi.org/10.13156/arac.2025.20.2.303
Lira AFA, DeSouza AM (2014) Microhabitat use by scorpion species (Arachnida: Scorpiones) in the montane Atlantic rainforest, Brazil. Revista Ibérica de Aracnología 24(1):107–108

Lira AFA, DeSouza AM, Silva-Filho AAC, Albuquerque CMR (2013) Spatio-temporal microhabitat use by two co-occurring species of scorpions in Atlantic rainforest in Brazil. Zoology 116(3):182–185. https://doi.org/10.1016/j.zool.2013.01.002

Lira AFA, Rego FNAA, Albuquerque CMR (2015) How important are environmental factors for the population structure of co-occurring scorpion species in a tropical forest? Can J Zool 93(1):15–19. https://doi.org/10.1139/cjz-2014-0238

Lira AFA, Araújo VLN, DeSouza AM, Rego FNAA, Albuquerque CMR (2016) The effect of habitat fragmentation on the scorpion assemblage of a Brazilian Atlantic Forest. J Insect Conserv 20:457–466. https://doi.org/10.1007/s10841-016-9878-6

Lira AFA, DeSouza AM, Albuquerque CMR (2017a) Report of intraguild predation and cannibalism in scorpions (Scorpiones: Buthidae) in the Brazilian Atlantic Forest. North West J Zool 13:356–358

Lira AFA, Pordeus LM, Albuquerque CMR (2017b) A new species of *Ananteris* (Scorpiones: Buthidae) from Caatinga biome, Brazil. Acta Arachnol 66(1):9–15. https://doi.org/10.2476/asjaa.66.9

Lira AFA, DeSouza AM, Albuquerque CMR (2018a) Environmental variation and seasonal changes as determinants of the spatial distribution of scorpions (Arachnida: Scorpiones) in Neotropical forests. Can J Zool 96(9):963–972. https://doi.org/10.1139/cjz-2017-0251

Lira AFA, Damasceno EM, Silva-Filho AAC, Albuquerque CMR (2018b) Linking scorpion (Arachnida: Scorpiones) assemblage with fragment restoration in the Brazilian Atlantic Forest. Stud Neotropical Fauna Environ 53(2):107–112. https://doi.org/10.1080/01650521.2017.1413823

Lira AFA, Pordeus LM, Salomão RP, Badillo-Montaño R, Albuquerque CMR (2019) Effects of anthropogenic land-use on scorpions (Arachnida: Scorpiones) in Neotropical forests. Int J Trop Insect Sci 39:211–218. https://doi.org/10.1007/s42690-019-00029-2

Lira AFA, Badillo-Montaño R, Lira-Noriega A, Albuquerquer CMR (2020) Potential distribution patterns of scorpions in North-Eastern Brazil under scenarios of future climate change. Austral Ecol 45(2):215–228. https://doi.org/10.1111/aec.12849

Lira AFA, Foerster SÍA, Salomão RP, Porto TJ, Albuquerque CMR, Moura GJB (2021a) Understanding the effects of human disturbance on scorpion diversity in Brazilian tropical forests. J Insect Conserv 25:147–158. https://doi.org/10.1007/s10841-020-00292-6

Lira AFA, Rego FNAA, Salomão RP, Albuquerque CMR (2021b) Effects of habitat quality on body size of the litter dwelling scorpion *Tityus pusillus* in fragmented rainforests of Brazil. J Arachnol 48(3):295–299. https://doi.org/10.1636/JoA-S-19-081

Lira AFA, Foerster SÍA, DeSouza AM, de Armas LF (2022) Disentangling diversity patterns in Cuban scorpions (Arachnida: Scorpiones). Novit Caribaea 19:72–91. https://doi.org/10.33800/nc.vi19.290

Mace GM (2004) The role of taxonomy in species conservation. Philos Trans R Soc B Biol Sci 359(1444). https://doi.org/10.1098/rstb.2003.1454

Martinez PA, Andrade MA, Bidau CJ (2018) Potential effects of climate change on the risk of accidents with poisonous species of the genus *Tityus* (Scorpiones, Buthidae) in Argentina. Spat Spatio-temporal Epidemiol 25:67–72. https://doi.org/10.1016/j.sste.2018.03.002

Martins JG, Almeida MRN, Procópio REL, Lira AFA (2024) Cannibalism in *Tityus metuendus* Pocock, 1897 (Scorpiones: Buthidae) from the Brazilian Amazon. Rev Chil Entomol 50(1):67–61

Mcdonald RI, Kareiva P, Forman RTT (2008) The implications of current and future urbanization for global protected areas and biodiversity conservation. Biol Conserv 141(6):1695–1703. https://doi.org/10.1016/j.biocon.2008.04.025

McKinney ML (2002) Urbanization, biodiversity, and conservation: the impacts of urbanization on native species are poorly studied, but educating a highly urbanized human population about these impacts can greatly improve species conservation in all ecosystems. Bioscience 52(10):883–890. https://doi.org/10.1641/0006-3568(2002)052[0883:UBAC]2.0.CO;2

McKinney ML (2006) Urbanization as a major cause of biotic homogenization. Biol Conserv 127(3):247–260. https://doi.org/10.1016/j.biocon.2005.09.005

Melo-dos-Santos G, Melo-dos-Santos G, Leite TM, Rocha AM, Jati SR, Frezarin-da-Silva E, Cerni FA, Sibajev A, Lira AFA, Pucca MB (2025) Cannibalistic behaviour in *Rhopalurus laticauda* Thorell, 1876 and *Tityus silvestris* Pocock, 1897 (Scorpiones: Buthidae) under captivity in the extreme north of the Brazilian Amazon. Rev Chil Entomol 51(2):169–175

Moreira MOM, Araújo VLN, Foerster SÍA, Moura GJB, Lira AFA (2022) Relationship between body size and habitat heterogeneity on cannibalism and intraguild predation in scorpions. Biologia 77:2867–2873. https://doi.org/10.1007/s11756-022-01154-z

Moreira MOM, Moura GJB, Lira AFA (2024) Sexual receptivity of females on Neotropical scorpion *Tityus pusillus* Pocock, 1893 in response to male reproductive advances. Biologia 79:2119–2124. https://doi.org/10.1007/s11756-024-01691-9

Moreno-Gonzalez JA, Hazzi NA (2012) Intraguild predation case: *Tityus forcipula* Gervais, 1843 (Scorpiones, Buthidae) feeding on *Chactas vandenedeni* Gervais, 1843 (Scorpiones, Chactidae) in Colombia. Rev Ibér Aracnol 20:17–20

Moura TA, Ojanguren-Affilastro AA, Sasa M, Guitérrez JM, Silva FF, Siqueira-Silva T, Martinez PA (2025) AI models uncover factors influencing scorpionism in Northern Brazil. Toxicon 258:108342. https://doi.org/10.1016/j.toxicon.2025.108342

Nagy DD, Magura T, Horváth R, Debnár Z, Tóthmérész B (2018) Arthropod assemblages and functional responses along an urbanization gradient: a trait-based multi-taxa approach. Urban For Urban Green 30:157–168. https://doi.org/10.1016/j.ufug.2018.01.002

Nime MF, Casanoves F, Mattoni CI (2014) Scorpion diversity in two different habitats in the Arid Chaco, Argentina. J Insect Conserv 18:373–384. https://doi.org/10.1007/s10841-014-9646-4

Nime MF, Casanoves F, Mattoni CI (2016) Microhabitat use and behavior differ across sex-age classes in the scorpion *Brachistosternus ferrugineus* (Scorpiones: Bothriuridae). J Arachnol 44(2):235–244. https://doi.org/10.1636/J15-63

Oliveira RF, Lira AFA, Zaffaroni-Caorsi V, Feitosa MLB, Moura GJB (2025) Wind farm noise negatively impacts the calling behavior of three frogs in Caatinga dry forest. PLoS One 20(3):e0318517. https://doi.org/10.1371/journal.pone.0318517

Ott R, Ott AP (2014) Abundance and seasonality of *Bothriurus signatus* (Arachnida, Scorpiones) in different vegetal formations in São Francisco de Paula, RS, Brazil. Iheringia 104:92–98. https://doi.org/10.1590/1678-4766201410419298

Pinto AV, Hansson B, Patramanis I, Morales HE, van Oosterhout C (2024) The impact of habitat loss and population fragmentation on genomic erosion. Conserv Genet 25:49–57. https://doi.org/10.1007/s10592-023-01548-9

Ponce-Saavedra J, Francke O, Suzán H (2006) Actividad superficial y utilización del hábitat por *Centruroides balsasensis* Ponce y Francke (Scorpiones: Buthidae). Biológicas 8:130–137

Porto TJ, Pinto-da-Rocha R, Rocha PLB (2018) Regional distribution patterns can predict the local habitat specialization of arachnids in heterogeneous landscapes of the Atlantic Forest. Divers Distrib 24(3):375–386. https://doi.org/10.1111/ddi.12685

Pucca MB, Cavalcante JS, Jati SR, Cerni FA, Ferreira RS Jr, Arantes EC (2025) Scorpions are taking over: the silent and escalating public health crisis in Brazil. Front Public Health 13. https://doi.org/10.3389/fpubh.2025.1573767

Puig-Gironès R, Real J (2022) A comprehensive but practical methodology for selecting biological indicators for long-term monitoring. PLoS One 17(3):e0265246. https://doi.org/10.1371/journal.pone.0265246

Quijano-Ravell AF, Ponce-Saavedra J, Francke O (2012) Densidad, distribución espacial y biomasa de *Hadrurus gertschi* Soleglad (Scorpiones: Iuridae) en una localidad de Guerrero, Mexico. Rev Ibér Aracnol 20:35–43

Rosenfield MF, Jakovac CC, Vieira DLM, Poorter L, Brancalion PHS, Vieira ICG, Almeida DRA, Massoca P, Schietti J, Albernaz ALM, Ferreira MJ, Mesquita RCG (2023) Ecological integrity of tropical secondary forests: concepts and indicators. Biol Rev 98(2):662–676. https://doi.org/10.1111/brv.12924

Sandall EL, Maureaud AA, Guralnick R, McGeoch MA, Sica YV, Rogan MS, Booher DB, Edwards R, Franz N, Ingenloff K, Lucas M, Marsh CJ, McGowan J, Pinkert S, Ranipeta A, Uetz P, Wieczorek J, Jetz W (2023) A globally integrated structure of taxonomy to support biodiversity science and conservation. Trends Ecol Evol 38(12):P1143–P1153. https://doi.org/10.1016/j.tree.2023.08.004

Santos GCSG, Dionisio-da-Silva W, Souza-Alves JP, Albuquerque CMR, Lira AFA (2018) Random or clumped: how litter dwelling scorpions are distributed in a fragment of Brazilian Atlantic. Eur J Entomol 115:445–449. https://doi.org/10.14411/eje.2018.045

Saout SL, Hoffmann M, Shi Y, Hughes A, Bernard C, Brooks TM, Bertzky B, Butchart SHM, Stuart SN, Badman T, Rodrigues ASL (2013) Protected areas and effective biodiversity conservation. Science 342(6160):803–805. https://doi.org/10.1126/science.1239268

Saunders ME, Lees AC, Grames EM (2025) Understanding and counteracting the denial of insect biodiversity loss. Curr Opin Insect Sci 68:101338. https://doi.org/10.1016/j.cois.2025.101338

Schwerdt L, Copperi S, Pompozzi G, Ferretti N (2016) Diversity and seasonal composition of the scorpion fauna from a mountainous system on pampean grasslands in central Argentina. Stud Neotropical Fauna Environ 51(3):169–175. https://doi.org/10.1080/01650521.2016.1215039

Sgarlata GM, Maié T, Zoeten T, Salmona J, Rasteiro R, Chikhi L (2025) The effect of habitat loss and fragmentation on isolation by distance and divergence. Proc Natl Acad Sci 122(31):e2410951122. https://doi.org/10.1073/pnas.2410951122

Silva-Júnior AO, Barbosa-da-Silva HR, Feitosa MLB, Lira AFA (2021) Intraguild predation in Brazilian Atlantic scorpions: report of *Tityus brazilae* Lourenço & Eickstedt, 1984 preying on *Tityus pusillus* Pocock, 1893 (Scorpiones: Buthidae). Rev Ibér Aracnol 39(2):133–134

Silva-Júnior AO, Brito-Almeida TR, Feitosa MLB, Jansen RG, Peres M, Azevedo R, Moura GJB, Barbosa AR, Lira AFA (2023) Can size make a difference? Cross-predation occurrences between lizards and scorpions in the Brazilian seasonal dry tropical forest. Invertebr Zool 20(3):343–347. https://doi.org/10.15298/invertzool.20.3.06

Simkin RD, Seto KC, McDonald RI, Jetz W (2022) Biodiversity impacts and conservation implications of urban land expansion projected to 2050. Proc Natl Acad Sci 119(12):e2117297119. https://doi.org/10.1073/pnas.2117297119

Sushinsky JR, Rhodes JR, Possingham HP, Gill TK, Fuller RA (2012) How should we grow cities to minimize their biodiversity impacts? Glob Chang Biol 19(2):401–410. https://doi.org/10.1111/gcb.12055

Tee SL, Samantha LD, Kamarudin N, Akbar Z, Lechner AM, Ashton-Butt A, Azhar B (2018) Urban forest fragmentation impoverishes native mammalian biodiversity in the tropics. Ecol Evol 8(24):12506–12521. https://doi.org/10.1002/ece3.4632

Theobald DM, Oakleaf JR, Moncrieff G, Voigt M, Kiesecker J, Kennedy CM (2025) Global extent and change in human modification of terrestrial ecosystems from 1990 to 2022. Sci Data 12(606). https://doi.org/10.1038/s41597-025-04892-2

Toranza C, Arim M (2010) Cross-taxon congruence and environmental conditions. BMC Ecol 10(18). https://doi.org/10.1186/1472-6785-10-18

Toscano-Gadea CA, Costa FG (2006) Is *Tityus uruguayensis* (Buthidae) an araneophagic scorpion? An experimental analysis of its predatory behavior in spiders and insects. Bull Br Arachnol Soc 13(7):256–264

Tourinho AL, Benchimol M, Porto W, Peres CA, Storck-Tonon D (2020) Marked compositional changes in harvestmen assemblages in Amazonian forest islands induced by a mega dam. Insect Conserv Divers 13(5):432–444. https://doi.org/10.1111/icad.12398

Trew BT, Maclean IMD (2021) Vulnerability of global biodiversity hotspots to climate change. Glob Ecol Biogeogr 30(4):768–783. https://doi.org/10.1111/geb.13272

Ureta C, González EJ, Ramírez-Barrón M, Contreras-Félix GA, Santibáñez-López CE (2020) Climate change will have an important impact on scorpion's fauna in its most diverse country, Mexico. Perspect Ecol Conserv 18(2):116–123. https://doi.org/10.1016/j.pecon.2020.04.003

Vellend M, Verheyen K, Jacquemin KA, van Calster H, Peterken G, Hermy M (2006) Extinction debt of forest plants persists for more than a century following habitat fragmentation. Ecology 87(3):542–548. https://doi.org/10.1890/05-1182

Venter O, Fuller FA, Segan DB, Carwardine J, Brooks T, Butchart SHM, Di Marco M, Iwamura T, Joseph L, O'Grady D, Possingham HP, Rondinini C, Smith RJ, Venter M, Watson JEM (2014) Targeting global protected area expansion for imperiled biodiversity. PLoS Biol 12(6):e1001891. https://doi.org/10.1371/journal.pbio.1001891

Villanueva-Bonilla GA, Azevedo R, Araújo KF, Lira AFA, Sobczak JF (2025) Biotic rather than abiotic factors influence ground-dwelling mygalomorph spider assemblages along an altitudinal gradient in the Brazilian semi-arid domain. Papéis Avulsos Zool 65:e2025650001. https://doi.org/10.11606/1807-0205/2025.65.001

Wallington TJ, Hobbs RJ, Moore SA (2005) Implications of current ecological thinking for biodiversity conservation: a review of the salient issues. Ecol Soc 10(1):1–16

Yamaguti HY, Pinto-da-Rocha R (2006) Ecology of *Thestylus aurantiurus* of the parquet estadual da serra da Cantareira, São Paulo, Brazil (Scorpiones, Bothriuridae). J Arachnol 34(1):214–220. https://doi.org/10.1636/T04-37.1

Yuan R, Zhang N, Zhang Q (2024) The impact of habitat loss and fragmentation on biodiversity in global protected areas. Sci Total Environ 931:173004. https://doi.org/10.1016/j.scitotenv.2024.173004

Zanetta S, Germino C, Rodrigues I, Waragaia A (2020) Urbanization and increased cases of scorpionism in Brazilian cities. Eur J Pub Health 30(5):163–166. https://doi.org/10.1093/eurpub/ckaa166.163

Chapter 14
Traditional Treatment of Scorpion Stings in the Neotropics

Jonas Gama Martins, Bruno Rafael Ribeiro de Almeida, Pedro Pereira de Oliveira Pardal, and Rudi Emerson de Lima Procópio

Abstract Scorpion stings are common in rural tropical regions, which often have limited access to health services. In this context, remote communities, both Amerindian and non-Amerindian, often employ traditional plant-based treatments, such as substances obtained from seeds, sap, leaves, and roots, to alleviate symptoms. In rural Amazon, traditional healers, often without medical training, treat scorpion stings at home. This practice contributes to the underreporting of cases, since many incidents are not formally recorded in health systems.

Plant extracts contain bioactive metabolites that are widely used in traditional medicine for treating venomous animal stings and bites. However, their clinical efficacy remains largely unproven due to a lack of rigorous scientific studies, and their use may pose risks such as allergic reactions or toxicity. In contrast, commercial antivenoms, subject to strict quality control and composed of functional immunoglobulins, are globally recognized as safe and effective. Cultural barriers, such as individual concepts of health and deeply rooted ethnic beliefs in rural communities, often prevent the adoption of appropriate treatments that could save lives. Although traditional treatments for scorpion stings are used as first aid and definitive measures in different rural areas of the Amazon, their use should be discouraged, especially among non-indigenous communities, which are generally more receptive to health education programs. To this end, regional health authorities should prioritize

J. G. Martins (✉)
Pós-Graduação em Genética, Conservação e Biologia Evolutiva (PPG GCBEv), Instituto Nacional de Pesquisas da Amazônia (INPA), Amazonas, Brazil
e-mail: jonasgama83@gmail.com

B. R. R. de Almeida
Universidade do Estado do Pará (UEPA), Cametá, Pará, Brazil

P. P. de Oliveira Pardal
Laboratório de Entomologia Médica e Animais Peçonhentos, Núcleo de Medicina Tropical, Universidade Federal do Pará (UFPA), Belém, PA, Brazil

R. E. de Lima Procópio
Pós-Graduação em Biotecnologia e Recursos Naturais da Amazônia, Universidade do Estado do Amazonas (UEA), Manaus, AM, Brazil

A. F. de Araujo Lira, E. González-Santillán (eds.), *The Biology of Neotropical Scorpions*, https://doi.org/10.1007/978-3-032-23194-9_14

the implementation of medical centers with antivenom stockpiles in high-risk regions, ensuring rapid access to safe and effective treatments.

Keywords Scorpion - Scorpion sting · Traditional treatment · Public health

14.1 Neotropical Scorpions of Medical Importance

The New World is rich in medically relevant scorpion species (Fet and Lowe 2000; Ythier 2018; Martins et al. 2021). In the Neotropics, these animals are found in diverse ecosystems, ranging from arid and semiarid areas to flooded forests and savannas (Fet and Lowe 2000; Ythier and Dupre 2024). The species of most concern are members of the genera *Centruroides* Max, 1890 and *Tityus* C.L Koch, 1836, which inhabit rural tropical regions (Fet and Lowe 2000; Ythier 2018; Batista et al. 2018; Martins et al. 2021, 2023, 2024). This proximity to areas inhabited by humans often results in envenoming (Pardal et al. 2014; Martins et al. 2023).

In South America, species of the genera *Tityus* and *Centruroides* are responsible for most of the officially reported incidents (Martins et al. 2026; Chippaux et al. 2020). While cases of *Centruroides* envenomation predominate in Central America (Ward et al. 2018), the genus *Tityus* is more relevant to public health in South America, especially in Brazil, Argentina, and Venezuela (Pardal et al. 2003; Martins et al. 2023). In Brazil, for instance, epidemiological data from 2021 reflect the magnitude of the problem, with 159,934 reported cases of incidents and 133 deaths (Brasil 2022), constituting one of the highest rates of scorpion envenomation in South America (Brasil 2022).

Neotropical scorpion envenomation can cause symptoms ranging from mild to fatal (Martins et al. 2026; Pardal et al. 2003). Among the most frequently reported clinical manifestations are tachycardia, vomiting, muscle spasms, dysphagia, and pulmonary edema, with high risks of respiratory and cardiac complications (Pardal et al. 2003; Cupo et al. 2007; Chippaux 2015; Chippaux et al. 2020). Children under 10 years of age constitute the most vulnerable group, due to their greater sensitivity to the effects of toxins (Pardal et al. 2014). In addition, systemic inflammatory processes can increase the risk of serious complications (Reis et al. 2023; Martins et al. 2024).

Clinical management of scorpion stings involves analgesia for pain control in mild cases, where symptoms are localized. In moderate to severe cases, characterized by systemic symptoms such as tachycardia, vomiting, and/or respiratory distress, antivenom is administered to neutralize the toxins and prevent life-threatening complications. These strategies are guided by health manuals but are not always applicable in remote rural communities in South America (Davis and Yost 1983). The residents of indigenous communities, for example, typically have their own concepts of disease and traditional treatments for envenomation (Grenand et al. 2004; Martins et al. 2023). Therefore, understanding the traditional practices associated with scorpion stings in high-risk rural areas is crucial for guiding both indigenous and non-indigenous communities on the importance of seeking medical treatment at health facilities.

14.2 Indigenous Cosmology on Scorpions in the Amazon

The indigenous cosmology of the Amazon reveals a diverse view of venomous animals, integrating myths and self-care practices against the symptoms of envenomation (Prance 1972; Kamen-Kaye 1975; Wilbert and Simoneau 1990; Posey and Elisabetsky 1991; Milliken and Albert 1996; Santos and Coimbra 1994). In the Northwest Amazon, for example, the term for "scorpion" varies among indigenous peoples, as do their conceptions of disease and health (Schultes and Raffauf 1990; Schultes et al. 2001).

Among the Tukano people, for example, scorpions are seen as instruments of revenge, used to attack villages (Reichel-Dolmatoff 1969). In ceremonial narratives, Amerindian myths describe a scorpion that, when drinking human blood, inflicts stings that cause intense pain (Wilbert and Simoneau 1990). Interestingly, this description aligns with reports in the medical literature, which identify pain at the site of the sting as a hallmark of scorpion envenomation (Pardal et al. 2003).

Members of the Yanomami communities find scorpions mainly in the "*urihi a*," an expression that, in their language, means "forest-land" (Wilbert and Simoneau, 1990) and can be understood as equivalent to the concept of "nature" for non-indigenous people. Among the myths of this ethnic group, which inhabits the upper Río Orinoco region in Venezuela, there is a black scorpion called "*uhutima*" (Wilbert and Simoneau 1990). The Kayapó people of the Gorotire village (7°45'58.5"S 51°07'55.1"W), Pará State, Brazil interpret the stings as causing the disease "kaàk kanê" (Posey and Elisabetsky 1991). Among many indigenous communities in the Amazon, venomous animals are believed to be associated with harmful spiritual forces (Lyon 1974; Wilbert and Simoneau 1990), subject to treatment by shamans, who act as spiritual mediators and healers (Posey and Elisabetsky 1991).

The Campa indigenous people of the Peruvian Amazon believe that a giant arthropod controls the water levels of the rivers where they fish, influencing their livelihoods (Lyon 1974). On the other hand, the Makú people of the upper Rio Negro in Brazil observe the sky to guide their agricultural practices. They recognize the constellation 'Bupbeh', which represents a snake or scorpion, as a sign of seasonal changes in flora and fauna (Silverwood-Cope 1990). This perceived connection between scorpions and natural cycles reflects the integration of these animals into the daily lives and worldview of this community.

In addition to health risks, scorpions carry various cultural meanings. In the upper Rio Negro, for example, they are associated with aphrodisiac properties (Eickstedt 1989), while in the state of Pará, they appear as supernatural symbols in Marajoara indigenous ceramics from 1500 BC (Schaan 1997). These representations suggest that ancient ethnic communities in the region already lived with these animals and were victims of their stings, since scorpions in the Amazon can infest villages (Chagnon 1968). The symbolism of scorpions has been recorded beyond the Amazon (Duke 2009). In Durango, Mexico, scorpions are part of local myths (González-Ponce et al. 2023), while in Venezuela, Andean communities use plant extracts to repel these animals (Kames-Kaye 1971).

14.3 Amerindians use Plants as First aid for Scorpion Stings

In the rural areas of South America, substances from various plant species are traditionally used to alleviate pain and other symptoms caused by scorpion stings (Fig. 14.1). Notable examples include species that produce edible fruits such as the genera *Bactris* spp. and *Mauritia* spp., both members of the Arecaceae family (Fig. 14.2). Studies show that Andean communities in Peru have been using therapeutic plants for at least 8000 years (Plowmann 1984). A significant example is the topical application of *Erythroxylum coca* leaves on areas of the body affected by scorpion stings (Duke 2009).

In the Americas, there are more than 100,000 species of vascular plants (Ulloa et al. 2017), meaning high potential for traditional use of natural resources to treat health problems, including scorpion stings. In Brazil and Mexico, which concentrate most of this flora, the utilization of such plants is a practice that sustains many rural communities (Kahn and de Granville 1992; Smith 2015). Interaction between botanists and indigenous peoples, such as the Paumari of the Purus River, Amazonas,

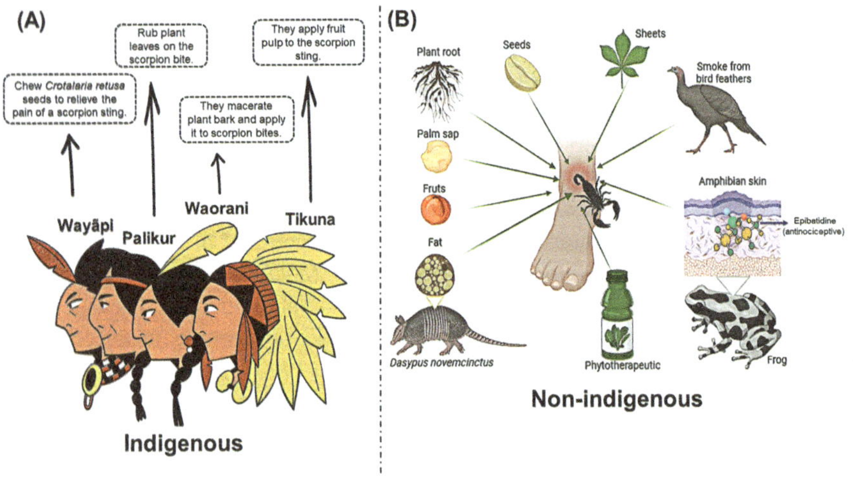

Fig. 14.1 Amerindian and non-Amerindian communities in the Amazon and the Andes use natural resources to combat scorpion stings. (**a**) Ethnic communities generally use plant parts at the site of the sting; (**b**) Non-indigenous communities in the Amazon use different natural resources to combat pain and other symptoms caused by stings

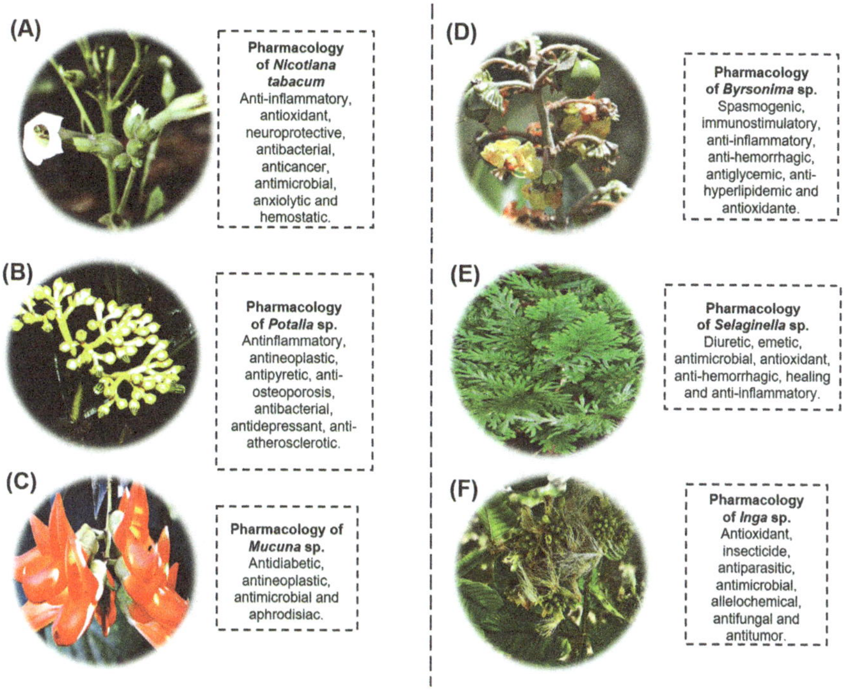

Fig. 14.2 Pharmacological potential of plant species used in the traditional treatment of scorpion stings. (Photos by G. Prance (**a**–**d**), J. Martins (**e**–**f**)

Brazil, helped to record the cultivation of medicinal plants, including those used against envenomation by venomous reptiles (Prance et al. 1977; Prance 1986). This traditional knowledge, still alive in many rural communities, highlights the resilience of ethnobotanical practices in mitigating pain and other symptoms caused by wild animals that threaten human life (Duke and Vasquez 1994; Grenand et al. 2004).

In the Amazon, plants are central to the traditional treatment of envenomation by animals. For example, the Waimiri Atroari people use the inner bark of *Guarea scabra* to relieve pain from scorpion stings, while the Palikur people apply mucilage from the *Bactris gasipaes* palm as a topical antidote (Fig. 14.2) (Rutter 1990; Grenand et al. 2004). Similarly, Kichwa communities in Ecuador apply a mixture of extracts from *Potalia amara* to treat wounds caused by these venomous animals, although the pharmacological basis of this plant still needs to be studied (Duke and Vasquez 1994).

In French Guiana, several plants are utilized to treat scorpion stings, including the following species: *Aristolochia trilobata*, *Ayapana triplinervis*, *Bactris gasipaes*, *Citrus aurantifolia*, *Commelina benghalensis*, *Commelina erecta*, *Crotalaria retusa*, *Dieffenbachia seguine*, *Euphorbia hirta*, *Inga alba*, *Inga bourgoni*, *Palhinhaea cernua*, *Mauritia flexuosa*, *Nautilocalyx pictus*, *Omphalea diandra*, and *Piper marginatum* var. *marginatum* (Fig. 14.3) (Grenand et al. 2004).

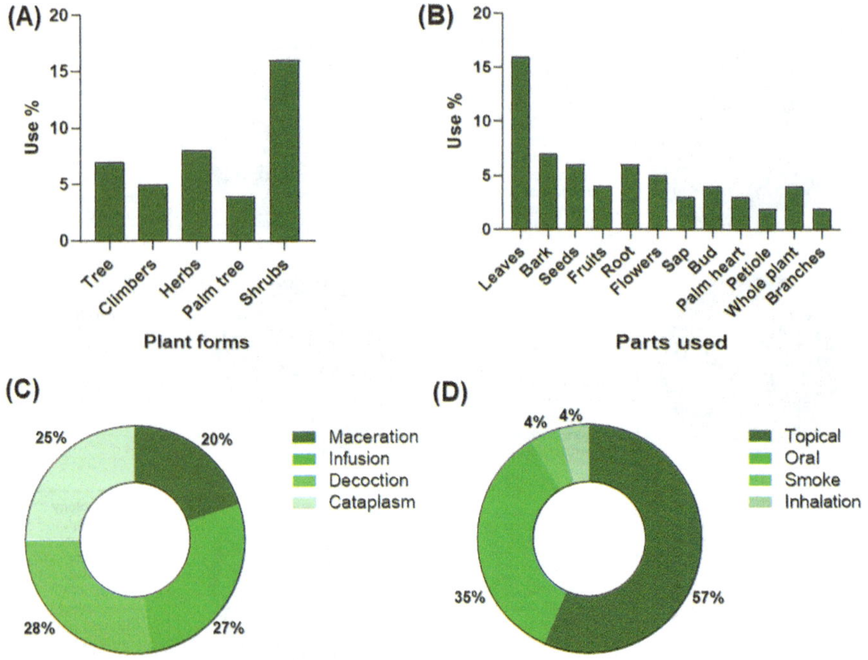

Fig. 14.3 Forms and parts of plants used to treat scorpion stings. (**a**) Parts used; (**b**) Methods of preparing the plants; (**c**) Routes of administration (**d**)

Hyperdominant plants in the Amazon, such as palms of the Arecaceae family, are widely used in the region. The "inajá" palm (*Attalea maripa*), for example, contains parts rich in bioactive compounds that are processed and applied topically to arachnid stings (Smith 2015). Other species, such as *Aristolochia trilobata* and *Citrus aurantifolia*, also have traditional uses for the treatment of scorpion stings (Grenand et al. 2004). Communities outside the Amazon also maintain herbal practices. In Baja California Sur, Mexico, tea or extract of twigs of *Wislizenia refracta* is used to treat scorpion stings, while leaves of *Bebbia atriplicifolia* are applied topically (Dimayuga and Agundez 1986). These traditional practices reflect the rich diversity of plants and local knowledge adapted to regional health needs, especially in areas where access to antivenoms is limited.

The traditional pharmacology of South American plants is a promising field for research (Roth and Lindorf 2002). Many plants used by indigenous communities, such as *Crotalaria retusa*, whose raw seeds are chewed to relieve pain caused by scorpion stings (Duke and Vasquez 1994), have potential for pharmacological studies (Roth and Lindorf, 2002). At the same time, species such as *Petiveria alliacea*, with anxiolytic and antidepressant properties, are used in the treatment of animal envenoming (Luz et al. 2016).

Despite advances in occidental medicine, many impoverished rural communities continue to rely on medicinal plants due to limited access to healthcare services and the high cost of conventional treatments. This is especially true in the case of neglected tropical diseases, where traditional remedies are often the only available option (Grenand et al. 2004; Ming 2006). This dependence on forests reinforces the importance of health education programs in regions at high risk of exposure to venomous animals.

Table 14.1 lists the plants most used to treat scorpion stings in rural communities. The use of these plants reflects deeply rooted cultural practices that contrast sharply with modern medicine's emphasis on antivenoms as scientifically validated therapies (Martins et al. 2026). While Amerindian communities often rely on natural remedies or treatments urged by shamans, modern medicine emphasizes the use of antivenom as the only scientifically validated therapy. This is because antivenom directly neutralizes toxins, whereas traditional remedies may only provide symptomatic relief without addressing the root cause of envenomation.

Table 14.1 Plants used in the traditional treatment of scorpion stings in Neotropical regions

Family/species	Location	Part used	Use	Reference
Araceae				
Philodendron solimoesense	Roraima—Brazil	Roots	Topical	Milliken and Albert (1996)
Monstera adansonii	Guyana	Buds	Topical	van Andel (2000)
Montrichardia arborescens	Pará—Brazil	Stems	Topical	Branch and da Silva (1983)
Philodendron brevispathum	Northern Guayana	Buds	Topical	van Andel (2000)
Arecaceae				
Attalea maripa	Peru	Palm hearts	Topical	Smith (2015)
Bactris gasipaes	Guyana	Sap	Topical	Grenand et al. (2004)
Euterpe edulis	–	Palm hearts	Topical	Plotkin and Balick (1984)
Mauritia flexuosa	Guyana	Petioles	–	Grenand et al. (2004)
Euterpe oleracea	Suriname	Palm hearts	Topical	DeFilipps et al. (2004)
Araliaceae				
Schefflera morototoni	Guyana	Bark	–	van Andel (2000)
Astereaceae				
Tagetes lucida	México	–	Tea	Duke (2009)
Eclipta albar	Brazil	Leaves	Juice	Lorenzi and Matos (2002)
Bebbia atriplicifolia	Mexico	Leaves	Decoction	Dimayuga and Agundez (1986)
Sphagneticola trilobata	Honduras	Leaves	Oral	Duke (2009)
Mikania guaco	Mexico	Leaves	Juice	Duke (2009)
Aristolochiaceae				
Aristolochia grandifolia	Mexico	Roots	–	Morton (1981)

(continued)

Table 14.1 (continued)

Family/species	Location	Part used	Use	Reference
Apocynaceae				
Tabernaemontana disticha	Guyana	Bark	–	DeFilipps et al. (2004)
Mandevilla tenuifolia	Northwest Amazon	Leaves	Cataplasm	Schultes and Raffauf (1990)
Boraginaceae				
Heliotropium indicum	South America	Flowers	Cataplasm	Roth and Lindorf (2002)
Caricaceae				
Carica papaya	South America	Seeds	Topical	
Cleomaceae				
Wislizenia refracta	Mexico	Branches	Decoction	Dimayuga and Agundez (1986)
Combretaceae				
Combretum karijonorum	Amazon	Leaves	Cataplasm	Schultes and Raffauf (1990)
Erythroxylaceae				
Erythroxylum coca	Peru	Leaves	Cataplasm	Duke (2009)
Euphorbiaceae				
Croton ciliatoglanduliferus	Mexico	Sap	–	Morton (1981)
Fabaceae				
Crotalaria retusa	Guyana	Seeds	Oral	Duke and Vasquez (1994)
Desmondium uncinatum	Venezuela	Roots	–	Morton (1981)
Acacia hindsii	Guatemala	Stems	Decoction	Morton (1981)
Lonchocarpus chrysophyllus	French Guiana	Bark	–	DeFilipps et al. (2004)
Pentaclethra macroloba	French Guiana, Suriname	Bark	–	DeFilipps et al. (2004)
Indigofera suffruticosa	Peru	Leaves	Topical	Duke (2009)
Gesneriaceae				
Nautilocalyx pictus	Colombia, Peru, Ecuador	Leaves	–	Grenand et al. (2004)
Lamiaceae				
Ocimum micranthum	Pará—Brazil	Whole plant	Topical	Branch and da Silva (1983)
Leguminosae				
Inga alba	Guyana	Leaves	Topical	Grenand et al. 2004
Meliaceae				
Guarea scabra	Amazonas, Brazil	Bark	Topical	Milliken et al. (1992)
Menispermaceae				
Cissampelos pareira	Nicaragua	Leaves	Decoction	Duke (2009)
Leguminosae				

(continued)

Table 14.1 (continued)

Family/species	Location	Part used	Use	Reference
Pentaclethra macroloba	Guyana	Bark	–	Duke (2009)
Lamiaceae				
Ocimum selloi	Brazil	Leaves	Decoction	Lorenzi and Matos (2002)
Peltodon radicans	Brazil	Leaves	Cataplasm	Lorenzi and Matos (2002)
Papilionaceae				
Clathrotropis brachypetala	Guyana, Suriname	Bark	Topical	van Andel (2000)
Phytolaccaceae				
Petiveria alliacea	Guyana	Bark	–	Duke and Vasquez (1994)
Polygalaceae				
Bredemeyera floribunda	Brazil	Roots	Topical	Lorenzi and Matos (2002)
Piperaceae				
Piper hostmannianum	Guyana	Leaves	Oral	van Andel (2000)
Rubiaceae				
Pentagonia spathicalyx	Ecuador	Fruits	–	Davis and Yost (1983)
Simaroubaceae				
Quassia amara	Honduras	Stems	Decoction	Duke (2009)
Rutaceae				
Citrus aurantiifolia	South America	Bark	Topical	Grenand et al. (2004)
Solanaceae				
Nicotiana tabacum	Northwest Venezuela	Cataplasm	Topical	Kames-Kaye (1971)
Solanum pectinatum	Ecuador	Fruits	Topical	Davis and Yost (1983)
Solanum sessiliflorum	Amazon	Whole plant	Drink	Schultes and Raffauf (1990)
Urticaceae				
Urera caracasana	upper Rio Negro (Amazonas—Brazil)	Stinging hairs	–	**Prance** (1972)
Cecropia palmatisecta	South America	Bark	Topical	Roth and Lindorf (2002)
		Animal		
Bradypodidae		**Parts**		
Urera caracasana	upper Rio Negro (Amazonas—Brazil)	Stinging hairs	–	**Prance** (1972)
Dasypodidae				
Dasypus novemcinctus		Fat		Martins et al. (2023)

(continued)

Table 14.1 (continued)

Family/species	Location	Part used	Use	Reference
Tinamidae				
Tinamus sp.		Feathers	**Smoke**	Martins et al. (2023)
Bufonidae				Martins et al. (2023)
Rhinella marina		Skin	**Topical**	Martins et al. (2023)
Tapiridae				
Tapirus sp.	Santarém-Pará-Brazil	Fat	–	Martins et al. (2023)

– unspecified

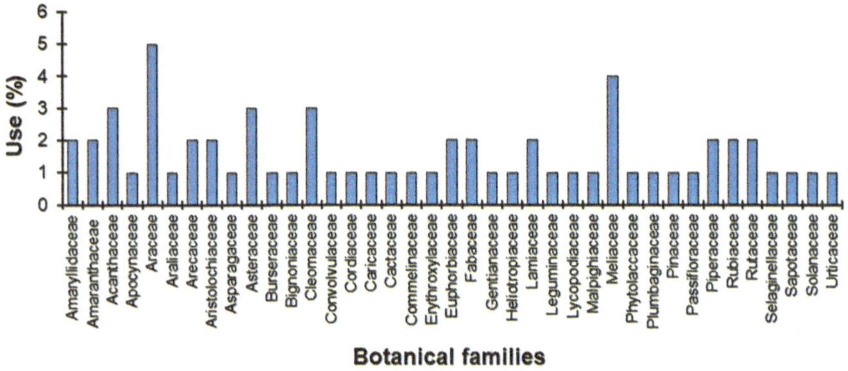

Fig. 14.4 Main botanical families of the Neotropical region used against symptoms of scorpion stings

14.4 How Plants Used as Antidotes are Selected by Rural Communities

In the recent past, Amerindian communities in South America have transferred plants from natural ecosystems to domestic cultivation areas (Fig. 14.4) (Lévi-Strauss 1986; Kerr 1986; Coradin and Lleras 1988; Clement et al. 2017), promoting their use both as food and as natural medicines (Lévi-Strauss 1986; Grenand et al. 2004; DeFilipps et al. 2004; Clement et al. 2010). One example is the palm tree *Bactris gasipaes* Kunth (Arecaceae), with stands domesticated by Native Americans (Clement et al. 2017). In addition to providing edible fruits, its sap is used as an antidote against the pain of scorpion stings in French Guiana (Table 14.1).

Indigenous communities in the Amazon have deep knowledge of the flowering and fruiting cycles of several useful plant species (Lévi-Strauss 1986; Prance 1986; Davis 1996; Clement et al. 2017). This knowledge is applied not only in the selection of medicinal plants, but also in traditional plant taxonomy, where roots, leaves, stems, and bark are given specific names in native languages (Davis 1996). This intimate relationship with the local flora reflects a detailed understanding of the properties and uses of plants.

Experienced indigenous groups are able to distinguish wild hallucinogenic, poisonous, and edible plants (Lévi-Strauss 1986; Prance 1986; Schultes et al. 2001). For example, rural communities in the Yucatan Peninsula, Mexico, rely on organoleptic properties, such as aroma and flavor, to identify species with therapeutic potential (Ankli et al. 1999). In this region, herbs and shrubs are the plant groups most frequently used in traditional treatments of scorpion stings.

The Ecological Appearance Hypothesis suggests that visible plant species, such as trees and shrubs, are often managed by local communities (Feeny 1976), while less visible herbs may be undervalued (Phillips and Gentry 1993). Before adopting a plant for medicinal use, indigenous people observe natural characteristics, such as pollinator interactions with aerial parts of the plant (Elisabetsky 1986). Furthermore, aspects such as the orientation of leaves in relation to the sun are considered significant by ethnic groups (Elisabetsky 1986). These specific criteria help Amerindians to identify plants with medicinal potential (Davis and Yost 1983; Schultes et al. 2001).

Furthermore, rural communities often associate specific morphological characteristics of plants with the animals that cause incidents. For example, in French Guiana, plants with structures resembling spiders are believed to be effective in treating symptoms caused by spider bites (Grenand et al. 2004). Likewise, in the indigenous cosmology of the natives of French Guiana, for example, plants with structures that resemble venomous animals, such as spiders, are considered effective in alleviating symptoms caused by these same animals (DeFilipps et al. 2004; Grenand et al. 2004).

The abundance of a plant species also influences its preference for medicinal purposes. Hyperdominant plants, such as palms (Arecaceae), are more commonly used as medicines, foods, and in religious rituals, in part because of their greater accessibility (Kahn and de Granville 1992). However, further studies are needed to deepen understanding of the belief systems of rural communities, as well as their methods of diagnosing and treating envenoming by animals.

The leaves of plants used as antidotes of scorpion stings are processed manually to intensify their aroma and perceived therapeutic activity (Rutter 1990; Schultes and Raffauf 1990; Santos and Coimbra 1994; Schultes et al. 2001). The Kulina Indians, from the Arawá linguistic group, who inhabit areas close to the Juruá and Purus rivers in the Brazilian state of Acre, choose aromatic herbs to treat arthropod stings, believing that the aroma penetrates the skin and exerts therapeutic effects (Santos and Coimbra 1994). Similarly, non-indigenous rubber tappers in Acre evaluate the smell and taste of plants before using them as natural remedies (Ming 2006).

In the Lower Amazon, Pará, Brazil, non-indigenous communities believe that the latex of the "sucuba" tree (*Himatanthus sucuuba*), used as medicine, can be toxic if extracted out of season (Amorozo and Gély 1988). Furthermore, teas prepared with its leaves should be made before sunset to ensure their effectiveness (Amorozo and Gély 1988). In Andean Venezuela, "chimó" (Fig. 14.5d), a paste made from tobacco (*Nicotiana* spp.), is applied to scorpion stings and used as protection against evil spirits and poisonous animals (Fig. 14.5) (Kames-Kaye 1971). *Nicotiana tabacum*, often called "God's plant" by rural communities (Schultes et al. 2001), has

Fig. 14.5 The figures illustrate the strong belief systems of local communities, which may influence the choice of treatment for people envenomed by animals. A home-based community pharmacy in Barcelos, Meddle Rio Negro, Amazonas, Brazil, where the women who produce natural remedies are known as "witches of God" (**a**); Bark and leaves of endemic plants stored in containers for the preparation of home remedies, including those against venomous animal bites (**b**–**c**); "Chimó", a paste composed of *Nicotiana tabacum* leaves, brown sugar, bicarbonate and tree ash used to treat scorpion bites in the Andean region of Venezuela (**d**)

extensive uses in traditional medicine and religious rituals among Amerindians (Schultes and Raffauf 1990; Duke and Vasquez 1994; Duke 2009).

People of Andean communities in Venezuela believe it is important to keep a perennial herbaceous plant (e.g., *N. tabacum*) inside their homes because its odor can repel both poisonous animals and evil spirits (Kames-Kaye 1971). Some of these plants have hallucinogenic properties whose consumption alters human perception, causing mood changes as well as auditory and tactile hallucinations (Elisabetsky 1986; Schultes et al. 2001). However, indigenous people carefully regulate the dosage of these substances (Posey and Elisabetsky 1991), adjusting the quantities according to the age of the users: smaller doses for children and larger doses for adults (de Oliveira and Hamú 1992).

Residents of rural communities tend to value home remedies. For example, the group of women who work in a community pharmacy in the municipality of Barcelos, Middle Rio Negro, Amazonas, Brazil, are called by the local population "little witches of God" (Fig. 14.5). Local studies show that plants used in ethnomedicine have not only a botanical identity, but also a therapeutic history deeply connected to local cultural values (Rutter 1990; Duke 2009).

14.5 Can Plant Extracts Replace Therapeutic Antibodies?

Scorpion envenomation syndrome has been safely treated in referral hospitals for decades (Ismail 1994; Pardal et al. 2003; Cupo 2015; Chippaux et al., 2020; Martins et al. 2026). Neotropical countries such as Brazil, Mexico, Argentina, Venezuela, and Bolivia stand out for their self-sufficiency in the production of effective and safe antivenoms that can be stored anywhere in the world at appropriate temperatures (Ismail 1994; Chippaux et al. 2020). These antivenoms consist mainly of functional antibodies capable of neutralizing circulating toxins in the deep tissues affected by scorpion toxins (Reis et al. 2023; Bordon et al. 2025; Ismail 1994).

Parenteral administration of antivenoms, indicated for moderate and severe cases, results in a significant reduction in the lethality risk (Martins et al. 2026), as observed in patients with symptoms of scorpionism in Mexico and Brazil (Pardal et al. 2003; Cupo 2015; Chippaux et al. 2020). This therapy is primarily responsible for reversing the mechanisms of animal envenoming. Systemic symptoms, such as nausea, hypertension, and agitation, regress with the neutralization of the toxins inoculated by the scorpion (Bordon et al. 2025; Ismail 1994).

The majority of plant extracts applied at the site of the sting or ingested (Table 14.1) do not guarantee relief from pain and other symptoms triggered by scorpion envenomation. Although some plants may have analgesic activity, from a clinical point of view, serum therapy administered directly into the bloodstream is the most effective treatment against the lethal effects of toxins. It is not safe to use plants in mild, moderate, or severe cases. This type of self-care should be discouraged, especially among the people of non-indigenous communities, who tend to be more receptive to modern treatments prescribed by health professionals.

However, in Amazonian and Andean regions, limited access to health services causes indigenous and non-indigenous communities to resort to traditional treatments without clinical evidence (Rutter 1990; Martins et al. 2023). Medicinal plants based on ancestral knowledge constitute the first line of care for some neglected tropical diseases in the Amazon (Duke and Vasquez 1994; Grenand et al. 2004). This practice, although questionable, reflects an adaptation of remote rural communities, where medicinal plant species are more accessible than pharmacies and hospitals.

14.6 Concluding Remarks

Research in the Amazon indicates that antivenoms against scorpion toxins have limited availability in most of the region. The authorities in French Guiana, Guyana, Suriname, Ecuador, and Colombia, for example, do not keep records of national production of antivenoms specific for scorpionism. In these places, the victims of scorpion stings often resort to symptomatic treatment.

Amazonian plants, which are deeply integrated into the lives of indigenous and non-indigenous peoples, are often considered remedies with potential equivalent to that of conventional pharmaceuticals. The residents of many riverside communities in the state of Amazonas, Brazil, for example, use extracts from plants grown near their homes in lieu of commercial medicines. These self-care practices are especially common in places with limited access to health services.

However, our review indicates that many plant species traditionally used as antidotes, such as *Crotalaria retusa* and *Petiveria alliacea*, warrant further scientific research to identify their active phytocompounds and validate their therapeutic efficacy against animal toxins. Furthermore, since the ingestion of plant extracts or phytotherapeutics can cause adverse reactions, there is a need for caution in the use of these resources.

Given this scenario, it is essential to invest in health education programs in high-risk areas, focusing on raising awareness about the dangers of venomous animal bites and promoting good prevention practices, such as the use of protective clothing and avoiding areas with a high concentration of scorpions. At the same time, it is important to expand the network of health centers with antivenom stocks to ensure timely and safe treatment of the victims.

References

Amorozo MCM, Gély A (1988) Uso de plantas medicinais por caboclos do Baixo Amazonas, Barcarena, Pará, Brasil. Bol do Mus Paraense Emílio Goeldi 4(1):47–131

Ankli A, Sticher O, Heinrich M (1999) Yucatec Maya Medicinal Plants Versus Nonmedicinal Plants: Indigenous Characterization and Selection. Hum Ecol 27:557–580. https://doi.org/10.1023/A:1018791927215

Batista CVF, Martins JG, Restano-Cassulini R, Coronas FIV, Zamudio FZ, Procópio R, Possani LD (2018) Venom characterization of the Amazonian scorpion *Tityus metuendus*. Toxicon 143:51–58. https://doi.org/10.1016/j.toxicon.2018.01.006

Branch LC, da Silva MF (1983) Folk medicine of Alter do Chão, Pará, Brazil. Acta Amazon 13:737–797. https://doi.org/10.1590/1809-4392135737

Brasil (2022) Ministério da Saúde/SVS. Bol Epidemiol 53:1–48

Bordon KCF, Santos GC, Martins JG, Wiezel GA, Amorim FG, Crasset T, Redureau D, Quinton L, Procópio REL, Arantes EC (2025) Pioneering comparative proteomic and enzymatic profiling of Amazonian scorpion venoms enables the isolation of their first α-ktx, metalloprotease, and phospholipase A2. Toxins (Basel)17(8):411

Chagnon NA (1968) Yanomamo: the fierce people, 1st edn. Holt, Rinehart and Winston, New York

Chippaux JP (2015) Epidemiology of envenomations by terrestrial venomous animals in Brazil based on case reporting: from obvious facts to contingencies. J Venom Anim Toxins Incl Trop Dis 21:1–13. https://doi.org/10.1186/s40409-015-0011-1

Chippaux JP, Alfredo C, Leslie B, Alejandro A (2020) Factors involved in the resilience of incidence and decrease of mortality from scorpion stings in Mexico. Toxicon 188:65–75. https://doi.org/10.1016/j.toxicon.2020.10.011

Clement CR, Cristo-Araújo M, Coppens d'Eeckenbrugge G, Alves Pereira A, Picanço-Rodrigues D (2010) Origin and domestication of native Amazonian crops. Diversity 2:72–106. https://doi.org/10.3390/d2010072

Clement CR, Cristo-Araújo M, Coppens d'Eeckenbrugge G, Reis VM, Lehnebach R, Picanço-Rodrigues D (2017) Origin and dispersal of domesticated peach palm. Front Ecol Evol 5:148. https://doi.org/10.3389/fevo.2017.00148

Coradin L, Lleras E (1988) Overview of palm domestication in Latin America. Adv Econ Bot 6:175–189

Cupo P (2015) Clinical update on scorpion envenoming. Rev Soc Bras Med Trop 48(6):642–646. https://doi.org/10.1590/0037-8682-0237-2015

Cupo P, Figueiredo AB, Filho AP, Pintya AO, Tavares Junior GA, Caligaris F, Marin-Neto JÁ, Hering SE, Simões MV (2007) Acute left ventricular dysfunction of severe scorpion envenomation is related to myocardial perfusion disturbance. Int J Cardiol 116:98–106. https://doi.org/10.1016/j.ijcard.2006.02.015

Davis W (1996) One River: explorations and discoveries in the Amazon rain forest. Simon and Schuster Paperbacks, New York

Davis EW, Yost JA (1983) The ethnomedicine of the Waorani of Amazonian Ecuador. J Ethnopharmacol 9(2–3):273–297. https://doi.org/10.1016/0378-8741(83)90036-3

de Oliveira AE, Hamú E (1992) Ciência Kayapó: alternativas contra destruição. Editora Museu Paraense Emílio Goeldi, Belém

DeFilipps RA, Maina SL, Crepin J (2004) Medicinal Plants of the Guianas (Guyana, Surinam, French Guiana). Medicinal plants of the Guianas, Guyana, Surinam, French Guiana. https://nmnh.typepad.com/the_plant_press/2004/01/plant-press-2004-vol-7-issue-1.html. Accessed 12 Mar 2024

Dimayuga RE, Agundez J (1986) Traditional medicine of Baja California Sur (Mexico) I. J Ethnopharmacol 7(2):183–193. https://doi.org/10.1016/0378-8741(86)90057-7

Duke JA (2009) Medicinal plants of Latin America. CRC Press, Boca Raton

Duke JA, Vasquez R (1994) Amazonian Ethnobotanical Dictionary. CRC Press, Boca Raton

Eickstedt VR (1989) O que você sabe sobre escorpiões? Rev Ensino Ciências 22:35–40

Elisabetsky, E (1986) Etnofarmacologia de algumas tribos brasileiras. In: Ribeiro, D. Suma etnológica brasileira. Petrópolis: Vozes

Feeny P (1976) Plant apparency and chemical defense. In: Biochemical interactions between plants and insects. Recent advances in phytochemistry. Springer, Boston

Fet V, Lowe G (2000) Family Buthidae C.L. Koch, 1837. In: Catalog of the scorpions of the world. The New York Entomological Society, New York, pp 54–286

Grenand P, Moretti C, Jacquemin H (2004) Pharmacopées traditionelles en Guyane. Créoles, Palikur, Wayãpi. Editions l'Orstom, Paris

Gonzalez-Ponce E, Rodríguez-Rangel S, Martinez R, Alvarado A, Ruiz-Baca, E Miranda, P Sánchez-Rodríguez JE, Lopez-Rodriguez A (2023) Scorpions, Science and Folklore in Durango City. Diversity 2023, 15(6):743. https://doi.org/10.3390/d15060743

Ismail M (1994) The treatment of the scorpion envenoming syndrome: the Saudi experience with serotherapy. Toxicon 32(9):1019–1026. https://doi.org/10.1016/0041-0101(94)90384-0

Kahn F, de Granville JJ (1992) Palm Communities in the forest ecosystems of Amazonia. Berlin: Springer

Kamen-Kaye D (1975) Chimó: Why Not? A Primitive Form of Tobacco Still in Use in Venezuela. Economic Botany 29(1):47–68

Kerr W (1986) Agricultura e selecões genéticas de plantas. In: Ribeiro D, Ribeiro BG (eds) Suma Etnológica Brasileira. Vozes, Petrópolis

Lévi-Strauss C (1986) O uso de plantas silvestres da América do Sul Tropical. In: Ribeiro D, Ribeiro BG (eds) Suma Etnológica Brasileira. Vozes, Petrópolis, pp 27–46

Lorenzi H, Matos FJA (2002) Plantas medicinais no Brasil: nativas e exóticas. Plantarum, Nova Odessa

Luz DA, Pinheiro AM, Silva ML, Monteiro MC, Prediger RD, Ferraz Maia CS, Fontes-Júnior EA (2016) Ethnobotany, phytochemistry and neuropharmacological effects of Petiveria alliacea L. (Phytolaccaceae): A review. J Ethnopharmacol 5;185:182–201. https://doi.org/10.1016/j.jep.2016.02.053

Lyon PJ (1974) Native South Americans: ethnology of the least known continent. University of Michigan, Ann Arbor

Martins JG, Santos GC, Procopio REL, Arantes EC, Bordon KCF (2021) Scorpion species of medical importance in the Brazilian Amazon: a review to identify knowledge gaps. J Venom Anim Toxins Incl Trop Dis 27:1–32. https://doi.org/10.1590/1678-9199-JVATITD-2021-0012

Martins JG, Almeida BR, Pardal PPO, Procópio REL (2026) Scorpion Stings and Antivenom Coverage in the Brazilian Amazon. Wilderness Environ Med 23:10806032261417950

Martins JG, de Castro Figueiredo Bordon K, Moreno-González JA, de Almeida BRR, Pardal PPO, Lira AFA, Cândido DM, Arantes EC, Procópio REL (2023) On the noxious black Amazonian scorpion, *Tityus obscurus* (Scorpiones, Buthidae): Taxonomic notes, biology, medical importance and envenoming treatment. Toxicon 228:107125. https://doi.org/10.1016/j.toxicon.2023.107125

Martins JG, Reis MB, Zuanazzi BA, Bordon KCF, Zoccal KF, Procópio REL, Arantes EC (2024) Clinical outcomes in a murine model after envenoming by the Amazonian scorpions *Tityus strandi* and *Tityus dinizi*. Toxicon 246:107797. https://doi.org/10.1016/j.toxicon.2024.107797

Milliken W, Miller RP, Pollard SR, Wandelli E (1992) Ethnobotany of the Waimiri-Atroari Indians of Brazil. Royal Botanic Gardens, London

Milliken W, Albert B (1996) The use of medicinal plants by the Yanomami indians of Brazil. Economic Botanic 50(1):10–25

Ming LC (2006) Plantas medicinais na reserva extrativista Chico Mendes, Acre: uma visão etnobotânica. UNESP, Botucatu-São Paulo

Morton JF (1981) Atlas of Medicinal Plants of Middle America. Charles C. Thomas, Illinois

Pardal PPO, Castro LC, Jennings E, Pardal JS, Monteiro MRC (2003) Aspectos epidemiológicos e clínicos do escorpionismo na região de Santarém, estado do Pará, Brasil. Rev Soc Bras Med Trop 36:349–353. https://doi.org/10.1590/S0037-86822003000300006

Pardal PPO, Ishikawa EA, Vieira JL, Coelho JS, Dórea RC, Abati PA, Quiroga MMM, Chalkidis HM (2014) Clinical aspects of envenomation caused by *Tityus obscurus* (Gervais, 1843) in two distinct regions of Pará state, Brazilian Amazon basin: a prospective case series. J Venom Anim Toxins Incl Trop Dis 20:1–7. https://doi.org/10.1186/1678-9199-20-3

Phillips O, Gentry AH (1993) The useful plants of Tambopata, Peru: II. Additional hypothesis testing in quantitative ethnobotany. Econ Bot 47:33–43. https://doi.org/10.1007/BF02862204

Plotkin MJ, Balick MJ (1984) Medicinal uses of South American palms. J Ethnopharmacol 10:157–179. https://doi.org/10.1016/0378-8741(84)90001-1

Plowman T (1984) The ethnobotany of coca (Erythroxylum spp., Erythroxylaceae). Advances Econ. Bot 1:62–111

Posey DA, Elisabetsky E (1991) Conceito de animais e seus espíritos em relação a doenças e curas entre os índios Kayapó da Aldeia Gorotire, Pará. Bletim do Museu Paraense Emílio Goeldi série Antropologia 7(1):21–36

Prance, G (1972) An ethnobotanical comparison of four tribes of Amazonian Indians. Acta Amazon 2(2):12–32

Prance GT (1986) Etnobotânica de algumas tribos brasileiras. In: Ribeiro D, Ribeiro BG (eds) Suma Etnológica Brasileira. Vozes, Petrópolis, pp 119–133

Prance GT, Campbell DG, Nelson BW (1977) The ethnobotany of the Paumarí Indians. Econ Bot 31:129–139

Reichel-Dolmatoff G (1969) El contexto cultural de un alucinógeno aborigen: *Banisteriopsis caapi*. Rev Acad Colomb Cienc Exact Fís Nat 13(51):327–345

Reis MB, Martins JG, Bordon KCF, Fraga-Silva TFC, Procópio REL, de Almedia BRR, Bonato VLD, Arantes EC (2023) Pioneering in vitro characterization of macrophage response induced by scorpion venoms from the Brazilian Amazon. Toxicon 230(7):107171

Roth I, Lindorf H (2002) South American Medicinal Plants. Springer, Berlin

Rutter RA (1990) Catalogo de plantas utiles de la Amazonia Peruana. Instituto Linguistico de Verano, Yarinacocha

Santos RV, Coimbra JR (1994) Saúde e povos indígenas. Rio de Janeiro: Editora FIOCRUZ

Schaan DP (1997) A linguagem iconográfica da cerâmica Marajoara. Um estudo da arte pré-histórica na Ilha de Marajó, Brasil (400–1300 A.D.). Coleção Arqueologia, Porto Alegre
Schultes RE, Raffauf RF (1990) The healing forest. Dioscorides Press, Portland
Schultes RE, Hofmann A, Rätsch C (2001) Plants of the Gods: their sacred, healing and hallucinogenic powers. McGraw-Hill, New York
Silverwood-Cope PL (1990) Os Makú: povo caçador do noroeste da Amazônia. Editora Universidade de Brasília, Brasília
Smith N (2015) Palms and people in the Amazon. Springer, Heidelberg
Ulloa UC, Acevedo-Rodríguez P, Beck S, Belgrano MJ, Bernal R, Berry PE, Brako L, Celis M, Davidse G, Forzza RC, Gradstein SR, Hokche O, León B, León-Yánez S, Magill RE, Neill DA, Nee M, Raven PH, Stimmel H, Strong MT, Villaseñor JL, Zarucchi JL, Zuloaga FO, Jørgensen PM (2017) An integrated assessment of the vascular plant species of the Americas. Science 22;358(6370):1614–1617. https://doi.org/10.1126/science.aao0398
van Andel TR (2000) Non-timber forest products of the North-West district of Guyana. Part I., Tropenbos-Guyana Series 8a. Tropenbos-Guyana Programme, Georgetown
Ward MJ, Ellsworth SA, Nystrom GS (2018) A global accounting of medically significant scorpions: Epidemiology, major toxins, and comparative resources in harmless counterparts. Toxicon 151:137–155. https://doi.org/10.1016/j.toxicon.2018.07.007
Wilbert J, Simoneau K (1990) Folk literature of the Yanomami Indians. UCLA Latin American Center Publications, Los Angeles
Ythier E (2018) A synopsis of the scorpion fauna of French Guiana, with description of four new species. ZooKeys 764:27–90. https://doi.org/10.3897/zookeys.764.25108
Ythier E, Dupre G (2024) Les scorpions de la Guyane Française (Arachnida: Scorpiones). Arachnides 118:1–7
Zoccal KF, Bordon KCF, Reis MB, Chini PBRNS, Martins JG, Zuanazzi BA, Wizel GA, Santos ACC, Procópio ERL, Arantes EC (2025) Divergent clinical, inflammatory, and histopathological responses induced by Amazonian Tityus venoms: insights and limitations of current antivenom therapy. Biochimie 238(Pt B):159–171

Chapter 15
Neotropical Scorpion Venom

Juana María Jiménez-Vargas, Jimena Isaias Cid-Uribe, and Rita Restano-Cassulini

Abstract Scorpion venom is a cocktail of hundreds of compounds, including metals, enzymes, protease inhibitors, mucopolysaccharides, salts, amino acids, ions, neurotransmitters, peptides, lipids, and different types of protein. Thanks to decades of studies on scorpion venom, it is now known that the cause of poisoning symptoms is the peptide fraction, which consists of toxins capable of affecting cell membrane receptors such as ion channels. Recently, the application of high-throughput technologies to the inquiry of biochemical and physiological effects of the components of venoms allowed us to peek through the immense diversity of non-peptide compounds and their possible function. This chapter describes the state of the art of the principal categories of scorpion venom compounds, their function, and possible application.

Keywords Ion channel toxin • Host defense peptides • Scorpion venom

15.1 Introduction

Scorpion venom contains a complex mixture of components (Fig. 15.1), though not all have been fully characterized. This chapter will focus on the components found in the scorpion venoms from the Neotropical region, including all the species located within the neovolcanic belt in Mexico (Morrone et al. 2022). Following the limits of the Neotropics suggested by Morrone et al. (2022), we will discuss the components identified in the scorpion venom from the families Buthidae (which are

J. M. Jiménez-Vargas
Facultad de Ciencias Químicas, Universidad de Colima, Coquimatlan, Colima, Mexico
e-mail: jjimenez45@ucol.mx

J. I. Cid-Uribe (✉) · R. Restano-Cassulini
Instituto de Biotecnología, Universidad Autónoma de México, Cuernavaca, Morelos, Mexico
e-mail: jimena.cid@ibt.unam.mx; rita.restano@ibt.unam.mx

A. F. de Araujo Lira, E. González-Santillán (eds.), *The Biology of Neotropical Scorpions*, https://doi.org/10.1007/978-3-032-23194-9_15

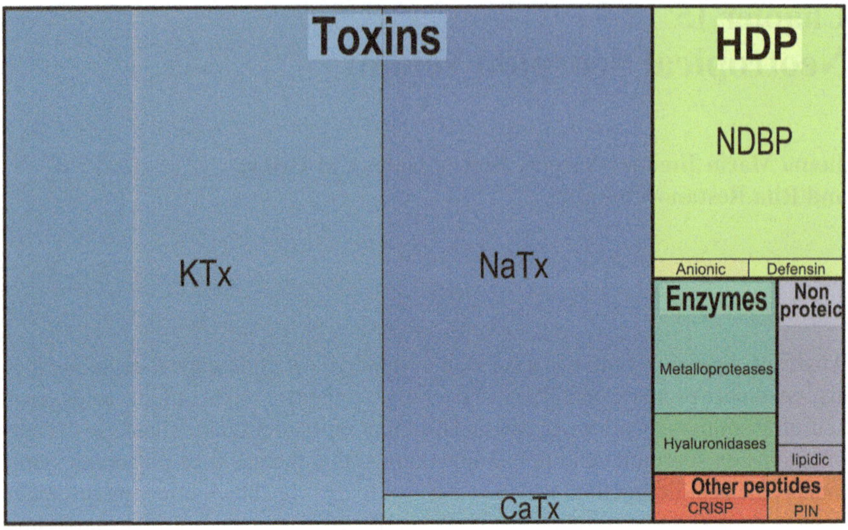

Fig. 15.1 Venom components isolated/identified from neotropical scorpion venoms, including NaTx (30.5%), KTx (45.1%), CaTx (1.7%), Metalloproteases (3.9%), Hyaluronidases (1.7%), Defensin (0.4%), NDBP (11.2%), Anionic (0.4%), CRISP (1.3%), PIN (0.8%), Heterocyclic (2.6%), and Lipidic (0.4%)

the genera *Centruroides, Tityus,* and *Rhopalurus*), Vaejovidae (including the genera *Thorellius* and *Vaejovis*), and Hadruridae (genus *Hoffmannihadrurus*).

15.2 Sodium Scorpion Toxins (NaTx)

Sodium channels consist of a transmembrane protein arranged into four domains (DI—DIV), which delimit a central pore. Each domain contains six transmembrane segments (S1–S6) connected by extracellular and intracellular loops and voltage--gated ion channels respond upon depolarization, opening the pore and permitting the ion flow through the membrane according to the diffusion principle (Yu and Catterall 2003). During membrane depolarization, sodium channels are first activated, but immediately after, the same depolarization promotes the channels transition to the inactivated state, which means channels are open but nonconductive (Fig. 15.2). The channels will remain unable to open again until the membrane repolarizes. This mechanism ensures perfect timing in the signal propagation that is necessary for the effective intercellular communication (Catterall 2014). In mammals, nine genes encode different voltage-gated sodium channels, known as Nav 1.1 to Nav 1.9. These channels are expressed in different tissues. Specifically, Nav 1.1–1.3 are found in the central nervous system; Nav 1.4 is present in skeletal muscles; Nav 1.5 is expressed in cardiac muscle, Nav 1.6 is found in the central nervous system, and ganglia, Nav 1.7, Nav1.8, and Nav 1.9 are expressed in the peripheral

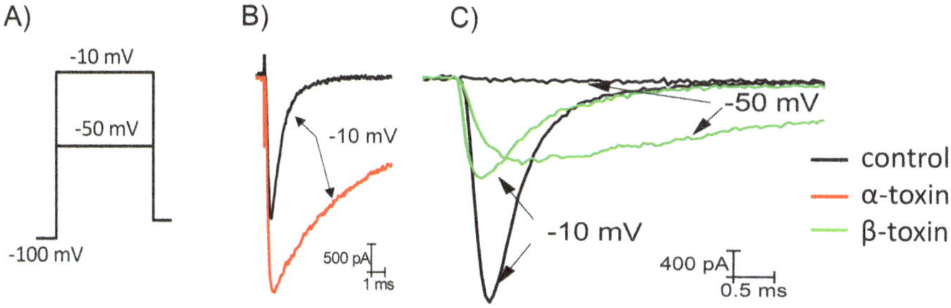

Fig. 15.2 Effect of alpha (red) and beta scorpion toxins (green) on sodium currents. (**a**) Subthreshold (−50 mV) and full activating potential (−10 mV) which elicited currents shown in (**b**) and (**c**). At −50 mV, in the control condition, sodium channels are closed, and no current is recorded, while at −10 mV, all channels are open, and current is maximal (black traces in **b** and **c**). In (**b**), the red trace is sodium current after application of α-NaTx that slows down the inactivation kinetics. In (**c**) green traces are sodium currents after application of β-NaTx that promotes the channel activation at more negative potential (green trace at −50 mV) and reduces the total current (green trace at −10 mV)

nervous system, playing a role in pain signalization (Alexander et al. 2023). Alterations in the functioning of sodium channels are responsible for a wide range of pathologies including epilepsy, arrhythmias, muscular dystrophies, and cancer, among others (Comini et al. 2024).

Among the peptide components of scorpion venom, the toxins that bind sodium channels are mainly responsible for the symptoms associated with their sting (Mendes et al. 2023). These toxins range from 58 to 76 amino acids in length (Possani et al. 1999) and contain three or four conserved disulfide bridges that stabilize their structural motif alpha-beta. To date, 71 toxins modulating sodium channels have been identified in the venom of neotropical scorpions. Based on their effect on sodium channels, NaTx were classified into two subtypes: alpha sodium toxins (α-NaTx) and beta sodium toxins (β-NaTx).

15.2.1 Alpha Sodium Toxins

The alpha sodium toxins (α-NaTx) are the most diverse subfamily in scorpion venoms from Europe, Asia, and Africa. However, only eight toxins with this effect have been identified in neotropical scorpion venoms, mainly in *Tityus* and *Centruroides* venoms (Vandendriessche et al. 2010; Tibery et al. 2024). These toxins bind to the extracellular S3–S4 segment of domain IV, known as site 3 (Cestèle 2000), increasing the neuronal excitation (Zilberberg et al. 1997) and prolonging the time the channel remains open (Cestèle et al. 2001) by slowing down the inactivation process (Fig. 15.2a, b).

The toxin Tst3, isolated from *T. stigmurus* venom, is one of the few alpha toxins characterized from the neotropical region. Structurally, Tst3 shows an α-helix

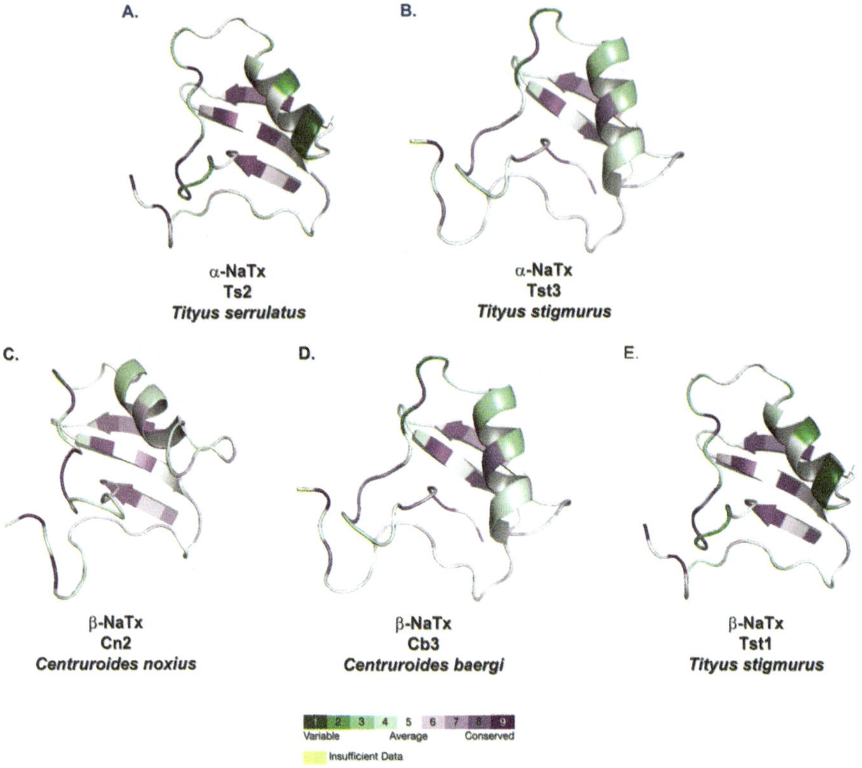

Fig. 15.3 Structures of NaTxs toxins. (**a**) Mammalian α-NaTx, Ts2 isolated *T. serrulatus* showing a typical motif that affects Nav1.2, Nav1.3, Nav1.5, Nav1.6, and Nav1.7 channels (Cologna et al. 2012). However, it shows a characteristic effect of the beta-toxins, causing a significant shift in the voltage dependence of activation for the Nav1.3 channel. (**b**) Structure of Tst3 toxin, derived from *T. stigmurus,* which has a typical action of α-NaTx on the Nav1.7 and Nav 1.5 channels. (**c**) Mammalian β-NaTxs: Cn2 isolated from *Centruroides noxius* and (**d**) Cb3 isolated from *Centruroides baergi*. Cn2 toxin is selective and highly active on the Nav1.6 channel (Schiavon et al. 2006). In contrast, the structure of the Cb3 toxin contributes to its high toxicity as it simultaneously affects the activity of Nav1.1, Nav1.2, Nav1.4, Nav1.5, and Nav1.6 channels (Gómez--Ramírez et al. 2020). (**e**) Mammalian and insect β-NaTx, Ts1 from *Tityus serrulatus*, its structure contributes to its highly active against mammals (Nav 1.6) and insect (DmNav) channels. The structures of Tst3, Ts2, and Cb3 toxins were generated using ColabFold v1.5.5 (Mirdita et al. 2022), and for the Ts1 and Cn2 toxins the PDB data 1B7D and 1CN2 were used. The Consurf web server represented evolutionarily conserved regions of residues in these toxins. The amino acids were colored based on their conservation grades, with a color-coding bar of green to dark purple indicating variable to conserved residues (Yariv et al. 2023)

connected to three β-sheets antiparallel (Figs. 15.2b and 15.3b). Remarkably, there is electrophysiological evidence that the delay of the fast inactivation kinetics of sodium channels, mainly on Nav1.7 and Nav1.5, results from the interaction between Tst3 and domain IV. However, the Nav1.2 and Nav1.7 channels showed the highest binding affinity for the toxin (Tibery et al. 2024).

15.2.2 Beta Sodium Toxins

The beta sodium toxins (β-NaTx) bind to site 4 at the S1–S2 and S3–S4 linkers in domains II and III of the channels (Gurevitz et al. 2015). These toxins modify the voltage threshold required for activation, allowing the channel to activate at more negative potentials and simultaneously reducing the total current by enhancing closed-state inactivation (Fig. 15.2c). The proposed mechanism for the β-NaTx is the so-called "voltage sensor trapping model," where the beta toxin encounters the channel and, during the first depolarization, "traps" the voltage sensor, facilitating the subsequent channel opening (Cestèle et al. 1998). At the end, beta-scorpion toxins work in the same direction as alpha-scorpion toxins, inducing a massive cellular depolarization.

Buthidae scorpions venom from neotropical regions is enriched in β-NaTx in contrast with the venom from Buthidae scorpions from Europe, Africa, and Asia. There are some toxins that are worth mentioning from the β-NaTx, Cn2, Cb3, and Ts1, these displayed alpha-beta structural motifs stabilized by four disulfide bridges (Fig. 15.3c–e). The most lethal one against mammals, isolated from the venom of *Centruroides noxius*, is the Cn2 (Pintar et al. 1999). Cn2 exhibits high selectivity against Nav1.6 and can induce a resurgent current in this channel (Schiavon et al. 2006). In contrast, Cb3, isolated from the *Centruroides baergi* venom, interacts with the majority of sodium channel subtypes, shifting the activation threshold to more negative potential in Nav1.1, Nav1.2, Nav1.4, and Nav1.6 (Gómez-Ramírez et al. 2020). It is interesting to notice that along with the classical beta effect, this toxin slows down the inactivation in Nav1.6 and Nav1.4. A similar activity is reported for Cbo4, isolated from *Centruroides bonito* and Cl13 from *C. limpidus* in Nav1.6 and Nav1.5 channels (Olamendi-Portugal et al. 2017; Restano-Cassulini et al. 2024).

Beta sodium toxins isolated from the venom of the *Tityus* scorpions are similar in function to those isolated from *Centruroides*. Ts1 represents the most abundant toxin in the venom of *Tityus serrulatus* and is promiscuous like Cbo4. Interestingly, Ts1 is highly active upon different subtypes of mammal sodium channels and also recognizes insect sodium channels (Bosmans and Tytgat 2007). The increasing interest in sodium channel toxins and their electrophysiological characterization has led to the identification of toxins that are difficult to categorize as alpha or beta. This is the case of Ts2, which originally was described as beta-scorpion toxin for the primary sequence similarity with this group of toxins (Fig. 15.3a), but more recently was functionally characterized, showing that is able to slow down the inactivation kinetic (a typical effect for an alpha-scorpion toxins), while shifts the activation process to more negative potentials (a typical effect for a beta-scorpion toxin) (Cologna et al. 2012). As the number of characterized sodium channel toxins increases, the current classification method for NaTx appears inadequate.

15.3 Potassium Scorpion Toxins (KTx)

Potassium channels are ion channels that regulate the movement of potassium ions in and out of cells. They play a critical role in many cellular processes, such as regulating neuronal excitability, the duration of action potentials, hormone secretion, smooth muscle relaxation, and cardiac muscle repolarization (Suppiramaniam et al. 2010; Kuang et al. 2015). There are several types of potassium channels, including voltage-gated, calcium-activated, inward rectifiers, and sodium-activated, among others (Suppiramaniam et al. 2010).

Potassium scorpion toxins (KTxs) are powerful molecular probes that have significantly advanced our understanding of various potassium channels. These toxins have been instrumental in uncovering structural and functional diversity, particularly of voltage-dependent (Kv) and calcium-dependent channels (K_{Ca}). The remarkable selectivity of KTxs for specific channels arises from their sequence homology and unique conformational structures. KTxs can be classified into seven distinct subfamilies: α-KTx, β-KTx, γ-KTx, δ-KTx, ε-KTx, κ-KTx, and λ-KTx (Fig. 15.4). This classification highlights the complexity and richness of these molecules. The λ-KTx toxin is distinctive in that it is the sole type absent in the venoms of Neotropical scorpions. Conversely, the presence of all other peptide types has been identified in these scorpions (Tabakmakher et al. 2019; Naseem et al. 2023). The potassium channel activity of one hundred and four different toxins derived from the venom of neotropical scorpions belonging to the genera *Centruroides* (e.g., Corona et al. 2002; Olamendi-Portugal et al. 2005, 2016; Cid-Uribe et al. 2017; Beltrán-Vidal et al. 2021), *Heteroctenus* (Rodríguez-Ravelo et al. 2014), *Hoffmannihadrurus* (Schwartz et al. 2006), *Opisthacanthus* (Silva et al. 2009; Vandendriessche et al. 2012), *Rhopalurus* (Rodríguez-Ravelo et al. 2015), *Tityus* (e.g., Batista et al. 2006; Diego-García et al. 2007; D'Suze et al. 2009; Almeida et al. 2013; Freire et al. 2019; Oliveira et al. 2022), and *Vaejovis* (Varga et al. 2012; Quintero-Hernández et al. 2015; Csoti et al. 2022) have been characterized, highlighting the fascinating biochemical diversity within these species.

15.3.1 The Alpha Potassium Channel Toxins (α-KTxs)

The alpha potassium channel toxins peptides are short molecules consisting of 23–40 amino acids that block potassium channels. They present one or two short alpha helices connected to anti-parallel beta-sheets stabilized by three or four disulfide bridges, exhibiting a structural motif known as cysteine-stabilized α-helix/β-sheet (CSα/β) (Fig. 15.4). This family of toxins is highly diverse, comprising 32 different subfamilies (Saucedo et al. 2012), and primarily inhibits Kv channels. When α-KTxs bind to the channel, a lysine residue penetrates deep into the selectivity filter, blocking the pore and closing off the K^+ ion pathway. This mechanism showcases the potency and sophistication of these peptides in

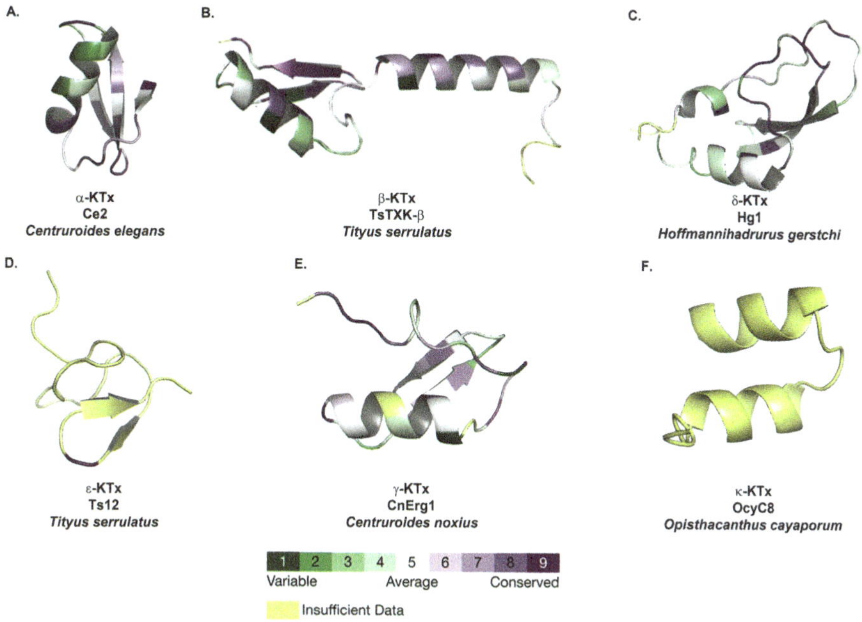

Fig. 15.4 Illustrates the structures of KTxs toxins. (**a**) α-KTx: Ce2, isolated from *C. elegans*, features a Cysteine-Stabilized α/β motif, which includes a triple-stranded β-sheet connected to an α-helix. (**b**) β-KTx: TsTXK-β, derived from *T. serrulatus*, exhibits an α-helix at the N-terminal connected to a double-stranded β-sheet, which is bonded to an α-helix at the C-terminal. (**c**) δ-KTx: Hg1 from *H. gertschi* showcases an α-helix linked to a double-stranded β-sheet bonded to another α-helix. (**d**) ε-KTx: Ts12, also from *T. serrulatus*, possesses two α-helixes in its structure. (**e**) γ-KTx: CnErg1, obtained from *C. noxius*, presents a double-stranded β-sheet linked to an α-helix. (**f**) κ-KTx: OcyC8, from *O. cayaporum*, features a structure comprising two α-helixes in their structure. The structures of Ce2 (P0C162), TsTXK-β (P69940), Hg1 (P0C8W3), Ts12 (P0C175), and OcyC8 (P86110) were generated using ColabFold v1.5.5 (Mirdita et al. 2022). For the CnErg1 toxin (Q86QT3), the PDB data 1NE5 was used. The Consurf web server was employed to identify evolutionarily conserved regions of residues in these toxins. The amino acids were color-coded based on their conservation levels, with a color-coding bar of green to dark purple indicating variable to conserved residues (Yariv et al. 2023)

regulating ion flow (Goldstein and Miller 1993; Banerjee et al. 2013; Selvakumar et al. 2022). To date, 67 α-KTx have been identified in the venoms of Neotropical scorpions, comprising the genera *Centruroides* (Corona et al. 2002; Olamendi-Portugal et al. 2005, 2016; Cid-Uribe et al. 2017; Beltrán-Vidal et al. 2021), *Heteroctenus* (Rodríguez-Ravelo et al. 2014), *Hoffmannihadrurus* (Schwartz et al. 2006), *Opisthacanthus* (Silva et al. 2009; Vandendriessche et al. 2012), *Rhopalurus* (Rodríguez-Ravelo et al. 2015), *Tityus* (e.g., Batista et al. 2007; D'Suze et al. 2009; Papp et al. 2009; Saucedo et al. 2012; Oliveira-Mendes et al. 2018; Freire et al. 2019; Titaux-Delgado et al. 2023), and *Vaejovis* (Varga et al. 2012; Quintero-Hernández et al. 2015; Csoti et al. 2022). The α-KTx peptides demonstrate significant diversity in their affinities for the different Kv channels, with a notable selectivity for the Kv1.1, Kv1.2, and Kv1.3 subtypes. Some of these peptides also interact with calcium-activated K+ channels (KCa). For instance, the Vm24 peptide (P0DJ31), derived from the venom of the *Vaejovis mexicanus* scorpion, consists of 24 amino acids and shows a high selectivity for Kv1.3 channels found in human T-lymphocytes, with a dissociation constant (Kd) of 2.9 pM (Varga et al. 2012). It selectivity by Kv1.3 channel is also observed with the peptides Ce2 (P0C162) and Ce4 (P0C164) from *Centruroides elegans*, which show a Kd of 0.25 nM and 0.98 nM, respectively (Olamendi-Portugal et al. 2005). In contrast, the Tst26 peptide (P0CB56), sourced from *Tityus trivittatus*, blocks the Kv1.2 channel with high affinity, exhibiting a Kd of 1.9 nM and demonstrating tenfold greater selectivity over the Kv1.3 channel (Papp et al. 2009).

15.3.2 Beta Potassium Channel Toxins (β-KTxs)

The beta potassium channel toxins subfamily consists of long-chain peptides, each containing between 47 and 67 amino acids. Sequence analysis and structural predictions have revealed that β-KTxs share a typical folding pattern characterized by two distinct domains. The N-terminal helical domain is understood to mediate antimicrobial and cytolytic activities. In contrast, the C-terminal domain contains the cysteine-stabilized α/β structural motif (CSαβ) (Fig. 15.4b), which is believed to exhibit potassium channel-blocking activity (Oliveira et al. 2022). As of 2024, researchers have identified ten distinct peptides in the venom of Neotropical scorpions (Diego-García et al. 2007; Oliveira et al. 2022; Titaux-Delgado et al. 2023), categorizing them into three groups:

- Peptides related to TsTxK-β are known for their antimicrobial properties and ability to inhibit potassium channels. TsTxK-β is a toxin derived from the venom of the Brazilian scorpion *T. serrulatus*. This toxin has shown effectiveness against Gram-positive and Gram-negative bacterial strains, fungi, and yeast while exhibiting low toxicity to human erythrocytes (Oliveira et al. 2022). Furthermore, homologous peptides in the venom of other neotropical scorpions of the same genus were identified, including *T. stigmurus* (TstbetaKTx), *T. discrepans* (TdibetaKTx), and *T. trivittatus* (TtrbetaKTx) (e.g., Batista et al. 2006, 2007; Diego-García et al. 2007; D'Suze et al. 2009; Titaux-Delgado et al. 2023).
- BmTxK-β-related peptides. Four peptides with high identity were found in the venom from *T. costatus* (TcoKIK), *T. trivittatus* (TtrKIK), *T. discrepans* (TdiKIK), and *T. stigmurus* (TstKMK) (Diego-García et al. 2007; Almeida et al. 2013; Titaux-Delgado et al. 2023).
- Scorpine-related peptides. These present two structural domains, an N-terminal domain that adopted α-helical folding and is responsible for the antimicrobial, cytolytic, and parasitic activities. In contrast, the C-terminal domain inhibits potassium channels. Two scorpines have been found in the venom of Neotropical scorpions, such as TcobetaKTx and OcyC7, in *T. costatus* and *Opisthacanthus*

cayaporum, respectively (Batista et al. 2007; Silva et al. 2009; López-Giraldo et al. 2023).

15.3.3 Potassium Channel Toxins Delta-KTx (δ-KTxs)

The δ-KTxs subfamily is recognized for two primary functions: acting as protease inhibitors and blocking potassium channels. These peptides contain long chains of 50–70 amino acids and adopt a Kunitz-type structural conformation. It presents two antiparallel β-sheets flanked by an α-helix at the C-terminus and N-terminus (Fig. 15.4c). Three disulfide bridges stabilize this structural motif. The first bridge connects the C-terminal α-helix to one strand of the β-sheet, while the second and third bridges link the C-terminal α-helix to the ends of both β-sheet domains (Zhao et al. 2011; Chen et al. 2012, 2013a, b).

Only one δ-KTx from the neotropical scorpion venom of *Hoffmannihadrurus gertschi* was isolated. This peptide is named Kunitz-type serine protease inhibitor Hg1 (δ-KTx 1.1). Several studies have shown that it has a dual function: it inhibits trypsin activity with a Kd of 107 nM and reduces the currents of mKv1.3 channels by 80% at a concentration of 1 μM (Schwartz et al. 2007; Chen et al. 2012).

15.3.4 Potassium Channel Toxins Epsilon-KTx (ε-KTxs)

The ε-KTxs subfamily consists of peptides that are fewer than 29 amino acids in length, distinguished by a unique structural conformation known as the inhibitor cystine knot (ICK). This remarkable motif features three compact, antiparallel β-sheets stabilized by four disulfide bridges (Fig. 15.4d) (Zhu et al. 2003; Smith et al. 2011; Cremonez et al. 2016). In the venom of Neotropical scorpions, two remarkable peptides stand out: Kappa-KTx 1.1 (Ts11) and Kappa-KTx 1.2 (Ts12). Both peptides were isolated from the venom of *T. serrulatus* (Cremonez et al. 2016).

Extensive testing of these peptides on various potassium (Kv) channels using advanced electrophysiological techniques has revealed significant differences in their effectiveness. At a concentration of 3 μM, Ts11 effectively inhibits currents by 25% for Kv1.2, 27% for Kv1.3, 25% for Kv4.2, 15% for Kv10.1, 12% for hERG, and 10% for Shaker IR. Although Ts12 is less potent, it demonstrates notable blocking effects: 5% on Kv1.2, 10% on Kv1.3, 20% on Kv1.4, 24% on hERG, and 27% on Shaker IR at the same concentration. These findings highlight the potential of ε-KTxs in understanding ion channel modulation and their applications in pharmacology (Cremonez et al. 2016).

15.3.5 Potassium Channel Toxins Gamma-KTx (γ-KTxs)

In the ɣ-KTx subfamily, Erg toxins specifically target voltage-gated potassium channels known as ERG (*Ether-à-go-go-related gene*) channels. In mammals, there are three subtypes of ERG channels: ERG1 (Kv11.1), ERG2 (Kv11.2), and ERG3 (Kv11.3). These channels have a high degree of amino acid sequence identity, although they differ subtly in their voltage-dependence of activation (Perry and Sanguinetti 2008). ERG channels are expressed in cardiac tissues, the brain, and the central nervous system (Shi et al. 1997; Wymore et al. 1997; Shepard et al. 2006). They play a crucial role in repolarizing the cardiac action potential and regulating neuronal excitability (Sanchez-Conde et al. 2022).

The ɣ-KTx are peptides of 42–47 amino acids, forming an alpha/beta structural conformation stabilized by four disulfide bonds (Fig. 15.4e). These toxins are contained mainly in the scorpion venoms belonging to the genus *Centruroides.* Among the neotropical species, notable examples include *C. elegans* (which has five identified Erg toxins: CeErg1, CeErg2, CeErg3, CeErg4, and CeErg5 (Corona et al. 2002, Restano-Cassulini et al. 2008)), *C. limpidus* (CllErg1, CllErg2, CllErg3, and CllErg4 (Corona et al. 2002)), *C. margaritatus* (CmErg1 (Beltrán-Vidal et al. 2021)), *C. noxius* (CnErg1, CnErg2, CnErg3, CnErg4, and CnErg5 (Gurrola et al. 1999, Bottiglieri et al. 2000, Corona et al. 2002, Lecchi et al. 2002)), *C. sculpturatus* (CsEKerg1, CsErg2, CsErg3, CsErg4, and CsErg5 (Corona et al. 2002)), and *C. tecomanus* (Toxin II.10.9 (Olamendi-Portugal et al. 2016)).

Twenty-two gamma toxins have been characterized in the venom of these species, demonstrating their ability to block approximately 90% blockage of potassium ion passage through the ERG channel. The Erg toxins that showed more affinity by the Kv11.1 channels were CmErg1 (3.4 nM; C0HLM3 (Beltrán-Vidal et al. 2021)), CnErg1 (4.5 nM; Q86QT3 (Gurrola et al. 1999)), CnErg2 (418 nM; P59939 (Restano-Cassulini et al. 2008)), and CsEKerg1 (231.7 nM; P59940 (Corona et al. 2002)).

15.3.6 Potassium Channel Toxins Kappa-KTx (κ-KTxs)

The κ-KTxs subfamily is composed of 22–28 amino acids, showcasing a remarkable structural conformation characterized by two α-helix stabilized by two disulfide bridges, commonly referred to as the cystine-stabilized helix-loop-helix (CSα/α) fold (Fig. 15.4f) (Mouhat et al. 2004; Chagot et al. 2005; Camargos et al. 2011; Vandendriessche et al. 2012). So far, researchers have successfully isolated two incredible neotropical scorpion peptides, OcyC8 (P86110) and OcyC9 (C5J893), from *Opisthacanthus cayaporum* (Camargos et al. 2011; Vandendriessche et al. 2012; Saucedo et al. 2012). These peptides exhibit a compelling ability to inhibit potassium flux through channels. For example, the OcyC8 has an IC_{50} of 217 and 71 μM, reducing the current by 50 percent in the Kv1.1 and Kv1.4 channels, respectively (Camargos et al. 2011). It highlights the potential of these peptides for further research and application in biomedical fields.

15.4 Calcium Channel Scorpion Toxins (CaTx)

Calcines are a family of highly basic peptides composed of 32–35 amino acids and three disulfide bridges (C1–C4, C2–C5, C3–C6). They present an inhibitor cystine knot (ICK) as a structural motif (Fig. 15.5) (Xiao et al. 2016; Haji-Ghassemi et al. 2023). Two peptides in the venom from two neotropical scorpion species have been identified: Hadrucalcin (B8QG00) of *Hoffmannihadrurus gertschi* (Schwartz et al. 2009) and in the *Vaejovis mexicanus smithi,* the Vejocalcin (P0DPT1) (Xiao et al. 2016). These calcins have the membrane-permeating capacity, causing subconducting states in the ryanodine receptors (RyR) of skeletal muscle (RyR1), causing the release of Ca^{+2} in the sarcoplasmic reticulum. Conserved basic residues (for instance, Arg25 and Arg26 in the Hadrucalcin and Arg31 and Arg33 in the Vejocalcin), are essentials in the bind to RyR1 (Schwartz et al. 2009; Xiao et al. 2016). Additionally, the venom gland of *Thorellius intrepidus* and *T. serrulatus* identified the genes of

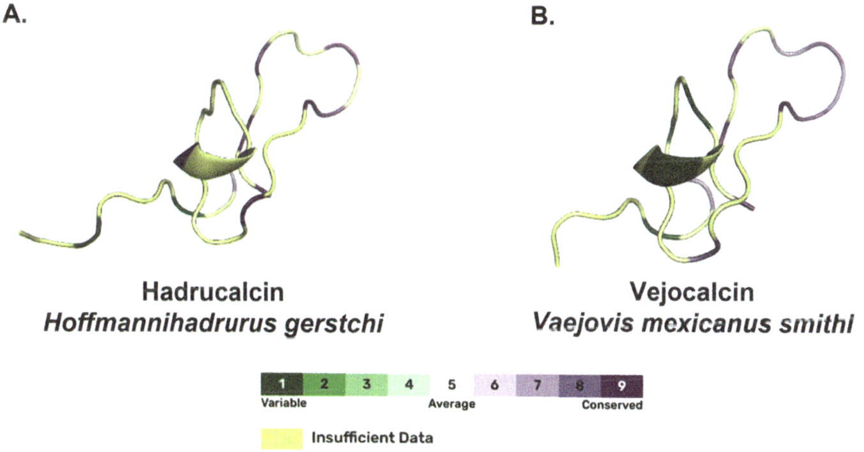

Fig. 15.5 Illustrates the structures of CaTxs toxins. (**a**) Structure of the Hadrucalcin (B8QG00), isolated from *H. gertschi,* features an ICK motif, which includes a triple-stranded β-sheet connected to an α-helix. (**b**) The Vejocalcin (P0DPT1) derived from *V. mexicanus smithi* also contains its structural motif. Both structures were generated using ColabFold v1.5.5 (Mirdita et al. 2022). The Consurf web server was employed to identify evolutionarily conserved regions of residues in these toxins. The amino acids were color-coded based on their conservation levels, with a color--coding bar of green to dark purple indicating variable to conserved residues (Yariv et al. 2023)

Intrepicalcin and CCP-Ts, respectively, through transcriptomic analysis (Quintero-Hernández et al. 2015; Oliveira-Mendes et al. 2018). Heterologous expression of the intrepicalcin gene revealed evaluating its activity on RyR, demonstrating a *Kd* of 17.4 nM. Furthermore, intrepicalcin enhanced the characteristic bell-shaped calcium-[3H]ryanodine binding curve, whose maximal activity occurs at approximately 10 μM calcium concentration (Vargas-Jaimes et al. 2017).

15.5 Enzymes from Neotropical Scorpions

Studies of scorpion venom have revealed that it contains several enzymes, with the hydrolases being the most extensively researched (e.g., Díaz-García et al. 2015; Cid-Uribe et al. 2020; Delgado-Prudencio et al. 2022). The purification of venom enzymes from neotropical scorpions has primarily focused on hyaluronidases, antareases, and angiotensin-converting enzymes. These enzymes are isolated from Buthidae scorpions, particularly from the genera *Tityus* and *Rhopalurus*.

15.5.1 Hyaluronidases

Hyaluronidases have been found in several neotropical scorpions. Recently, research has suggested that they are ubiquitous enzymes found in scorpion venoms. These enzymes hydrolyze chondroitin sulfate and hyaluronic acid, both connective tissue components (Bordon et al. 2015). This characteristic allows them to act as a spreading factor within different venoms, enhancing the dispersion of other venom components throughout the tissues and into the bloodstream (Silva de França and Tambourgi 2023). Invertebrate hyaluronidases have a structure like the TIM barrel (Marković-Housley et al. 2000), with active sites at the D and E residues of the barrel's interior. The predicted models for scorpion hyaluronidase demonstrate a TIM barrel structure (beta/alpha fold) with active sites in positions 101 and 103, corresponding to D and E residues, respectively (Horta et al. 2014). Figure 15.6a shows a predicted model performed with AlphaFold v, based on a sequence (P85841) previously identified from *T. serrulatus*. This model highlights the active site residues. Hyaluronidases have been purified from two neotropical scorpions, *Rhopalurus crassicauda* (Abreu et al. 2020) and *T. serrulatus* (Horta et al. 2014). The role of this enzyme in envenomation and the biodistribution of *Tityus serrulatus* venom has been evaluated. When the enzyme was immunoneutralized, there was a notable reduction in venom biodistribution, resulting in delayed envenomation symptoms and decreased toxicity (Horta et al. 2014; Oliveira-Mendes et al. 2019).

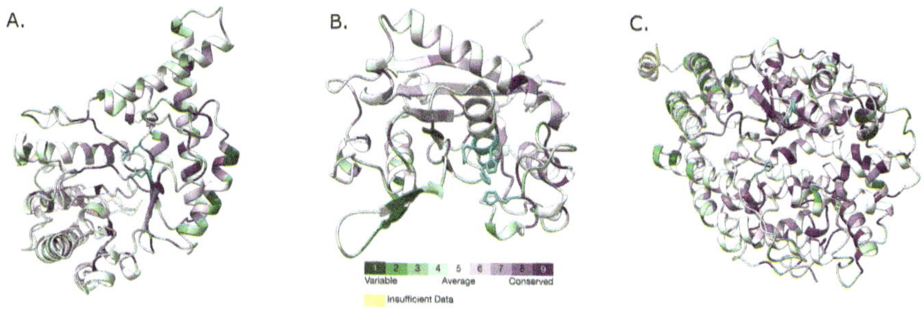

Fig. 15.6 Illustrates the predicted structure of enzymes isolated from neotropical scorpion venoms. (**a**) Hyaluronidase (P85841); (**b**) Metalloprotease known as antarease (P86392), and (**c**) Angiotensin-Converting Enzyme (A0A218QXF0). All enzymes were derived from the venom of *Tityus serrulatus*. The models were generated using AlphaFold and submitted to the Consurf web server (Yariv et al. 2023). The residues located at the catalytic site are in cyan

15.5.2 Metalloproteases

Metalloproteases are the largest family of protease, which requires divalent ions as part of their catalytic mechanism (Bode et al. 1993; Stöcker et al. 1995). This family includes enzymes that range widely in size, from 20 to 100 kDa. Metalloproteases are present in several animal venoms (Delgado-Prudencio et al. 2022). Until now, only two types of metalloproteases have been purified: antareases and angiotensin-converting enzymes, both derived from the venom of *Tityus* scorpions (Cajado-Carvalho et al. 2019; Fletcher et al. 2010). The first, known as antareases, belong to the metzincins and contain a zinc-binding domain and a conserved methionine (Ortiz et al. 2014). Figure 15.6b illustrates a predicted model generated by AlphaFold using a previously reported antarease sequence (P86392) from *Tityus serrulatus*. This model highlights the zinc-binding residues, with glutamic acid and methionine as the active site. Antareases are ubiquitous in *Tityus* scorpions (Ortiz et al. 2014). It has been observed that these enzymes cleave the vesicle-associated membrane protein (VAMP) in pancreatic tissue. Consequently, the proteolytic activity of the venom disrupts vesicular trafficking in the pancreas, leading to acute pancreatitis (Fletcher et al. 2010). Furthermore, it has been suggested that this enzyme could bind to a calcium channel (Zornetta et al. 2017). Recently, the angiotensin-converting enzyme (ACE) has been found and isolated from scorpion venoms (Cajado-Carvalho et al. 2016). Structurally, ACE belongs to the zincins family. Figure 15.6c shows a predicted model obtained from AlphaFold, using an ACE sequence (A0A218QXF0) previously reported from *T. serrulatus*. These enzymes play a crucial role in the hypertensive symptoms that emerge during envenomation (Cajado-Carvalho et al. 2016). However, further research into this enzyme is essential to unravel its complexities fully.

In summary, the most isolated enzymes in neotropical scorpion venoms are hyaluronidases and metalloproteases. This shows relevance as a spreading factor or an etiology for acute pancreatitis associated with envenomation from *Tityus* scorpions.

From the Buthidae family, the *Tityus* venoms are the most diverse in enzymes and the best-studied venoms in this subject. However, several reports show evidence of other enzymes in scorpion venoms or their activity, and they need further research (Schwartz et al. 2007; Amorim et al. 2019; Solano-Godoy et al. 2021; Delgado--Prudencio et al. 2022).

15.6 Host Defense Peptides

This chapter focuses on the components of category host defense peptides (HDP), which have been suggested as the response against pathogens (Mookherjee et al. 2020). Notably, defensins, anionic peptides, and non-disulfide bridge peptides (NDBP) are members of this category. The antimicrobial properties of peptides derived from scorpion venoms are fascinating and well-documented (Zeng et al. 2005; Cao et al. 2012). Several peptides reported as antimicrobial in scorpion venoms belong to the non-disulfide-bridged peptides. Furthermore, these peptides are recognized for their outstanding anti-inflammatory (Daniele-Silva et al. 2016), hypotensive (Verano-Braga et al. 2010), and antineoplastic effects (Guo et al. 2013). Their consistent effectiveness against a diverse range of pathogens, spanning bacteria, fungi, and viruses, underscores their impressive therapeutic potential (Guo et al. 2013; Daniele-Silva et al. 2016; Pedron et al. 2017, 2020; Mechkarska et al. 2023).

Defensins are well-known as defense peptides from several organisms (Adhikary and Poget 2024). Among these, scorpion defensins—also known as insect defensins—are part of the superfamily of cysteine-stabilized alpha-beta (CS-αβ) peptides (Tarr 2016). This superfamily includes both long-chain and short-chain scorpion toxins. Recent studies suggest an evolutionary connection between scorpion toxins and insect defensins, indicating that scorpion toxins may have originated from these defensins (Zhang et al. 2016). Furthermore, research has shown that defensins derived from non-venomous tissues and those from the venom could block potassium channels, like the action of short scorpion toxins (Meng et al. 2020). Defensins have been found in scorpion venoms (Zhu et al. 2000). However, in neotropical scorpions, the presence of defensins in venoms has only been confirmed through high-throughput analysis (D'Suze et al. 2009; Cid-Uribe et al. 2019; García-Villalvazo et al. 2023).

On the other hand, the non-disulfide bridged peptides have only been described in venoms from scorpions and ants (Orivel et al. 2001; Zeng et al. 2005; Lima et al. 2018; Nixon et al. 2021). These short peptides between 11 and 56 amino acids form a short helix. NDBP has been classified into five categories according to length in NDBP1-NDBP5 (Almaaytah and Albalas 2014). However, the length range between categories overlaps, and sometimes the classification is hard to assign. However, recent studies show relationships between the net charge and their activity against pathogens (Jiang et al. 2008). The antimicrobial activity of these peptides depends on various factors including physicochemical features, such as hydrophobicity, amphipathicity, hydrophobic moment, electrostatic potential, dipolar moment, helicity, and amidation, which significantly influence their varied potencies (Fong--Coronado et al. 2024). Regarding the mechanism of these peptides, research suggested that electrostatic interactions between peptides and membrane lipids perturb the membrane, which leads to pore formation (Lee et al. 2015).

Anionic peptides are antimicrobial peptides with a negative net charge (Harris et al. 2009). In neotropical scorpion venoms only one peptide has been isolated from *Tityus stigmurus* venom and is known as TanP, which shows immunomodulatory properties (Melo et al. 2022).

Until now, several peptides from the HDP category in the Hadruridae, Vaejovidae, and Buthidae families have been found in neotropical scorpions. Table 15.1 summarizes the NDBP found in neotropical scorpion venoms. Figure 15.7 shows two peptides, A. Hadrurin (NDBP2) and B. TsAP-1 (NDBP4), which are effective against Gram (+) and Gram (−) bacteria. TsAP-1 even shows a broader spectrum of activity affecting fungi and neoplastic cells (Guo et al. 2013).

The high-throughput analysis conducted on venom glands enhances the diversity of host defense peptides across various scorpion families (Cid-Uribe et al. 2020).

Table 15.1 Summary of NDBP form Neotropical scorpion venoms

UniProt entry	Name	Scorpion species
P0DQT1	Stigmurin	*Tityus stigmurus*
S6CWV8	Antimicrobial peptide TsAP-1 (non-disulfide--bridged peptide 4.22) (NDBP-4.22)	*Tityus serrulatus*
S6D3A7/Q5G8B5	Antimicrobial peptide TsAP-2 (non-disulfide--bridged peptide 4.23) (NDBP-4.23)	*Tityus serrulatus*
	Putative antimicrobial peptide clone 4 (non-disulfide-bridged peptide 4.23-like) (NDBP-4.23-like)	*Tityus costatus*
A0A1D3IXJ5	Antimicrobial peptide ToAP2	*Tityus obscurus*
A0A1D3IXR7	Antimicrobial peptide ToAp1	*Tityus obscurus*
P0DRB5	Peptide TtAP-1	*Tityus trinitatis*
P0DRB6	Peptide TtAP-2	*Tityus trinitatis*
P0DRB7/ A0A1E1WVR9/ Q5G8B3/	Peptide TtAP-3/Peptide ToAP4/antimicrobial peptide 6 (putative antimicrobial peptide clone 6)	*Tityus trinitatis*/ *Tityus obscurus*/ *Tityus costatus*
P0DQT2	Antimicrobial peptide ToAP3	*Tityus obscurus*
P0C8W1	Non-disulfide-bridged peptide 5.5 (NDBP-5.5)	*Hoffmannihadrurus gertschi*
P0C8W2	Non-disulfide-bridged peptide 5.6 (NDBP-5.6)	*Hoffmannihadrurus gertschi*
P82656	Hadrurin (non-disulfide-bridged peptide 2.1) (NDBP-2.1) (non-disulfide bridged peptide 3.1) (NDBP-3.1)	*Hoffmannihadrurus aztecus*
I0DEB3	Amphipathic peptide VmCT1 (CT1) (non--disulfide-bridged peptide 4.13) (NDBP-4.13) (non-disulfide-bridged peptide 5.13) (NDBP-5.13)	*Vaejovis smithi*

(continued)

Table 15.1 (continued)

UniProt entry	Name	Scorpion species
F1AWB0	Vejovine (non-disulfide-bridged peptide 2.7) (NDBP-2.7)	*Vaejovis mexicanus*
I0DEB4	Amphipathic peptide CT2 (VmCT2) (non--disulfide-bridged peptide 4.14) (NDBP-4.14) (non-disulfide-bridged peptide 5.14) (NDBP-5.14)	*Vaejovis smithi*
I0DEB5	Amphipathic peptide CT1 (VsCT1) (non--disulfide-bridged peptide 5.11) (NDBP-5.11)	*Mesomexovis subcristatus*
I0DEB6	Amphipathic peptide CT2 (VsCT2) (non--disulfide-bridged peptide 5.12) (NDBP-5.12)	*Mesomexovis subcristatus*
P84189	Hypotensin-1 (anti-hypertensive peptide) (Hypotensin I) (TsHpt-I) (Tityustoxin-14 1) (Hypotensin toxin Ts14.1) (Ts14 1) [Cleaved into: Hypotensin-3 (Hypotensin III) (TsHpt-III) (Toxin Ts14 3)]	*Tityus serrulatus*
P84190	Hypotensin-2 (Hypotensin II) (TsHpt-II) (Toxin Ts14 2) [Cleaved into: Hypotensin-4 (Hypotensin IV) (TsHpt-IV) (Toxin Ts14 4)]	*Tityus serrulatus*
A0A218QWY8	Hypotensin toxin Ts31 (anti-hypertensive peptide) (Tityustoxin-31)	*Tityus serrulatus*
Q9TWR4	Bradykinin-potentiating peptide T (BPP-T) (non-disulfide-bridged peptide 1.1) (NDBP-1.1) (Ts10)	*Tityus serrulatus*
P86828	Venom peptide 7.1 (venom peptide 7) [Cleaved into: Venom peptide 7.2]	*Tityus serrulatus*
C0HJT2	Venom peptide 8	*Tityus serrulatus*
No entry	TanP	*Tityus stigmurus*

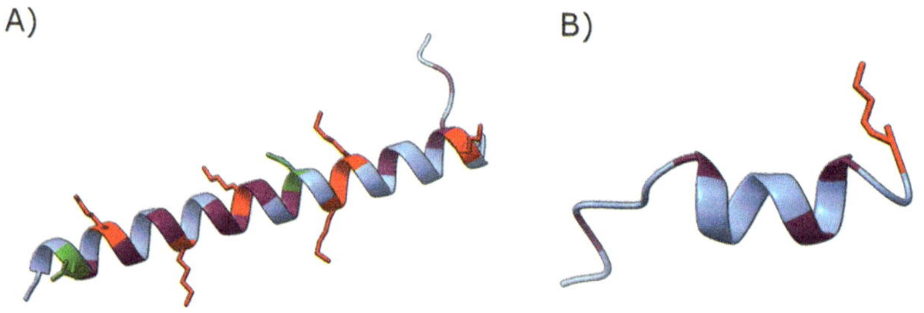

Fig. 15.7 Predicted structure from host defense peptides isolated from neotropical scorpion venoms. (**a**) Hadrurin (P82656) isolated from *Hoffmannihadrurus aztecus* venom, (**b**) TsAp-1 (S6CWV8) obtained from *Tityus serrulatus* venom. Hydrophobic residues are in blue, hydrophilic in purple, with positive charges in red and negative in green

Although anionic peptides have not been identified in other neotropical scorpion venoms, the transcriptomic analysis revealed precursors of these peptides. This finding has led to the synthesis of mature peptides, which exhibit chelating, hemolytic, and immunomodulatory properties (Melo et al. 2022).

Remarkably, two particular peptides are worth mentioning in this section: Scorpines and TS1. These toxins affect ion channels but also show antimicrobial activity. The first structurally belongs to the β-KTx but also has a cecropin domain (Zhu and Tytgat 2004). The presence of both domains provides two different activities: an ion channel blocker and an antimicrobial peptide (López-Giraldo et al. 2023). These peptides have been isolated and tested from the Neotropical scorpion venoms from *Hoffmannihadrurus* (Flores-Solis et al. 2016) and *Tityus* (Titaux--Delgado et al. 2023). The second case refers to the toxin TS1, which structurally is a sodium channel toxin that acts on channels Nav1.2 to Nav1.6 and Nav1.9 of mammals and the DmNav1 from insects (Campos et al. 2007; Bosmans et al. 2011; Peigneur et al. 2015). However, this peptide has an antifungal activity (Santussi et al. 2017), whose mechanism remains unclear. However, this toxin may bind the voltage-dependent sodium channel from the yeasts (Gustin et al. 1986).

15.7 Other Components

Scorpion venoms are complex mixtures, and most research has focused on peptidic and proteic compounds. However, a few studies have explored non-protein compounds. Notable findings include the identification of two benzoquinones from *Diplocentrus melici* and *Didymocentrus krausi* (Carcamo-Noriega et al. 2019; Rojas-Azofeifa et al. 2019), an alkaloid from *Megacormus gertschi* venom (Banerjee et al. 2018), three Indole alkylamines from *Thorellius intrepidus* (Ibarra-Vega et al. 2023), and the lipidomic profile of *Tityus serrulatus* venom (Acunha et al. 2023). The function of these compounds as venom components is still unknown. However, it is well documented that p-benzoquinones are highly reactive against proteins; even the evidence shows they are toxic to cells (Yang and Zhou 2010). *Didymocentrus* venom shows hemorrhagic and myotoxic effects, which authors explain could be related to the presence of these benzoquinones (Rojas-Azofeifa et al. 2019). On the other hand, it has recently been mentioned that venom components of *Megacormus* males, such as alkaloids, pass into the hemolymph through sexual stinging (Olguin--Perez et al. 2021).

Lastly, other components found in scorpion venomsScorpion venoms include La1 peptides, CAP proteins, and protease inhibitors (Chen et al. 2013c; Ma et al. 2016; Martinez et al. 2016). Unfortunately, these compounds have not been identified in neotropical scorpion venoms. Additionally, only a few studies have focused on characterizing those components; further research is needed to understand their role in the venom. The composition of scorpion venom is a dynamic mixture, continually shaped by selective pressure (Sunagar et al. 2013). Researchers hypothesized that the remarkable diversity and specificity of its components, a broad spectrum of toxins, enzymes, and antimicrobial peptides, reflect a continuous co-evolutionary arms race driven prey-predator dynamics and adaptations to specific ecological niches (Dietl and Kelley 2002). This selective pressure, exerted by both predator-prey interactions and intraspecific competition, has likely driven the diversification and functional optimization of venom toxins (Sharma et al. 2015; Dietl and Kelley 2002). Consequently, the genomic architecture underlying venom production, including gene duplication events and subsequent functional divergence, has facilitated this evolutionary process. While ion channel toxins, crucial for paralyzing prey or deterring predators, are the most studied peptides, recent evidence also points to the presence of compounds that could be involved in digestion, antimicrobial protection, and even mating (Sunagar et al. 2013; Simone and van der Meijden 2021). Despite these advances, researchers are currently unable to predict the function of all venom components, indicating we still have a long way to go in fully understanding scorpion venoms.

Acknowledgments We are very grateful to the CONAHCyT-FORDECYT grant [303045] awarded to the consortium Alagon-Becerril-Corzo-Possani for the postdoctoral fellowship granted to Rita Restano-Cassulini.

References

Abreu CB, Bordon KCF, Cerni FA, Oliveira IS, Balenzuela C, Alexandre-Silva GM, Zoccal KF, Reis MB, Wiezel GA, Peigneur S, Pinheiro-Júnior EL, Tytgat J, Cunha TM, Quinton L, Faccioli LH, Arantes EC, Zottich U, Pucca MB (2020) Pioneering study on *Rhopalurus crassicauda* scorpion venom: isolation and characterization of the major toxin and hyaluronidase. Front Immunol 11. https://doi.org/10.3389/fimmu.2020.02011

Acunha T, Rocha BA, Nardini V, Barbosa F Jr, Faccioli LH (2023) Lipidomic profiling of the Brazilian yellow scorpion venom: new insights into inflammatory responses following *Tityus serrulatus* envenomation. J Toxic Environ Health A 86(9):283–295. https://doi.org/10.1080/15287394.2023.2188896

Adhikary K, Poget SF (2024) Expanding the insect defensin landscape. Structure 32(9):1294–1296. https://doi.org/10.1016/j.str.2024.08.009

Alexander SPH, Mathie AA, Peters JA, Veale EL, Striessnig J, Kelly E, Armstrong JF, Faccenda E, Harding SD, Davies JA, Aldrich RW, Attali B, Baggetta AM, Becirovic E, Biel M, Bill RM, Caceres AI, Catterall WA, Conner AC et al (2023) The concise guide to PHARMACOLOGY 2023/24: ion channels. Br J Pharmacol 180(S2). https://doi.org/10.1111/bph.16178

Almaaytah A, Albalas Q (2014) Scorpion venom peptides with no disulfide bridges: a review. Peptide 5:35–45. https://doi.org/10.1016/j.peptides.2013.10.021

Almeida DD, Torres TM, Barbosa EG, Lima JPMS, Fernandes-Pedrosa MF (2013) Molecular approaches for structural characterization of a new potassium channel blocker from *Tityus stigmurus* venom: cDNA cloning, homology modeling, dynamic simulations and docking. Biochem Biophys Res Commun 430(1):113–118. https://doi.org/10.1016/j.bbrc.2012.11.044

Amorim FG, Longhim HT, Cologna CT, Degueldre M, Pauw E, Quinton L, Arantes EC (2019) Proteome of fraction from *Tityus serrulatus* venom reveals new enzymes and toxins. J Venom Anim Toxins Incl Trop Dis 25. https://doi.org/10.1590/1678-9199-jvatitd-1482-18

Banerjee A, Lee A, Campbell E, MacKinnon R (2013) Structure of a pore-blocking toxin in complex with a eukaryotic voltage-dependent K+ channel. ELife 2. https://doi.org/10.7554/eLife.00594

Banerjee S, Gnanamani E, Lynch SR, Zuñiga FZ, Jiménez-Vargas JM, Possani LD, Zare RN (2018) Correction to an alkaloid from scorpion venom: chemical structure and synthesis. J Nat Prod 81(9):2161–2161. https://doi.org/10.1021/acs.jnatprod.8b00722

Batista C, D'Suze G, Gómez-Lagunas F, Zamudio FZ, Encarnación S, Sevcik C, Possani L (2006) Proteomic analysis of *Tityus discrepans* scorpion venom and amino acid sequence of novel toxins. Proteomics 6(12):3718–3727. https://doi.org/10.1002/pmic.200500525

Batista C, Román-González SA, Salas-Castillo SP, Zamudio FZ, Gómez-Lagunas F, Possani LD (2007) Proteomic analysis of the venom from the scorpion *Tityus stigmurus*: biochemical and physiological comparison with other *Tityus* species. Comp Biochem Physiol C Toxicol Pharmacol 146(1–2):147–157. https://doi.org/10.1016/j.cbpc.2006.12.004

Beltrán-Vidal J, Carcamo-Noriega E, Pastor N, Zamudio-Zuñiga F, Guerrero-Vargas JA, Castaño S, Possani LD, Restano-Cassulini R (2021) Colombian scorpion *Centruroides margaritatus*: purification and characterization of a gamma Potassium toxin with full-block activity on the hERG1 channel. Toxins 13(6):407. https://doi.org/10.3390/toxins13060407

Bode W, Gomis-Rüth F-X, Stöckler W (1993) Astacins, serralysins, snake venom and matrix metalloproteinases exhibit identical zinc-binding environments (HEXXHXXGXXH and Met-turn) and topologies and should be grouped into a common family, the 'metzincins'. FEBS Lett 331(1–2):134–140. https://doi.org/10.1016/0014-5793(93)80312-1

Bordon KCF, Wiezel GA, Amorim FG, Arantes EC (2015) Arthropod venom Hyaluronidases: biochemical properties and potential applications in medicine and biotechnology. J Venom Anim Toxins Incl Trop Dis 21(1):43. https://doi.org/10.1186/s40409-015-0042-7

Bosmans F, Tytgat J (2007) Voltage-gated sodium channel modulation by scorpion α-toxins. Toxicon 49(2):142–158. https://doi.org/10.1016/j.toxicon.2006.09.023

Bosmans F, Puopolo M, Martin-Eauclaire M-F, Bean BP, Swartz KJ (2011) Functional properties and toxin pharmacology of a dorsal root ganglion sodium channel viewed through its voltage sensors. J Gen Physiol 138(1):59–72. https://doi.org/10.1085/jgp.201110614

Bottiglieri C, Ferrara L, Corona M, Gurrola GB, Batista C, Wanke E, Possani LD (2000) Disulfide bridges of Ergtoxin, a member of a new sub-family of peptide blockers of the ether-a-go-go -related K + channel. FEBS Lett 479(3):156–157. https://doi.org/10.1016/S0014-5793(00)01891-3

Cajado-Carvalho D, Kuniyoshi A, Duzzi B, Iwai L, Oliveira Ú, Azevedo IJ, Kodama R, Portaro F (2016) Insights into the Hypertensive effects of *Tityus serrulatus* scorpion venom: purification of an Angiotensin-Converting Enzyme-Like peptidase. Toxins 8(12):348. https://doi.org/10.3390/toxins8120348

Cajado-Carvalho D, da Silva CCF, Kodama RT, Mariano DOC, Pimenta D C, Duzzi B, Kuniyoshi AK, Portaro FV (2019) Purification and biochemical characterization of TsMS 3 and TsMS 4: neuropeptide-degrading metallopeptidases in the tityus serrulatus venom. Toxins 11(4):194. https://doi.org/10.3390/toxins11040194

Camargos TS, Restano-Cassulini R, Possani LD, Peigneur S, Tytgat J, Schwartz CA, Alves EMC, de Freitas SM, Schwartz EF (2011) The new kappa-KTx 2.5 from the scorpion *Opisthacanthus cayaporum*. Peptides 32(7):1509–1517. https://doi.org/10.1016/j.peptides.2011.05.017

Campos FV, Chanda B, Beirão PSL, Bezanilla F (2007) β-scorpion toxin modifies gating transitions in all four voltage sensors of the Sodium channel. J Gen Physiol 130(3):257–268. https://doi.org/10.1085/jgp.200609719

Cao L, Dai C, Li Z, Fan Z, Song Y, Wu Y, Cao Z, Li W (2012) Antibacterial activity and mechanism of a scorpion venom peptide derivative in vitro and in vivo. PLoS One 7(7):e40135. https://doi.org/10.1371/journal.pone.0040135

Carcamo-Noriega EN, Sathyamoorthi S, Banerjee S, Gnanamani E, Mendoza-Trujillo M, Mata-Espinosa D, Hernández-Pando R, Veytia-Bucheli JI, Possani LD, Zare RN (2019) 1,4-Benzoquinone antimicrobial agents against *Staphylococcus aureus* and *Mycobacterium tuberculosis* derived from scorpion venom. Proc Natl Acad Sci 116(26):12642–12647. https://doi.org/10.1073/pnas.1812334116

Catterall WA (2014) Structure and function of voltage-gated sodium channels at atomic resolution. Exp Physiol 99(1):35–51. https://doi.org/10.1113/expphysiol.2013.071969

Cestèle S (2000) Molecular mechanisms of neurotoxin action on voltage-gated sodium channels. Biochimie 82(9–10):883–892. https://doi.org/10.1016/S0300-9084(00)01174-3

Cestèle S, Qu Y, Rogers JC, Rochat H, Scheuer T, Catterall WA (1998) Voltage sensor–trapping. Neuron 21(4):919–931. https://doi.org/10.1016/S0896-6273(00)80606-6

Cestèle S, Scheuer T, Mantegazza M, Rochat H, Catterall WA (2001) Neutralization of gating charges in domain II of the Sodium Channel α subunit enhances voltage-sensor trapping by a β-scorpion toxin. J Gen Physiol 118(3):291–302. https://doi.org/10.1085/jgp.118.3.291

Chagot B, Pimentel C, Dai L, Pil J, Tytgat J, Nakajima T, Corzo G, Darbon H, Ferrat G (2005) An unusual fold for potassium channel blockers: NMR structure of three toxins from the scorpion *Opisthacanthus madagascariensis*. Biochem J 388(1):263–271. https://doi.org/10.1042/BJ20041705

Chen Z, Hu Y-T, Yang W-S, He Y-W, Feng J, Wang B, Zhao R-M, Ding J-P, Cao Z-J, Li W-X, Wu Y-L (2012) Hg1, novel peptide inhibitor specific for Kv1.3 channels from first scorpion Kunitz-type Potassium channel toxin family. J Biol Chem 287(17):13813–13821. https://doi.org/10.1074/jbc.M112.343996

Chen Z, Cao Z, Li W, Wu Y (2013a) Cloning and characterization of a novel Kunitz-type inhibitor from scorpion with unique cysteine framework. Toxicon 72:5–10. https://doi.org/10.1016/j.toxicon.2013.05.022

Chen Z, Luo F, Feng J, Yang W, Zeng D, Zhao R, Cao Z, Liu M, Li W, Jiang L, Wu Y (2013b) Genomic and structural characterization of Kunitz-type peptide LmKTT-1a highlights diversity and evolution of scorpion potassium channel toxins. PLoS One 8(4):e60201. https://doi.org/10.1371/journal.pone.0060201

Chen Z, Wang B, Hu J, Yang W, Cao Z, Zhuo R, Li W, Wu Y (2013c) SjAPI, the First Functionally characterized ascaris-type protease inhibitor from animal venoms. PLoS One 8(3):e57529. https://doi.org/10.1371/journal.pone.0057529

Cid-Uribe JI, Jiménez-Vargas JM, Ferreira-Batista CV, Zamudio-Zuñiga F, Possani LD (2017) Comparative proteomic analysis of female and male venoms from the Mexican scorpion *Centruroides limpidus*: novel components found. Toxicon 125:91–98. https://doi.org/10.1016/j.toxicon.2016.11.256

Cid-Uribe JI, Meneses EP, Batista CVF, Ortiz E, Possani LD (2019) Dissecting toxicity: the venom gland transcriptome and the venom proteome of the highly venomous scorpion *Centruroides limpidus* (Karsch, 1879). Toxins 11(5):247. https://doi.org/10.3390/toxins11050247

Cid-Uribe JI, Veytia-Bucheli JI, Romero-Gutierrez T, Ortiz E, Possani LD (2020) Scorpion venomics: a 2019 overview. Expert Rev Proteomics 17(1):67–83. https://doi.org/10.1080/14789450.2020.1705158

Cologna CT, Peigneur S, Rustiguel JK, Nonato MC, Tytgat J, Arantes EC (2012) Investigation of the relationship between the structure and function of Ts2, a neurotoxin from *Tityus serrulatus* venom. FEBS J 279(8):1495–1504. https://doi.org/10.1111/j.1742-4658.2012.08545.x

Comini M, Themistocleous AC, Bennett DLH (2024) Human pain channelopathies. Handb Clin Neurol:89–109. https://doi.org/10.1016/B978-0-323-90820-7.00004-5

Corona M, Gurrola GB, Merino E, Cassulini RR, Valdez-Cruz NA, García B, Ramírez-Domínguez ME, Coronas FIV, Zamudio FZ, Wanke E, Possani LD (2002) A large number of novel Ergtoxin-like genes and ERG K + -channels blocking peptides from scorpions of the genus *Centruroides*. FEBS Lett 532(1–2):121–126. https://doi.org/10.1016/S0014-5793(02)03652-9

Cremonez C, Maiti M, Peigneur S, Cassoli J, Dutra A, Waelkens E, Lescrinier E, Herdewijn P, De Lima M, Pimenta A, Arantes E, Tytgat J (2016) Structural and functional elucidation of peptide Ts11 shows evidence of a novel subfamily of scorpion venom toxins. Toxins 8(10):288. https://doi.org/10.3390/toxins8100288

Csoti A, Meza RCN, Bogár F, Tajti G, Szanto TG, Varga Z, Gurrola GB, Tóth GK, Possani LD, Panyi G (2022) sVmKTx, a transcriptome analysis-based synthetic peptide analogue of Vm24, inhibits Kv1.3 channels of human T cells with improved selectivity. Biochem Pharmacol 199:115023. https://doi.org/10.1016/j.bcp.2022.115023

D'Suze G, Schwartz E, García-Gómez BI, Sevcik C, Possani LD (2009) Molecular cloning and nucleotide sequence analysis of genes from a cDNA library of the scorpion *Tityus discrepans*. Biochimie 91(8):1010–1019. https://doi.org/10.1016/j.biochi.2009.05.005

Daniele-Silva A, Machado RJA, Monteiro NKV, Estrela AB, Santos ECG, Carvalho E, Araújo Júnior RF, Melo-Silveira RF, Rocha HAO, Silva-Júnior AA, Fernandes-Pedrosa MF (2016) Stigmurin and TsAP-2 from *Tityus stigmurus* scorpion venom: assessment of structure and therapeutic potential in experimental sepsis. Toxicon 121:10–21. https://doi.org/10.1016/j.toxicon.2016.08.016

Delgado-Prudencio G, Cid-Uribe JI, Morales JA, Possani LD, Ortiz E, Romero-Gutiérrez T (2022) The enzymatic core of scorpion venoms. Toxins 14(4):248. https://doi.org/10.3390/toxins14040248

Díaz-García A, Ruiz-Fuentes JL, Yglesias-Rivera A, Rodríguez-Sánchez H, Riquenes Garlobo Y, Fleitas Martinez O, Fraga Castro JA (2015) Enzymatic analysis of venom from Cuban scorpion Rhopalurus junceus. J Venom Res 6:11–8

Diego-García E, Schwartz EF, D'Suze G, González SAR, Batista CVF, García BI, de la Vega RCR, Possani LD (2007) Wide phylogenetic distribution of Scorpine and long-chain β-KTx-like peptides in scorpion venoms: identification of "orphan" components. Peptides 28(1):31–37. https://doi.org/10.1016/j.peptides.2006.06.012

Dietl GP, Kelley PH (2002) The fossil record of predator-prey arms races: coevolution and escalation hypotheses. Paleontol Soc Pap 8:353–374. https://doi.org/10.1017/S1089332600001157

Fletcher PL, Fletcher MD, Weninger K, Anderson TE, Martin BM (2010) Vesicle-associated membrane protein (VAMP) cleavage by a new metalloprotease from the brazilian scorpion *Tityus serrulatus*. J Biol Chem 285(10):7405–7416. https://doi.org/10.1074/jbc.M109.028365

Flores-Solis D, Toledano Y, Rodríguez-Lima O, Cano-Sánchez P, Ramírez-Cordero BE, Landa A, de la Vega RCR, del Rio-Portilla F (2016) Solution structure and antiparasitic activity of scorpine-like peptides from *Hoffmannihadrurus gertschi*. FEBS Lett 590(14):2286–2296. https://doi.org/10.1002/1873-3468.12255

Fong-Coronado PA, Ramirez V, Quintero-Hernández V, Balleza D (2024) A critical review of short antimicrobial peptides from scorpion venoms, their physicochemical attributes, and potential for the development of new drugs. J Membr Biol 257(3–4):165–205. https://doi.org/10.1007/s00232-024-00315-2

Freire MCLC, Menezes YAS, Ferraz MVF, Cruz CHB, Ferreira LS, Fernandes-Pedrosa MF, Barbosa EG (2019) Molecular basis of *Tityus stigmurus* alpha toxin and potassium channel kV1.2 interactions. J Mol Graph Model 87:197–203. https://doi.org/10.1016/j.jmgm.2018.11.012

García-Villalvazo PE, Jiménez-Vargas JM, Lino-López GJ, Meneses EP, Bermúdez-Guzmán MJ, Barajas-Saucedo CE, Enciso ID, Possani LD, Valdez-Velazquez LL (2023) Unveiling the Protein components of the secretory-venom gland and venom of the scorpion *Centruroides possanii* (Buthidae) through Omic Technologies. Toxins 15(8):498. https://doi.org/10.3390/toxins15080498

Goldstein SA, Miller C (1993) Mechanism of charybdotoxin block of a voltage-gated K+ channel. Biophys J 65(4):1613–1619. https://doi.org/10.1016/S0006-3495(93)81200-1

Gómez-Ramírez IV, Riaño-Umbarila L, Olamendi-Portugal T, Restano-Cassulini R, Possani LD, Becerril B (2020) Biochemical, electrophysiological and immunological characterization of the venom from *Centruroides baergi*, a new scorpion species of medical importance in Mexico. Toxicon 184:10–18. https://doi.org/10.1016/j.toxicon.2020.05.021

Guo X, Ma C, Du Q, Wei R, Wang L, Zhou M, Chen T, Shaw C (2013) Two peptides, TsAP-1 and TsAP-2, from the venom of the Brazilian yellow scorpion, *Tityus serrulatus*: evaluation of their antimicrobial and anticancer activities. Biochimie 95(9):1784–1794. https://doi.org/10.1016/j.biochi.2013.06.003

Gurevitz M, Gordon D, Barzilai MG, Kahn R, Cohen L, Moran Y, Zilberberg N, Froy O, Altman-Gueta H, Turkov M, Dong K, Karbat I (2015) Molecular description of scorpion toxin interaction with voltage-gated sodium channels. In: Scorpion venoms. Springer Netherlands, pp 471–491. https://doi.org/10.1007/978-94-007-6404-0_10

Gurrola GB, Rosati B, Rocchetti M, Pimienta G, Zaza A, Arcangeli A, Olivotto M, Possani LD, Wanke E (1999) A toxin to nervous, cardiac, and endocrine ERG K + channels isolated from *Centruroides noxius* scorpion venom. FASEB J 13(8):953–962. https://doi.org/10.1096/fasebj.13.8.953

Gustin MC, Martinac B, Saimi Y, Culbertson MR, Kung C (1986) Ion channels in yeast. Science 233(4769):1195–1197. https://doi.org/10.1126/science.2426783

Haji-Ghassemi O, Chen YS, Woll K, Gurrola GB, Valdivia CR, Cai W, Li S, Valdivia HH, Van Petegem F (2023) Cryo-EM analysis of scorpion toxin binding to Ryanodine Receptors reveals subconductance that is abolished by PKA phosphorylation. Sci Adv 9(21). https://doi.org/10.1126/sciadv.adf4936

Harris F, Dennison S, Phoenix D (2009) Anionic antimicrobial peptides from eukaryotic organisms. Curr Protein Pept Sci 10(6):585–606. https://doi.org/10.2174/138920309789630589

Horta CCR, Magalhães BF, Oliveira-Mendes BBR, Carmo AO, Duarte CG, Felicori LF, Machado-de-Ávila RA, Chávez-Olórtegui C, Kalapothakis E (2014) Molecular, immunological, and biological characterization of *Tityus serrulatus* venom hyaluronidase: new insights into its role in envenomation. PLoS Negl Trop Dis 8(2):e2693. https://doi.org/10.1371/journal.pntd.0002693

Ibarra-Vega R, Jiménez-Vargas JM, Pineda-Contreras A, Martínez-Martínez FJ, Barajas-Saucedo CE, García-Ortega H, Magaña-Vergara NE, Possani LD, Corzo G, Gaitan-Hinojosa MA, Vázquez-Vuelvas OF, Zamudio F, Valdez-Velazquez LL (2023) Indolealkylamines in the venom of the scorpion *Thorellius intrepidus*. Toxicon 233:107232. https://doi.org/10.1016/j.toxicon.2023.107232

Jiang Z, Vasil AI, Hale JD, Hancock REW, Vasil ML, Hodges RS (2008) Effects of net charge and the number of positively charged residues on the biological activity of amphipathic α-helical cationic antimicrobial peptides. Pept Sci 90(3):369–383. https://doi.org/10.1002/bip.20911

Kuang Q, Purhonen P, Hebert H (2015) Structure of potassium channels. Cell Mol Life Sci 72(19):3677–3693. https://doi.org/10.1007/s00018-015-1948-5

Lecchi M, Redaelli E, Rosati B, Gurrola G, Florio T, Crociani O, Curia G, Cassulini RR, Masi A, Arcangeli A, Olivotto M, Schettini G, Possani LD, Wanke E (2002) Isolation of a long-lasting eag -related gene-type K + current in MMQ lactotrophs and its accommodating role during slow firing and prolactin release. J Neurosci 22(9):3414–3425. https://doi.org/10.1523/JNEUROSCI.22-09-03414.2002

Lee T-H, Hall KN, Aguilar M-I (2015) Antimicrobial peptide structure and mechanism of action: a focus on the role of membrane structure. Curr Top Med Chem 16(1):25–39. https://doi.org/10.2174/1568026615666150703121700

Lima DB, Mello CP, Bandeira ICJ, Menezes RRPB, Sampaio TL, Falcão CB, Morlighem J-ÉRL, Rádis-Baptista G, Martins AMC (2018) The dinoponeratoxin peptides from the giant ant Dinoponera quadriceps display in vitro antitrypanosomal activity. Biol Chem 399(2):187–196. https://doi.org/10.1515/hsz-2017-0198

López-Giraldo E, Carrillo E, Titaux-Delgado G, Cano-Sánchez P, Colorado A, Possani LD, Río-Portilla F (2023) Structural and functional studies of scorpine: a channel blocker and cytolytic peptide. Toxicon 222:106985. https://doi.org/10.1016/j.toxicon.2022.106985

Ma H, Xiao-Peng T, Yang S-L, Lu Q-M, Lai R (2016) Protease inhibitor in scorpion (*Mesobuthus eupeus*) venom prolongs the biological activities of the crude venom. Chin J Nat Med 14(8):607–614. https://doi.org/10.1016/S1875-5364(16)30071-1

Marković-Housley Z, Miglierini G, Soldatova L, Rizkallah PJ, Müller U, Schirmer T (2000) Crystal structure of hyaluronidase, a major allergen of bee venom. Structure 8(10):1025–1035. https://doi.org/10.1016/S0969-2126(00)00511-6

Martinez G, Hograindleur J-P, Voisin S, Nahed RA, Aziz TMAE, Escoffier J, Bessonnat J, Fovet C-M, De Waard M, Hennebicq S, Aucagne V, Ray PF, Schmitt E, Bulet P, Arnoult C (2016) Spermaurin, an La1-like peptide from the venom of the scorpion *Scorpio maurus palmatus*, improves sperm motility and fertilization in different mammalian species. Mol Hum Reprod 23(2):116–131. https://doi.org/10.1093/molehr/gaw075

Mechkarska M, Cunning TS, Taggart MG, Ternan NG, Leprince J, Coquet L, Jouenne T, Tena-Garcés J, Calvete JJ, Conlon JM (2023) Identification of an Antimicrobial peptide from the venom of the Trinidad thick-tailed scorpion *Tityus trinitatis* with potent activity against ESKAPE pathogens and *Clostridioides difficile*. Antibiotics 12(9):1404. https://doi.org/10.3390/antibiotics12091404

Melo MMA, Oliveira VS, Queiroz-Neto MF, Paiva WS, Torres-Rêgo M, Silva SRB, Pontes DL, Rocha HAO, Souza MÂF, Silva-Júnior AA, Fernandes-Pedrosa MF (2022) TanP: a multifunctional anionic peptide from *Tityus stigmurus* scorpion vVenom. Front Mol Biosci 8. https://doi.org/10.3389/fmolb.2021.785316

Mendes LC, Viana GMM, Nencioni ALA, Pimenta DC, Beraldo-Neto E (2023) Scorpion peptides and ion channels: an insightful review of mechanisms and drug development. Toxins 15(4):238. https://doi.org/10.3390/toxins15040238

Meng L, Zhao Y, Qu D, Xie Z, Guo X, Zhu Z, Chen Z, Zhang L, Li W, Cao Z, Tian C, Wu Y (2020) Ion channel modulation by scorpion hemolymph and its defensin ingredients highlights origin of neurotoxins in telson formed in Paleozoic scorpions. Int J Biol Macromol 148:351–363. https://doi.org/10.1016/j.ijbiomac.2020.01.133

Mirdita M, Schütze K, Moriwaki Y, Heo L, Ovchinnikov S, Steinegger M (2022) ColabFold: making protein folding accessible to all. Nat Methods 19(6):679–682. https://doi.org/10.1038/s41592-022-01488-1

Mookherjee N, Anderson MA, Haagsman HP, Davidson DJ (2020) Antimicrobial host defence peptides: functions and clinical potential. Nat Rev Drug Discov 19(5):311–332. https://doi.org/10.1038/s41573-019-0058-8

Morrone JJ, Escalante T, Rodriguez-Tapia G, Carmona A, Arana M, Mercado-Gómez JD (2022) Biogeographic regionalization of the Neotropical region: new map and shapefile. An Acad Bras Cienc 94(1). https://doi.org/10.1590/0001-3765202220211167

Mouhat S, Jouirou B, Mosbah A, De Waard M, Sabatier J-M (2004) Diversity of folds in animal toxins acting on ion channels. Biochem J 378(3):717–726. https://doi.org/10.1042/bj20031860

Naseem MU, Gurrola-Briones G, Romero-Imbachi MR, Borrego J, Carcamo-Noriega E, Beltrán-Vidal J, Zamudio FZ, Shakeel K, Possani LD, Panyi G (2023) Characterization and chemical synthesis of Cm39 (α-KTx 4.8): a scorpion toxin that inhibits voltage-gated K+ channel KV1.2 and small- and intermediate-conductance Ca2+-activated K+ channels KCa2.2 and KCa3.1. Toxins 15(1):41. https://doi.org/10.3390/toxins15010041

Nixon SA, Robinson SD, Agwa AJ, Walker AA, Choudhary S, Touchard A, Undheim EAB, Robertson A, Vetter I, Schroeder CI, Kotze AC, Herzig V, King GF (2021) Multipurpose peptides: the venoms of Amazonian stinging ants contain anthelmintic ponericins with diverse predatory and defensive activities. Biochem Pharmacol 192:114693. https://doi.org/10.1016/j.bcp.2021.114693

Olamendi-Portugal T, Somodi S, Fernández JA, Zamudio FZ, Becerril B, Varga Z, Panyi G, Gáspár R, Possani LD (2005) Novel α-KTx peptides from the venom of the scorpion *Centruroides elegans* selectively blockade Kv1.3 over IKCa1 K+ channels of T cells. Toxicon 46(4):418–429. https://doi.org/10.1016/j.toxicon.2005.06.001

Olamendi-Portugal T, Bartok A, Zamudio-Zuñiga F, Balajthy A, Becerril B, Panyi G, Possani LD (2016) Isolation, chemical and functional characterization of several new K+-channel blocking peptides from the venom of the scorpion *Centruroides tecomanus*. Toxicon 115:1–12. https://doi.org/10.1016/j.toxicon.2016.02.017

Olamendi-Portugal T, Restano-Cassulini R, Riaño-Umbarila L, Becerril B, Possani LD (2017) Functional and immuno-reactive characterization of a previously undescribed peptide from the venom of the scorpion *Centruroides limpidus*. Peptides 87:34–40. https://doi.org/10.1016/j.peptides.2016.11.008

Olguin-Perez L, Francke OF, Carbajal-Saucedo A (2021) Evidence of piercing and sexual differences in venom composition in a sexual stinging scorpion (Scorpiones: Euscorpiidae). J Arachnol 49:98–107. https://doi.org/10.1636/JoA-S-19-056

Oliveira TJ, Pereira N, Riciluca KCT, Nascimento SM, Castro de Oliveira U, Silva Junior PI (2022) TsTX-κ beta (Ts8), a neurotoxin from the venom of the scorpion *Tityus serrulatus*: antimicrobial and hemolytic activity. bioRxiv. https://doi.org/10.1101/2022.10.09.511467

Oliveira-Mendes BBR, Horta CCR, do Carmo AO, Biscoto GL, Sales-Medina DF, Leal HG, Brandão-Dias PFP, Miranda SEM, Aguiar CJ, Cardoso VN, Barros ALB, Chávez-Olortégui C, Leite MF, Kalapothakis E (2018) CPP-Ts: a new intracellular calcium channel modulator and a promising tool for drug delivery in cancer cells. Sci Rep 8(1):14739. https://doi.org/10.1038/s41598-018-33133-3

Oliveira-Mendes BBR, Miranda SEM, Sales-Medina DF, Magalhães BF, Kalapothakis Y, Souza RP, Cardoso VN, Barros ALB, Guerra-Duarte C, Kalapothakis E, Horta CCR (2019) Inhibition of *Tityus serrulatus* venom hyaluronidase affects venom biodistribution. PLoS Negl Trop Dis 13(4):e0007048. https://doi.org/10.1371/journal.pntd.0007048

Orivel J, Redeker V, Le Caer J-P, Krier F, Revol-Junelles A-M, Longeon A, Chaffotte A, Dejean A, Rossier J (2001) Ponericins, new antibacterial and insecticidal peptides from the venom of the ant *Pachycondyla goeldii*. J Biol Chem 276(21):17823–17829. https://doi.org/10.1074/jbc.M100216200

Ortiz E, Rendón-Anaya M, Rego SC, Schwartz EF, Possani LD (2014) Antarease-like Zn-metalloproteases are ubiquitous in the venom of different scorpion genera. Biochim Biophys Acta 1840(6):1738–1746. https://doi.org/10.1016/j.bbagen.2013.12.012

Papp F, Batista CVF, Varga Z, Herceg M, Román-González SA, Gaspar R, Possani LD, Panyi G (2009) Tst26, a novel peptide blocker of Kv1.2 and Kv1.3 channels from the venom of *Tityus stigmurus*. Toxicon 54(4):379–389. https://doi.org/10.1016/j.toxicon.2009.05.023

Pedron CN, Torres MDT, Lima JAS, Silva PI, Silva FD, Oliveira VX (2017) Novel designed VmCT1 analogs with increased antimicrobial activity. Eur J Med Chem 126:456–463. https://doi.org/10.1016/j.ejmech.2016.11.040

Pedron CN, Freire KA, Torres MDT, Lima DB, Monteiro ML, Menezes RRPPB, Martins AMC, Oliveira VX (2020) Arg-substituted VmCT1 analogs reveals promising candidate for the development of new antichagasic agent. Parasitology 147(14):1810–1818. https://doi.org/10.1017/S0031182020001882

Peigneur S, Cologna CT, Cremonez CM, Mille BG, Pucca MB, Cuypers E, Arantes EC, Tytgat J (2015) A gamut of undiscovered electrophysiological effects produced by *Tityus serrulatus* toxin 1 on NaV-type isoforms. Neuropharmacology 95:269–277. https://doi.org/10.1016/j.neuropharm.2015.03.027

Perry M, Sanguinetti MC (2008) A single amino acid difference between ether-a-go-go- related gene channel subtypes determines differential sensitivity to a small molecule activator. Mol Pharmacol 73(4):1044–1051. https://doi.org/10.1124/mol.107.043018

Pintar A, Possani LD, Delepierre M (1999) Solution structure of toxin 2 from *Centruroides noxius* Hoffmann, a β-scorpion neurotoxin acting on sodium channels 1 1Edited by P. E Wright. Journal of Molecular Biology 287(2):359–367. https://doi.org/10.1006/jmbi.1999.2611

Possani LD, Becerril B, Delepierre M, Tytgat J (1999) Scorpion toxins specific for Na + -channels. Eur J Biochem 264(2):287–300. https://doi.org/10.1046/j.1432-1327.1999.00625.x

Quintero-Hernández V, Ramírez-Carreto S, Romero-Gutiérrez MT, Valdez-Velázquez LL, Becerril B, Possani LD, Ortiz E (2015) Transcriptome analysis of scorpion species belonging to the *Vaejovis* genus. PLoS One 10(2):e0117188. https://doi.org/10.1371/journal.pone.0117188

Restano-Cassulini R, Olamendi-Portugal T, Zamudio F, Becerril B, Possani LD (2008) Two novel ergtoxins, blockers of K+-channels, purified from the mexican scorpion *Centruroides elegans elegans*. Neurochem Res 33(8):1525–1533. https://doi.org/10.1007/s11064-008-9634-8

Restano-Cassulini R, Olamendi-Portugal T, Riaño-Umbarila L, Zamudio FZ, Delgado-Prudencio G, Becerril B, Possani LD (2024) Characterization of sodium channel peptides obtained from the venom of the scorpion *Centruroides bonito*. Toxins 16(3):125. https://doi.org/10.3390/toxins16030125

Rodríguez-Ravelo R, Restano-Cassulini R, Zamudio FZ, Coronas FIV, Espinosa-López G, Possani LD (2014) A K+ channel blocking peptide from the Cuban scorpion *Rhopalurus garridoi*. Peptides 53:42–47. https://doi.org/10.1016/j.peptides.2013.10.010

Rodríguez-Ravelo R, Batista CVF, Coronas FIV, Zamudio FZ, Hernández-Orihuela L, Espinosa-López G, Ruiz-Urquiola A, Possani LD (2015) Comparative proteomic analysis of male and female venoms from the Cuban scorpion *Rhopalurus junceus*. Toxicon 107:327–334. https://doi.org/10.1016/j.toxicon.2015.06.026

Rojas-Azofeifa D, Sasa M, Lomonte B, Diego-García E, Ortiz N, Bonilla F, Murillo R, Tytgat J, Díaz C (2019) Biochemical characterization of the venom of central American scorpion *Didymocentrus krausi* Francke, 1978 (Diplocentridae) and its toxic effects in vivo and in vitro. Comp Biochem Physiol C Toxicol Pharmacol 217:54–67. https://doi.org/10.1016/j.cbpc.2018.11.021

Sanchez-Conde FG, Jimenez-Vazquez EN, Auerbach DS, Jones DK (2022) The ERG1 K+ channel and its role in neuronal health and disease. Front Mol Neurosci 15. https://doi.org/10.3389/fnmol.2022.890368

Santussi WM, Bordon KCF, Alves APNR, Cologna CT, Said S, Arantes EC (2017) Antifungal activity against filamentous fungi of Ts1, a multifunctional toxin from *Tityus serrulatus* scorpion venom. Front Microbiol 8. https://doi.org/10.3389/fmicb.2017.00984

Saucedo AL, Flores-Solis D, Rodríguez de la Vega RC, Ramírez-Cordero B, Hernández-López R, Cano-Sánchez P, Navarro RN, García-Valdés J, Coronas-Valderrama F, de Roodt A, Brieba LG, Possani LD, Río-Portilla F (2012) New tricks of an old pattern. J Biol Chem 287(15):12321–12330. https://doi.org/10.1074/jbc.M111.329607

Schiavon E, Sacco T, Cassulini RR, Gurrola G, Tempia F, Possani LD, Wanke E (2006) Resurgent current and voltage sensor trapping enhanced activation by a β-scorpion toxin solely in Nav1.6 channel. J Biol Chem 281(29):20326–20337. https://doi.org/10.1074/jbc.M600565200

Schwartz EF, Schwartz CA, Gómez-Lagunas F, Zamudio FZ, Possani LD (2006) HgeTx1, the first K+-channel specific toxin characterized from the venom of the scorpion *Hadrurus gertschi* Soleglad. Toxicon 48(8):1046–1053. https://doi.org/10.1016/j.toxicon.2006.08.009

Schwartz EF, Diego-Garcia E, Rodríguez de la Vega RC, Possani LD (2007) Transcriptome analysis of the venom gland of the Mexican scorpion *Hadrurus gertschi* (Arachnida: Scorpiones). BMC Genomics 8(1):119. https://doi.org/10.1186/1471-2164-8-119

Schwartz EF, Capes EM, Diego-García E, Zamudio FZ, Fuentes O, Possani LD, Valdivia HH (2009) Characterization of hadrucalcin, a peptide from *Hadrurus gertschi* scorpion venom with pharmacological activity on ryanodine receptors. Br J Pharmacol 157(3):392–403. https://doi.org/10.1111/j.1476-5381.2009.00147.x

Selvakumar P, Fernández-Mariño AI, Khanra N, He C, Paquette AJ, Wang B, Huang R, Smider VV, Rice WJ, Swartz KJ, Meyerson JR (2022) Structures of the T cell potassium channel Kv1.3 with immunoglobulin modulators. Nature. Communications 13(1):3854. https://doi.org/10.1038/s41467-022-31285-5

Sharma PP, Fernández R, Esposito LA, González-Santillán E, Monod L (2015) Phylogenomic resolution of scorpions reveals multilevel discordance with morphological phylogenetic signal. Proc R Soc B 282(1804):20142953. https://doi.org/10.1098/rspb.2014.2953

Shepard PD, Canavier CC, Levitan ES (2006) Ether-a-go-go related gene potassium channels: what's all the buzz about? Schizophr Bull 33(6):1263–1269. https://doi.org/10.1093/schbul/sbm106

Shi W, Wymore RS, Wang H-S, Pan Z, Cohen IS, McKinnon D, Dixon JE (1997) Identification of two nervous system-specific members of the erg potassium channel gene family. J Neurosci 17(24):9423–9432. https://doi.org/10.1523/JNEUROSCI.17-24-09423.1997

Silva de França F, Tambourgi DV (2023) Hyaluronan breakdown by snake venom hyaluronidases: from toxins delivery to immunopathology. Front Immunol 14. https://doi.org/10.3389/fimmu.2023.1125899

Silva ÉCN, Camargos TS, Maranhão AQ, Silva-Pereira I, Silva LP, Possani LD, Schwartz EF (2009) Cloning and characterization of cDNA sequences encoding for new venom peptides of the Brazilian scorpion *Opisthacanthus cayaporum*. Toxicon 54(3):252–261. https://doi.org/10.1016/j.toxicon.2009.04.010

Simone Y, van der Meijden A (2021) Armed stem to stinger: a review of the ecological roles of scorpion weapons. J Venom Anim Toxins Incl Trop Dis 3:e20210002. https://doi.org/10.1590/1678-9199-JVATITD-2021-0002

Smith JJ, Hill JM, Little MJ, Nicholson GM, King GF, Alewood PF (2011) Unique scorpion toxin with a putative ancestral fold provides insight into evolution of the inhibitor cystine knot motif. Proc Natl Acad Sci 108(26):10478–10483. https://doi.org/10.1073/pnas.1103501108

Solano-Godoy JA, González-Gómez JC, Torres-Bonilla KA, Floriano RS, Miguel ATSF, Murillo-Arango W (2021) Comparison of biological activities of *Tityus pachyurus* venom from two Colombian regions. J Venom Anim Toxins Incl Trop Dis 27. https://doi.org/10.1590/1678-9199-jvatitd-2021-0005

Stöcker W, Grams F, Reinemer P, Bode W, Baumann U, Gomis-Rüth F, Mckay DB (1995) The metzincins — topological and sequential relations between the astacins, adamalysins, serralysins, and matrixins (collagenases) define a super family of zinc-peptidases. Protein Sci 4(5):823–840. https://doi.org/10.1002/pro.5560040502

Sunagar K, Undheim EA, Chan AH, Koludarov I, Muñoz-Gómez SA, Antunes A, Fry BG (2013) Evolution stings: the origin and diversification of scorpion toxin peptide scaffolds. Toxins 5(12):2456–2487. https://doi.org/10.3390/toxins5122456

Suppiramaniam V, Abdel-Rahman EA, Buabeid MA, Parameshwaran K (2010) Ion channels. In: Comprehensive toxicology. Elsevier, pp 129–171. https://doi.org/10.1016/B978-0-08-046884-6.01310-5

Tabakmakher VM, Krylov NA, Kuzmenkov AI, Efremov RG, Vassilevski AA (2019) Kalium 2.0, a comprehensive database of polypeptide ligands of potassium channels. Sci Data 6(1):73. https://doi.org/10.1038/s41597-019-0074-x

Tarr DEK (2016) Establishing a reference array for the CS-αβ superfamily of defensive peptides. BMC Res Notes 9(1):490. https://doi.org/10.1186/s13104-016-2291-0

Tibery DV, Nunes JAA, da Mata DO, Menezes LFS, de Souza ACB, Fernandes-Pedrosa MF, Treptow W, Schwartz EF (2024) Unveiling Tst3, a multi-target gating modifier scorpion α toxin from *Tityus stigmurus* venom of northeast Brazil: evaluation and comparison with well-studied Ts3 toxin of *Tityus serrulatus*. Toxins 16(6):257. https://doi.org/10.3390/toxins16060257

Titaux-Delgado G, Lopez-Giraldo AE, Carrillo E, Cofas-Vargas LF, Carranza LE, López-Vera E, García-Hernández E, del Rio-Portilla F (2023) Beta-KTx14.3, a scorpion toxin, blocks the human potassium channel KCNQ1. Biochim Biophys Acta (BBA) Proteins Proteom 1871(4):140906. https://doi.org/10.1016/j.bbapap.2023.140906

Vandendriessche T, Olamendi-Portugal T, Zamudio FZ, Possani LD, Tytgat J (2010) Isolation and characterization of two novel scorpion toxins: the α-toxin-like CeII8, specific for Nav1.7 channels and the classical anti-mammalian CeII9, specific for Nav1.4 channels. Toxicon 56(4):613–623. https://doi.org/10.1016/j.toxicon.2010.06.008

Vandendriessche T, Kopljar I, Jenkins DP, Diego-Garcia E, Abdel-Mottaleb Y, Vermassen E, Clynen E, Schoofs L, Wulff H, Snyders D, Tytgat J (2012) Purification, molecular cloning and functional characterization of HelaTx1 (*Heterometrus laoticus*): the first member of a new κ-KTX subfamily. Biochem Pharmacol 83(9):1307–1317. https://doi.org/10.1016/j.bcp.2012.01.021

Varga Z, Gurrola-Briones G, Papp F, Rodríguez de la Vega RC, Pedraza-Alva G, Tajhya RB, Gaspar R, Cardenas L, Rosenstein Y, Beeton C, Possani LD, Panyi G (2012) Vm24, a natural immunosuppressive peptide, potently and selectively blocks Kv1.3 potassium channels of human T cells. Mol Pharmacol 82(3):372–382. https://doi.org/10.1124/mol.112.078006

Vargas-Jaimes L, Xiao L, Zhang J, Possani LD, Valdivia HH, Quintero-Hernández V (2017) Recombinant expression of Intrepicalcin from the scorpion *Vaejovis intrepidus* and its effect on skeletal ryanodine receptors. Biochim Biophys Acta Gen Subj 1861(4). https://doi.org/10.1016/j.bbagen.2017.01.032

Verano-Braga T, Figueiredo-Rezende F, Melo MN, Lautner RQ, Gomes ERM, Mata-Machado LT, Murari A, Rocha-Resende C, Lima ME, Guatimosim S, Santos RAS, Pimenta AMC (2010) Structure–function studies of *Tityus serrulatus* Hypotensin-I (TsHpt-I): a new agonist of B2 kinin receptor. Toxicon 56(7):1162–1171. https://doi.org/10.1016/j.toxicon.2010.04.006

Wymore RS, Gintant GA, Wymore RT, Dixon JE, McKinnon D, Cohen IS (1997) Tissue and species distribution of mRNA for the i kr -like K + channel, ERG. Circulation Research 80(2):261–268. https://doi.org/10.1161/01.RES.80.2.261
Xiao L, Gurrola GB, Zhang J, Valdivia CR, SanMartin M, Zamudio FZ, Zhang L, Possani LD, Valdivia HH (2016) Structure-function relationships of peptides forming the calcin family of ryanodine receptor ligands. J Gen Physiol 147(5). https://doi.org/10.1085/jgp.201511499
Yang F, Zhou JH (2010) Cytotoxicity and DNA damage induced by 1, 4-benzoquinone in v79 Chinese hamster lung cells. J Toxicol Environ Health A 73(7):483–489. https://doi.org/10.1080/15287390903523402
Yariv B, Yariv E, Kessel A, Masrati G, Ben CA, Martz E, Mayrose I, Pupko T, Ben-Tal N (2023) Using evolutionary data to make sense of macromolecules with a "face-lifted" ConSurf. Protein Sci 32(3). https://doi.org/10.1002/pro.4582
Yu FH, Catterall WA (2003) Overview of the voltage-gated sodium channel family. Genome Biol 4(3). https://doi.org/10.1186/gb-2003-4-3-207
Zeng X-C, Corzo G, Hahin R (2005) Scorpion venom peptides without disulfide bridges. IUBMB life (International Union of Biochemistry and Molecular Biology: Life) 57(1):13–21. https://doi.org/10.1080/15216540500058899
Zhang S, Gao B, Zhu S (2016) Independent origins of scorpion toxins affecting potassium and sodium channels. In: Evolution of venomous animals and their toxins. Springer Netherlands, pp 1–16. https://doi.org/10.1007/978-94-007-6727-0_12-2
Zhao R, Dai H, Qiu S, Li T, He Y, Ma Y, Chen Z, Wu Y, Li W, Cao Z (2011) SdPI, the first functionally characterized kunitz-type trypsin inhibitor from scorpion venom. PLoS One 6(11):e27548. https://doi.org/10.1371/journal.pone.0027548
Zhu S, Tytgat J (2004) The scorpine family of defensins: gene structure, alternative polyadenylation and fold recognition. Cell Mol Life Sci 61(14). https://doi.org/10.1007/s00018-004-4149-1
Zhu S, Li W, Jiang D, Zeng X (2000) Evidence for the existence of insect defensin-like peptide in scorpion venom. IUBMB Life 50(1):57–61. https://doi.org/10.1080/15216540050176601
Zhu S, Darbon H, Dyason K, Verdonck F, Tytgat J (2003) Evolutionary origin of inhibitor cystine knot peptides. FASEB J 17(12):1765–1767. https://doi.org/10.1096/fj.02-1044fje
Zilberberg N, Froy O, Loret E, Cestele S, Arad D, Gordon D, Gurevitz M (1997) Identification of structural elements of a scorpion α-neurotoxin important for receptor site recognition. J Biol Chem 272(23):14810–14816. https://doi.org/10.1074/jbc.272.23.14810
Zornetta I, Scorzeto M, Mendes Dos Reis P, De Lima M, Montecucco C, Megighian A, Rossetto O (2017) Electrophysiological characterization of the antarease metalloprotease from *Tityus serrulatus* Venom. Toxins 9(3):81. https://doi.org/10.3390/toxins9030081

Chapter 16
Scorpionism in Neotropics: Public Health Issues and Prevention

Clara Guerra-Duarte and Ana Luiza Bittencourt Paiva

Abstract Scorpionism is a significant but often neglected public health issue in the Neotropical region, characterized by high scorpion biodiversity. The most medically relevant genera are *Centruroides* and *Tityus*, and hotspot areas include mainly Mexico, Brazil, and Venezuela. However, climate change, urbanization, and deforestation are contributing to the spread of affected regions. Underreporting is a major challenge, hindering accurate assessment of the problem's scope. Antivenoms, the primary treatment, are often mismatched to the specific venoms involved, leading to poor therapeutic outcomes. Mexico has a high incidence but low mortality due to improved treatment, while Brazil sees rising cases and fatalities, despite antivenom availability. Venezuela, despite lacking systematic tracking, produces effective antivenom, though its efficacy varies by species. Regardless of the country, the profile of scorpionism victims is quite similar. Severe cases are more common in young children and typically occur in urban settings or in areas surrounding homes. Additionally, there is a strong correlation with economically disadvantaged communities. Prevention requires environmental conservation, urban control, housing security, and improved medical care. Furthermore, addressing taxonomic complexities and venom variations is crucial for effective antivenom development and management.

Keywords Scorpionism • Neotropics • Public health

C. Guerra-Duarte (✉) · A. L. B. Paiva
Research and Development Department, Fundação Ezequiel Dias, Belo Horizonte, Minas Gerais, Brazil
e-mail: clara.duarte@funed.mg.gov.br; analuiza.paiva@funed.mg.gov.br

A. F. de Araujo Lira, E. González-Santillán (eds.), *The Biology of Neotropical Scorpions*, https://doi.org/10.1007/978-3-032-23194-9_16

16.1 Introduction

Scorpions are fascinating creatures that have thrived for over 400 million years, showing little change in their physical form throughout evolution. This remarkable consistency highlights their adaptability to life on Earth (Wendruff et al. 2020; Lourenço 2023). While scorpions are found in nearly every region across various biomes and exhibit diverse living strategies, only a few species, members of the Buthidae family, possess highly toxic venom to humans, posing a significant health risk in the Americas. Many scorpion species have specific ecological needs for survival. Unfortunately, some of the most toxic species demonstrate remarkable adaptability, successfully colonizing human-altered environments (Lourenço 2018). These species tend to be concentrated in certain geographical areas that frequently experience scorpion stings, creating "hot spots" for such incidents (White 2016). These regions include the Middle East, North and South Africa, India, China, and Central and South America.

The Neotropical region is renowned for its biodiversity, especially considering its variety of scorpion species, including those that pose a risk to human health (Chippaux and Goyffon 2008; Lourenço and Ythier 2013; Ward et al. 2018). Historically, these noxious species have been more abundant in Mexico, Venezuela, and Brazil, where the health impact of scorpion stings has long been recognized (Chippaux and Goyffon 2008).

However, climate change, disorganized urbanization, and deforestation are altering this scenario, leading to increased scorpionism in the Neotropical region (Rafinejad et al. 2020; Lacerda et al. 2022a; Guerra-Duarte et al. 2023). Possibly due to these phenomena, species previously considered less relevant to human health are becoming more involved in harmful incidents, as interactions between them and the population are becoming more frequent (Borges et al. 2015; Silva et al. 2017). Deforestation is a major contributing factor, as it expands human encroachment into forested areas and allows scorpions to invade urban environments where they find food, water, and shelter while also facing fewer natural predators (Vilarinho et al. 2023). Furthermore, scorpion species already adapted to urban environments are expanding into new areas, as human mobility and commercial trades enable the introduction of foreign species into new habitats (Mario da Rosa et al. 2015; Costa et al. 2020).

There are additional factors that raise concerns regarding this issue. Although rising incidents of scorpionism have been reported in many countries, reporting these incidents is often not mandatory, leading to significant underestimation in the number of envenomations (Santos et al. 2016). As a result, there is no clear understanding of the true impact of scorpion stings in the Neotropical region. In addition, severe cases occurring in areas where such incidents were previously uncommon face challenges due to the lack of experience among medical staff at health centers, which can lead to unfavorable outcomes (Borges and Román 2023). It is also unclear whether some of these accidents were caused by native species, previously disregarded as potentially dangerous to humans, or by species that may have started to colonize new territories. Moreover, the taxonomy of the relevant species remains controversial, making it difficult to grasp the real contribution of each one of them to the medical problem (Ojanguren-Affilastro et al. 2021; Pardal et al. 2023).

Antivenoms are the primary treatment recommended for severe scorpion envenomation (World Health Organization 2024). However, these antidotes must be specific to the venom of the species responsible for the envenomation to be effective. Unfortunately, this crucial aspect of managing victims envenomed by scorpions is often overlooked both by the health professionals and by the policy makers responsible for their acquisition, as the available antivenoms are directed toward only a few harmful scorpion species found in the Neotropical region (Borges et al. 2020). Indeed, the antivenoms currently produced do not adequately reflect the biodiversity of the region, leading to frequent reports of poor responses to antivenom therapy (de Roodt et al. 2009; Borges et al. 2020; Silva de Oliveira et al. 2020; Gomes et al. 2020; Martins et al. 2021).

Scorpionism certainly poses a significant health issue in the Neotropical region. In hotspot areas such as Brazil and Mexico, where this problem has persisted for over a century, several measures to contain scorpion colonization in urban areas and prevent accidents have been proposed over time (Dias et al. 1924). However, the high number of envenomation cases still occurring in these countries indicates that additional strategies must be implemented for effective management (Guerra-Duarte et al. 2023).

To help address this public health problem, this chapter compiles available information regarding scorpion envenomation cases in various parts of the Neotropical region to illustrate the scope of this issue. The second part also suggests potential strategies to prevent the further spread of affected areas and reduce the number of incidents in regions already impacted by scorpions.

16.2 Epidemiology

The first step in addressing a health issue is to acknowledge the extent of the damage. While the problem of scorpion stings in the Neotropical region has been known for decades, a comprehensive understanding of this issue has only recently begun to emerge in a few countries.

To our knowledge, scorpion stings are classified as a disease of mandatory reporting to health authorities by law or regulation in only three countries: Mexico, Brazil, and Argentina. With an increasing situation, Paraguay has begun to hold official records of the accidents. Other countries, such as Ecuador and Panama, monitor the situation, but do so inconsistently. Uruguay and Peru have established monitoring systems for snake and spider bites, but not for scorpions. In Guatemala, scorpion sting notifications are categorized alongside other types of poisonings, including drugs and medicinal substances. Conversely, some countries, like Colombia and surprisingly Venezuela—one of the countries most affected by scorpions—do not keep official records of these envenomations.

As a result, it is challenging to obtain a clear picture of the real impact of scorpions on human health in the Neotropical region. Nevertheless, by gathering information from scientific papers, dissertations, and other publicly available data, we have attempted to provide a general overview of the situation (Fig. 16.1), which will be discussed further in the sections below.

In this chapter, we will focus on accidents caused by scorpions of the genera *Centruroides* and *Tityus,* which are more frequently reported as the etiological agent of medically relevant accidents. Chile is the sole country that was not included in our analysis, as these two genera are not found, to date, in the region. Nevertheless, it is important to mention that scorpion stings do occur in the country, but as they are mostly related to scorpions of the Bothriuridae family, the victims present only mild symptoms (Schenone and Fontecilla 1998; Chippaux and Goyffon 2008; Pizarro-Araya and Ojanguren-Affilastro 2021).

Regardless of the country, the profile of scorpionism victims is remarkably consistent. Stings predominantly affect the extremities (hands and feet) and tend to be more severe in young children under 10 years of age. These incidents commonly occur in urban or peridomiciliary environments and are strongly correlated with impoverished communities. Unlike snakebites, scorpionism affects men and women almost equally, without significant gender-based distinctions. However, regional specificities exist. For example, envenomations involving certain Amazonian scorpions resemble the profile of snakebites, disproportionately affecting men during occupational activities. Additionally, most scorpion stings occur during the warmest months of the year and are correlated with periods of increased precipitation (Santos et al. 2016).

16.2.1 Mexico

Mexico has the greatest diversity of scorpions in the world, with 38 genera and nearly 300 species, representing over 12% of all known scorpion species (Santibáñez-López et al. 2015). Among these, the Buthidae family holds particular importance for public health, especially the genus *Centruroides* Marx, 1860, which is by far the most diverse scorpion genus of this family in Mexico. This genus comprises around 42 species distributed across almost the entire territory, and approximately 20 species are considered dangerous to humans (Riaño-Umbarila et al. 2017; González-Santillán and Possani 2018; Chippaux et al. 2020). Medically significant scorpions are predominantly found in the central region of the country extending to the Pacific coastal states (Hernández-Muñoz et al. 2024). One of the most venomous species is *Centruroides noxius*, whose venom has an LD_{50} of 5 μg per 20 g of mouse weight (Dent et al. 1980). Other dangerous species include *Centruroides sculpturatus, Centruroides limpidus*, and *Centruroides infamatus* (Dehesa-Dávila et al. 1996; Corona et al. 2001; Valdez-Cruz et al. 2004; Santibáñez-López et al. 2015). As new *Centruroides* species are being continuously identified, including harmful ones, this number is likely to increase (García-Villalvazo et al. 2023).

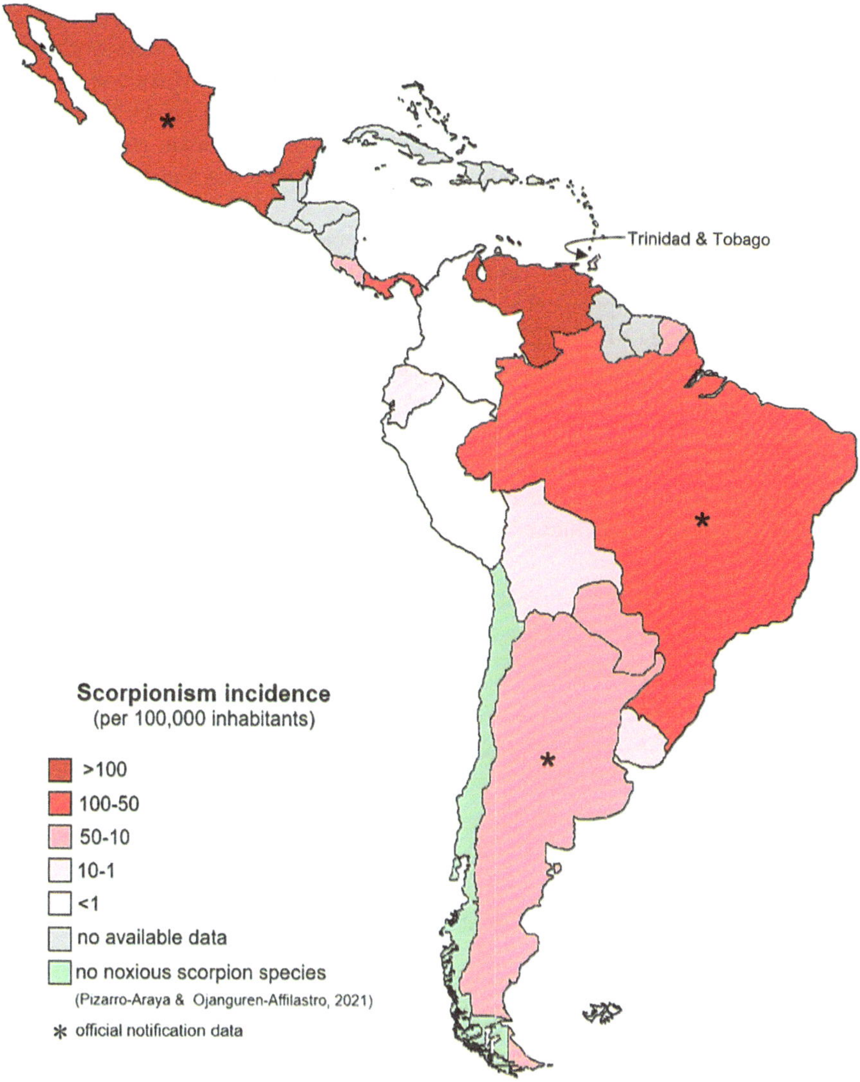

Fig. 16.1 Estimated scorpionism incidence in the Neotropical region in 2020's. Publicly available data (government websites, articles, dissertations, news) on annual numbers of scorpionism for each country, from 2020 until 2023, were gathered and incidence per 100.000 inhabitants was estimated considering the country's population reported in https://worldpopulationreview.com/. References: Benmosbah et al. (2013), Telemundo (2017), Versión Final (2017), Universidad de Costa Rica (2018), Restrepo et al. (2020), Ministerio de Salud Ecuador (2021), Ministerio de Salud Argentina - Direccion epidemiologia (2023), Ministerio de Salud Mexico (2023), Rocío Arjona 2023, Instituto Nacional de Laboratorios de Salud (2023), Ministerio de Salud Paraguay (2024), Ministério da Saúde Brasil (2024)

Scorpionism has long been a public health concern in Mexico (Chippaux et al. 2020), which accounts for 25% of all scorpion sting cases reported worldwide (Hernández-Muñoz et al. 2024). The reality of scorpionism in Mexico is so common that the population of the most affected areas, like Durango, has even embraced scorpions as an important symbol, having their local soccer team named after it ("Los Alacranes"), picturing the animal in souvenirs, and using it as food and drink (Gonzalez-Ponce et al. 2023).

Cases of scorpion stings are systematically reported to the state epidemiological surveillance system by health centers, which then relay the data to the Federal Ministry of Health (Chippaux et al. 2020). The incidence of scorpionism in Mexico over the past decade (2014–2023) was 228.3 per 100,000 inhabitants (Secretaría de Salud México 2023), with yearly cases fluctuating around 300.000, without any noticeable increase or decrease over these years (Fig. 16.2a). Although the overall incidence has remained relatively stable since the mid-2000s, four states report significantly higher rates. For instance, in the last decade, Colima state recorded an average incidence of 1383.2 per 100,000 people, followed by Morelos with 1374.2, Nayarit with 1226.1, and Guerrero with 1177.7 (Secretaría de Salud México 2023).

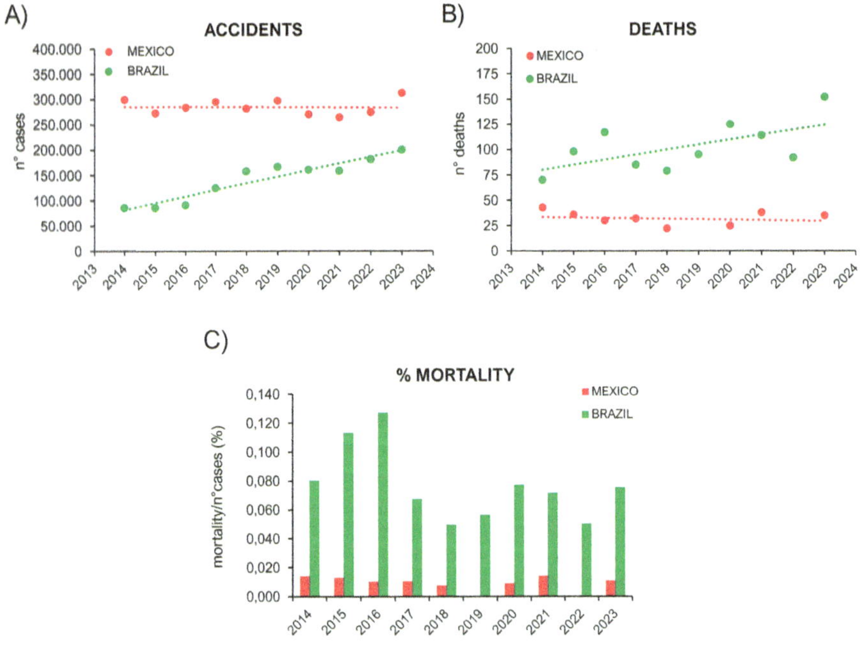

Fig. 16.2 Comparison of scorpionism between Mexico and Brazil. The number of accidents and deaths due to scorpionism was obtained from the official Ministry of Health websites of each country. The dotted lines represent the calculated tendency of each dataset. (**a**) Number of reported scorpion incidents per year from 2014 to 2023. (**b**) Number of reported deaths related to scorpionism. (**c**) Case/fatality rate, considering the number of cases per year

Within Guerrero, the municipality of Chilapa stands out, reporting the state's highest incidence. The associated factors to the high frequency of scorpion stings in seven highly marginalized communities of Chilapa were reported by Trinidad-Porfirio et al. (2023). Likewise, at the end of 2023, an outbreak of scorpion sting envenomation cases was reported in Villa Unión, Sinaloa state, with an incidence rate of 1364 per 100,000 inhabitants (Hernández-Muñoz et al. 2024). Among the 231 reported cases, 151 (65.3%) had a previous history of scorpion sting, although the case fatality rate remained stable. Scorpion stings envenomation has been the second leading cause of morbidity in the Villa Unión Hospital for five consecutive years, surpassed only by acute upper respiratory tract infections. The highest incidence of scorpion stings in Mexico is often concentrated in regions with lower socioeconomic development and inadequate housing conditions. Field investigations revealed that many homes in these areas lack basic protective features, such as tiled floors, glass windows, doors, and fly screens. Notably, in these impoverished regions, most accidents occur inside dwellings, a pattern distinct from other parts of the country (Trinidad-Porfirio et al. 2023; Hernández-Muñoz et al. 2024).

Although Mexico records one of the highest incidences of scorpion stings worldwide, and some states and localities report rates well above the national average, it is noteworthy that the mortality rate has decreased significantly over the years and has remained stable since the early 2000s (Fig. 16.2b). Between 1940 and 1958, the average annual deaths from scorpion sting envenoming were estimated at 1775, corresponding to a mortality rate of 7.1 per 100,000 inhabitants (Mazzotti and Bravo-Becherelle 1961). By the late 1970s, this had dropped to fewer than 500 deaths annually (0.9 per 100,000) and has since fallen to fewer than 50 deaths per year, approximately 0.03 per 100,000 population. A statistically significant decline in mortality occurred between 1979 and 2003, with an 86.5% total reduction during 2001–2003 compared to 1979–1982 (Celis et al. 2007). This dramatic reduction in mortality is attributed to improved treatment protocols and expanded healthcare access. Advances in intensive care and early immunotherapy, including the introduction of highly purified IgG fragments in the late 1990s, enhanced treatment safety and reduced the severity of envenomations. These medical advancements, coupled with proactive health policies, assured greater access to care in underserved populations. Over the past 40 years, these efforts have drastically reduced fatalities. However, approximately 30 deaths per year still occur, predominantly among young children (Chippaux et al. 2020). Within this age group, severe cases can represent almost 60% the total number of reported cases (Robledo-Aceves et al. 2024).

16.2.2 Brazil

Brazil ranks second in the number of scorpion incidents in South America, with over 200,000 cases reported in 2023, according to SINAN (Notifiable Diseases Information System) (Ministério da Saúde Brasil 2024). This represents a dramatic increase compared to the 8000 annual cases reported in the early 1990s (Reckziegel and Pinto 2014). While the incidence remains lower than that reported in Mexico, the number of accidents has continued to rise steadily over the past decade (Fig. 16.2a). Despite the availability of high-quality antivenom, which has been freely provided to envenoming victims through the Brazilian Unified Health System (SUS) since the 1980s, the number of fatalities has also increased, reaching approximately 100 deaths annually (Fig. 16.2b). Notably, scorpion-related fatalities have even surpassed those caused by snakebites in recent years. However, the case-fatality rate has shown a decline over the past few years (Fig. 16.2c).

Tityus serrulatus is the primary species responsible for scorpion incidents in Brazil. Its ubiquitous presence in urban areas and its status as a national public health concern can be attributed to several factors: parthenogenesis (Lourenço 2008), low requirements for food and water (Pimenta et al. 2019), a broad thermal tolerance range (Hoshino et al. 2006), adaptability to anthropized environments (Lourenço 2022), and the highly toxic nature of its venom (Pucca et al. 2015). The type locality of *T. serrulatus* is the city of Belo Horizonte, located in the state of Minas Gerais in the Southeast region of Brazil (Lourenço 2015). Historically, this region has recorded the highest incidence of scorpionism in the country. However, over the past decade, the state of São Paulo has emerged as a significant focus in this scenario (Chiaravalloti-Neto et al. 2023; Silva et al. 2023). Originally, this state was predominantly inhabited by *Tityus bahiensis*, another species of medical importance. The introduction of *T. serrulatus*, however, has led to a species replacement, resulting in an increase in both the number of accidents and their severity (von Eickstedt et al. 1996). The expansion of *T. serrulatus* continues, with the neighboring South region of Brazil also reporting a growing number of cases involving this species (Ortiz et al. 2021; Seemann et al. 2023; Gaviraghi et al. 2023). Alongside the Southeast region, the Northeast region also reports the highest number of scorpionism cases (Guerra-Duarte et al. 2023). In addition to *T. serrulatus*, another yellow scorpion species, *Tityus stigmurus*, is responsible for a significant number of envenomation incidents in this region (Furtado et al. 2020).

The role of scorpions as a public health concern in the Southeast and Northeast regions has long been recognized. However, in recent years, the Middle West region of Brazil has also drawn significant attention. Brasília, the country's capital, is an artificially constructed city located in this region. Established around 1960, it has only recently been invaded by *T. serrulatus*. Within a span of 15–20 years, the sporadic occurrence of this species has led to the establishment of a stable population, resulting in frequent accidents and invasions of public buildings, including the Parliament and the Senate (Lourenço et al. 1996). In 2024, a previously overlooked species, *Tityus confluens*, was elevated to medical concern status after being linked to life-threatening envenomations in the region. This development has established a new hotspot for scorpionism in Brazil (Borges et al. 2024).

The North region of Brazil, which encompasses the Amazon rainforest, is also emerging as a new area of concern for scorpionism (Monteiro et al. 2019). This region is home to approximately 13% of the world's scorpion diversity (Borges et al. 2021). The demographic expansion in the Amazon has disrupted local ecosystems, leading to more frequent encounters between human populations and endemic scorpion species, which were previously not considered harmful (Monteiro et al. 2019). Amazonian scorpions exhibit distinct envenomation profiles and are not effectively neutralized by currently available antivenoms (Gomes et al. 2020), raising significant public health concerns.

Among the species in the region, *Tityus obscurus* is commonly regarded by health authorities as the primary threat. This species exhibits regional variations in venom composition and in the clinical symptoms it causes, which have recently been linked to its genetic diversity, a phenomenon that has just begun to be unveiled (Pardal et al. 2023). In addition to *T. obscurus*, several other species are emerging as problematic in the North region, including *Tityus metuendus, Tityus matthieseni, Tityus bastosi, Tityus silvestris* (Costa et al. 2016), *Tityus raquelae*, and *Tityus apiacas* (Gomes et al. 2020), all of which have been associated with human hospitalizations. Furthermore, the potential colonization of the region by *T. serrulatus* is a growing concern, as commercial trade between Brazilian states frequently facilitates the unintentional transport of non-native species, which can establish stable populations, considering the potential of this species to do so (Costa et al. 2020).

The profile of scorpion accidents in the North region has traditionally differed from other parts of the country, with most incidents occurring during field activities in natural areas. Additionally, the expansion of agricultural frontiers has contributed to the rise in scorpionism in Northern states, as rural areas dominated by monocultures report a comparatively higher number of accidents (Pardal et al. 2003). With increasing deforestation, cities such as Manaus are now experiencing scorpion invasions, with accidents being reported in peridomiciliary areas and even inside homes (Gomes et al. 2020; Benzaken et al. 2024).

16.2.3 Venezuela

Venezuela has the highest incidence of scorpionism in the Neotropical region, a situation partly attributed to the remarkable diversity of *Tityus* scorpions within the country, with at least 54 species documented (González-Sponga 1996). Among these, 17 species have been associated with fatal envenomations: *T. discrepans, T. isabelceciliae, T. pittieri*, and *T. carabobensis* in the North Central region; *T. caripitensis, T. quirogae, T. arellanoparrai, T. monaguensis, T. surorientalis, and T. nororientalis* in the North Eastern region; *T. falconensis, T. ivic-nancor*, and *T. sanarensis* in the West Central region; *T. valerae* and *T. zulianus* in the Andean region; *T. breweri* in the Southern region; and *T. perijanensis* in the Perijá mountain in the Colombian border (D'Suze et al. 2015). Despite the severity of the issue, health authorities in Venezuela do not systematically track scorpionism epidemiology. As a result, reported cases are likely underestimated, painting an even more concerning picture of the public health challenge posed by scorpion envenomations in the country.

Fortunately, Venezuela is one of the few countries in the Neotropical region that produces scorpion antivenom. Although the antivenom uses only *Tityus discrepans* as the immunogen, its general effectiveness in preventing lethality has been documented (De Davila et al. 1997), even in severe cases (Fuentes Fiore 2024). *Tityus discrepans* was the chosen species possibly because it was among the first Venezuelan species linked to severe envenomations. Typically, envenoming by this species can cause gastrointestinal toxicity and pancreatitis (Borges et al. 2006). In contrast, *T. zulianus*, another medically significant species, highly prevalent in the country, primarily induces cardiorespiratory complications and is not fully neutralized by the anti-*T. discrepans* antivenom, requiring higher dosages to be effective (Mazzei de Dàvila et al. 2002). Indeed, further studies have demonstrated the molecular diversity of the toxins from *T. discrepans* and *T. zulianus* venoms, accounting for the limited cross-neutralization observed (Borges et al. 2006). Species more closely related to *T. discrepans,* such as *T. breweri,* are likely to respond better to the antivenom treatment (Borges et al. 2010b), whereas those from other groups tend to respond only partially. Gathering information on the venom composition and the implication of the different Venezuelan species in the reported accidents is paramount to understand the reality of the scorpionism problem in the country and to provide better treatment to the victims.

The antigenic complexity of Venezuelan scorpion venoms seems not to be adequately addressed by using a single species for antivenom production (Borges et al. 2008, 2010a). Dr. Adolfo Borges, a Venezuelan toxinologist who has dedicated his career to this field, has advocated for the establishment of antigenic groups to develop antivenoms that better reflect the phylogenetic diversity of scorpion venoms in the Neotropical region (Borges et al. 2020).

16.2.4 Panama, Central America, and Caribbean

Panama reports the highest incidence of scorpion envenomations in Central America. There, scorpionism is considered a public health issue, and scorpion stings are mandatory notifiable events. Most stings are attributed to species within the *Centruroides* genus. However, identifying these scorpions at the species level remains challenging due to significant intraspecific morphological variation within the genus (Escudero-Sanjur et al. 2022). *Centruroides margaritatus* is the most common species with the highest population density in urban areas of Panama. It is the scorpion most frequently associated with stings in the Panamanian population, typically without significant consequences (Salazar et al. 2018; Rocío Arjona 2023). *Tityus* is the genus of greatest medical significance, and the species *Tityus pachyurus, T. asthenes, T. festae*, and *T. cerroazul* have been linked to severe and fatal cases. Among these, *T. cerroazul* is particularly critical, as its venom's toxic effects are not neutralized by any existing antivenoms (Borges et al. 2020).

By the late 1990s, reports of severe envenomations and deaths caused by scorpion stings began to emerge in Panama. The habitats of most toxic species coincide with tropical humid forest areas, which were previously sparsely populated and underdeveloped. However, as urban centers expanded into forested regions, waste accumulation, deforestation, and vegetation removal created conditions conducive to the appearance of toxic scorpions in or near human dwellings, as observed in other countries (Borges et al. 2012). Between 2000 and 2016, there was a significant increase in scorpionism cases, rising from 385 cases in 2000 to 4399 in 2016. In this period, the Department of Epidemiology of the Ministry of Health registered 33 623 cases with a total of 47 deaths (Rocío Arjona 2023). A particularly sharp rise occurred after 2007, a year marked by an unusual increase in scorpionism-related mortality. In 2007, Panama recorded an incidence of 52 cases per 100,000 inhabitants, the highest in Latin America after Mexico. This upward trend over 17 years is attributed in part to underreporting in earlier years and heightened awareness following the 2007 mortality spike. Between 1998 and 2006, 28 scorpionism-related deaths were reported, predominantly in children aged 1–14 years (Borges et al. 2012). Similar to other countries, the 1–4-year age group is at the highest risk of mortality. From 2000 to 2019, 122 deaths were recorded in this group, with a mortality rate of 7.9 deaths per 100,000 inhabitants (Rocío Arjona 2023).

In Costa Rica, scorpion stings have been noted as a recurring issue, with increasing numbers since 1992. For instance, 24 cases were reported in 1992, rising to 206 cases in 1998 and 405 in 1999 (Leveridge 2000). Surprisingly, according to the Costa Rican Ministry of Health, no cases of scorpion envenomation occurred between 2005 and 2010. Limited information exists about envenomation by *Tityus* species and its clinical effects in Costa Rica, despite the presence of this genus in the country (Borges et al. 2012).

In contrast to previously noted, envenoming by *Tityus* scorpions inhabiting Costa Rica does not seem to reach the same clinical severity as observed in Panama (Díaz et al. 2023). Although there is one reported fatality linked to scorpion envenomation in the southern region of Costa Rica, where *T. championi* and *T. dedoslargos* are found (Leveridge 2000), most scorpion sting cases in the country are considered mild. These cases are predominantly attributed to species of the *Centruroides* genus, with only a few confirmed incidents involving *Tityus* species. The low incidence of *Tityus* envenomation in Costa Rica may be influenced by various factors. One possibility is differences in venom composition at the species or population level, as has been observed in *T. pachyurus* from different countries (Borges et al. 2020). This variation could play a role in the reduced severity of envenomation. On the other hand, unlike countries with rapid urban expansion, Costa Rican *Tityus* species primarily inhabit secondary rainforests and rarely coexist with humans. Moreover, *Centruroides* scorpions are more likely to be found inside premises and in their roundabouts, which may displace *Tityus*. Thus, the limited medical emergencies caused by *Tityus* scorpions in Costa Rica may stem more from ecological and demographic factors than from the venom's characteristics (Díaz et al. 2023).

Cases of scorpionism in the other Central American countries (Guatemala, Belize, El Salvador, Nicaragua) are generally underreported and regional morbidity rates are low. Among the *Centruroides* species, *C. margaritatus* and *C. gracilis* are the most widespread throughout the area. Meanwhile, a study of the medical relevance of species of *Tityus* in Central America has determined that they are present only in Panama and Costa Rica. The most abundant and widely distributed species in the genus is *T. pachyurus* (Borges et al. 2012).

The true extent of the scorpionism problem in most Central American countries is difficult to estimate since most countries lack official records of scorpion incidents. The most recent epidemiological data available in the literature was reported by Borges et al. (2012), who gathered information through personal communication with local health authorities in each country. In El Salvador, during the period 2006–2011, the Ministry of Health reported 61 cases of scorpion envenomation, all of which were non-fatal. In Guatemala, 38 cases were documented between 2005 and 2010, with 57.8% of these occurring in the Izabal department, near the borders with Belize and Honduras. Scorpion envenomation in Honduras is categorized under "sting by unknown animal" in health records, making it challenging to determine the true scale of the problem. However, based on statistics from neighboring Nicaragua and El Salvador, it can be inferred that scorpion stings are likely common. In Nicaragua, 392 scorpion sting cases were recorded between 2005 and 2011, making it the leading cause of envenomation by invertebrates in the country. However, no fatalities were reported during this period.

As in most Central American countries, there is a lack of comprehensive reports and official records regarding scorpion incidents on Caribbean islands. This is likely because most scorpion species found in these regions are not of significant medical concern. However, since the early twentieth century, numerous cases of pancreatitis (Waterman 1938; Bartholomew 1970; Bartholomew et al. 1974; Angus et al. 1995), as well as myocarditis and fatalities (Daisley et al. 1999), have been attributed to stings by *T. trinitatis,* in Trinidad and Tobago. This species is endemic to Trinidad and Tobago and has a limited range of occurrence, extending into Venezuela. Recently, cases with this species seem to be increasing, with 24 cases reported in 4 months in 2023 (Trinidad Express 2023), including a case report of acute coronary syndrome developed after *T. trintattis* sting in a HIV-positive man (Katwaroo et al. 2024).

The sole documented cases of envenomation by a *Centruroides* species in the Caribbean occurred on the island of Guadeloupe. The first reported incident involved *Centruroides pococki*, a species endemic to the island (Schmitt et al. 2017). As is typical with most *Centruroides* stings, the envenomation did not result in serious consequences. A second case with the same species reported later (Caré et al. 2021) described a class III envenoming with neuromuscular toxicity. The patient was not treated with antivenom but reported full recovery after 2 days. These case reports illustrate that, even if scorpion envenomations are not so common in this region, their potential to cause harm cannot be underestimated.

16.2.5 Paraguay, Argentina, and Uruguay

The Southern Cone of South America has experienced significant changes in its scorpionism landscape. Argentina, which has an official surveillance system in place, has documented a notable increase in scorpion sting incidents. In Paraguay, although scorpion sting reporting is not mandatory, official data available on government websites highlight the growing public health concern posed by scorpions (Cáceres et al. 2022). In contrast, information on scorpionism in Uruguay is scarce. The country is home to *T. uruguayensis* (Toscano-Gadea 2015), a species not associated with significant envenomation cases. However, given Uruguay's shared borders and active trade with Brazil and Argentina—both endemic for scorpionism—there is a risk of invasive, medically significant species establishing themselves in the country.

In Paraguay, several *Tityus* species have been reported, including *T. trivitatus, T. confluens, T. bahiensis, T. paraguayensis*, and *T. mattogrossensis. Tityus trivitatus* is the species more frequently found in urban areas of Asunción, the country's capital (Rojas et al. 2022). Although the individuals collected in this area seem to exhibit more limited toxicity, scorpion stings are common, with a local hospital reporting at least one case per day throughout 2019 (Borges and Rojas de Arias 2019). More recently, *T. confluens* appears to be displacing *T. trivitatus* in Asunción, raising concerns due to its higher toxicity (Borges et al. 2024). In 2023, 1259 scorpion sting cases were reported in Paraguay, resulting in four deaths (ABC Color 2024). The rise in scorpion incidents in Paraguay is recent, and many medical professionals are not yet familiar with the symptoms and optimal treatment protocols. Additionally, Paraguay lacks a locally produced antivenom and relies on imports from Argentina and Brazil when needed (Rojas et al. 2022).

In Argentina, scorpionism has been a recognized issue for a longer period, and a locally produced antivenom is available. However, the situation is deteriorating, with a 32% increase in reported cases between 2021 and 2022 (Ministerio de Salud Argentina - Direccion epidemiologia 2023). Prior to this, the Argentinian Health Ministry had already noted a significant rise in both the number of cases and fatalities, alongside an expansion of areas reporting severe envenomations (de Roodt et al. 2009, 2010, 2019). Advances in scientific knowledge have also revealed new challenges. Historically, *T. trivitatus* was considered the primary species responsible for scorpionism in Argentina, but its venom differs from that of Paraguayan populations. In 2021, Ojanguren-Affilastro et al. (2021) redefined the taxonomy of this species, identifying the southernmost population as a distinct, more toxic species, *Tityus carrilloi*. Argentina also faces threats from *T. confluens*, which has caused fatal envenomations (de Roodt et al. 2009), and *T. bahiensis*, both of which are expanding their ranges of occurrence (Giambelluca et al. 2021b; Jannello and López 2024). Climate change is expected to further exacerbate this expansion (Martinez et al. 2018). While scorpions in Buenos Aires, the country's most populous city, were previously considered less toxic compared to other regions (Blanco et al. 2016), severe cases have been reported in recent years (de Roodt et al. 2019).

Additionally, the first confirmed case involving *T. serrulatus* in Argentina was reported in 2020 (López et al. 2020), raising significant public health concerns. The augmented complexity of the Argentinean scorpiofauna of medical concern raises questions around the adequacy of the currently produced antivenom, which relies solely on *T. carrilloi* venom (Lopez 2021).

16.2.6 Suriname, Guiana, French Guiana

The Guianas, located in the northeastern part of South America, is a sparsely populated region characterized by dense primary forests, with most inhabitants concentrated along the coast. This region boasts significant biodiversity, including a rich scorpiofauna (Benmosbah et al. 2013). While epidemiological data on scorpionism are primarily available for French Guiana, Guyana and Suriname likely exhibit similar profiles.

Reported scorpion sting incidents are relatively few but represent a significant incidence rate given the region's small population (Chippaux et al. 1984). In French Guiana, the official National Health Agency reports an incidence of approximately 14 cases per 100,000 inhabitants, though some studies suggest rates as high as 90 stings per 100,000 inhabitants annually (Vaucel et al. 2020). When considering only the population of Cayenne, the capital, the incidence can reach up to 40 cases per 100,000 inhabitants (Benmosbah et al. 2013). Despite discrepancies in reported numbers, it is evident that scorpionism is a relevant public health issue in French Guiana. Additionally, most documented cases likely occur in urbanized areas, while incidents in forest-dwelling communities remain largely unreported.

The primary causative agent of scorpion stings in the Guianas appears to be *Tityus obscurus* or related species within the Black Amazonian scorpion group. *Tityus silvestris* has also been implicated in severe envenomations in the region (Vaucel et al. 2022). While a significant proportion of cases are mild, some require medical intervention. In one case series, 16.6% of scorpion sting victims were hospitalized, with 4% admitted to intensive care units (Benmosbah et al. 2013). Another study focusing on pediatric patients found that 30% of cases were classified as grade III envenomations (Vaucel et al. 2020). Although no fatalities were reported in these studies, at least two deaths attributed to *T. obscurus* envenomations have been documented in French Guiana (Ganteaume and Imbert 2014). Severe cases of pancreatitis have also been reported in the region (Kallel et al. 2016).

Data on scorpionism in Suriname are scarce, but reports from Guyana highlight a concerning situation. Three deaths were recorded within a single year, all of which began with localized pain but progressed to systemic symptoms and fatalities despite medical intervention and the absence of preexisting comorbidities in the patients (Iserson and Jagit Ramcharran 2019). Notably, antivenom specific to the causative species was unavailable, and other related antivenoms were also absent in the Guianas region. The authors of the report suggest that the venom of *T. obscurus* may be evolving toward greater toxicity, as no similar case clusters have been historically documented in Guyana.

16.2.7 Other South American Countries: Colombia, Ecuador, Bolivia, and Peru

The remaining Neotropical countries—Colombia, Ecuador, Bolivia, and Peru—report a relatively low incidence of scorpionism, with official data indicating 10 to 1 case per 100,000 inhabitants (Fig. 16.1). However, none of these countries maintain organized surveillance systems for scorpionism, suggesting that the reported numbers likely underestimate the true scale of the issue. Given that Lourenço (1984) identified this region as one of the most diverse areas for scorpions globally, there is a considerable risk of human encounters with medically significant species.

In Colombia, *T. pachyurus* is the species most frequently involved in human incidents and is associated with severe envenomations. Other species, such as *T. asthenes, T. fuhrmanni*, and *Centruroides gracilis*, are also present and cause less severe envenomations compared to *T. pachyurus* (Otero et al. 2004). Additionally, *Tityus colombianus, Tityus forcipula, T.* aff. *metuendus* sp. n., *Centruroides edwardsii*, and *Centruroides margaritatus* have been documented in Colombia (Guerrero-Vargas et al. 2015). Recently, a grade II envenomation caused by *Tityus vaissadei* was reported (Buitrago et al. 2020), highlighting the incomplete understanding of the local scorpionism landscape. The number of scorpion species recorded in Colombia has doubled in the last two decades to 80 species (Guerrero-Vargas et al. 2015), and the discovery of a new *Tityus achilles* capable of spraying its venom further underscores the complexity of scorpionism in the country (Laborieux 2024).

Despite numerous reports of severe envenomations in Colombia, including cases of coronary artery dissection (Arrieta-Mendoza et al. 2022), cardiac arrest (Izquierdo and Rodríguez Buitrago 2012), and interstitial nephritis (Cortés et al. 2022), there is a widespread perception that local scorpions are harmless. Medical staff are often inadequately trained to identify and manage scorpion envenomations (Reyes-Vega et al. 2021). This perception, combined with the belief that most scorpion stings are mild, has led to approximately 82.5% of cases going unregistered, as many individuals do not seek medical care (Gómez et al. 2010). Nevertheless, fatal cases have been reported in various departments (Otero et al. 2004; Gómez and Otero 2007; Restrepo et al. 2020). A study by Carrasquilla et al. (2023) found that 24% of scorpion envenomation cases treated at a hospital in the Norte de Santander region between 2018 and 2023 were severe, with a perceived increase in cases over time. Furthermore, species like *T. fuhrmanni* are known to invade dwellings in populous cities such as Medellín (Cardona et al. 2002), and hospitals in the city have treated envenomations caused by this species in newborns, children, and adults (Estrada-Gómez et al. 2017).

Although Colombia has two national antivenom producers, no specific antivenom for scorpion envenomations has been manufactured to date. Available antivenoms from Mexico, Brazil, and Venezuela have shown efficacy against local venoms. However, motivated by the potential escalation of scorpionism in Colombia, an experimental lab-scale antivenom targeting the country's primary scorpion species has been proposed (Estrada-Gomez et al. 2022).

In Ecuador, the situation mirrors that of Colombia, with no mandatory reporting of scorpionism cases and limited knowledge of the species responsible for envenomations and their public health impact. The higher prevalence of mild cases caused by *Centruroides margaritatus*, which is commonly found in urban areas, has led to an underestimation of the medical significance of scorpionism in the country. However, Ecuador is home to at least 16 *Tityus* species, many of which have the potential to cause harm, as they inhabit domiciliary and peridomiciliary areas in both forested and coastal regions (Brito and Borges 2015). Severe cases do occur, particularly in children, as highlighted by recent local press reports (Primicias 2024). Scientific journals have also documented severe envenomations with cardiorespiratory symptoms, including fatalities, attributed to species such as *Tityus asthenes* and others (Borges et al. 2015; Gaona et al. 2021). Notably, no scorpion antivenom is currently available in Ecuador, and none of the patients in these reports received specific antivenom therapy.

A fatal case involving a boy from the Shuar ethnic group in Morona Santiago Province was caused by *Tityus cisandinus*, a species related to the Atreus subgenus. Despite the administration of Mexican antivenom, the treatment failed to reverse the clinical outcome, underscoring the need for species-specific antivenom to ensure efficacy (Borges and Román 2023). A transcriptomic study of the venom gland of *T. cisandinus* revealed toxin similarities with *Tityus obscurus* and demonstrated limited reactivity of Venezuelan antivenom with these venoms in Western Blot assays. This suggests that an antivenom tailored to Amazonian scorpion species may be necessary, given the rising number of severe envenomations caused by this group (Kalapothakis et al. 2023). Previous epidemiological studies in Morona Santiago Province reported that 12.7% of scorpionism cases were severe (Ochoa-Andrade et al. 2022), with another study documenting two fatalities (Román et al. 2018). However, it is suspected that the number of accidents in this region is significantly underreported. The Shuar people believe that envenomation by *T. cisandinus* is invariably fatal, which may discourage victims from seeking medical help. Additionally, the region's remote and difficult-to-access terrain further complicates accurate reporting and medical intervention (Borges and Román 2023).

In Bolivia, studies on scorpions remain limited, but several *Tityus* species have been documented, including *T. argentinus*, *T. birabens*, *T. bolivianus*, *T. confluens*, *T. gasci*, *T. metuendus*, *T. proseni*, and *T. soratensis* (Acosta and Ochoa 2002). The presence of species associated with severe envenomations, such as *T. confluens* (Borges et al. 2024) and *T. metuendus*, is particularly noteworthy. *Tityus confluens* was the most frequently identified species in the collection of the Centro Nacional de Enfermedades Tropicales de Bolivia. Among 18 analyzed specimens, the majority originated from local health centers. Sixteen were identified as *T. confluens*, with three linked to reported accidents, one of which involved systemic symptoms. One case was attributed to *T. argentinus*, presenting only mild symptoms (Montaño et al. 2024).

However, *T. soratensis*, previously considered of minor medical importance, was implicated in a recent fatal envenomation of a child in 2023 (Auza-Santiváñez and Franco Lacato 2023). In the following weeks, additional scorpion sting cases were reported in the Cochabamba region (Laura Valencia, pers. comm.) These incidents prompted Bolivia's Ministry of Health to announce the development of a species-specific antivenom by the national laboratory INLASA (Instituto Nacional de Laboratorios de Salud "Dr. Néstor Morales Villazón") (INLASA 2024).

Peru shares borders and part of its scorpiofauna with Colombia, Ecuador, Brazil, and Bolivia. At least 16 *Tityus* species have been documented in the country (Ochoa 2005), though no reports of severe envenomations or fatalities attributed to scorpions have been identified. In 2010, a press release from Peru's Ministry of Health (Ministerio de Salud Perú 2010) reported 41 scorpion sting cases in Lima, all of which were classified as mild. These cases were attributed to scorpions of the genus *Hadruroides*, which belong to the less harmful family Caraboctonidae. However, a series of research studies revealed that the venom of *Hadruroides lunatus* can cause damage in experimental in vivo models, including cardiac dysfunction (Costal-Oliveira et al. 2012, 2015, 2017). The diversity of Buthidae scorpions in Peru continues to be uncovered, with new species being described and known species expanding their ranges (Kovařík et al. 2015). Experimental studies have demonstrated that the venom of *Tityus kaderkai* and its purified fractions are toxic to mice (Escobar et al. 2013). Among the species found in Peru, *Tityus footei* is one of the most widespread and has been reported inside dwellings as well as in the Machu Picchu Sanctuary and its surrounding areas (Ochoa 2024). Given the high volume of tourism in this region, the presence of *T. footei* could pose a potential public health risk, but its potential to cause relevant human accidents is still unknown.

16.3 Prevention Strategies

As demonstrated in the previous section, scorpionism is a significant public health concern in the Neotropical region, accounting for a substantial portion of the global burden of scorpion envenomations. Despite underreporting, the mortality rate associated with scorpion stings in the region may be considered relatively low. However, it is crucial to recognize that scorpion envenomation is a preventable injury that disproportionately affects children which is, per se, problematic. Also, the traumatic loss of a child can have profound and lasting impacts on families and communities, underscoring the need for targeted attention and action.

The factors contributing to the rise in scorpion incidents are complex, necessitating multifaceted solutions that address various dimensions of the problem, requiring coordinated actions across multiple fronts. It is worth noting that efforts to address scorpionism have been ongoing for over a century, yet a definitive solution remains elusive (Dias et al. 1924), as the number of envenoming cases continues to rise. Importantly, the complete eradication of scorpions is neither achievable nor desirable, as it could disrupt ecological balances, particularly affecting species predated by scorpions (de Roodt et al. 2021). Therefore, prevention strategies must consider this broader environmental context. While individual and community actions are vital in reducing the risk of harmful encounters with scorpions, it is essential to recognize that individuals cannot be solely responsible for their health outcomes. Social determinants of health, such as socioeconomic status, access to education, infrastructure, and basic sanitation, play a critical role in shaping the risk of scorpion envenomations (Braveman and Gottlieb 2014; Almeida et al. 2021b; Araújo et al. 2024).

Successful initiatives, such as those implemented in Mexico, which dramatically reduced scorpion-related fatalities (Chippaux et al. 2020), and the recent restructuring of health procedures in the state of São Paulo, Brazil, which curbed the rise in scorpionism lethality (Eloy et al. 2021), serve as valuable models. However, the persistence of outbreaks without clear, methodologically guided responses (Hernández-Muñoz et al. 2024) and the ongoing extremely high number of incidents in these regions highlight the need for continued efforts. Recent works have proposed some strategies for prevention and control (Guerra-Duarte et al. 2023; Hernández-Muñoz et al. 2024). In this context, we discuss six levels of action to mitigate the problem of scorpionism, not only applicable to the Neotropical region but globally (Fig. 16.3).

16.3.1 Minimize Disturbance to Scorpion's Natural Habitats: Environmental Conservation and Ecological Studies

It has been recognized that scorpionism cases are closely linked to environmental factors (Hurtado-Díaz et al. 2018; Ghorbani et al. 2021; Chiaravalloti-Neto et al. 2023). Climate change and deforestation significantly alter the natural habitats of scorpions (Lacerda et al. 2022a), often forcing them into closer contact with human communities. In these areas, scorpions find shelter and abundant prey while facing fewer natural predators. Indeed, numerous reports document envenomations by scorpion species previously considered of minor medical concern, occurring in recently urbanized regions adjacent to green areas (Benzaken et al. 2024). Preserving natural vegetation not only helps maintain scorpions in their natural habitats but also mitigates rising global temperatures, a factor directly correlated with increased scorpion invasions.

Knowledge of the life dynamics of most scorpion species remains limited (Lourenço 2018). Ecological studies aimed at understanding scorpion life history, behavior, taxonomy, and the factors driving their incursion into human dwellings are essential for shaping effective action plans (Giambelluca et al. 2021b). For example, the Amazonian species *Tityus raquelae*, implicated in human

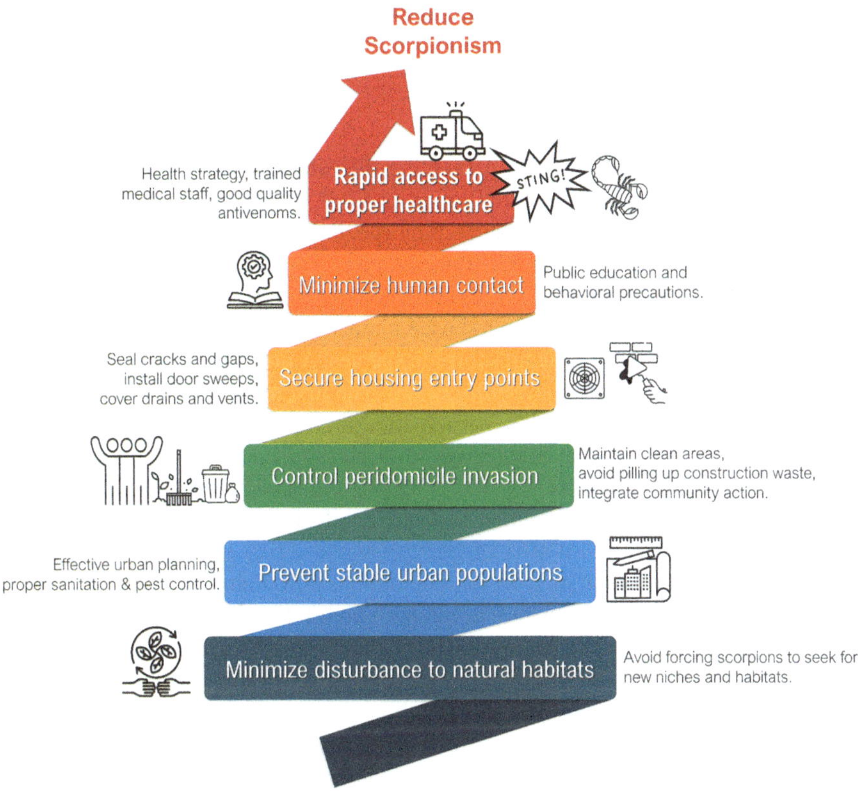

Fig. 16.3 Different level strategies to mitigate the scorpionism problem. Icons used in this figure were obtained from and are currently available at the Noun Project (https://thenounproject.com/), used under a CC BY 3.0 license. Icons creator's attributions: environmental protection by Eri Setiawan, urban planning by Faridah, garbage by Suhatidja, construction material by Arkinasi, rake by Creative Mania, drain by Iconiyo, wall plaster by ProSymbols, knowledge by Fina Arfiana, ambulance by Rakhmat Setiawan, scorpion by Simon Child, group by Hui wen

envenomation cases, was initially thought to be almost exclusively arboreal. However, most reported stings caused by *T. raquelae* occurred on the victims' feet, contradicting the presumed habits of this species. Understanding the drivers behind this apparent shift toward more terrestrial behavior can aid in preventing related incidents (Gomes et al. 2020). Cataloging the scorpiofauna diversity and identifying which species pose a potential threat to humans is also an important strategy for designing appropriate prevention plans (Jimenez-Flores et al. 2022). Additionally, preserving natural predators compatible with urban environments can be beneficial. For instance, hens and frogs, known for their voracious appetite for scorpions and minimal toxicity, could be harnessed as biological control agents (Jared et al. 2020; Murayama et al. 2022).

16.3.2 Prevent the Formation of Stable Scorpion Urban Populations: Effective Urban Planning and Pest Control

A second level of action involves preventing the establishment of stable scorpion communities once they have entered urban areas. In addition to natural migration, there are numerous reports of alien scorpion species being introduced to new locations through terrestrial or even air transportation (Freitas and Vasconcelos 2008; Lobo et al. 2011; Mario da Rosa et al. 2015; Restrepo et al. 2020). These incidents should be thoroughly documented and used to develop monitoring plans to prevent the spread of medically significant scorpion species. A notable example is the introduction of *T. stigmurus* into São Paulo, Brazil. The first record of this species in the area dates back to 2012, and within 3 years, 86 additional occurrences were reported, indicating the establishment of three distinct scorpion communities in the region (Bertani et al. 2018). *Tityus stigmurus* is capable of causing severe envenomations, though the extent of its contribution to local accidents in São Paulo city remains unclear.

Regardless of the origin of scorpion populations colonizing new areas, hotspots should be identified, and related accidents should be systematically recorded (Kotviski and Barbola 2013). The georeferencing of hotspots for scorpionism allows for the efficient allocation of resources to contain the spread. Areas with higher incidence rates should be prioritized in public policies (Brites-Neto et al. 2023). Identifying these hotspots and investing in them, rather than taking a diffused approach, helps to reduce costs, time, and human resource demands in public policies while improving the efficiency of control (Almeida et al. 2016). Many of these urban scorpionism hotspots arise from rapid, disorganized urbanization processes. Addressing this issue requires providing urban cleaning and basic sanitation services, as well as proper drainage for rainfall and sewage systems (Cruz et al. 1995). Implementing such infrastructure is the responsibility of public authorities, underscoring the need to sensitize governments and decision-makers to the challenges posed by scorpionism.

The One Health concept offers a valuable framework for promoting the health and well-being of urban communities and ecosystems. While numerous studies have applied the One Health approach to combat vector-borne diseases (Murray et al. 2022) and urban rodent infestations (Awoniyi et al. 2024), scorpions have been largely overlooked in these discussions. Nevertheless, adopting the One Health approach to provide basic urban infrastructure can improve population health and quality of life holistically, addressing multiple health challenges while fostering sustainable urban environments (Bruno et al. 2024).

Pest control is another strategic priority for preventing the stable colonization of urban areas by scorpions. In addition to the previously mentioned biological control using natural predators, methods such as manual collection assisted by UV lamps (Brites-Neto and Brasil 2012), the use of specific scorpion pathogens like fungi (Brites-Neto et al. 2021) and nematodes (de Roodt et al. 2021), and chemical control can form a comprehensive arsenal against infestations. However, chemical control remains controversial. Some studies suggest that insecticides used for scorpion control may have an excitatory and dislodging effect rather than a lethal one (Albuquerque et al. 2009). As a result, the Brazilian Ministry of Health specifically recommends avoiding this method. Nevertheless, given the urgency of controlling urban scorpion populations, other researchers advocate for increased investment in identifying effective substances and application strategies to enable responsible chemical control protocols (Brites-Neto 2023).

16.4 Control Peridomicile Scorpion Invasion: Investment in Maintaining Clean Areas and Community Engagement

The third level of proposed action focuses on controlling peridomiciliary areas to prevent scorpions from approaching residential premises. This can be achieved by maintaining clean surroundings around homes and ensuring proper storage of garbage prior to collection (Cruz et al. 1995). While individual responsibility is crucial, it is equally important to emphasize that common areas and vacant lots, which often become illegal garbage disposal sites, must be managed by the public sector. These measures reduce the availability of shelter and food sources for scorpions, thereby discouraging their establishment in residential areas.

Ensuring scorpion-free neighborhoods is a shared responsibility that requires collaboration between public sector agents and local communities. Residents possess invaluable knowledge of their local environment and can play an active role in maintaining clean and safe surroundings. Neighborhood associations can be instrumental in this effort, while modern approaches, such as leveraging social networks, can also be employed to raise awareness and combat scorpion infestations. A notable example is a Facebook community established in Sudan for data collection and health promotion related to venomous animals, even during wartime. This initiative demonstrated the power of social media to disseminate accurate information and engage citizens in monitoring venomous animal encounters (Baleela et al. 2024). The organizers, affiliated with Khartoum University, observed that inquiries about scorpions were nearly as frequent as those about snakes—a well-documented issue in the region—highlighting the growing but neglected problem of scorpion stings. This initiative utilized a simple social media feature to foster community integration and engagement. More advanced tools, such as smartphone applications, can further enhance these efforts by providing integrated information and resources. For instance, the use of artificial intelligence (AI) to detect and identify harmful scorpion species, distinguishing them from non-threatening ones, has been proposed as a tool to bolster vigilance against scorpion invasions in peridomiciliary areas (Giambelluca et al. 2021a).

16.5 Secure Housing Entry Points: Construction Recommendations

The fourth level of action involves improving household structures to prevent the unmonitored entry of scorpions. As early as 1924, Ezequiel Dias highlighted the role of construction features in providing suitable shelter for scorpions, which exploit small gaps in poorly finished buildings (Dias et al. 1924). Among the long-standing recommendations for scorpion-proof construction are ensuring adequate isolation from the soil, eliminating empty spaces beneath floors and above ceilings, sealing cracks and gaps in walls, and using smooth wall finishes to prevent climbing by these unwanted visitors. The Brazilian Ministry of Health further recommends sealing doors and windows and covering drains and vents (Ministério da Saúde Brasil 2025).

The importance of adhering to these construction guidelines has been demonstrated in the field. During a recent scorpionism outbreak in Villa Unión, Sinaloa, Mexico, researchers observed that many of the affected homes did not comply with these recommendations. Consequently, they proposed control and prevention strategies aligned with those outlined here (Hernández-Muñoz et al. 2024). Like other preventive measures discussed earlier, housing improvements are recognized as a critical strategy for controlling urban vector-borne diseases. Although this approach requires significant financial investment, it offers a long-term solution to structural issues that impact multiple aspects of urban health (Vazquez-Prokopec et al. 2016).

16.6 Minimize Human Contact: Education and Behavior Precautions

A fifth level of action to prevent scorpionism involves educating individuals to adopt safe behaviors, even in environments where scorpions are present. Education is a critical factor that inversely correlates with the risk of scorpion envenomation. According to a study by Almeida et al. (2021b), increasing the "expected number of years of schooling" in a municipal population by 1 year was associated with a 12% reduction in the incidence of scorpion stings.

While education is essential, school curricula in regions where scorpionism is prevalent often lack comprehensive information about venomous animals. Consequently, even in highly educated populations, myths and misinformation about scorpions can persist and spread, overshadowing accurate and useful information, as highlighted in a survey conducted in Argentina (Gil et al. 2020).

Understanding the behavior, habits, and risks associated with venomous animals helps individuals adopt more cautious behaviors. Educating the population about high-risk situations can further reduce incidents. For example, individuals should be encouraged to inspect clothing and shoes before wearing them, keep beds and cribs away from walls, ensure bed linens do not touch the floor, and wear gloves and closed-toe shoes when gardening (Ministério da Saúde Brasil 2025).

Equally important is educating the public on how to respond in the event of a scorpion sting. Seeking immediate medical care is the most critical step, as the time elapsed between the sting and receiving appropriate treatment is the most significant factor in ensuring a positive outcome for the victim, as a 1 h delay in providing antivenom is associated with 9% increase in the fatality rate related to scorpionism (Almeida et al. 2021a).

16.7 Rapid Access to Proper Medical Care: Strategy, Trained Staff and Good Quality Antivenoms

The sixth and final strategy to mitigate the impact of scorpionism is ensuring that victims seek immediate medical care following a sting, as previously emphasized. The designated health services must be clearly communicated to the population, and these facilities should be strategically located to serve high-incidence areas, enabling victims to receive care within one hour. In São Paulo, Brazil, for example, monitoring and georeferencing clusters of cases and vulnerable areas helped redesign the state's health strategy, establishing new strategic points and expanding access to antivenom treatment (Eloy et al. 2021).

However, timely arrival at a medical facility does not guarantee prompt and adequate treatment. This requires well-defined protocols, particularly for children under 10, who are at higher risk of developing severe symptoms, as well as a trained medical staff capable of recognizing signs of severity (Almeida et al. 2021a). A descriptive report from a reference hospital in Pará, Brazil, highlighted complaints from scorpionism victims regarding delays in care and shortages of medication—both critical factors for effective treatment. Equally concerning were reports of a lack of empathy and solidarity from medical staff, leaving victims in pain and distress without the humanized care they needed (Da Silva et al. 2020).

In addition to timely access and well-trained staff, high-quality, species-specific antivenoms must be available in sufficient quantities. Over a century ago, Vital Brazil demonstrated the specificity of antivenoms, showing that they are effective only against the venom of the species used in their production or closely related species (Lira-da-Silva et al. 2016). For instance, Borges et al. (1999) found that Brazilian antivenom produced against *Tityus serrulatus* failed to fully neutralize the effects of Venezuelan scorpion venoms, a finding supported by other studies (Monteiro et al. 2019; Campbell et al. 2024). Currently, only four scorpion antivenoms are produced in the Neotropical region (Fan et al. 2019): in Mexico, targeting *Centruroides limpidus limpidus, C. suffusus suffusus*, and *C. noxius*; in Brazil, targeting *T. serrulatus*; in Venezuela, targeting *T. discrepans*; and in Argentina, targeting *T. carrilloi*. Although Bolivia has initiated efforts to produce an antivenom for *T. soratensis*, the limited availability of antivenoms for the region's diverse and medically significant scorpion species remains a critical concern (Borges et al. 2020). This gap undermines the ability to provide adequate treatment for victims, highlighting the urgent need for expanded antivenom development and distribution.

16.8 Concluding Remarks

Scorpionism is a significant yet underreported and underaddressed public health issue in the Neotropical region, disproportionately affecting vulnerable populations. As one of the most biodiverse regions globally, the Neotropical scorpion fauna remains understudied and challenging to control (Chippaux and Goyffon 2008). Scorpionism meets all the criteria to be recognized as a neglected tropical disease (NTD), as it primarily impacts impoverished communities in tropical and subtropical regions, causing substantial morbidity and mortality despite being preventable and treatable, receiving limited research attention (WHO- World Health Organization 2025). As highlighted by recent studies (Guerra-Duarte et al. 2023; Kumar et al. 2023; Hernández-Muñoz et al. 2024), granting scorpionism NTD status could galvanize global attention and resources to address this pressing issue.

Effective management of scorpionism in the Neotropical region requires accurate data collection on its incidence. Mandatory reporting of cases has proven to be a valuable tool for improving the understanding of and response to envenomations (Chippaux 2015; Lacerda et al. 2022b). For instance, the introduction of Brazil's surveillance system in 2007 significantly enhanced the precision and operationalization of scorpionism notifications, providing a robust framework for studying these accidents (Brito et al. 2023). However, while gathering data for this chapter, we encountered significant gaps in scorpionism reporting, even in countries known for their high risk of scorpion envenomations. Replicating successful surveillance strategies across the region would be instrumental in addressing these risks. Beyond understanding the extent and context of accidents, identifying the species involved in envenomations is critical for shaping targeted prevention and treatment strategies—an approach that has yet to be systematically implemented anywhere.

Scorpions do not recognize political borders; their distribution is shaped by geographic and ecological boundaries. This underscores the importance of regional collaboration, particularly in shared biomes, to exchange information and develop region-specific antivenoms (Borges et al. 2020). Such efforts could address current gaps in both the epidemiological understanding of scorpionism and the availability of effective antivenoms, ensuring that treatments are tailored to the unique venom profiles of local scorpion species.

In conclusion, addressing scorpionism requires a multifaceted approach that integrates environmental conservation, urban planning, community education, and improved healthcare access. By leveraging data-driven strategies, fostering regional cooperation, and advocating for NTD recognition, the Neotropical region can mitigate the burden of scorpionism and protect its most vulnerable populations.

References

ABC Color (2024) Cuidado, se inicia la epoca de los alacranes. In: https://www.abc.com.py/nacionales/2024/09/23/cuidado-inicia-la-epoca-de-los-alacranes/

Acosta LE, Ochoa JA (2002) Lista de los escorpiones bolivianos (Chelicerata: Scorpiones), con notas sobre su distribución. Rev Soc Entomol Argent 61:15–23

Albuquerque CMR, Barbosa MO, Iannuzzi L (2009) *Tityus stigmurus* (Thorell, 1876) (Scorpiones; Buthidae): response to chemical control and understanding of scorpionism among the population. Rev Soc Bras Med Trop 42:255–259. https://doi.org/10.1590/S0037-86822009000300004

Almeida TSO, Fook SML, França FOS, Monteiro TMR, Silva EL, Gomes LCG, Farias AMG (2016) Spatial distribution of scorpions according to the socioeconomic conditions in Campina Grande, State of Paraíba, Brazil. Rev Soc Bras Med Trop 49:477–485. https://doi.org/10.1590/0037-8682-0128-2016

Almeida ACC, Carvalho FM, Mise YF (2021a) Risk factors for fatal scorpion envenoming among Brazilian children: a case–control study. Trans R Soc Trop Med Hyg 115:975–983. https://doi.org/10.1093/trstmh/trab120

Almeida ACC, Mise YF, Carvalho FM, Silva RML (2021b) Associação ecológica entre fatores socioeconômicos, ocupacionais e de saneamento e a ocorrência de escorpionismo no Brasil, 2007–2019. Epidemiol Serv Saúde 30:1–11. https://doi.org/10.1590/s1679-49742021000400021

Angus LDG, Salzman S, Fritz K, Yaman RM, Gintautas J (1995) Chronic relapsing pancreatitis from a scorpion sting in Trinidad. Ann Trop Paediatr 15:285–289. https://doi.org/10.1080/02724936.1995.11747785

Araújo KAM, Torres-Rêgo M, Gurgel-Medeiros T, Silva-Júnior AA, Leite RS, Daniela-Silva A, Fernandes-Pedrosa MF (2024) Epidemiological study in Brazil: scorpion sting cases in Natal, Rio Grande do Norte. Heliyon 10:e24190. https://doi.org/10.1016/j.heliyon.2024.e24190

Arrieta-Mendoza ME, Zamora A, Domínguez JD, Solano I, Álvarez R (2022) Disección coronaria espontánea posterior a envenenamiento por escorpión. Rev Colomb Cardiol 29:66–71. https://doi.org/10.24875/RCCAR.21000132

Auza-Santiváñez JC, Franco Lacato AO (2023) Picadura de escorpión en Bolivia: una revisión crítica de la literatura. Gaceta Médica Boliviana 46:108–113. https://doi.org/10.47993/gmb.v46i2.583

Awoniyi AM, Barreto AM, Argibay HD, Santana JO, Palma FAG, Riviere-Cinnamond A, Dobigny G, Bertherat E, Fergunson L, Belmain S, Costa F (2024) Systematic surveillance tools to reduce rodent pests in disadvantaged urban areas can empower communities and improve public health. Sci Rep 14:4503. https://doi.org/10.1038/s41598-024-55203-5

Baleela RMH, Mohammad A, Saeed SAK (2024) The role of social media in public health awareness during times of war in Sudan: snakebites and scorpion stings. BMC Public Health 24:1752. https://doi.org/10.1186/s12889-024-19156-8

Bartholomew C (1970) Acute scorpion pancreatitis in Trinidad. BMJ 1:666–668. https://doi.org/10.1136/bmj.1.5697.666

Bartholomew C, Murphy J, McGeeney K, Fitzgerald O (1974) Pancreatic response to *Tityus trinitatis* venom. West Indian Med J 23:41–43

Benmosbah M, Gueguenniat P, Mayence C, Egmann G, Narcisse E, Gono S, Hommel D, Kallel H (2013) Epidemiological and clinical study on scorpionism in French Guiana. Toxicon 73:56–62. https://doi.org/10.1016/j.toxicon.2013.05.025

Benzaken Z, Silva LL, Silva-Neto A, Sachett J, Fé N, Guerra MGVB, Monteiro W (2024) Urbanization and scorpion stings in Manaus, the largest metropolis in the Brazilian Amazon. Toxicon 247:107823. https://doi.org/10.1016/j.toxicon.2024.107823

Bertani R, Bonini RK, Toda MM, Isa LS, Figueiredo JVA, Santos MR, Ferraz SC (2018) Alien scorpions in the Municipality of São Paulo, Brazil—evidence of successful establishment of *Tityus stigmurus* (Thorell, 1876) and first records of *Broteochactas parvulus* Pocock, 1897, and *Jaguajir rochae* (Borelli, 1910). Bioinvasions Rec 7:89–94. https://doi.org/10.3391/bir.2018.7.1.14

Blanco GAC, Laskowicz RD, Lanari LC, Scarlatto E, Damin CF, de Titto EH, de Roodt AR (2016) Distribución de los hallazgos de escorpiones en la Ciudad de Buenos Aires en el período 2001–2012 y sus implicancias sanitarias. Arch Argent Pediatr 114:177–183

Borges A, Rojas de Arias A (2019) El accidente por escorpiones tóxicos en el Paraguay: mito y realidad en el contexto de la emergencia por escorpionismo en el sudeste de la América del Sur. Rev Soc Cient Parag 24:27–35. https://doi.org/10.32480/rscp.2019-24-1.27-35

Borges A, Román JP (2023) Case Report: fatal scorpion envenomation in a shuar child by *Tityus cisandinus* from Amazonian Ecuador: a call for specific antivenom availability in the amazon basin. Am J Trop Med Hyg 108:807–810. https://doi.org/10.4269/ajtmh.22-0681

Borges A, Tsushima RG, Backx PH (1999) Antibodies against *Tityus discrepans* venom do not abolish the effect of *Tityus serrulatus* venom on the rat sodium and potassium channels. Toxicon 37:867–881. https://doi.org/10.1016/S0041-0101(98)00201-3

Borges A, García CC, Lugo E, Alfonzo MJ, Jowers MJ, Op den Camp HJM (2006) Diversity of long-chain toxins in *Tityus zulianus* and *Tityus discrepans* venoms (Scorpiones, Buthidae): Molecular, immunological, and mass spectral analyses. Comp Biochem Physiol C Toxicol Pharmacol 142:240–252. https://doi.org/10.1016/j.cbpc.2005.10.011

Borges A, De Sousa L, Espinoza J, Santos RG, Kalapothakis E, Valadares D, Chávez-Olórtegui C (2008) Characterization of *Tityus* scorpion venoms using synaptosome binding assays and reactivity towards Venezuelan and Brazilian Antivenoms. Toxicon 51:66–79. https://doi.org/10.1016/j.toxicon.2007.08.006

Borges A, Bermingham E, Herrera N, Herrera N, Alfonzo MJ, Sanjur OI (2010a) Molecular systematics of the neotropical scorpion genus *Tityus* (Buthidae): The historical biogeography and venom antigenic diversity of toxic Venezuelan species. Toxicon 55:436–454. https://doi.org/10.1016/j.toxicon.2009.09.011

Borges A, Rojas-Runjaic FJM, Diez N, Faks JG, Op den Camp HJM, Souza L (2010b) Envenomation by the scorpion *Tityus breweri* in the Guayana shield, Venezuela: report of a case, efficacy and reactivity of antivenom, and proposal for a toxinological partitioning of the Venezuelan scorpion Fauna. Wilderness Environ Med 21:282–290. https://doi.org/10.1016/j.wem.2010.06.008

Borges A, Miranda R, Pascale J (2012) Scorpionism in Central America, with special reference to the case of Panama. J Venom Anim Toxins Incl Trop Dis 18:130–143. https://doi.org/10.1590/S1678-91992012000200002

Borges A, Morales M, Loor W, Delgado M (2015) Scorpionism in Ecuador: First report of severe and fatal envenoming cases from northern Manabí by *Tityus asthenes* Pocock. Toxicon 105:56–61. https://doi.org/10.1016/j.toxicon.2015.08.019

Borges A, Lomonte B, Angulo Y, Patiño HA, Pascale JM, Otero R, Miranda RJ, Sousa L, Graham MR, Gómez A, Pardal PPO, Ishikawa E, Bonilla F, Castillo A, Avila RAM, Gómez JP, Caro-López JA (2020) Venom diversity in the Neotropical scorpion genus *Tityus*: implications for antivenom design emerging from molecular and immunochemical analyses across endemic areas of scorpionism. Acta Trop 204:105346. https://doi.org/10.1016/j.actatropica.2020.105346

Borges A, Graham MR, Cândido DM, Pardal PPO (2021) Amazonian scorpions and scorpionism: integrating toxinological, clinical, and phylogenetic data to combat a human health crisis in the world's most diverse rainfores. J Venom Anim Toxins Incl Trop Dis 27:1–20. https://doi.org/10.1590/1678-9199-jvatitd-2021-0028

Borges A, Rojas de Arias A, Montaño AM, de Souza CMV (2024) Scorpion envenoming as an emerging public health problem in Paraguay, Bolivia, and midwest Brazil: involvement of *Tityus confluens* and the need for a panregional evaluation of available antivenoms. Am J Trop Med Hyg 111:1166–1172. https://doi.org/10.4269/ajtmh.24-0424

Braveman P, Gottlieb L (2014) The Social determinants of health: it's time to consider the causes of the causes. Public Health Rep 129:19–31. https://doi.org/10.1177/00333549141291S206

Brites-Neto J (2023) Controle químico de escorpiões em áreas de risco para acidentes escorpiônicos. Revista Estudos - Revista de Ciências Ambientais e Saúde 49:1–14

Brites-Neto J, Brasil J (2012) Estratégias de controle do escorpionismo no município de Americana, SP. BEPA 9:4–15

Brites-Neto J, Maimone NM, Piedade SMDS, Andrino FG, Andrade PAM, Baroni FA, Gomes LH, Lira SP (2021) Scorpionicidal activity of secondary metabolites from *Paecilomyces* sp. CMAA1686 against *Tityus serrulatus*. J Invertebr Pathol 179:107541. https://doi.org/10.1016/j.jip.2021.107541

Brites-Neto J, Dell Ducas V, Figueiredo FS (2023) Spatial analysis in areas with high occurrence of accidents caused by *Tityus serrulatus* and *Tityus bahiensis* (Scorpiones: Buthidae) in Brazil. Wilderness Environ Med 34:63–71. https://doi.org/10.1016/j.wem.2022.11.001

Brito G, Borges A (2015) A checklist of the scorpions of Ecuador (Arachnida: Scorpiones), with notes on the distribution and medical significance of some species. J Venom Anim Toxins Incl Trop Dis 21:1–23. https://doi.org/10.1186/s40409-015-0023-x

Brito M, de Almeida ACC, Cavalcante F, Mise YF (2023) Completeness of notifications of accidents involving venomous animals in the Information System for Notifiable Diseases: a descriptive study, Brazil, 2007–2019. Epidemiol Serv Saúde 32:e2022666. https://doi.org/10.1590/s2237-96222023000100002

Bruno A, Arnoldi I, Barzaghi B, Boffi M, Casiraghi M, Colombo B, Di Gennaro P, Epis S, Facciotti F, Ferrari N, Fesce E, Ficetola GF, Fumagalli S, Galimberti A, Ghisleni G, Nissim WG, Mainardi L, Manenti R, Messina V, Negri A, Palm E, Piga BEA, Rainisio N, Tommasi N, Labra M (2024) The One Health approach in urban ecosystem rehabilitation: an evidence-based framework for designing sustainable cities. iScience 27:110959. https://doi.org/10.1016/j.isci.2024.110959

Buitrago JR, Esguerra IS, NMR A (2020) Scorpion envenoming by *Tityus* (*Atreus*) *vaissadei* (Buthidae) in Santander, Colombia. Toxicon 177:S30–S31. https://doi.org/10.1016/j.toxicon.2019.12.034

Cáceres E, Medina M, Olmedo C, Rojas V, Melo A, Borges A (2022) La importancia médica del escorpionismo en Paraguay: revisión de la falla multisistémica asociada al envenenamiento y contribución a su diagnóstico diferencial. Rev Soc Cient Parag 27:122–152. https://doi.org/10.32480/rscp.2022.27.2.122

Campbell SID, Chow CY, Neri-Castro E, Alagón A, Gómez A, Soria R, King GF, Fry BG (2024) Taking the sting out of scorpions: Electrophysiological investigation of the relative efficacy of three antivenoms against medically significant *Centruroides* species. Comp Biochem Physiol C Toxicol Pharmacol 283:109977. https://doi.org/10.1016/j.cbpc.2024.109977

Cardona JPG, Otero-Patiño R, Rangel VN, Córdoba MMS, Cadavi AD, Sandino MPV (2002) Aspectos toxinológicos, clínicos y epidemiológicos del envenenamiento producido por el escorpión *Tityus fuhrmanni* Kraepelin. MedUnab 5:159–165

Caré W, Larréché S, Busser P, Dufayet L, Vodovar D, Haro L, Langrand J (2021) Envenomation by *Centruroides pococki* scorpion with neuromuscular toxicity. Toxicon 190:39–40. https://doi.org/10.1016/j.toxicon.2020.12.005

Carrasquilla A, Pérez V, Torres M, Samper W (2023) Aculiadura escorpionica en pediatría en los últimos 5 años en el hospital universitario Erasmo Meoz

Celis A, Gaxiola-Robles R, Sevilla-Godínez E, Valerio MJO, Armas J (2007) Tendencia de la mortalidad por picaduras de alacrán en México, 1979–2003. Rev Panam Salud Publica 21:373–380. https://doi.org/10.1590/S1020-49892007000500005

Chiaravalloti-Neto F, Lorenz C, Lacerda AB, de Azevedo TS, Càndido DM, Eloy LJ, Wen FH, Blangiardo M, Pirani M (2023) Spatiotemporal bayesian modelling of scorpionism and its risk factors in the state of São Paulo, Brazil. PLoS Negl Trop Dis 17:e0011435. https://doi.org/10.1371/journal.pntd.0011435

Chippaux J-P (2015) Epidemiology of envenomations by terrestrial venomous animals in Brazil based on case reporting: from obvious facts to contingencies. J Venom Anim Toxins Incl Trop Dis 21:1–13. https://doi.org/10.1186/s40409-015-0011-1

Chippaux J-P, Goyffon M (2008) Epidemiology of scorpionism: a global appraisal. Acta Trop 107:71–79. https://doi.org/10.1016/j.actatropica.2008.05.021

Chippaux JP, Galtier J, Lefait JF (1984) Epidemiology of envenomation in French Guiana. Bull Soc Pathol Exot:206–2015

Chippaux J-P, Celis A, Boyer L, Alagón A (2020) Factors involved in the resilience of incidence and decrease of mortality from scorpion stings in Mexico. Toxicon 188:65–75. https://doi.org/10.1016/j.toxicon.2020.10.011

Corona M, Valdez-Cruz NA, Merino E, Zurita M, Possani LD (2001) Genes and peptides from the scorpion Centruroides sculpturatus Ewing, that recognize Na+-channels. Toxicon 39:1893–1898. https://doi.org/10.1016/S0041-0101(01)00174-X

Cortés A, Castaño M, Medina S, Berrouet MC (2022) Lesión renal aguda y nefritis intersticial aguda debido a picadura por escorpión: reporte de caso. Medicina UPB 41:80–84. https://doi.org/10.18566/medupb.v41n1.a11

Costa CLSO, Fé NF, Sampaio I, Tadei WP (2016) A profile of scorpionism, including the species of scorpions involved, in the State of Amazonas, Brazil. Rev Soc Bras Med Trop 49:376–379. https://doi.org/10.1590/0037-8682-0377-2015

Costa GG, Serejo LFM, Coelho JS, Cândido DM, Gadelha MAC, Pardal PPO (2020) First report of scorpionism caused by *Tityus serrulatus*, described by Lutz and Mello, 1922 (Scorpiones, Buthidae), a species non-native to the state of Pará, Brazilian Amazon. Rev Soc Bras Med Trop 53:e20190285. https://doi.org/10.1590/0037-8682-0285-2019

Costal-Oliveira F, Duarte CG, Machado de Avila RA, Melo MM, Bordon KCF, Arantes EC, Paredes NC, Tintaya B, Bonilla C, Bonilla RE, Suarez WS, Yarleque A, Fernadez JM, Kalapothakis E, Chávez-Olórtegui C (2012) General biochemical and immunological characteristics of the venom from Peruvian scorpion *Hadruroides lunatus*. Toxicon 60:934–942. https://doi.org/10.1016/j.toxicon.2012.06.013

Costal-Oliveira F, Guerra-Duarte C, Castro KLP, Tintaya B, Bonilla C, Silva W, Yarlequé A, Fujiwara R, Melo MM, Chávez-Olórtegui C (2015) Serological, biochemical and enzymatic alterations in rodents after experimental envenomation with *Hadruroides lunatus* scorpion venom. Toxicon 103:129–134. https://doi.org/10.1016/j.toxicon.2015.06.022

Costal-Oliveira F, Guerra-Duarte C, Oliveira MS, Castro KLP, Lopes-de-Sousa L, Lara A, Gomes ERM, Bonilla C, Guatimosim S, Melo MM, Chávez-Olórtegui C (2017) Cardiorespiratory alterations in rodents experimentally envenomed with *Hadruroides lunatus* scorpion venom. J Venom Anim Toxins Incl Trop Dis 23:1–9. https://doi.org/10.1186/s40409-016-0076-5

Cruz EFS, Yassuda CRW, Jim J, Barraviera B (1995) Programa de controle de surto de escorpião *Tityus serrulatus*, Lutz e Mello 1922, no município de Aparecida, SP (Scorpiones, Buthidae). Rev Soc Bras Med Trop 28:123–128. https://doi.org/10.1590/S0037-86821995000200007

D'Suze G, Castillo C, Sevcik C, Brazón J, Malave C, Hernandez D, Zerpa N (2015) Scorpionism and dangerous species of Venezuela. In: Scorpion venoms. Springer Netherlands, Dordrecht, pp 273–298

Da Silva EMR, Bentes MB, De Oliveira SMS, Dos Santos MED (2020) Assistência de enfermagem: a visão de vítimas de escorpionismo em um hospital de referência no oeste do Pará. Revista Eletrônica Acervo Enfermagem 6:e5338. https://doi.org/10.25248/reaenf.e5338.2020

Daisley H, Alexander D, Pitt-Miller P (1999) Acute myocarditis following *Tityus trinitatis* envenoming: morphological and pathophysiological characteristics. Toxicon 37:159–165. https://doi.org/10.1016/S0041-0101(98)00174-3

De Davila CAM, Parra M, Fuenmayor A, Salgar N, Gonzalez Z, Davila DF (1997) Scorpion envenomation in Merida, Venezuela. Toxicon 35:1459–1462. https://doi.org/10.1016/S0041-0101(97)00009-3

de Roodt AR, Lago NR, Salomón OD, Laskowicz RD, Román LEN, López RA, Montero TE, Vega VV (2009) A new venomous scorpion responsible for severe envenomation in Argentina: *Tityus confluens*. Toxicon 53:1–8. https://doi.org/10.1016/j.toxicon.2008.10.003

de Roodt AR, Coronas FIV, Lago N, González ME, Laskowicz RD, Beltramino JC, Saavedra S, López RA, Reati GJ, Vucharchuk MG, Bazán E, Varni L, Salomón OD, Possani LD (2010)

General biochemical and immunological characterization of the venom from the scorpion *Tityus trivittatus* of Argentina. Toxicon 55:307–319. https://doi.org/10.1016/j.toxicon.2009.08.014
de Roodt AR, Lanari LC, Remes-Lenicov M, Cargnel E, Damin CF, Greco V, Orduna TA, Lloveras S, Desio MA, van Grootheest JH, Casas N, Ojanguren-Affilastro A (2019) Expansión de la distribución de escorpiones del género *Tityus* C. L. Koch 1836 en Argentina: Implicancias sanitarias. Acta Toxicol Argent 27:109–119
de Roodt AR, Lanari LC, Ojanguren-Affilastro A, Goñi FM, Malinovsky V, Dozoretz D, Cargnel E, Titto EH, Damin CF (2021) Control y prevención de los accidentes causados por Tityus trivittatus (Scorpiones: Buthidae). Acta Toxicol Argent 29:11–29
Dehesa-Dávila M, Ramfrez AN, Zamudio FZ, Gurrola-Briones G, Liévano A, Darszon A, Possani LD (1996) Structural and functional comparison of toxins from the venom of the scorpions *Centruroides infamatus* infamatus, *Centruroides limpidus limpidus* and *Centruroides noxius*. Comp Biochem Physiol B Biochem Mol Biol 113:331–339. https://doi.org/10.1016/0305-0491(95)02031-4
Dent MAR, Possani LD, Ramírez GA, Fletcher PL (1980) Purification and characterization of two mammalian toxins from the venom of the Mexican scorpion *Centruroides noxius* Hoffmann. Toxicon 18:343–350. https://doi.org/10.1016/0041-0101(80)90015-X
Dias E, Libanio S, Lisbôa M (1924) Lucta contra os Escorpiões. Mem Inst Oswaldo Cruz 17:5–44. https://doi.org/10.1590/S0074-02761924000100001
Díaz C, Serna-Gonzalez M, Chang-Castillo A, Lamonte B, Bonilla F, Alfaro-Chinchilla A, Triana F, Sasa M (2023) Proteomic profile of the venom of three dark-colored *Tityus* (Scorpiones: Buthidae) from the tropical rainforests of Costa Rica. Acta Trop 248:107031. https://doi.org/10.1016/j.actatropica.2023.107031
Eloy L, Camargo J, Spinola R, Paulo E, Malaque C, Gallafrio C, Piorelli R, ucaretchi F, Hui F, Cupo P, França F, Cladeira C, Medeiros C (2021) Escorpionismo no estado de São Paulo: reestruturação operacional para o atendimento oportuno às vítimas. BEPA Boletim Epidemiológico Paulista 18:16–30. https://doi.org/10.57148/bepa.2021.v.18.36653
Escobar E, Tincopa R, Ochoa JA (2013) Estudio bioquímico del veneno de *Tityus kaderkai* (Scorpiones: Buthidae) con notas sobre su distribución y hábitat en el Perú. Rev Peru Biol 20:151–158
Escudero-Sanjur S, Castro-Perez E, De Patiño HA, Rastogi I, Ramos CW (2022) Genetic diversity of medically important scorpions of the genus *Centruroides* (Buthidae) from Panama including two endemic species. J Genet 101:1–12. https://doi.org/10.1007/s12041-022-01374-x
Estrada-Gómez S, Gomez-Rave L, Vargas-Muñoz LJ, van der Meijden A (2017) Characterizing the biological and biochemical profile of six different scorpion venoms from the Buthidae and Scorpionidae family. Toxicon 130:104–115. https://doi.org/10.1016/j.toxicon.2017.02.007
Estrada-Gomez S, Núñez V, Vargas-Muñoz LJ, Madrid-Bracamonte CA, Preciado LM (2022) Characterization of a lab-scale process to produce whole IgG antivenom covering scorpion stings by genus *Tityus* and *Centruroides* of Colombia. Pharmaceuticals 15:1047. https://doi.org/10.3390/ph15091047
Fan HW, Natal Vigilato MA, Augusto Pompei JC, Gutiérrez JM (2019) Situación de los laboratorios públicos productores de antivenenos en América Latina. Rev Panam Salud Publica 43:1. https://doi.org/10.26633/RPSP.2019.92
Freitas GCC, Vasconcelos SD (2008) Scorpion fauna of the island of Fernando de Noronha, Brazil: first record of *Tityus stigmurus* (Thorell 1877) (Arachnida, Buthidae). Biota Neotrop 8:235–237. https://doi.org/10.1590/S1676-06032008000200019
Fuentes Fiore VF (2024) Escorpionismo severo; riesgo pediátrico en Venezuela. Reporte de un caso. Gac Med Caracas 132:820–824. https://doi.org/10.47307/GMC.2024.132.3.22
Furtado AA, Daniele-Silva A, Silva-Júnior AA, Fernandes-Pedrosa MF (2020) Biology, venom composition, and scorpionism induced by Brazilian scorpion *Tityus stigmurus* (Thorell, 1876) (Scorpiones: Buthidae): a mini-review. Toxicon 185:36–45. https://doi.org/10.1016/j.toxicon.2020.06.015

Ganteaume F, Imbert C (2014) Le point en 2013 sur les envenimations liées aux arthropodes en Guyane française. Bull Soc Pathol Exot 107:31–38. https://doi.org/10.1007/s13149-014-0331-8
Gaona O, Quishpe C, Sotamba J, De La Cruz G (2021) Alacranismo en el paciente pediátrico: a propósito de un caso clínico. Revista UNIANDES Ciencias de la Salud 4:663–674
García-Villalvazo PE, Jiménez-Vargas JM, Lino-López GJ, Meneses EP, Bermúdez-Guzmán MJ, Barajas-Saucedo CE, Enciso IV, Possani LD, Valdez-Velazquez LL (2023) Unveiling the protein components of the secretory-venom gland and venom of the scorpion Centruroides possanii (Buthidae) through Omic Technologies. Toxins 15:498. https://doi.org/10.3390/toxins15080498
Gaviraghi É, Saraiva L, Henz Giovelli GM, Bassani C (2023) Perfil epidemiológico de acidentes escorpiônicos no estado do Rio Grande do Sul. Rev Ciênc Humaniz Hosp Clín Passo Fundo 3:64–82. https://doi.org/10.61085/rechhc.v3i1.131
Ghorbani A, Mansouri B, Baradaran M (2021) Effects of climate variables on the incidence of scorpion stings in Iran for five years. J Venom Anim Toxins Incl Trop Dis 27:e20200110. https://doi.org/10.1590/1678-9199-jvatitd-2020-0110
Giambelluca FL, Cappelletti MA, Osio JR, Giambelluca LA (2021a) Novel automatic scorpion-detection and -recognition system based on machine-learning techniques. Mach Learn Sci Technol 2:025018. https://doi.org/10.1088/2632-2153/abd51d
Giambelluca LA, González SE, Rodríguez Gil SG, González A (2021b) Distribución del género *Tityus* Koch, 1836 (Scorpiones: Buthidae) en la ciudad de La Plata (Argentina). Rev Peru Biol 28:e21154. https://doi.org/10.15381/rpb.v28i4.21154
Gil SGR, Scioscia CL, Giambelluca L, Reboredo G, Gonzalez A (2020) Evaluación del conocimiento de mitos y verdades sobre sobre arácnidos en Argentina y su posible incidencia en salud. Physis: Rev Saúde Coletiva 30:1–22. https://doi.org/10.1590/s0103-73312020300116
Gomes JV, Fé NF, Santos HLR, Jung B, Bisneto PF, Sachett A, Moura VM, Silva IM, Melo GC, Paradal PPO, Lacerda M, Sampaio V, Wen FH, Sachett JAG, Monteiro WM (2020) Clinical profile of confirmed scorpion stings in a referral center in Manaus, Western Brazilian Amazon. Toxicon 187:245–254. https://doi.org/10.1016/j.toxicon.2020.09.012
Gómez CJP, Otero PR (2007) Ecoepidemiología de los escorpiones de importancia médica en Colombia. Rev Fac Nac Salud Pública 25:50–60
Gómez JP, Castillo JCQ, Montoya MPA, Fernández J, Acevedo MJB, Gutiérrez JC, Cadavid AD, Patiño RO (2010) Picaduras por escorpión *Tityus asthenes* en Mutatá, Colombia: aspectos epidemiológicos, clínicos y toxinológicos. Biomedica 30:126–139
Gonzalez-Ponce E, Rodríguez-Rangel S, Martinez R, Alvarado A, Ruiz-Baca E, Miranda P, Sánchez-Rodríguez JE, Lopez-Rodrígues A (2023) Scorpions, science and folklore in Durango city. Diversity 15:743. https://doi.org/10.3390/d15060743
González-Santillán E, Possani LD (2018) North American scorpion species of public health importance with a reappraisal of historical epidemiology. Acta Trop 187:264–274. https://doi.org/10.1016/j.actatropica.2018.08.002
González-Sponga MA (1996) Guía para identificar escorpiones de Venezuela. Caracas
Guerra-Duarte C, Saavedra-Langer R, Matavel A, Oliveira-Mendes BBR, Chavez-Olortegui C, Paiva ALB (2023) Scorpion envenomation in Brazil: Current scenario and perspectives for containing an increasing health problem. PLoS Negl Trop Dis 17:e0011069. https://doi.org/10.1371/journal.pntd.0011069
Guerrero-Vargas JA, Rodríguez Buitrago JR, Ayerbe S, Daza EF, Vidal JTB (2015) Scorpionism and dangerous species of Colombia. In: Scorpion venoms. Springer Netherlands, Dordrecht, pp 245–272
Hernández-Muñoz EA, Zavala-Sánchez EV, Rubio-Atondo A, López-Chávez JA, Arredondo-Rojas V, Reyes-Reyes JS (2024) Scorpion sting envenomation outbreak in Mexico. Strategies for prevention and control. Toxicon 237:107549. https://doi.org/10.1016/j.toxicon.2023.107549
Hoshino K, Moura ATV, De Paula HMG (2006) Selection of environmental temperature by the yellow scorpion *Tityus serrulatus* Lutz & Mello, 1922 (Scorpiones, Buthidae). J Venom Anim Toxins Incl Trop Dis 12:59–66. https://doi.org/10.1590/S1678-91992006000100005

Hurtado-Díaz M, Guzmán-Ontiveros J, Arias-Medellín LA, Hernández-Cadena L, Moreno-Banda GL, Rodríguez-Dozal SL, Texcalac-Sangrador JL, Zúñiga-Bello PE, Riojas-Rodríguez H (2018) Influence of increasing temperature on the scorpion sting incidence by climatic regions. Int J Climatol 38:2167–2173. https://doi.org/10.1002/joc.5318

INLASA Instituto Nacional de Laboratorios de Salud (2024) Innovación científica: sueros y antídotos 'made in Bolivia.' https://www.inlasa.gob.bo/2024/06/innovacion-cientifica%E2%80%88sueros-y-antidotos-made-in-bolivia/

Instituto Nacional de Laboratorios de Salud (2023) Sedes reporta de 20 a 40 picaduras por semana. https://www.inlasa.gob.bo/2023/07/sedes-reporta-de-20-a-40-picaduras-por-semana/

Iserson KV, Jagit Ramcharran SD (2019) Black scorpion (*Tityus obscurus*) fatalities in Guyana and a literature review. J Emerg Med 57:554–559. https://doi.org/10.1016/j.jemermed.2019.07.018

Izquierdo LM, Rodríguez Buitrago JR (2012) Cardiovascular dysfunction and pulmonary edema secondary to severe envenoming by *Tityus pachyurus* sting. Case report. Toxicon 60:603–606. https://doi.org/10.1016/j.toxicon.2012.05.021

Jannello JM, López CA (2024) Primer registro de *Tityus bahiensis* Perty 1834 en la provincia de Mendoza Argentina. Bol Soc Zool Urug 33:e33.1.2

Jared C, Alexandre C, Mailho-Fontana PL, Pimenta DC, Brodie JED, Antoniazzi MM (2020) Toads prey upon scorpions and are resistant to their venom: a biological and ecological approach to scorpionism. Toxicon 178:4–7. https://doi.org/10.1016/j.toxicon.2020.02.013

Jimenez-Flores I, Pingarrón-Ríos RA, Arredondo-Armenta JM (2022) Claves en la identificación de alacranes de importancia para la salud pública en México: mitos y realidades. Arch Med Urg Méx 14:20–28

Kalapothakis Y, Miranda K, Molina DAM, Conceição IMCA, Larangote D, Op den Camp HJM, Kalapothakis E, Chávez-Olórtegui C, Borges A (2023) An overview of *Tityus cisandinus* scorpion venom: transcriptome and mass fingerprinting reveal conserved toxin homologs across the Amazon region and novel lipolytic components. Int J Biol Macromol 225:1246–1266. https://doi.org/10.1016/j.ijbiomac.2022.11.185

Kallel H, Mayence C, Guegueniat P, Hommel D (2016) Acute necrotizing pancreatitis after *Tityus obscurus* scorpion envenomation in French Guiana. J Clin Toxicol 6:1–3. https://doi.org/10.4172/2161-0495.1000328

Katwaroo A, Austin K, Bharat A, Chatoo V, Ramcharan P, Seecheran V, Seecheran R, Giddings S, Seecheran NA (2024) Scorpion-induced acute coronary syndrome: a stinging complication. J Investig Med High Impact Case Rep 12:1–7. https://doi.org/10.1177/23247096241261255

Kotviski BM, Barbola IF (2013) Aspectos espaciais do escorpionismo em Ponta Grossa, Paraná, Brasil. Cad Saude Publica 29:1843–1858. https://doi.org/10.1590/0102-311X00043712

Kovařík F, Teruel R, Lowe G, Friedrich S (2015) Four new scorpion species (Scorpiones: Buthidae) from Amazonian Peru. Euscorpius 2015:1–40. https://doi.org/10.18590/euscorpius.2015.vol2015.iss210.1

Kumar A, Goyal S, Garg MK, Gopalakrishnan M (2023) Scorpion sting envenomation, a neglected tropical disease: a nationwide survey exploring perspectives and attitudes of resident doctors from India. Am J Trop Med Hyg 109:957–964. https://doi.org/10.4269/ajtmh.23-0194

Laborieux L (2024) Biomechanics of venom delivery in South America's first toxungen-spraying scorpion. Zool J Linn Soc 202:zlae161. https://doi.org/10.1093/zoolinnean/zlae161

Lacerda AB, Lorenz C, Azevedo TS, Cândido DM, Wen FH, Eloy LJ, Chiaravalloti-Neto F (2022a) Detection of areas vulnerable to scorpionism and its association with environmental factors in São Paulo, Brazil. Acta Trop 230:106390. https://doi.org/10.1016/j.actatropica.2022.106390

Lacerda AB, Lorenz C, De Azevedo TS, Cândido DM, Wen FH, Eloy LJ, Bersusa AAS, Chiaravalloti-Neto F (2022b) Scorpion envenomation in the state of São Paulo, Brazil: Spatiotemporal analysis of a growing public health concern. PLoS One 17:e0266138. https://doi.org/10.1371/journal.pone.0266138

Leveridge Y (2000) Accidente y cuadro clínico por la "picadura" de alacranes en Costa Rica. Rev Costarric Cienc Med 21:161–167

Lira-da-Silva RM, Lourenço M, Bochner R, Brazil EV, Brazil TK, Cunha LER, Castro AJW (2016) A contribuição de Vital Brazil para a medicina tropical: dos envenenamentos à especificidade da soroterapia. 2º Encontro Luso-Brasileiro de História da Medicina Tropical 15:27–32

Lobo RA, Goldoni PAM, de Souza CAR, Medeiros CR (2011) Accident caused by *Centruroides testaceus* (DeGeer, 1778) (Scorpiones, Buthidae), native to the Caribbean, in Brazilian airport. Rev Soc Bras Med Trop 44:789–791. https://doi.org/10.1590/S0037-86822011000600028

Lopez CA (2021) Escorpiones de la provincia de Misiones, Argentina: Un elenco de interés médico. Rev Argent Salud Pública 13:e54

López CA, Couto E, Gularte A (2020) Escorpionismo y primeros registros de *Tityus trivittatus* y *Tityus serrulatus* en Puerto Iguazú, Provincia de Misiones. Rev Argent Salud Pública 11:51–54

Lourenço WR (2008) Parthenogenesis in scorpions: some history—new data. J Venom Anim Toxins Incl Trop Dis 14:19–44. https://doi.org/10.1590/S1678-91992008000100003

Lourenço WR (2015) What do we know about some of the most conspicuous scorpion species of the genus *Tityus*? A historical approach. J Venom Anim Toxins Incl Trop Dis 21:1–12. https://doi.org/10.1186/s40409-015-0016-9

Lourenço WR (2018) Scorpions and life-history strategies: from evolutionary dynamics toward the scorpionism problem. J Venom Anim Toxins Incl Trop Dis 24:1–12. https://doi.org/10.1186/s40409-018-0160-0

Lourenço WR (2022) Back to *Tityus serrulatus* Lutz & Mello, 1922 (Scorpiones: Buthidae): new comments about an old species. J Venom Anim Toxins Incl Trop Dis 28:1–12. https://doi.org/10.1590/1678-9199-jvatitd-2022-0016

Lourenço WR (2023) Scorpions trapped in amber: a remarkable window on their evolution over time from the Mesozoic period to present days. J Venom Anim Toxins Incl Trop Dis 29:e20230040. https://doi.org/10.1590/1678-9199-jvatitd-2023-0040

Lourenço W, Ythier E (2013) The remarkable scorpion diversity in the Ecuadorian Andes and description of a new species of *Tityus* C. L. Koch, 1836 (Scorpiones, Buthidae). Zookeys 307:1–13. https://doi.org/10.3897/zookeys.307.5334

Lourenço WR, Cloudsley-Thompson J, Cuellar O, von Eickstedt CRD, Barraviera B, Knox MB (1996) The evolution of scorpionism in Brazil in recent years. J Venom Anim Toxins 2:121–134. https://doi.org/10.1590/S0104-79301996000200005

Mario da Rosa C, Abegg AD, Borges LM, Bitencourt GSS, Di Mare RA (2015) New record and occurrence map of Tityus serrulatus Lutz & Mello, 1922 (Scorpiones, Buthidae) in the state of Rio Grande do Sul, southern Brazil. Check List 11:1556. https://doi.org/10.15560/11.1.1556

Martinez PA, Andrade MA, Bidau CJ (2018) Potential effects of climate change on the risk of accidents with poisonous species of the genus *Tityus* (Scorpiones, Buthidae) in Argentina. Spat Spatio-temporal Epidemiol 25:67–72. https://doi.org/10.1016/j.sste.2018.03.002

Martins JG, Santos GC, Procópio REL, Arantes EC, Bordon KCF (2021) Scorpion species of medical importance in the Brazilian Amazon: a review to identify knowledge gaps. J Venom Anim Toxins Incl Trop Dis 27:e20210012. https://doi.org/10.1590/1678-9199-jvatitd-2021-0012

Mazzei de Dàvila CA, Dàvila DF, Donis JH, Bellabarba GA, Villareal V, Barboza JS (2002) Sympathetic nervous system activation, antivenin administration and cardiovascular manifestations of scorpion envenomation. Toxicon 40:1339–1346. https://doi.org/10.1016/S0041-0101(02)00145-9

Mazzotti L, Bravo-Becherelle MA (1961) Scorpionism in the Mexican Republic. Rev Inst Salubr Enferm Trop 21:3–19

Ministério da Saúde Brasil (2024) Doenças e Agravos de Notificação - 2007 em diante (SINAN). https://datasus.saude.gov.br/

Ministério da Saúde Brasil (2025) Acidentes por Escorpiões. https://www.gov.br/saude/pt-br/assuntos/saude-de-a-a-z/a/animais-peconhentos/acidentes-por-escorpioes

Ministerio de Salud Argentina - Direccion epidemiologia (2023) Boletín epidemiológico nacional - numero 637

Ministerio de Salud Ecuador (2021) Dirección Nacional de Vigilancia Epidemiológica. https://www.salud.gob.ec/wp-content/uploads/2021/09/GACETA-GENERAL-OTRAS-INTOXICACIONES-SE-36.pdf

Ministerio de Salud Paraguay (2024) Boletín Epidemiológico Semanal. https://dgvs.mspbs.gov.py/boletin-epidemiologico-semanal/

Ministerio de Salud Perú (2010) Lima reportó 41 casos de personas picadas por alacranes durante el 2009. https://www.gob.pe/institucion/minsa/noticias/37604-lima-reporto-41-casos-de-personas-picadas-por-alacranes-durante-el-2009

Montaño AM, Baldiviezo CDV, de Souza CMV (2024) Nuevos registros geográficos y médicos de escorpiones de la familia Buthidae (Arachinida: Scorpiones) del sureste de Bolivia. Kempffiana 20:1–13

Monteiro WM, Gomes J, Fé N, Silva IM, Lacerda M, Alencar A, Farias AS, Val F, Sampaio VS, Melo GC, Pardal PPO, Silva AM, Bernarde PS, Ferreira LCL, Gutierrez JM, Sachett JAG, Fan HW (2019) Perspectives and recommendations towards evidence-based health care for scorpion sting envenoming in the Brazilian Amazon: a comprehensive review. Toxicon 169:68–80. https://doi.org/10.1016/j.toxicon.2019.09.003

Murayama GP, Pagoti GF, Guadanucci JPL, Willemart RH (2022) Voracity, reaction to stings, and survival of domestic hens when feeding on the yellow scorpion (*Tityus serrulatus*). J Venom Anim Toxins Incl Trop Dis 11:e20210050. https://doi.org/10.1590/1678-9199-JVATITD-2021-0050

Murray MH, Buckley J, Byers KA, Fake K, Lehrer EW, Magle SB, Stone C, Tuten H, Schell CJ (2022) One Health for all: advancing human and ecosystem health in cities by integrating an environmental justice lens. Annu Rev Ecol Evol Syst 53:403–426. https://doi.org/10.1146/annurev-ecolsys-102220-031745

Ochoa JA (2005) Patrones de distribución de escorpiones de la región andina en el sur peruano. Rev Peru Biol 12:49–68

Ochoa JA (2024) Los escorpiones (Arachnida) del Santuario Histórico de Machupicchu, Cusco, Perú. Cantua 19:11–15. https://doi.org/10.51343/cantu.v19i0.1358

Ochoa-Andrade MJ, Abril-López PA, Yanza FBM, Gaibor AAM, León JLG, Pacheco JN, Chávez LI, Imba AR, Ochoa-Andrade ME (2022) Escorpionismo en la población amazónica del cantón Taisha en Ecuador. Rev Cubana Med Trop 74:e817

Ojanguren-Affilastro A, Kochalka J, Guerrero-Orellana D, Garcete-Barrett B, de Roodt AR, Borges A, Ceccarelli FS (2021) Redefinition of the identity and phylogenetic position of *Tityus trivittatus* Kraepelin 1898, and description of *Tityus carrilloi* n. sp. (Scorpiones; Buthidae), the most medically important scorpion of southern South America. Rev Mus Argent Cienc Nat 23:27–55. https://doi.org/10.22179/REVMACN.23.714

Ortiz MM, Lima PKGC, Sales CCF, Elvira IKS, Barbosa AO, Rocha KF, Guedes MRJ, Oliveira SRL, Oliveira MLF (2021) Acidentes por escorpião *Tityus* sp. (scorpiones: Buthidae) em crianças: discussão de dois óbitos. Res Soc Dev 10:e24110212457. https://doi.org/10.33448/rsd-v10i2.12457

Otero R, Navío E, Céspedes FA, Núñez MJ, Lozano L, Moscoso ER, Matallana C, Arsuza NB, García J, Fernández D, Rodas JH, Rodríguez OJ, Barona J, Vaderrama R, Lourenço WR (2004) Scorpion envenoming in two regions of Colombia: clinical, epidemiological and therapeutic aspects. Trans R Soc Trop Med Hyg 98:742–750. https://doi.org/10.1016/j.trstmh.2003.12.018

Pardal PPO, Castro LC, Jennings E, Pardal JSO, Monteiro MRCC (2003) Aspectos epidemiológicos e clínicos do escorpionismo na região de Santarém, Estado do Pará, Brasil. Rev Soc Bras Med Trop 36:349–353. https://doi.org/10.1590/S0037-86822003000300006

Pardal PPO, Coelho JS, Silva JM, Almeida BRR, Chalkidis HM, Borges A, Ishikawa EAY, Diniz CG, Santos-Neto GC, Melo MAD (2023) Interpopulational genetic diversity in the medically important scorpion *Tityus obscurus* (Scorpiones: Buthidae) from northeastern Brazilian Amazonia. Acta Amazon 53:215–222. https://doi.org/10.1590/1809-4392202202441

Pimenta RJG, Brandão-Dias PFP, Leal HG, Carmo AO, Oliveira-Mendes BBR, Chávez-Olórtegui C, Kalapothakis E (2019) Selected to survive and kill: *Tityus serrulatus*, the Brazilian yellow scorpion. PLoS One 14:e0214075. https://doi.org/10.1371/journal.pone.0214075

Pizarro-Araya J, Ojanguren-Affilastro A (2021) El orden Scorpiones en Chile: Estado del conocimiento, desde la taxonomía a la conservación. In: Mauricio Canals (ed) Arácnidos comunes de Chile, 2nd edn

Primicias (2024) Niña de 4 años sufrió taquicardia tras picadura de alacrán en Pichincha, ⊵hay otros riesgos a la salud? https://www.primicias.ec/sociedad/picadura-alacran-taquicardia-nina-puerto-quito-76169

Pucca MB, Cerni FA, Pinheiro Junior EL, Bordon KCF, Amorim FG, Cordeiro FA, Longhim HT, Cremonez CM, Oliveira GH, Arantes EC (2015) *Tityus serrulatus* venom—a lethal cocktail. Toxicon 108:272–284. https://doi.org/10.1016/j.toxicon.2015.10.015

Rafinejad J, Shahi M, Navidpour S, Jahanifard E, Hanafi-Bojd AA (2020) Effect of climate change on spatial distribution of scorpions of significant public health importance in Iran. Asian Pac J Trop Med 13:503. https://doi.org/10.4103/1995-7645.295361

Reckziegel G, Pinto V (2014) Scorpionism in Brazil in the years 2000 to 2012. J Venom Anim Toxins Incl Trop Dis 20:1–8. https://doi.org/10.1186/1678-9199-20-46

Restrepo GJ, Lastra SM, Gutiérrez DH, Giraldo MAM, Zuluaga AF (2020) Envenenamientos por escorpiones reportados al centro de información y estudio de medicamentos y tóxicos (CIEMTO) de Medellín, Colombia. Medicina 42:19–27. https://doi.org/10.56050/01205498.1484

Reyes-Vega DF, Bermúdez JF, Buitrago-Toro K, Jiménez-Salazar S, Zamora-Suárez A (2021) Aspectos epidemiológicos, clínicos y paraclínicos del accidente escorpiónico en el Hospital Universitario de Neiva, Colombia. Iatreia 34:295–306

Riaño-Umbarila L, Rodríguez-Rodríguez ER, Santibañez-López CE, Guereca L, Uribe-Romero SJ, Gómez-Ramírez IV, Cárcamo-Noriega EN, Possani LD, Becerril B (2017) Updating knowledge on new medically important scorpion species in Mexico. Toxicon 138:130–137. https://doi.org/10.1016/j.toxicon.2017.08.022

Robledo-Aceves M, Barrón-Balderas A, Corona-Gutiérrez AA, Camarena-Pulido EE, Rodríguez-Rendón SJ (2024) Pediatric scorpion envenoming in Western Mexico. Salud Publica de México 66:331–332. https://doi.org/10.21149/15417

Rocío Arjona (2023) Situación Epidemiológica de la Picadura de Alacrán y Mordedura de Ofidio, en la República de Panamá. Años: 2020 y 2021

Rojas VJR, Rojas EAC, Gimenéz CSO, Godoy MMM, López EAA, Hetter MEA, Zaya LSA (2022) Evaluación del conocimiento del personal médico acerca del envenenamiento sistémico tras la picadura de escorpión en el Paraguay. Rev Parag Biofís 2:20–23

Román JP, García F, Medina D, Vásquez M, García J, Graham MR, Romero-Alvarez D, Pardal PPO, Ishikawa EAY, Borges A (2018) Scorpion envenoming in Morona Santiago, Amazonian Ecuador: Molecular phylogenetics confirms involvement of the *Tityus obscurus* group. Acta Trop 178:1–9. https://doi.org/10.1016/j.actatropica.2017.10.014

Salazar MH, Arenas I, Corrales-García LL, Miranda R, Vélez S, Sánchez J, Mendoza K, Cleghorn J, Zamudio FZ, Castillo A, Possani LD, Corzo G, Acosta H (2018) Venoms of *Centruroides* and *Tityus* species from Panama and their main toxic fractions. Toxicon 141:79–87. https://doi.org/10.1016/j.toxicon.2017.11.013

Santibáñez-López C, Francke O, Ureta C, Possani L (2015) Scorpions from Mexico: from species diversity to venom complexity. Toxins 8:2. https://doi.org/10.3390/toxins8010002

Santos MSV, Silva CGL, Neto BS, Grangeiro JCRP, Lopes VHG, Teixeira JAG, Bezerra DA, Luna JVCP, Cordeira JB, Gonçalves JJ, Lima MAP (2016) Clinical and epidemiological aspects of scorpionism in the world: a systematic review. Wilderness Environ Med 27:504–518. https://doi.org/10.1016/j.wem.2016.08.003

Schenone H, Fontecilla J (1998) Scorpion stings outbreaks in recently constructed urban dwellings inhabitants. Bol Chil Parasitol 53:35–37

Schmitt C, Torrents R, Simon N, de Haro L (2017) First described envenomation by *Centruroides pococki* scorpion in the French Caribbean Island Guadeloupe. Wilderness Environ Med 28:159–160. https://doi.org/10.1016/j.wem.2017.03.005

Secretaría de Salud México (2023) Anuario epidemiologia. https://epidemiologia.salud.gob.mx/

Seemann TM, Rocha AP, Cremonese F, Resener MC, Petry A, Costa ACC (2023) Evolution of the occurrence of *Tityus serrulatus* (Lutz & Mello, 1992) in the state of Santa Catarina. Rev Soc Bras Med Trop 56:e0434. https://doi.org/10.1590/0037-8682-0434-2022

Silva de Oliveira SM, Bertani R, Quispe Torrez PP, Sousa PRL, Quiroga MMM, Bertolozzi MR, França FOS (2020) Electric shock sensation in the first reports of envenomations by *Tityus strandi* in the Brazilian Amazon. Toxicon 178:8–12. https://doi.org/10.1016/j.toxicon.2020.01.005

Silva BAJ, Fé NF, Gomes AAS, Souza AS, Sachett JAG, Fan HW, Melo GC, Monteiro WM (2017) Implication of *Tityus apiacas* (Lourenco, 2002) in scorpion envenomations in the Southern Amazon border, Brazil. Rev Soc Bras Med Trop 50:427–430. https://doi.org/10.1590/0037-8682-0490-2016

Silva GV, Trandafilov GC, Polidoro BR, Antonangelo LR, Oliveira CF (2023) Escorpionismo no estado de São Paulo: um problema de saúde pública em ascensão. Braz J Health Rev 6:26973–26982. https://doi.org/10.34119/bjhrv6n6-034

Telemundo (2017) Picaduras de escorpiones y arañas y mordidas de serpientes (en Uruguay). https://www.teledoce.com/telemundo/nacionales/picaduras-de-escorpiones-y-aranas-y-mordidas-de-serpientes-en-uruguay/

Toscano-Gadea CA (2015) Biología reproductiva, gestación y cantidad de crías de seis especies de escorpiones de Uruguay (Scorpiones: Bothriuridae, Buthidae, Euscorpiidae). Bol Soc Zool Urug 24:130–145

Trinidad Express (2023) 27 stung by scorpions in 4 months. https://trinidadexpress.com/newsextra/27-stung-by-scorpions-in-4-months/article_2a1f124a-3d2d-11ee-a228-6b4e201ba760.html#google_vignette

Trinidad-Porfirio BE, Morales-Pérez A, Nava-Aguilera E, Flores-Moreno M, Morales-Nava L, García-Leyva J, Silva-Domínguez R, Cortés-Guzmán AJ, Fernández-Salas I, Andersson N (2023) Occurrence of scorpion sting and associated factors in a highly marginalized municipality in Guerrero, Mexico: a cross-sectional study. PLoS Negl Trop Dis 17:e0011271. https://doi.org/10.1371/journal.pntd.0011271

Universidad de Costa Rica (2018) Costa Rican scorpions are not lethal for humans. https://vinv.ucr.ac.cr/en/news/costa-rican-scorpions-are-not-lethal-humans

Valdez-Cruz NA, Dávila S, Licea A, Zamudio FZ, García-Valdes J, Boyer L, Possani LD (2004) Biochemical, genetic and physiological characterization of venom components from two species of scorpions: *Centruroides exilicauda* Wood *and Centruroides sculpturatus* Ewing. Biochimie 86:387–396. https://doi.org/10.1016/j.biochi.2004.05.005

Vaucel J, Mutricy R, Hoarau M, Pujo J-M, Elenga N, Labadie M, Kallel H (2020) Pediatric scorpionism in northern Amazonia: a 16-year study on epidemiological, environmental and clinical aspects. J Venom Anim Toxins Incl Trop Dis 26:e202000038. https://doi.org/10.1590/1678-9199-jvatitd-2020-0038

Vaucel J-A, Larréché S, Paradis C, Courtois A, Pujo J-M, Elenga N, Résiere D, Caré W, Haro L, Gallart J-C, Torrents R, Schmitt C, Chevalier J, French PCC Research Group, Labadie M, Kallel H (2022) Les envenimations scorpioniques en France métropolitaine et d'Outremer: mise au point. Toxicol Anal Clin 34:68–85. https://doi.org/10.1016/j.toxac.2022.02.003

Vazquez-Prokopec GM, Lenhart A, Manrique-Saide P (2016) Housing improvement: a novel paradigm for urban vector-borne disease control? Trans R Soc Trop Med Hyg 110:567–569. https://doi.org/10.1093/trstmh/trw070

Versión Final (2017) Escorpiones pican a nueve mil venezolanos al ano. https://diarioversionfinal.com/ciudad/escorpiones-pican-a-nueve-mil-venezolanos-al-ano/escorpiones-pican-a-nueve-mil-venezolanos-al-ano

Vilarinho AC, Lopez VM, Tizo-Pedroso E, Oliveira CJF, Ferreira RG (2023) Scorpion envenomation in the Neotropical savannah: environmental predictors and years of lost life. Toxicon 234:107277. https://doi.org/10.1016/j.toxicon.2023.107277

von Eickstedt VRD, Ribeiro LA, Candido DM, Albuquerque MJ, Jorge MT (1996) evolution of scorpionism by *Tityus bahiensis* (Perty) and *Tityus serrulatus* Lutz and Mello and geographical distribution of the two species in the state of São Paulo—Brazil. J Venom Anim Toxins 2:92–105. https://doi.org/10.1590/S0104-79301996000200003

Ward MJ, Ellsworth SA, Nystrom GS (2018) A global accounting of medically significant scorpions: Epidemiology, major toxins, and comparative resources in harmless counterparts. Toxicon 151:137–155. https://doi.org/10.1016/j.toxicon.2018.07.007

Waterman JA (1938) Some notes on scorpion poisoning in Trinidad. Trans R Soc Trop Med Hyg 31:607–624. https://doi.org/10.1016/S0035-9203(38)90154-3

Wendruff AJ, Babcock LE, Wirkner CS, Kluessendorf J, Mikulic DG (2020) A Silurian ancestral scorpion with fossilised internal anatomy illustrating a pathway to arachnid terrestrialisation. Sci Rep 10:14. https://doi.org/10.1038/s41598-019-56010-z

White J (2016) Overview of scorpion envenoming. In: Critical care toxicology. Springer International Publishing, Cham, pp 1–15

World Health Organization (2024) Animal bites. https://www.who.int/news-room/fact-sheets/detail/animal-bites

World Health Organization (2025) Neglected tropical diseases. https://www.who.int/health-topics/neglected-tropical-diseases#tab=tab_1

Chapter 17
Managing Scorpionism: Envenomation, Diagnostics, and Therapeutic Approaches

Manuela B. Pucca, Isabela C. dos Santos, and Felipe A. Cerni

Abstract Scorpion envenomation represents a significant public health concern in tropical and subtropical regions, most often causing mild local pain but also capable of leading to severe systemic complications, particularly in children and the elderly. This chapter provides a comprehensive overview of scorpionism, covering the classification and clinical manifestations of envenomation, the biochemical complexity of scorpion venom, and both traditional and modern therapeutic strategies. Special attention is given to neurotoxic mechanisms of action and the multifactorial determinants of clinical severity. Additionally, the chapter highlights recent advances in antivenom development and addresses ongoing challenges related to treatment effectiveness.

Keywords Antivenom therapy · Clinical manifestations · Neurotoxicity · Supportive care · Venom composition

17.1 Introduction

This chapter offers a comprehensive overview of scorpion envenomation—commonly referred to as scorpionism—and the medical and therapeutic strategies developed to manage its effects.

The primary focus is on the pathophysiological effects of scorpion venom in humans, including clinical signs and symptoms, diagnostic approaches, and the therapeutic protocols used to mitigate these effects. Particular emphasis is placed on antivenom development and administration, symptomatic treatments, and

M. B. Pucca (✉) · I. C. dos Santos
Laboratório de Imunologia e Toxinologia ITox-Lab, Faculdade de Ciências Farmacêuticas, Universidade Estadual Paulista (UNESP), Araraquara, São Paulo, Brazil
e-mail: manuela.pucca@unesp.br

F. A. Cerni
Curso de Medicina, Universidade Federal de Roraima (UFRR), Boa Vista, Roraima, Brazil

A. F. de Araujo Lira, E. González-Santillán (eds.), *The Biology of Neotropical Scorpions*, https://doi.org/10.1007/978-3-032-23194-9_17

emergency response practices in regions where scorpion stings represent a significant public health concern.

The chapter also addresses challenges related to healthcare accessibility in the context of scorpionism, highlighting efforts to enhance diagnostic capacity, treatment availability, and public awareness in vulnerable communities. Although global in perspective, the discussion is rooted in examples from Central and South America.

17.2 Scorpion Envenomation

Scorpion stings are reported worldwide; however, clinically significant scorpion envenomation—referred to as scorpionism—is primarily concentrated in seven geographic regions: the northern Sahara, southern and eastern Africa, the Middle East, southern India, Mexico, Brazil, and the Amazon Basin, which includes the Guianas, Venezuela, and northern Brazil (Chippaux and Goyffon 2008; Santos et al. 2016). In these endemic areas, scorpion stings are frequent and pose a significant public health concern (Abroug et al. 2020), particularly in communities with limited access to prompt and adequate medical care. Indeed, due to its predominance in low- and middle-income countries, scorpion envenomation has historically received limited clinical research attention, contributing to its classification as a neglected disease (Abroug et al. 2019).

The vast majority of scorpion stings are classified as dry stings, typically resulting in mild, localized discomfort. However, in cases where systemic envenomation does occur, multiple physiological systems may be affected (Pucca et al. 2015; Abroug et al. 2019). The severity of envenomation can range from mild local reactions to life-threatening systemic complications, including cardiopulmonary dysfunction, neurological disturbances, and, in extreme cases, death (Cesaretli and Ozkan 2010; Godoy et al. 2021).

The outcome of scorpion envenomation is shaped by a complex interplay of factors related to both the scorpion and the human host. On the scorpion's side, important variables include species identity, venom composition and potency, the size and physiological condition of the animal, and the number and depth of stings. Venom yield and delivery may also vary depending on whether the sting was defensive or predatory (El Joud et al. 2024). On the human side, critical factors influencing the severity of symptoms include the anatomical location of the sting, the individual's age, weight, overall health status, and immune response. Children, elderly individuals, and those with pre-existing health conditions are at elevated risk of developing severe or fatal scorpion sting complications (Santos et al. 2016). Nonetheless, healthy adults are not immune to severe reactions, and fatal outcomes have also been documented in this group. Given the potentially rapid progression of symptoms, accurate and early assessment of envenomation severity is essential. This is especially true in pediatric and immunocompromised patients, for whom clinical deterioration may occur within hours (Chippaux and Goyffon 2008).

Despite the widespread nature of scorpionism, reliable epidemiological data remain scarce. Underreporting, lack of standardized diagnostic protocols, limited health surveillance systems, and disparities in healthcare access all contribute to an incomplete picture of scorpionism's true burden (Souza et al. 2019).

17.3 Clinical Effects of Envenomation

Scorpion envenomation manifests initially with intense, sharp pain at the sting site, often without visible tissue damage. While mild in most cases, certain species (e.g., *Tityus, Androctonus, Leiurus, Mesobuthus*, and *Centruroides*) produce potent venoms capable of inducing severe systemic responses. These effects stem primarily from neurotoxins that act on voltage-gated ion channels (see Chap. 15), leading to prolonged depolarization of excitable cells and a massive release of neurotransmitters, particularly catecholamines. This process triggers what is known as an "autonomic storm", characterized by simultaneous overstimulation of both the sympathetic and parasympathetic nervous systems (Fig. 17.1) (Chippaux and Goyffon 2008; Cupo 2015; Pimenta et al. 2019). This storm is central to the

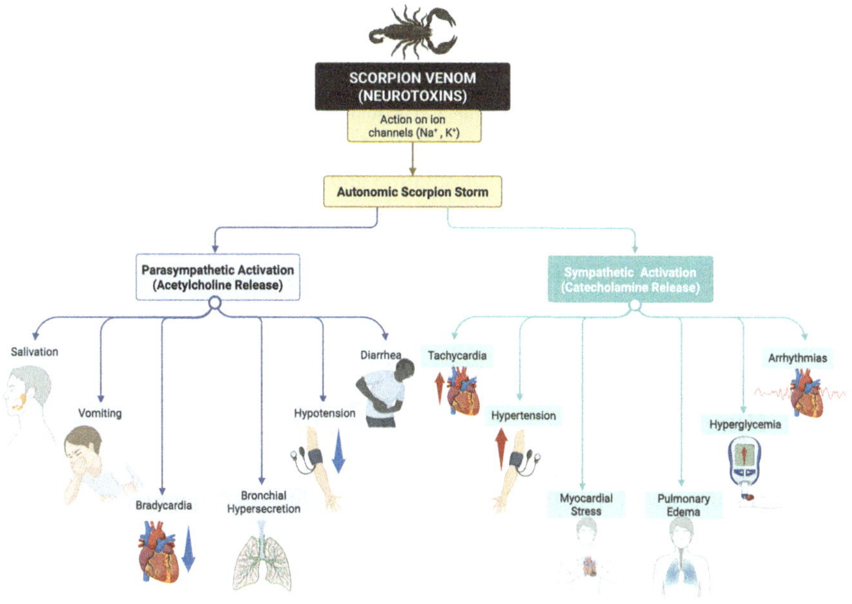

Fig. 17.1 Schematic representation of the scorpion venom-induced autonomic storm. The figure shows how scorpion venom causes overstimulation of the autonomic nervous system by affecting ion channels. The result is excessive activation of both parasympathetic and sympathetic responses, leading to several symptoms and potentially life-threatening complications like pulmonary edema. (Created in BioRender.com))

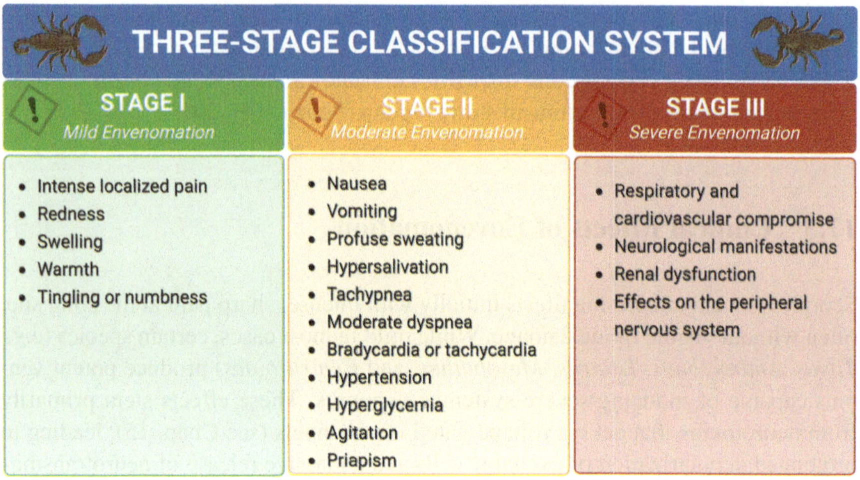

Fig. 17.2 Three-Stage Classification System of Scorpion Envenomation. Stage I (Mild): Symptoms are localized and self-limited, typically appearing within 5–30 min post-sting. Clinical signs are restricted to the sting site and may include pain, redness, swelling, and tingling, without systemic involvement. Stage II (Moderate): Systemic effects emerge due to autonomic dysregulation and inflammatory mediator release. These may include nausea, sweating, respiratory changes, and cardiovascular alterations. Prompt medical intervention can reverse symptoms. Stage III (Severe): Life-threatening systemic toxicity usually develops within 1–4 h. Multi-organ dysfunction may occur, including cardiovascular complications (e.g., myocarditis, arrhythmias, shock), respiratory failure (e.g., pulmonary edema, ARDS), neurological events (e.g., seizures, coma, strokes), renal impairment (e.g., hemolysis, AKI), and peripheral nervous system involvement (e.g., paralysis, Guillain-Barré-like syndromes). (Created in BioRender.com))

pathogenesis of severe scorpion envenomation and underlies many of its cardiovascular, respiratory, and neurological complications (Fig. 17.2).

To facilitate a clearer and more didactic understanding of the clinical manifestations of scorpion envenomation, the condition is often described using a three-stage (I-III) classification system.

17.4 Diagnosis

Currently, there are no specific laboratory tests or definitive diagnostic methods for scorpionism described in the scientific literature. Although symptoms may develop rapidly, diagnosis remains primarily clinical, based on the patient's history and presenting signs. In severe cases, laboratory findings—such as elevated blood glucose, creatinine, and cardiac enzymes—can help assess the extent of systemic involvement (Mohseni et al. 2013). Imaging may be necessary to confirm complications such as pulmonary edema (via chest X-ray or ultrasound) or cerebrovascular events (via CT or MRI). The toxicokinetics of scorpion venom vary depending on the species and host response. While neurotoxic effects may subside within hours, other venom components—especially those causing inflammation or endothelial damage—can persist longer and accumulate in organs such as the kidneys, brain, and heart, contributing to delayed complications (Abdel-Haleem et al. 2006; Chippaux and Goyffon 2008).

However, despite the scarcity of data regarding possible diagnostic methods for scorpion envenomation, some clinical studies have emerged over the years aiming to develop tests that address the existing gap in scorpionism diagnosis. A study by Puzari et al. (2025) reviewed immunodiagnostic methods developed over the years to detect venom antigens—such as toxins—from various venomous and poisonous species, focusing on the identification of specific antibodies against venoms. These tests are critical for more accurately determining the appropriate antivenom dosage by assessing the circulating venom levels in scorpionism patients (Puzari et al. 2025). The study also presents a global overview of research on scorpion venom detection, summarizing findings from studies conducted across the Americas, Asia, Africa, and Europe (Table 17.1).

17.5 Therapy

The management of scorpion envenomation is primarily supportive, focusing on symptom relief and the prevention of complications. In moderate to severe cases, the administration of antivenom may be recommended to neutralize circulating toxins (Cupo 2015). Nevertheless, clinical management of scorpion envenomation may vary considerably depending on the envenomation severity, the scorpion species implicated, and the local availability of medical infrastructure and antivenom.

17.5.1 Symptomatic Care

In cases of mild scorpion envenomation, typically characterized by localized pain, erythema, and paresthesia without systemic manifestations, antivenom is not indicated. Pain is usually the predominant symptom and is often described as a "burning sensation at the site, accompanied by redness and tingling." Management in these cases is primarily supportive and includes analgesics such as acetaminophen or nonsteroidal anti-inflammatory drugs (NSAIDs), antihistamines, and benzodiazepines when necessary for muscle spasms or anxiety. Local anesthetic blocks represent an effective strategy for controlling intense localized pain. This involves infiltrating a local anesthetic, such as 1% lidocaine (without vasoconstrictor), near the sting site or along the affected nerve pathway—for example, through a digital nerve block in finger stings. This technique provides rapid pain relief, reduces the need for systemic medications, and is safe, cost-effective, and widely applicable in emergency care. Vasoconstrictors should be avoided in extremities due to the risk of ischemia.

Table 17.1 Immunodiagnostic methods applied to scorpion envenomation

Family	Scorpion species	Geographical distribution				References
		America	Asia	Africa	Europe	
Buthidae	*Androctonus crassicauda*		Iran and Turkey	North Africa		Ozkan et al. (2006)
	Androctonus australis hector			North Africa		Mars et al. (2018)
	Androctonus australis garzoni			Middle East and North Africa		Krifi et al. (1998)
	Buthus occitanus tunetanus			Middle East and North Africa	Southern Europe	Mejri et al. (2003)
	Centruroides sculpturatus	Mexico				Riaño-Umbarila et al. (2017)
	Centruroides noxious	Western Mexico				Petricevich (2006)
	Centruroides suffusus	Mexico				Martin et al. (1987)
	Centruroides limpidus	Mexico				Dehesa-Dávila et al. (1996)
	Centruroides tecomanus	Mexico				Valdez-Velazquéz et al. (2016)
	Tityus serrulatus	Brazil				Barral-Netto et al. (1991), Chávez-Olórtegui et al. (1994), Nunan et al. (2001), Revelo et al. (1996), Rezende et al. (1995) and Rosário et al. (2024)

(continued)

Table 17.1 (continued)

Family	Scorpion species	Geographical distribution				References
		America	Asia	Africa	Europe	
Buthidae	*Tityus bahiensis*	Brazil				Lourenço et al. (2002)
	Tityus obscurus	Brazil				Santos-da-Silva et al. (2017)
	Tityus stigmurus	Brazil				Furtado et al. (2020)
	Tityus pachyurus	Colombia				Guerrero-Vargas et al. (2012)
	Tityus trivittatus	Argentina				De Roodt et al. (2003)
	Leiurus quinquestriatus			Middle East and North Africa		Salama (2013)
	Leiurus macroctenus		Oman			Gunas et al. (2024)
	Parabuthus transvaalicus			Southern Africa		Debont et al. (1998)
	Parabuthus granulatus			Southern Africa		Debont et al. (1998)
	Odontobuthus doriae		Iran and Iraq			Mazhdi and Hamidi (2021)
	Mesobuthus tamulus		India, Nepal, Pakistan, Sri Lanka			Tiwari and Deshpande (1993) and Puzari et al. (2024)
Hemiscorpiidae	*Hemiscorpius lepturus*		Iran, Iraq, Oman, Pakistan, Saudi Arabia, Yemen			Latifi and Tabatabai (1979)
Scorpionidae	*Scorpio maurustownsendi*		Iran			Nazari and Hassan (2016)

This approach may reduce the need for systemic opioids or sedatives, which can complicate monitoring in severe cases. Although warm or cold compresses are not recommended, many patients report symptomatic relief from warm water immersion, which may help alleviate pain and inflammation. All patients should be observed for a minimum of 6 h, particularly children and the elderly, due to the risk of symptom progression. If no clinical deterioration occurs during this period and symptoms remain limited to localized pain, the patient may be safely discharged with appropriate guidance for outpatient follow-up if necessary (Cupo 2015; Pucca et al. 2015; Santos et al. 2016).

17.5.2 Antivenom Therapy

Antivenom remains the only specific treatment for envenomation by venomous animals, including scorpion stings (Laustsen et al. 2018). It is composed of purified heterologous immunoglobulin (IgG) or its fragments, typically F(ab')$_2$, obtained from immunized animals, most commonly horses. When administered early, ideally within the first 1–2 h after the sting, antivenom can significantly improve clinical outcomes by neutralizing circulating venom before it binds to tissue receptors (Santos et al. 2016).

The type and availability of antivenom vary by region. In South America, available formulations primarily target *Tityus serrulatus*, *Tityus stigmurus*, and *Centruroides* species. In Africa and the Middle East, antivenoms are produced against *Androctonus* spp., *Buthus occitanus*, and *Leiurus quinquestriatus*. In South Asia, antivenoms specific to *Mesobuthus tamulus* are manufactured and used locally (Tiwari and Deshpande 1993; Pucca et al. 2014; Nath and Mukherjee 2023). For moderate envenomation, characterized by intense localized pain and early systemic manifestations such as diaphoresis, vomiting, or mild dyspnea, treatment usually consists of 2–3 vials of antivenom. In severe cases, marked by cardiopulmonary or neurological involvement, 4–6 vials or more may be required, depending on the patient's clinical response and local institutional protocols (Chippaux and Goyffon 2008).

17.5.3 Safety and Considerations in Antivenom Use

Although generally safe when properly purified, heterologous antivenoms still carry the risk of adverse reactions, including anaphylaxis, serum sickness, and, in rare cases, transmissible infections (Laustsen et al. 2018; Pucca and Camphora 2024). Thus, the availability of epinephrine and resuscitation equipment during administration remains essential. In high-risk patients or in settings with limited access to emergency care, premedication with antihistamines or corticosteroids may be considered as a preventive strategy (Feng et al. 2022).

Ongoing research into recombinant antivenoms, humanized monoclonal antibodies, and nanobody-based therapies seeks to improve antigen specificity while minimizing immunogenicity, offering promising alternatives for safer and more effective treatment (Carmo et al. 2015; Pucca et al. 2012, 2019).

17.5.4 Other Supportive and Adjunct Therapies

Supportive and adjunctive therapies play a critical role in the management of moderate to severe scorpion envenomation, particularly in settings where antivenom is unavailable, delayed, or insufficient to control systemic manifestations. These interventions aim to stabilize the patient, prevent complications, and improve overall clinical outcomes.

- **Hemodynamic support** should be prioritized in the presence of autonomic instability. Vasodilators, such as hydralazine, or other antihypertensive agents may be employed to manage sympathetic overdrive and severe hypertension. In cases of myocardial dysfunction or cardiogenic shock, inotropic agents like dopamine or dobutamine may be required to maintain adequate cardiac output. Cardiac arrhythmias should be treated with appropriate anti-arrhythmic medications, guided by continuous ECG monitoring.
- **Respiratory support** is essential in patients presenting with respiratory distress or failure, a complication more frequently observed in children due to their limited physiological reserve. Oxygen supplementation should be initiated promptly, and mechanical ventilation may be necessary in severe cases.
- **Fluid resuscitation** with intravenous crystalloids is also a cornerstone of supportive care, especially in pediatric patients, to correct dehydration, counteract capillary leak, and maintain tissue perfusion.
- **Infectious complications**, although infrequent, may arise from secondary bacterial infection at the sting site. When clinically indicated, empirical antibiotic therapy should be initiated and adjusted based on microbiological findings and local resistance patterns.
- **Metabolic disturbances** are not uncommon and should be monitored closely. Blood glucose levels must be assessed regularly, especially in children, older adults, and individuals with diabetes or severe systemic reactions, as both hypo- and hyperglycemia can occur.
- **Cardiac assessment** with transthoracic echocardiography is strongly recommended in all moderate to severe cases. This enables the early detection of ventricular dysfunction, myocardial depression, or pulmonary hypertension, which can significantly influence clinical management, including decisions about ICU admission and the need for advanced cardiovascular support.

17.6 Concluding Remarks

The clinical management of scorpion envenomation remains a multifaceted challenge that demands a multidisciplinary and context-sensitive approach. Although antivenom is currently the only specific treatment available, its accessibility and efficacy can be limited, particularly in regions with diverse scorpion fauna. This highlights the critical importance of early symptomatic management and personalized care based on the severity of envenomation and patient-specific factors. Recent advances in biotechnology (e.g., recombinant antivenoms and targeted toxin-neutralizing molecules) offer promising prospects for safer, more effective therapeutic alternatives.

References

Abdel-Haleem A-HA, Meki A-HMA, Noaman HA, Mohamed ZT (2006) Serum levels of IL-6 and its soluble receptor, TNF-α and chemokine RANTES in scorpion envenomed children: their relation to scorpion envenomation outcome. Toxicon 47(4):437–444. https://doi.org/10.1016/j.toxicon.2005.12.008

Abroug F, Ouanes-Besbes L, Bouchoucha S (2019) Scorpion envenomation: from a neglected to a helpful disease? Intensive Care Med 45(1):72–74. https://doi.org/10.1007/s00134-018-5226-5

Abroug F, Ouanes-Besbes L, Tilouche N, Elatrous S (2020) Scorpion envenomation: state of the art. Intensive Care Med 46(3):401–410. https://doi.org/10.1007/s00134-020-05924-8

Barral-Netto M, Vinhas V, Schriefer A, Barral A, Santos SB, Almeida AR, Novaes G (1991) Immunological studies with the venom of the scorpion *Tityus serrulatus*. Braz J Med Biol Res 24(2):171–180

Carmo AO, Chatzaki M, Horta CCR, Magalhães BF, Oliveira-Mendes BBR, Chávez-Olórtegui C, Kalapothakis E (2015) Evolution of alternative methodologies of scorpion antivenoms production. Toxicon 97:64–74. https://doi.org/10.1016/j.toxicon.2015.02.007

Cesaretli Y, Ozkan O (2010) Scorpion stings in Turkey: epidemiological and clinical aspects between the years 1995 and 2004. Rev Inst Med Trop Sao Paulo 52:215–220. https://doi.org/10.1590/S0036-46652010000400010

Chávez-Olórtegui C, Fonseca SCG, Campolina D, Amaral CFS, Diniz CR (1994) ELISA for the detection of toxic antigens in experimental and clinical envenoming by *Tityus serrulatus* scorpion venom. Toxicon 32(12):1649–1656. https://doi.org/10.1016/0041-0101(94)90323-9

Chippaux J-P, Goyffon M (2008) Epidemiology of scorpionism: a global appraisal. Acta Trop 107(2):71–79. https://doi.org/10.1016/j.actatropica.2008.05.021

Cupo P (2015) Clinical update on scorpion envenoming. Rev Soc Bras Med Trop 48(6):642–649. https://doi.org/10.1590/0037-8682-0237-2015

De Roodt AR, García SI, Salomón OD, Segre L, Dolab JA, Funes RF, Titto EH (2003) Epidemiological and clinical aspects of scorpionism by *Tityus trivittatus* in Argentina. Toxicon 41(8):971–977. https://doi.org/10.1016/S0041-0101(03)00066-7

Debont T, Swerts A, van der Walt JJ, Muller GJ, Verdonck F, Daenens P, Tytgat J (1998) Comparison and characterization of the venoms of three *Parabuthus* scorpion species occurring in southern Africa. Toxicon 36(2):341–352. https://doi.org/10.1016/S0041-0101(97)00099-8

Dehesa-Dávila M, Ramfrez AN, Zamudio FZ, Gurrola-Briones G, Liévano A, Darszon A, Possani LD (1996) Structural and functional comparison of toxins from the venom of the scorpions *Centruroides infamatus infamatus, Centruroides limpidus limpidus* and *Centruroides noxius*. Comp Biochem Physiol Part B Biochem Mol Biol 113(2):331–339. https://doi.org/10.1016/0305-0491(95)02031-4

El Joud Y, Bouazzaoui AE, El-Ghali S, Laghzaoui E-M, Toulon O, Larradia MA, Elmourid A, Hamdan YA, Rhazi M, Kahime K, El Mouden EH, Merzouki M, El Hidan MA (2024) Assessing the effects of temperature, diet and threat conditions on defensive behaviour and venom regeneration in scorpion (*Buthus atlantis*). J Therm Biol 124:103966. https://doi.org/10.1016/j.jtherbio.2024.103966

Feng J, Wu Z, Yu Q, Li H, Ji P, Yang Y, Zeng X, Zheng X, Zhao C, Zhang J (2022) Hydrocortisone for preventing adverse drug reactions to snake antivenom: a meta-analysis. Emerg Med Int 2022(1):6151206. https://doi.org/10.1155/2022/6151206

Furtado AA, Daniele-Silva A, Silva-Júnior AA, Fernandes-Pedrosa MF (2020) Biology, venom composition, and scorpionism induced by Brazilian scorpion *Tityus stigmurus* (Thorell, 1876) (Scorpiones: Buthidae): a mini-review. Toxicon 185:36–45. https://doi.org/10.1016/j.toxicon.2020.06.015

Godoy DA, Badenes R, Seifi S, Salehi S, Seifi A (2021) Neurological and systemic manifestations of severe scorpion envenomation. Cureus 13(4):e14715. https://doi.org/10.7759/cureus.14715

Guerrero-Vargas JA, Mourão CBF, Quintero-Hernández V, Possani LD, Schwartz EF (2012) Identification and phylogenetic analysis of *Tityus pachyurus* and *Tityus obscurus* novel putative Na+-channel scorpion toxins. PloS One 7(2):e30478. https://doi.org/10.1371/journal.pone.0030478

Gunas V, Maievskyi O, Raksha N, Vovk T, Savchuk O, Shchypanskyi S, Gunas I (2024) Study of the acute toxicity of scorpion *Leiurus macroctenus* venom in rats. Scientific World Journal 2024(1):9746092. https://doi.org/10.1155/2024/9746092

Krifi MN, Kharrat H, Zghal K, Abdouli M, Abroug F, Bouchoucha S, Dellagi K, El Ayeb M (1998) Development of an ELISA for the detection of scorpion venoms in sera of humans envenomed by *Androctonus australis garzonii* (Aag) and *Buthus occitanus tunetanus* (Bot): correlation with clinical severity of envenoming in Tunisia. Toxicon 36(6):887–900. https://doi.org/10.1016/S0041-0101(97)00136-0

Latifi M, Tabatabai M (1979) Immunological studies on Iranian scorpion venom and antiserum. Toxicon 17(6):617–620. https://doi.org/10.1016/0041-0101(79)90236-8

Laustsen AH, Guitérrez JM, Knudsen C, Johansen KH, Bermúdez-Méndez E, Cerni FA, Jurgensen JA, Ledsgaard L, Martos-Esteban A, Ohlenschlaeger M, Pus U, Andersen MR, Lomonte B, Engmark M, Pucca MB (2018) Pros and cons of different therapeutic antibody formats for recombinant antivenom development. Toxicon 146:151–175. https://doi.org/10.1016/j.toxicon.2018.03.004

Lourenço GA, Lebrun I, Dorce VAC (2002) Neurotoxic effects of fractions isolated from *Tityus bahiensis* scorpion venom (Perty, 1834). Toxicon 40(2):149–157. https://doi.org/10.1016/S0041-0101(01)00202-1

Mars A, Bouhaouala-Zahar B, Raouafi N (2018) Ultrasensitive sensing of *Androctonus australis hector* scorpion venom toxins in biological fluids using an electrochemical graphene quantum dots/nanobody-based platform. Talanta 190:182–187. https://doi.org/10.1016/j.talanta.2018.07.087

Martin MF, Garcia y Perez LG, el Ayeb M, Kopeyan C, Bechis G, Jover E, Rochat H (1987) Purification and chemical and biological characterizations of seven toxins from the Mexican scorpion, *Centruroides suffusus suffusus*. J Biol Chem 262(10):4452–4459

Mazhdi Y, Hamidi SM (2021) Detection of scorpion venom by optical circular dichroism method. Sci Rep 11(1):15854. https://doi.org/10.1038/s41598-021-95493-7

Mejri T, Borchani L, Srairi-Abid N, Benkhalifa R, Cestele S, Regaya I, Karoui H, Pelhate M, Rochat H, el Ayeb M (2003) BotIT6: a potent depressant insect toxin from *Buthus occitanus tunetanus* venom. Toxicon 41(2):163–171. https://doi.org/10.1016/S0041-0101(02)00246-5

Mohseni A, Vazirianzadeh B, Hossienzadeh M, Salehcheh M, Moradi A, Moravvej SA (2013) The roles of some scorpions, *Hemiscorpius lepturus* and *Androctonus crassicauda*, in a scorpionism focus in Ramhormorz, southwestern Iran. J Insect Sci 13(1):1–12. https://doi.org/10.1673/031.013.8901

Nath S, Mukherjee AK (2023) Ethnomedicines for the treatment of scorpion stings: a perspective study. J Ethnopharmacol 305(6):116078. https://doi.org/10.1016/j.jep.2022.116078

Nazari M, Hassan R (2016) Study on distribution of scorpions to provide prevention and interventions in combating scorpionism in Poldokhtar county, Lorestan province, Iran. J Clin Diagn Res 10(12):LC05–LC09. https://doi.org/10.7860/JCDR/2016/22759.8960

Nunan EA, Cardoso VN, Moraes-Santos T (2001) Lethal effect of the scorpion *Tityus serrulatus* venom: comparative study on adult and weaning rats. Braz J Pharm Sci 37(1):39–44

Ozkan O, Adiguzel S, Yakistiran S, Filazi A (2006) Study of the relationship between *Androctonus crassicauda* (Oliver, 1807; Scorpiones, Buthidae) venom toxicity and telson size, weight and storing condition. J Venom Anim Toxins Incl Trop Dis 12(2):297–309. https://doi.org/10.1590/S1678-91992006000200011

Petricevich VL (2006) Balance between pro- and anti-inflammatory cytokines in mice treated with *Centruroides noxius* scorpion venom. Mediators Inflamm 2006(6):054273. https://doi.org/10.1155/MI/2006/54273

Pimenta RJG, Brandão-Dias PFP, Leal HG, Carmo AO, Oliveira-Mendes BBR, Chávez-Olórtegui C, Kalapothakis E (2019) Selected to survive and kill: *Tityus serrulatus*, the Brazilian yellow scorpion. PloS One 14(4):e0214075. https://doi.org/10.1371/journal.pone.0214075

Pucca MB, Camphora AL (2024) The potential risks of equine serum therapy in transmitting new infectious diseases: lessons from a post-pandemic era. Front Public Health 12:1366929. https://doi.org/10.3389/fpubh.2024.1366929

Pucca MB, Zoccal KF, Roncolato EC, Bertolini TB, Campos LB, Cologna CT, Faccioli LH, Arantes EC, Barbosa JE (2012) Serrumab: a human monoclonal antibody that counters the biochemical and immunological effects of *Tityus serrulatus* venom. J Immunotoxicol 9(2):173–183. https://doi.org/10.3109/1547691X.2011.649220

Pucca MB, Oliveira FN, Schwartz EF, Arantes EC, Lira-da-Silva RM (2014) Scorpionism and dangerous species of Brazil. In: Gopalakrishnakone P (ed) Toxinology: scorpion venoms. Springer Netherlands, Dordrecht, pp 1–24

Pucca MB, Cerni FA, Pinheiro Júnior EL, Bordon KCF, Amorim FG, COrdeiro FA, Longhim HT, Cremonez CM, Oliveira GH, Arantes EC (2015) *Tityus serrulatus* venom—a lethal cocktail. Toxicon 108:272–284. https://doi.org/10.1016/j.toxicon.2015.10.015

Pucca MB, Cerni FA, Janke R, Bermúdez-Méndez E, Ledsgaard L, Barbosa JE, Laustsen AH (2019) History of envenoming therapy and current perspectives. Front Immunol 10:1598. https://doi.org/10.3389/fimmu.2019.01598

Puzari U, Khan MR, Mukherjee AK (2024) Development of a gold nanoparticle-based novel diagnostic prototype for *in vivo* detection of Indian red scorpion (*Mesobuthus tamulus*) venom. Toxicon: X 23:100203. https://doi.org/10.1016/j.toxcx.2024.100203

Puzari U, Das B, Mukherjee AK (2025) Advancements in diagnostic techniques for scorpion venom identification: a comprehensive review. Toxicon 253:108191. https://doi.org/10.1016/j.toxicon.2024.108191

Revelo MP, Bambirra EA, Ferreira AP, Diniz CR, Chávez-Olórtegui C (1996) Body distribution of *Tityus serrulatus* scorpion venom in mice and effects of scorpion antivenom. Toxicon 34(10):1119–1125. https://doi.org/10.1016/0041-0101(96)00074-8

Rezende NAD, Dias MB, Campolina D, Chávez-Olórtegui C, Amaral CFS (1995) Standardization of an enzyme linked immunosorbent assay (ELISA) for detecting circulating toxic venom antigens in patients stung by the scorpion *Tityus serrulatus*. Rev Inst Med Trop Sao Paulo 37(1):71–74. https://doi.org/10.1590/S0036-46651995000100011

Riaño-Umbarila L, Rodríguez-Rodríguez ER, Santibañez-Lopez CE, Guereca L, Uribe-Romero SJ, Gómez-Ramírez IV, Cárcamo-Noriega EN, Possani LD, Becerril B (2017) Updating knowledge on new medically important scorpion species in Mexico. Toxicon 138:130–137. https://doi.org/10.1016/j.toxicon.2017.08.022

Rosário VAM, Nappi GU, Matencio T, Carvalho TXT, Heneine LGD (2024) Detection of *Tityus serrulatus* venom using carbon immunosensor label-free impedimetric. BioNanoScience 14(2):1351–1361. https://doi.org/10.1007/s12668-024-01338-8

Salama WM (2013) The lethal effect of *Leiurus quinquestriatus* venom on adult and weanling albino mice with evaluating its effects on some biochemical parameters. Egypt J Zool 60(60):429–441. https://doi.org/10.12816/0003302

Santos MSV, Silva CGL, Neto BS, Grangeiro CRP Jr, Lopes VHG, Teixeira AG Jr, Bezerra DA, Luna JVCP, Cordeiro JB, Gonçalves J Jr, Lima MAP (2016) Clinical and epidemiological aspects of scorpionism in the world: a systematic review. Wilderness Environ Med 27(4):504–518. https://doi.org/10.1016/j.wem.2016.08.003

Santos-da-Silva AP, Candido DM, Nencioni ALA, Kimura LF, Prezotto-Neto JP, Barbaro KC, Chalkidis HM, Dorce VAC (2017) Some pharmacological effects of *Tityus obscurus* venom in rats and mice. Toxicon 126:51–58. https://doi.org/10.1016/j.toxicon.2016.12.008

Souza WMP, Silva GA, Cerni FA, Oliveira IS, Zottich U, Bassoli BK, Pucca MB (2019) Envenomings caused by venomous animals in Roraima: a neglected health problem in the Brazil's Northernmost state. Toxicol Curr Res 3(1):1–8. https://doi.org/10.24966/TCR-3735/100011

Tiwari AK, Deshpande SB (1993) Toxicity of scorpion (*Buthus tamulus*) venom in mammals is influenced by the age and species. Toxicon 31(12):1619–1622. https://doi.org/10.1016/0041-0101(93)90346-K

Valdez-Velazquéz LL, Romero-Gutierrez MT, Delgado-Enciso I, Dobrovinskaya O, Melnikov V, Quintero-Hernández V, Ceballos-Magaña SG, Gaitan-Hinojosa MA, Coronas FI, Puebla-Perez AM, Zamudio F, De la Cruz-García I, Vázquez-Vuelvas OF, Soriano-Hernandez AD, Possani LD (2016) Comprehensive analysis of venom from the scorpion *Centruroides tecomanus* reveals compounds with antimicrobial, cytotoxic, and insecticidal activities. Toxicon 118:95–103. https://doi.org/10.1016/j.toxicon.2016.04.046

Chapter 18
Perspectives on Biology and Interactions with Humans of Neotropical Scorpions

Edmundo González-Santillán

Abstract This chapter compiles overlooked topics on Neotropical scorpions presented in this book, integrating contributions from both emerging and seasoned scientists to advance regional knowledge. It synthesizes recent biological and evolutionary findings, highlighting the complex interaction between humans and scorpions, typically characterized by fear rooted in their toxic reputation. Emphasizing that knowledge reduces fear, the work advocates clearer identification of medically important species and reconsideration of traditional medicinal practices to improve public health responses and antivenom use. Key biological insights include reproductive behaviors, sexual dimorphism, and the ecological roles of scorpions within trophic webs, alongside morphophysiological adaptations to their environments. Geological history and evolutionary biology frame a comprehensive understanding of species diversity, supported by recent advances in molecular systematics and high-throughput genetic techniques. New research frontiers explored are troglomorphic adaptations in cave-dwelling species, fluorescence of the scorpion cuticle, and cytogenetics as a tool for phylogenetic study. The study underscores the potential of scorpions as models for conservation biology and urban ecology, particularly regarding public health challenges in South America. Finally, the work encourages ongoing innovation and collaboration in scorpion research, emphasizing collective scientific effort to expand knowledge in evolution, ecology, and human interactions within the Neotropical context.

Keywords Neotropical region • Arachnids • Taxonomy • Ecology

E. González-Santillán (✉)
Colección Nacional de Arácnidos, Departamento de Zoología, Instituto de Biología, Universidad Nacional Autónoma de México, Coyoacán, Mexico City, Mexico
e-mail: edmundo.gonzalez@ib.unam.mx

A. F. de Araujo Lira, E. González-Santillán (eds.), *The Biology of Neotropical Scorpions*, https://doi.org/10.1007/978-3-032-23194-9_18

18.1 Our Contribution

This contribution is a compilation of topics that have historically been overlooked in the study of scorpions in the Neotropics region. The assembly of young and highly experienced scientists contributed to the knowledge of this regional fauna. It is acknowledged that the contemporary world is characterized by dynamic and contrasting postures about these topics. Despite that, the present endeavor is regarded as a modest contribution to the broader field of scorpion research. However, we are convinced that this undertaking has the potential to serve as a foundational basis for individuals pursuing a career in evolutionary biology, particularly in the context of scorpions or simply fellow citizens looking for an entertaining reading.

In accordance with extant contributions to the field of scorpion biology, the present study focused on topics where new evidence and data have advanced our knowledge of the biology of Neotropical scorpions. It is acknowledged that the present approach was influenced by the work of Polis (1990), which was further enhanced by Brownell and Polis (2001), as well as by Fet and Selden (2001). The field of biodiversity studies has provided the impetus for our research initiative, which aims to synthesize the biology of the Neotropics—a region where scorpions and humans coexist (see Chap. 1). Furthermore, it was anticipated that individuals residing outside the Neotropics were enticed by the captivating biodiversity on display.

The interaction of humans with scorpions in the Neotropics has been characterized by a predominant presence of caution and fear in daily life. The underlying rationale for this phenomenon is the pervasive association of scorpions with toxicity. Nevertheless, we have endeavored to address this negative perspective implicitly. The fundamental principle of our philosophy is the belief that knowledge is the most effective remedy for fear. One of the objectives of this contribution was to broaden the existing body of knowledge concerning the interaction between humans and scorpions.

The identification of species that pose a medical threat to public health can mitigate the associated fear (see Chaps. 14, 15, 16 and 17). It is imperative to acknowledge that not all species possess the capacity to severely harm human health (see Chap. 5) in order to facilitate a reevaluation of our response to encounters with scorpions. Conversely, we emphasized the significance of transcending cultural limitations and preconceived notions concerning traditional medicine to enhance the well-being of isolated communities and ensure the judicious utilization of antivenoms (see Chap. 14).

An analysis of the mating habits and care mechanisms employed by scorpions in relation to their offspring provided a valuable opportunity to elucidate a significant aspect of their cryptic biology (see Chap. 3). We hoped that such description had helped to realize that we humans are not that different from scorpions. We provided insights into the tools that have facilitated understanding the factors contributing to the size disparity between male and female scorpions, including the larger size of female scorpions and the larger size of male scorpions' pincers. The fundamental hypothesis posits that females, in general, necessitate additional space in their body to carry and nurture their offspring. Whereas males have evolved to require larger pincers or chelae for successful mating due to sexual selection (see Chap. 2).

Scorpions, as wild animals, are engaged in the utilization of ecological resources. One of the numerous challenges that scorpions encounter as predators is survival, a topic that was explored in Chap. 13. This underscores their role as a crucial component of an intricate and delicate trophic web within the ecosystem. As delineated in Chap. 10, a comprehensive survey of the traits exhibited by scorpions is furnished, with an emphasis on the integrated adaptation of morphophysiological modifications to their habitat.

From the perspective of geological history and its core components in organismic evolution, time, species, and form, our understanding is just as important as understanding present relationships with scorpions. Chapters 5 through 8 presented a comprehensive compilation and in-depth exploration of the evolutionary biology of the ancestral Neotropics, with a focus on its global influence and its impact on the region. The most recent hypothesis regarding the patterns left by geological changes is an integration of abiotic and biotic factors. Recent advancements in systematics have profoundly impacted our understanding of scorpion classification. These advancements encompass both technological and conceptual aspects. Technological advances, including massive sequencing, have enabled the analysis of genetic variation in scorpions, resulting in a more comprehensive understanding of species diversity. Conceptual advances, such as the manipulation of large data sets reflected as the "omics" approach, have enabled novel insights into the taxonomy, biogeography, and evolution of scorpions.

In Chap. 13 (cf. Chap. 11 but in an utilitary form), a rationale for investigating the biodiversity of Neotropical scorpions from the perspective of a broader spectrum of opportunities to address current global problems was presented. The notion that scorpions can serve as models for conservation studies has been a subject of recent discussion, with several studies proposing the use of scorpions in this capacity. Urban ecology emerged as a particularly salient topic of interest, given its close association with the subject matter addressed in Chaps. 16 and 17. The comprehension of the dynamics governing species populations that are of public health significance holds considerable promise in addressing the escalating challenges plaguing South America.

A notable trait exhibited by a subset of Neotropical scorpions is their capacity to adopt a troglobiont lifestyle, which involves a comprehensive ecomorphological and physiological metamorphosis known as troglomorphism (see Chap. 9). Our reevaluation not only explored the current knowledge of their diversity but also delved into the conservation and preservation of these unique scorpion species.

Another topic that has garnered significant interest is the fluorescence emitted by the cuticle of scorpions. We presented new insights into this optical phenomenon and conducted a comparative study for the first time, which produced results that can be extrapolated beyond Neotropical scorpions (see Chap. 4).

Finally, but not least, there was a resurgence of interest in cytogenetics, which had not yet been thoroughly explored. This area has the potential to provide new evidence for phylogenetics and to enhance our understanding of cellular biology in scorpions (see Chap. 12).

18.2 What Is Next?

Our final recommendation, which is intended for both the readership and the scientific community, is to continue innovating and searching for facts, methods, theories, and elements to increase our knowledge. Evolution, ecology, and human interaction with scorpions may be areas of particular interest to young or enthusiastic veterans. The pursuit of scientific knowledge is not the exclusive domain of any individual or group; rather, it is a collective endeavor that requires openness to diverse perspectives and a commitment to rigorous inquiry. Consequently, those who demonstrate a willingness to engage with alternative proposals while maintaining a rigorous approach to their own work may be entitled to a platform through which their contributions can be acknowledged and disseminated. Scorpions may be regarded as a mere pretext for inclusion within the Neotropical scientific community; however, this dynamic can also be applied in reverse, with the Neotropical scientific community extending its embrace to any community.

References

Brownell P, Polis GA (2001) Scorpion biology and research. Oxford University Press, Oxford

Fet V, Selden PA (2001) Scorpions 2001: in memoriam Gary A. Polis. British Arachnological Society, Burnham Beeches, Bucks

Polis GA (1990) The biology of scorpions. Stanford University Press, Stanford

Index

A. F. de Araujo Lira, E. González-Santillán (eds.), *The Biology of Neotropical Scorpions*, https://doi.org/10.1007/978-3-032-23194-9

The manufacturer's authorised representative in the EU is Springer Nature Customer Service Centre GmbH, Europaplatz 3, 69115 Heidelberg, Germany. If you have any concerns regarding our products, please contact ProductSafety@springernature.com

Printed and bound by CPI Group (UK) Ltd, Croydon, CR0 4YY
05/10/2026
02228057-0010